Principle and Application of Blind Source Separation Technology

Weihong Fu

Principle and Application of Blind Source Separation Technology

Springer

Weihong Fu
Xidian University
Xi'an, Shaanxi, China

ISBN 978-981-96-0021-2 ISBN 978-981-96-0022-9 (eBook)
https://doi.org/10.1007/978-981-96-0022-9

Jointly published with Xidian University Press
The print edition is not for sale in China (Mainland). Customers from China (Mainland) please order the print book from: Xidian University Press.
ISBN of the Co-Publisher's edition: 978-7-5606-6798-0

Translation from the Chinese Simplified language edition: "盲源分离技术原理与应用" by Weihong Fu,

This Springer imprint is published by the registered company Springer Nature Singapore Pte Ltd.
The registered company address is: 152 Beach Road, #21-01/04 Gateway East, Singapore 189721, Singapore

Preface

With the rapid development of network, communication, and computer technology, we have entered a digital and information age, and digital signal processing is one of its key links. As the main content of blind signal processing, blind source separation refers to the technology of recovering or separating the source signal from the observed mixed signal when the prior information of the source signal and transmission channel is unknown. It is a traditional and challenging problem in signal processing. Blind source separation technology is widely used in speech signal processing, image processing, mechanical fault diagnosis, biomedical signal processing, communication signal processing, and so on.

Blind source separation has very important application value in the field of communication reconnaissance. In the actual military communication environment, the signals intercepted are often a mixture of enemy interference signals, enemy communication signals, our communication signals, and various electromagnetic interference and noise signals. In order to reconnaissance enemy communication signals, we must first separate the required enemy communication signals from the intercepted mixed signals. Due to the diversity of modulation styles of communication signals, we often cannot know any information about the enemy-transmitted signals. Coupled with the complexity of the electromagnetic environment, we have no way to know how the enemy communication signals are mixed together. In the case of uncertain prior information about the signal, it is very difficult to process such signals in time and frequency. Blind source separation technology is an effective way to solve this problem.

As a researcher of communication engineering, the author focuses on the blind source separation technology of communication signals. This book is divided into four parts. The first part is the theoretical and technical basis of blind source separation, the second part is the overdetermined/determined blind source separation technology, the third part is the underdetermined blind source separation technology and the fourth part is the application of blind source separation technology.

The first part of this book includes three chapters. The first chapter introduces the concept, development, and application of blind source separation. The second chapter introduces the mathematical model of blind source separation and the separable condition. The third chapter introduces the theoretical basis of blind source separation, including PCA, ICA, SCA, and other theoretical basic knowledge used in overdetermined, determined and underdetermined blind separation, common methods of overdetermined and underdetermined blind separation, and performance evaluation indexes of algorithms.

The second part includes five chapters. Chapter 4 is the preprocessing technology of overdetermined/determined blind separation, which mainly introduces the whitening technology of blind source separation and the source signal estimation technology. Chapter 5 introduces the overdetermined/determined blind separation algorithm based on gradient. It firstly introduces the fixed step-size stochastic gradient blind separation algorithm, natural gradient blind separation algorithm, and EASI blind separation algorithm. Secondly it introduces the variable step-size stochastic gradient algorithm, natural gradient algorithm, and EASI algorithm, and then analyzes the convergence of the algorithm. Finally, the performance simulation results of different algorithms are given. In Chap. 6, the fast fixed-point algorithm of overdetermined blind separation is introduced. The unit FastICA algorithm and multivariate FastICA algorithm are introduced respectively. In Chap. 7, the overdetermined blind separation algorithm based on joint diagonalization is mainly introduced, which is based on second-order cumulants and fourth-order cumulants. Chapter 8 introduces the convolutional mixed blind source separation technology, the mathematical model of convolutional mixed blind source separation and the algorithm of convolutional blind source separation in time domain and frequency domain.

The third part includes four chapters. Chapter 9 researches the mixing matrix estimation technology of fully sparse underdetermined blind source separation, introduces the K-means clustering algorithm, the fuzzy C-means clustering algorithm, and the principle and implementation steps of improved K-means clustering algorithm, and gives the performance simulation results of the algorithm. In Chap. 10, the estimation techniques of insufficiently sparse underdetermined mixing matrix are introduced, which are based on K-dimensional subspace, parameter estimation, potential function, polynomial and single source detection. In Chap. 11, the source signal recovery algorithm based on compression reconstruction is introduced. On the basis of introducing the compressed sensing mathematical model and the relationship between the source signal recovery and underdetermined blind separation, the source signal recovery algorithm based on greedy idea, L1 norm, L0 norm, and cascade network are introduced in detail. In Chap. 12, other algorithms of source signal recovery are studied, including source signal recovery algorithm based on shortest path, statistical sparse decomposition and its improved algorithm.

The third part includes two chapters. Chapter 13 is the application of blind source separation in communication reconnaissance. It introduces the DS-CDMA signal pseudo-code estimation and multi-user separation based on blind source separation, frequency hopping signal sorting and splicing based on blind source separation, communication anti-jamming technology based on blind source separation, etc. In Chap. 14, the experimental results of communication signal blind source separation are given.

The author has studied blind source separation technology for 17 years since 2005. The author led the graduate students to conduct in-depth research on overdetermined/determined BSS and underdetermined BSS. The author spent nearly 3 years to summarize and organize the research work in recent years on the basis of referring to the relevant literature of domestic and foreign. In this book, the key algorithm principle of blind source separation technology is discussed in detail, the implementation steps of the algorithm are summarized, and the performance simulation results of the algorithm are given, and finally the application examples of blind source separation technology and the measured results in real environment are given.

This book is written by Weihong Fu. In the process of writing this book, the author has referred to some literature materials, which is not listed in the literatures. I would like to express my sincere thanks to all the authors of the references. Thanks to Zhao Yichen, Zhao Wensheng, Chen Yue, Zhou Yufei, Zhang Xinyu, Li Haoyi, Zheng Peiyuan, and other graduate students who participated in the material arrangement and proofreading of this book, to the leaders and teachers of the School of Communication Engineering of Xidian University for their help and support, and to Xidian University Press for their support in publishing this book.

Due to the limited level of the author, coupled with the short time, the inadequacies in the book are inevitable, and I urge readers to criticize and correct.

Xi'an, China
August 2024

Weihong Fu

Introduction

This book comprehensively describes the principle, key algorithms, and applications of blind source separation technology. On the basis of introducing the principle of related algorithms of blind source separation, the simulation results of algorithms are given. The book is divided into four parts, a total of 14 chapters. This book first introduces the related theories and technical basis of blind source separation, then describes the overdetermined/determined blind source separation technology and underdetermined blind source separation technology, and finally introduces the application of blind source separation technology.

This book can be used as a teaching material for graduate students and senior undergraduates in signal processing-related majors, and also as a reference book for engineers and technicians in wireless communication and signal processing fields.

Contents

Chapter 1
The Proposal and Application of the Blind Source Separation

Blind source separation (BSS) refers to the signal processing technology that recovers or separates the source signal from the observed mixed signal according to the statistical characteristics of the source signal when the prior information of the source signal and transmission channel are unknown. It is a traditional and challenging problem in signal processing. In scientific research and engineering practice, many observed signals can be regarded as a mixture of multiple source signals. The main task of BSS is to recover the source signals that we are interested in from the observed signal data.

In this chapter, we first introduce the background of blind source separation, then summarizes the development of BSS technology, including overdetermined, determined and underdetermined BSS technology, and finally briefly introduce the application of BSS.

1.1 The Proposal of Blind Source Separation Problem

Suppose you're at a cocktail party and there are a variety of sound sources, such as chatter (and possibly in different languages), music, and maybe a steam-whistle outside the window. If there are enough microphones in different locations to record these sounds, the signal recorded by each microphone is a mixed signal of the speaker's speech signals with different weights. Despite there are many interferences on the scene, you can still focus on what your friend is saying, and you can even listen to music while talking. Without knowing any information about the sound source or the location of the microphone, how do you separate the desired speaker voice from the speech signal received by the microphone? Blind source separation comes into being to solve such problems [1].

Blind source separation has obvious advantages in the following situations:

The source signal is unobservable;

W. Fu, *Principle and Application of Blind Source Separation Technology*,
https://doi.org/10.1007/978-981-96-0022-9_1

The mixing mode of the source signal is unknown;
The transmission channel between receiver and transmitter is difficult to establish mathematical model or the prior knowledge about the transmission channel cannot be obtained.

The significant feature of BSS algorithm is "blind", which is especially suitable for signal separation in complex electromagnetic environment.

According to the relationship between the number of observed signals and source signals, blind source separation can be divided into overdetermined BSS, determined BSS and underdetermined BSS. When the number of observed signals (mixed signals) is greater than the number of source signals, the BSS problem is called overdetermined BSS. When the number of observed signals (mixed signals) is equal to the number of source signals, the blind source separation problem is called determined BSS. When the number of observed signals (mixed signals) is less than the number of source signals, the BSS problem is called underdetermined BSS.

At present, the basic theoretical framework adopted for overdetermined and determined BSS problem is the same, which is solved by Independent Component Analysis (ICA) according to the statistical independence between source signals. According to the sparse characteristic of source signal, Sparse Component Analysis (SCA) is used to solve the problem of underdetermined BSS.

1.2 The Development of Blind Source Separation Technology

1.2.1 Development of Overdetermined/Determined BSS Technology

In 1986, Herault and Jutten presented a feedback neural network model and a learning algorithm based on Hebb learning rules (H-J algorithm), at an international conference on neural networks held in the United States. When the linear mixed channel and the source signal are unknown, the H-J algorithm realizes the separation of two independent source signals only by using the mixed signal. Herault and Jutten's work opens up a new research topic, Blind Source Separation (BSS).

Since the only assumption of blind source separation is that the source signals are statistically independent from each other, blind source separation has become a widely used signal processing method. Especially when it is difficult or impossible to build a transmission channel model from the source to the sensor, blind source separation becomes the only feasible signal processing method.

Independent Component Analysis (ICA) was first proposed on the basis of neural network model, and can be regarded as a further development of principal component analysis and factor analysis, but there are great differences. ICA can also be seen as a technique for dealing with BSS.

At the Advanced Spectral Analysis Conference in 1989, Cardoso and Common published two early papers [2, 3] on ICA, and Cardoso proposed the so-called Joint Approximate Diagonalization of Eigen-matrix (JADE) algorithm.

In 1991, Tong [4] conducted a systematic study on the uncertainty and identifiability of blind source separation solutions. Because of the lack of signal transmission (i.e. channel) and source signal knowledge, there are two kinds of uncertainties or fuzziness in blind source separation: the arrangement and amplitude of the separated signal. Tong et al. pointed out that when the source signals are independent of each other, if the source signal vector is transformed, if and only if the transformed signals remain independent of each other, the transformation matrix can be decomposed into the product of a full-rank diagonal matrix and a transpose matrix. That is, the transformation only changes the amplitude and arrangement of the source signal, but does not change the waveform shape of the signal.

In 1994, Common [5] systematically explained the concept of independent component analysis, gave a strict mathematical definition, and proved that the blind source separation can be realized if the signals are independent of each other through appropriate linear transformation, which makes the realization problem of BSS into the solution problem of independent component analysis (ICA). Common defines the cost function of ICA and puts forward the famous ICA algorithm based on minimum mutual information. The cost function is a real valued scalar function used to measure the degree of mutual independence of each signal after transformation. If and only if each output signal is independent of each other, the cost function obtains the maximum or minimum value. In this way, the ICA problem of random vector can be transformed into an optimization problem by maximizing and minimizing the cost function. The work of Common gives a clear vein to the BSS problem, which makes the future algorithms have a clear theoretical basis.

In the case of low noise, Nadal and Parga can factor the output distribution by maximizing the mutual information of the network output and input that is, the joint probability density function can be decomposed into an edge probability density function. On this basis, Bell et al. derived the stochastic gradient learning criterion, established an information theory-based framework for BSS problem, and gave the Infomax algorithm, but the convergence and stability of the algorithm need to be improved. In 1996, Cardoso [6] proposed the relative gradient algorithm, and then Amari et al. [7–9] proposed the natural gradient algorithm, and illustrated the effective working principle of such algorithms from the perspective of Riemannian geometry. Because natural gradient algorithm eliminates the problem of matrix inversion, it makes it possible for many algorithms to process real signals. It has been verified that the relative gradient algorithm and the natural gradient algorithm are equivalent.

In 1997, Hyvarine and Oja et al. obtained a serial fixed-point algorithm [10–11] based on the negative entropy maximization principle of the output signal using Newton's method. The algorithm has a fast convergence speed, so it has been widely used in large-scale data processing.

In China, there are also many universities to carry out researches on blind source separation technology, such as Tsinghua University, Northwestern Polytechnical

University, Southeast University, Shanghai Jiao Tong University, Xidian University, South China University of Technology, Beijing Normal University and so on.

1.2.2 Development of Underdetermined Blind Separation Technology

Blind source separation has made great progress in the past 20 years, and overdetermined BSS technology has been widely used. However, in many practical applications, overdetermined conditions are difficult to satisfy. For example, in wireless sensor networks, the number of sensors is less than the number of source signals due to cost or environmental constraints, so underdetermined BSS has become a research focus in recent years, and it is also more challenging. In the early research work, scholars assumed that the source signal has the feature of segmented mutual exclusion in the time–frequency domain (the source signal is sufficiently sparse), that is, at any time–frequency point, only one source signal is non-zero. However, the constraint condition of time–frequency mutual exclusion is too ideal, which is difficult to satisfy in practical application. When the time–frequency distribution of the source signal overlaps slightly, the performance of the algorithm will decrease significantly. In order to relax the requirement of time–frequency distribution, one researcher proposed the time–frequency ratio of mixtures (TIFROM) algorithm [12], which allows the time–frequency distributions of source signals to have a certain degree of overlap, but still requires that there is a certain number of adjacent time–frequency windows in which only one source signal is non-zero. In order to solve the problem of underdetermined BSS, the two-step method is more effective among the underdetermined BSS methods with sparse source signals that is, the mixing matrix is estimated first, and then the source signal is recovered by using the mixing matrix and the observed signal.

Mixing matrix estimation is mainly achieved by clustering method. Some people combine k-means clustering and Hough transform to estimate mixing matrix. Firstly, k-means clustering method is used to obtain clustering center, and differential evolution method is used to solve the problem of selecting initial cluster center. And then Hough transform is used to modify clustering center. The accuracy of the mixing matrix estimation is improved, but the complexity of the algorithm is greatly increased. Similarly, some literature uses the swarm algorithm and k-means clustering algorithm to estimate the clustering center, and then uses the grid density method to correct the clustering center, which improves the robustness of the algorithm in the case of ideal sufficient sparsity, that is, there are a large number of single-source points (SSP), the mixing matrix estimating method has a good performance. However, when the number of multiple-source points (MSP) is equal to the number of single source points, the performance of the above method will deteriorate dramatically, and even the mixing matrix cannot be estimated. At present, a simple and effective method to solve this problem is the recognition time–frequency SSP method. This method first performs linear time–frequency analysis on the observed signal, and then identifies the single source point by comparing whether the real and imaginary parts of the observed signal time–frequency coefficient vector are in the same direction or opposite direction. Finally, the traditional clustering method is used to cluster all single source points to get each column vector of the mixing matrix. The mixing matrix estimation method based on recognition time–frequency SSP only retains single source points and eliminates multiple source points and interference points, which not only reduces the amount of clustering data, but also significantly improves the accuracy of clustering, and improves the accuracy and robustness of mixing matrix estimation while reducing complexity. The problem of current methods based on recognition time–frequency SSP is that the time–frequency resolution is low, and the accuracy of time–frequency single-source point detection is easily affected by the time–frequency resolution.

In the case that the source signal is not sufficiently sparse, there are a large number of multi-source points in the observed signal space and a few single source points, so the number of single source points obtained by using the single source point detection method is likely to fail to meet the requirements of clustering, and the mixing matrix cannot be completely estimated. Therefore, in the case of insufficiently sparse, we must make full use of the observed signal subspace composed of multiple source points to estimate the mixing matrix. Different from the linear clustering presented by single source points, the multi-source points in the observation signal space show the characteristics of subspace clustering or hyperplane clustering.

For source signal recovery in underdetermined BSS, the sparse recovery method is mainly adopted at present, because when the mixing matrix has been estimated, the underdetermined source signal recovery model is consistent with the mathematical model of sparse recovery, and the sparse recovery method generally has good stability. When the mixing matrix is known, the shortest path method can be used to recover the source signal, which is essentially to optimize the minimum ℓ_1 norm under the constraint of linear equations. By introducing non-negative variables, the minimization ℓ_1 norm problem under linear constraints can be transformed into a linear programming problem, such as basis pursuit algorithm.

In order to reduce the complexity of the source signal recovery algorithm and solve the problem that ℓ_0 norm is highly sensitive to noise, a sparse recovery algorithm based on the smoothed ℓ_0 norm (SL0) is proposed. This method obtains the minimum ℓ_0 norm solution by optimizing the function of the approximate ℓ_0 norm, and proves that the solution of the minimized approximate ℓ_0 norm can be approximated to the real solution infinitely, and also proves the robustness of the algorithm. At the same time, it is pointed out that SL0 algorithm can be used for blind source separation. Simulation results show that the running time of SL0 algorithm is significantly lower than that of the algorithm with minimum ℓ_1 norm. However, since the gradient descent method is adopted in each iteration of SL0 algorithm, the recovery accuracy is easily affected by the step size. Since then, more scholars have proposed some improved algorithms based on SL0 algorithm.

1.3 Application of Blind Source Separation

Applications of blind source separation include biomedical signal processing, image processing, communication signal processing, geophysical exploration, etc.

(1) Biomedical Signal Processing

Biomedical source signals are often weak, unstable, easily contaminated by a variety of noise and interference, and often superimpose on each other, using some traditional methods can not get very clear results. Since most biomedical signals are statistically independent of each other, ICA technology is used to separate or extract biomedical signals, so that each decomposed component is easier to have actual physical or physiological significance. Typical applications of ICA in bio-medicine include: separation or extraction of fetal ECG signals, isolation or extraction of atrial fibrillation signals, feature extraction for induced EEG signals, nuclear magnetic functional imaging, etc.

Blind source separation technique can be used to separate the eye movement and blinking artifacts recorded by EEG recording. Experimenter read text on a screen while their EEG signal were displayed via an electroencephalogram. The horizontal movement of the eye produces artifacts in the EEG, and the vertical movement of the eyelid caused by blinking also produces artifacts in the EEG. The blind source separation technique can be used to separate the eye movement and blinking artifacts, so as to distinguish the eye movement and blinking action.

The blind source separation method has also been used in the analysis of multi-channel electrogastrogram. Electrogastrogram analysis uses electrodes to record the myoelectric activity of stomach contraction on the surface of human abdomen. However, the waveform of electromyographic activity shown on electrogastrogram is not only from the stomach, but also from the interference of other organs. ICA can separate the interference waveform.

(2) Image Processing

With the in-depth study of BSS technology, it is applied in image processing. In the process of image shooting, camera shake, the skill level of the photographer, noise and other problems will cause image pollution, in the image recovery and reconstruction problems, we must extract the original image from the polluted image. BSS technology can not only solve these problems, but also retain the original information of the image. In recent years, BSS technology has been more and more applied to image feature extraction, image classification, image fusion and so on.

(3) Voice Signal Processing

Speech signal is an important medium of information exchange and emotional communication between human beings, is the carrier of information, and also is the most basic, most efficient and most convenient means of communication for human society. With the rapid progress of science and technology, in order to make the expression method of speech more widely used in every field of human life, speech signal processing technology has emerged.

In the actual environment, speech signals are often interfered with by various noises or other speakers, resulting in the reduction of speech quality. It is necessary to adopt different speech enhancement algorithms according to different interference. Speech enhancement is an important research content in the field of speech signal processing, which is to recover pure speech signals from polluted speech signals as much as possible, so as to improve speech quality, eliminate auditory fatigue and improve speech intelligibility. In many applications, voice enhancement technology is needed.

Speech enhancement includes two aspects: one is to remove noise from speech; The other is to remove the interference of other speech, which is the problem of separation of speech signals. From the point of view of signal separation, the problem of speech noise reduction can be regarded as the problem of separation of speech signal and noise, which can make full use of the difference between speech signal and noise in time domain, frequency domain, statistical characteristics and so on to achieve the enhancement of speech signal.

(4) Geophysical Exploration

Geophysical exploration is an algorithm and theory to explore underground minerals and study geological structures by applying the principles of physics. It is mainly used in resource exploration and environmental protection. This theory is the result of interpretation and inference of geological bodies or geological structures based on physical phenomena, the problem of solving the source body of the field according to the measured data or the observed geophysical field, and the problem of inversion of the geophysical field. And BSS technology is the theory of using the observed data to estimate the source signal and even the transmission channel. Therefore, some problems in the geophysical exploration belong to the application category of BSS theory, and it is theoretically feasible to apply BSS technology to solve some problems encountered in the geophysical exploration.

(5) Communication Signal Processing

With the development of the new generation of wireless communication technology and the increasing complexity of the communication system, the number of radiation sources in wireless communication environment is also increasing, and the modulation style is becoming more and more complex. For the wireless communication detection system, the received signal is usually a mixed signal composed of many different source signals, which are often mixed and overlapping in the time–frequency domain and the modulation domain. In this case, it is difficult to separate the required signal by using traditional signal processing methods (such as frequency domain filtering), because traditional methods usually use frequency resources or power resources for signal detection, analysis and identification processing. At the same time, it is impossible for the receiving system to obtain all kinds of prior information of the signal. Especially for non-cooperative communication systems, the receiver cannot obtain the exact carrier and other signal parameters of the transmitter in advance. Therefore, if the receiver is required to complete the demodulation successfully, it is necessary to use blind signal processing technology to separate various signals from the received signals, extract the necessary signal parameters, and try to complete the data demodulating without knowing the channel. The main task of the non-cooperative communication receiving system is to sort and identify the received signals quickly and accurately, and obtain useful signals. However, in most cases, the prior information of the received signal is unknown in the non-cooperative communication. The signal detection, parameter estimation, recognition and demodulation processing are all have the characteristics of 'blind'. This makes the non-cooperative communication receiving system can only start from the 'blind' perspective to find the difference between the signals in time, frequency, space, code domain and other fields to separate and process the useful signal, which is actually the blind source separation and signal processing problem of non-cooperative communication systems.

Summary

This chapter briefly introduces the background of blind source separation problem, summarizes the development of blind source separation technology, and explains the application of blind source separation technology.

Chapter 2
The Mathematical Model and Separability Condition of BSS

There are two signal mixing modes in blind source separation (BSS) system: linear mixing and nonlinear mixing. It is difficult to solve the BSS problem of nonlinear mixing. At present, the research on BSS of linear mixing is more mature.

In this chapter, we first introduce the signal mixing model of BSS, including linear instantaneous mixing model and linear convolution mixing model. Then we give the hypothesis conditions to solve BSS problem, and finally analyze the inherent fuzziness of BSS technology.

2.1 Mathematical Model of Blind Source Separation

2.1.1 Linear Instantaneous Mixing Model

Assuming that there are N source signals and M receiving antennas (or sensors), only the attenuation of signal amplitude is considered, and the difference in time delay at different arrays (sensors) is not concerned. In this case, the mixing model is a linear instantaneous mixing model, and the mathematical expression is as follows:

$$x_j(t) = \sum_{i=1}^{N} a_{ji} s_i(t) \tag{2.1}$$

where, a_{ji} is the transmission coefficient of the i-th source signal to the j-th receiving antenna (sensor), also known as the mixing coefficient, $s_i(t)$ is the sampling value of the i-th source signal at time t, $x_j(t)$ is the sampling value of the j-th mixed signal (observation signal) at time t, $i \in [1, 2, \ldots, N]$, $j \in [1, 2, \ldots, M]$, and expands the formula (2.1):

W. Fu, *Principle and Application of Blind Source Separation Technology*,
https://doi.org/10.1007/978-981-96-0022-9_2

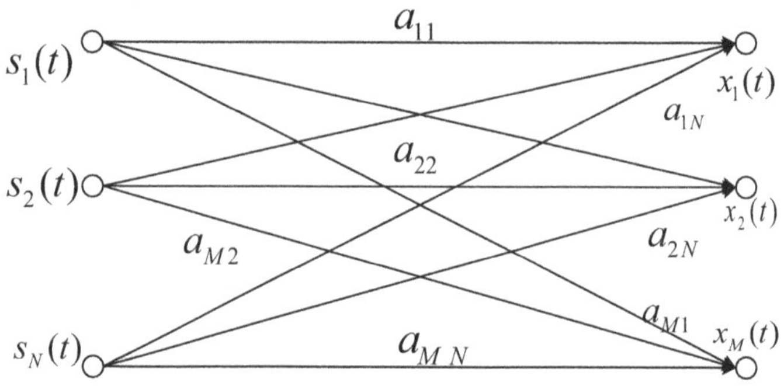

Fig. 2.1 Schematic diagram of instantaneous mixing model

$$\begin{pmatrix} x_1(t) \\ x_2(t) \\ \vdots \\ x_M(t) \end{pmatrix} = \begin{pmatrix} a_{11} \\ a_{21} \\ \vdots \\ a_{M1} \end{pmatrix} s_1(t) + \begin{pmatrix} a_{12} \\ a_{22} \\ \vdots \\ a_{M2} \end{pmatrix} s_2(t) + \ldots \begin{pmatrix} a_{1N} \\ a_{2N} \\ \vdots \\ a_{MN} \end{pmatrix} s_N(t) \tag{2.2}$$

The Eq. (2.2) is expressed as a vector:

$$\boldsymbol{x}(t) = \boldsymbol{A}\boldsymbol{s}(t) \tag{2.3}$$

where, $\boldsymbol{s}(t) = [s_1(t), s_2(t), \ldots, s_N(t)]^T$, $\boldsymbol{x}(t) = [x_1(t), x_2(t), \ldots, x_M(t)]^T$, $\boldsymbol{A}$ is the mixing matrix, and the linear instantaneous mixing model is shown in Fig. 2.1.

When $M > N$, it means the number of observed signals is more than that of the source signals, it is called overdetermined blind source separation. When $M = N$, it means the number of observed signals is equal to the number of the source signals, it is called determined blind source separation; When $M < N$, it means the number of observed signals is smaller than the number of source signal, it is called underdetermined blind source separation.

For overdetermined/determined blind separation, the goal is to find an $N \times M$ unmixing matrix also known as the separation matrix $\boldsymbol{W}$ that makes

$$\boldsymbol{y}(t) = \boldsymbol{W}\boldsymbol{x}(t) \tag{2.4}$$

is the estimation of the source signal. In general, overdetermined/determined BSS uses the statistical independence between source signals to find a separation matrix.

Underdetermined blind separation is an ill-conditioned problem. Therefore, to solve the problem of underdetermined blind source separation, it is necessary to make use of the sparsity of source signals. Firstly we estimate the mixing matrix, and then use the estimated mixing matrix to recover the source signals.

Since in practice, the transmission channel and the receiving antenna (sensor) array inevitably contain additive noise, additive noise must be considered in practical applications, and the mixing model in Eq. (2.3) becomes

$$x(t) = A(s(t) + v(t)) + r(t) \tag{2.5}$$

where $v(t) = [v_1(t), v_2(t), \ldots, v_N(t)]^T$ is the channel additive noise, $r(t) = [r_1(t), r_2(t), \ldots, r_M(t)]^T$ is the additive noise of the receiving antenna array. At present, the research on linear instantaneous mixing BSS generally only considers the influence of additive noise of the receiving antenna array.

2.1.2 Linear Convolution Mixing Model

In the actual system, the received signal is often a linear combination of the source signal after different delays, that is, the observed data is the convolution sum of the source signal, which is called the linear convolution mixing model. The linear convolution mixing model is relatively close to the reality, because: ① In reality, each source signal does not reach all sensors at the same time. Each sensor has a different delay for different sources, and the delay value depends on the relative position between the sensor and the source signal and the speed of signal propagation; ② The source signal reaches the receiving antenna after multipath propagation. Assuming that the signals from different propagation paths are linear combinations, the signals received by the receiving antenna are linear combinations of the source signals with different delays.

Suppose there are N sources $s_i(t), i \in [1, 2, \ldots, N]$, and their convolutional mixed signals $x_j(t), j \in [1, 2, \ldots, M]$ are received by M receivers. $x_j(t)$ can be expressed as:

$$x_j(t) = \sum_{i=1}^{N} \sum_{p=0}^{P} a_{ji,p} \cdot s_i(t-p) = \sum_{i=1}^{N} a_{ji} * s_i(t) \tag{2.6}$$

where, $h_{ji,p}(p = 1, 2, \ldots, P)$ denotes the transmission coefficient when the delay between the i-th ($i \in [1, 2, \ldots, N]$) source and the j-th ($j \in [1, 2, \ldots, M]$) receiving antenna is p. p is the order of the transmission function and $*$ denotes the convolution. For convenience, the Eq. (2.6) is expressed in vector form as follows:

$$x(t) = \sum_{p=0}^{P} A(p) s(t-p) \tag{2.7}$$

$$A(p) = \begin{pmatrix} a_{11,p} & \cdots & a_{1n,p} \\ \vdots & \ddots & \vdots \\ a_{m1,p} & \cdots & a_{mn,p} \end{pmatrix} \tag{2.8}$$

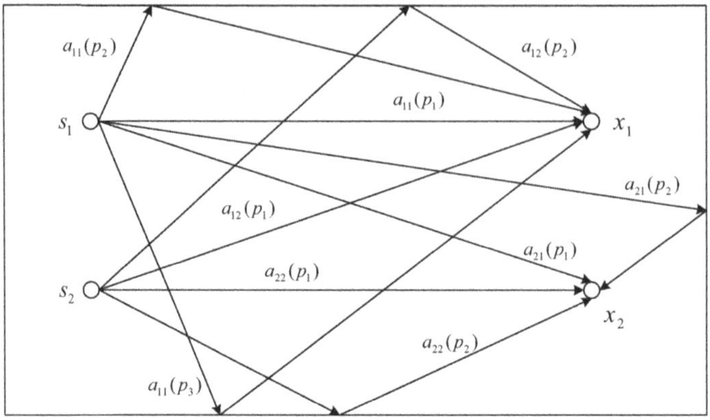

Fig. 2.2 Schematic diagram of linear convolution hybrid model

where $\boldsymbol{s}(t) = [s_1(t), s_2(t), \ldots, s_N(t)]^T, \boldsymbol{x}(t) = [x_1(t), x_2(t), \ldots, x_M(t)]^T$. When p is only 0, the Eq. (2.7) is a linear instantaneous mixing model. The schematic diagram of the linear convolution mixing model when $M = 2$ and $N = 2$ is shown in Fig. 2.2.

Under the convolution mixing model, the purpose of blind source separation is extended to solve a Q order $N \times M$ separation filter matrix $\boldsymbol{W}(q)$ that makes

$$\boldsymbol{y}(t) = \sum_{q=0}^{Q} \boldsymbol{W}(q)\boldsymbol{x}(t-p) \tag{2.9}$$

is the estimation of the source signal and $\boldsymbol{y}(t) = [y_1(t), y_2(t), \ldots, y_N(t)]^T$ is an N-dimensional column vector.

2.2 Assumptions of Linear Instantaneous Mixed Blind Source Separation

2.2.1 Assumptions of Overdetermined/Determined Blind Source Separation

In the absence of any prior information of the mixing matrix $\boldsymbol{A}$ and the source signal $\boldsymbol{s}(t)$, if there are no certain preconditions, the solution of the blind source separation problem must be multi-solution. Therefore, in order to make the blind source separation problem have a deterministic solution, it is necessary to give the assumptions and constraints of the blind source separation problem:

(1) Different source signals are statistically independent from each other.

Intuitively, if any sequence of random variables $y_1, y_2, \ldots, y_N$ is statistical independent of each other, this means that no information about the random variable y_j can be obtained from the information about the random variable y_i $(i \neq j)$. The statistical independence between random variables can be accurately characterized by the probability density function. Assume $p(y_1, y_2, \ldots, y_N)$ represents the joint probability density function of y_i $(i = 1, 2, \ldots, N)$, which $p_i(y_i)$ represents the edge probability density function of the random variable y_i $(i = 1, 2, \ldots)$, if

$$p(y_1, y_2, \ldots, y_N) = p_1(y_1)p_2(y_2)\ldots p_N(y_N) \tag{2.10}$$

we say y_i $(i = 1, 2, \ldots)$ is statistically independent of each other.

If the probability density function of the source signal s_j is $p_j(s_j)$, the probability density of the source signal vector $\boldsymbol{s} = [s_1, s_2, \ldots, s_N]$ is:

$$p_s(\boldsymbol{s}) = \prod_{j=1}^{N} p_j(s_j) \tag{2.11}$$

The assumption of independence is the basis of overdetermined/determined blind source separation, and it is also a reasonable assumption in many practical applications.

(2) Assume that only one of the N source signals is Gaussian signal.

Because the mixture of two statistically independent Gaussian signals is still a Gaussian signal, their independence is equivalent to being uncorrelated. It can be shown that the results obtained by any rotation $\boldsymbol{y} = \boldsymbol{W}\boldsymbol{x}$ ($\boldsymbol{W}$ is rotation matrix, that is $\boldsymbol{W}\boldsymbol{W}^{\mathrm{T}} = \boldsymbol{I}$) separation do not change the second-order irrelevance of Gaussian vectors, i.e., they always meet the requirement of statistical independence. Therefore, if more than one source signal follows the Gaussian distribution, the source signals are inseparable. In the real natural environment, the true pure Gaussian signal is very few, so the hypothesis (2) is reasonable and can be satisfied.

(3) Assume that the mixing matrix is full column rank.

In order to ensure that all source signal components can be separated or extracted, blind source separation requires that the mixing matrix be column full rank. In other words, if the mixing matrix is column deficient, only part of the source signal components can be extracted.

(4) the mean of source signal is zero.

If the zero mean condition is not satisfied, we can achieve this condition by preprocessing. In general, we use the technique of centralizing the observed signals, i.e. subtracting the mean of observed signals. This means that before processing the data with the BSS algorithm, the raw observed signal $\boldsymbol{x}'$ can be preprocessed by the following formula:

$$x = x' - E\{x'\} \tag{2.12}$$

Because $E\{s\} = A^{-1}E(x)$, the independent components also have zero mean. The mixing matrix remains unchanged after preprocessing, so we can centralize the observed signals without affecting the estimation of the mixing matrix.

Under the above assumptions, the task of BSS is to find a separation matrix W that makes the components of $y(t)$ in Eq. (2.13) independent of each other.

$$y(t) = Wx(t) = W(t)A(t)s(t) + \tilde{v}(n) = \Lambda(t)P(t)s(t) + \tilde{v}(n) \tag{2.13}$$

where $\tilde{v}(n)$ is noise, $\Lambda(t)$ is a real diagonal matrix, and $P(t)$ is an arbitrary exchange matrix. Then we can say that $y(t)$ is an estimation of $s(t)$, and the independent components of sources are separated. In ideal condition (i.e. without noise), if the separating matrix W we find is exactly the inverse of A, that is, the $\Lambda(t)P(t)$ is identity matrix, then

$$y(t) = Wx(t) = A^{-1}x(t) = A^{-1}As(t) = s(t) \tag{2.14}$$

In this case, the individual signals in s can be accurately separated.

2.2.2 Assumptions of Underdetermined Blind Source Separation

When the mixing matrix and source signal meet the following three conditions, the sparse component analysis method can be used to solve the underdetermined BSS problem [13]:

Condition 1: mixing matrix $A \in R^{M \times N}$, and the submatrix of any $M \times M$ dimension that constitute A is invertible matrix;

Condition 2: For any sampling time, the source signal $s(t)$ has at most $M - 1$ non-zero elements;

Condition 3: For any set containing $N - M + 1$ subscripts $I = \{i_1, i_2, \ldots, i_{N-M+1}\} \subset \{1, 2, \ldots, N\}$, there are at least M columns in the source signal matrix S such that the elements of the corresponding set of subscripts I in each column are 0, and any $M - 1$ column vectors are linearly independent.

In the above three conditions, condition (1) is the requirement for the mixing matrix, condition (2) is the condition that each signal column vector should meet, and condition (3) requires that the received signal during the sampling period should carry all the information of the mixing matrix A, so that the mixing matrix can be estimated. When an underdetermined BSS problem meets the above three conditions, the received signal can be used to recover the source signal accurately [14]. This theory has played a great role in promoting the development of underdetermined BSS.

2.3 Ambiguity of Linear Instantaneous Mixing Blind Source Separation (BSS)

2.3.1 Ambiguity of Overdetermined/Determined Blind Source Separation (BSS)

BSS technology estimates the mixing matrix and source signals according to the observed signal vector $\boldsymbol{x}$ by using independence between source signals. When two source signals are independent of each other, if their order is reversed, or both signals are multiplied by a constant, they are still independent. Therefore, BSS technology has the following two inherent uncertainties, or ambiguity, which cannot be eliminated in BSS technology.

(1) Uncertainty of the order

That is, it is not possible to determine which component of the source signal vector $\boldsymbol{s}(t)$ is corresponded to the individual component y_j of the recovered signal, because

$$\boldsymbol{x}(k) = \boldsymbol{A}\boldsymbol{s}(k) = \boldsymbol{A}\boldsymbol{P}^{-1}\boldsymbol{P}\boldsymbol{s}(k) = \tilde{\boldsymbol{A}}\tilde{\boldsymbol{s}}(k) \tag{2.15}$$

where $\boldsymbol{P}$ is the identity permutation matrix, for example

$$\boldsymbol{P}\boldsymbol{s} = \begin{bmatrix} 0\,0\,1\,0 \\ 0\,1\,0\,0 \\ 0\,0\,0\,1 \\ 1\,0\,0\,0 \end{bmatrix} \begin{bmatrix} s_1 \\ s_2 \\ s_3 \\ s_4 \end{bmatrix} = \begin{bmatrix} s_3 \\ s_2 \\ s_4 \\ s_1 \end{bmatrix} \tag{2.16}$$

Therefore, the sequence of source signals may be changed after blind separation.

(2) Uncertainty of signal (complex) amplitude.

Because

$$\boldsymbol{x} = \boldsymbol{A}\boldsymbol{s} = \sum_{j=1}^{n} \left(\frac{1}{\alpha_j} h_j \right) (s_j \alpha_j) \tag{2.17}$$

It can be seen that if s_j multiply by any non-zero complex factor α_j, and every elements in the *j*-th column of $\boldsymbol{A}$ is multiplied by $1/\alpha_j$, whatever value α_j takes, the value of $\boldsymbol{x}$ does not change. Therefore, when the source signal is obtained from $\boldsymbol{x}$, there is a scale uncertainty.

In view of the above reasons, it is generally agreed that each source signal has unit variance, that is $E\{|s_i|^2\} = 1$. The autocorrelation matrix of $\boldsymbol{s}$ is the identity matrix, i.e. $\boldsymbol{R}_{ss} = E(\boldsymbol{s}\boldsymbol{s}^{\mathrm{T}}) = \boldsymbol{I}_n$. For complex signals, we have $\boldsymbol{R}_{ss} = E(\boldsymbol{s}\boldsymbol{s}^{\mathrm{H}}) = \boldsymbol{I}_n$. In other words, after doing such constraints, the covariance matrix of recovered signal is identity matrix. For complex signals, in addition to the above two ambiguities, there is also phase ambiguity, because if α_j is complex, it contains both amplitude and phase.

In practice, the covariance matrix of source signal $\boldsymbol{R}_{ss}$ is often not equal to $\boldsymbol{I}_n$, but $\boldsymbol{R}_{ss} = \boldsymbol{D} = \mathrm{diag}(d_1, d_2, \ldots, d_n)$. At this time, $\boldsymbol{s}$ can be decomposed into $\boldsymbol{s} = \boldsymbol{D}^{1/2}\tilde{\boldsymbol{s}}$, then $\boldsymbol{R}_{\tilde{s}\tilde{s}} = \boldsymbol{I}_n$, it will be regarded $\tilde{\boldsymbol{s}}$ as the source signal vector ($\tilde{\boldsymbol{s}}$ is only different from $\boldsymbol{s}$ in amplitude), and $\tilde{\boldsymbol{A}} = \boldsymbol{A}\boldsymbol{D}^{1/2}$ will be regarded as $\boldsymbol{A}$, then observed signal also meets the linear mixing model of BSS. So, in the subsequent chapter, we assume $\boldsymbol{R}_{ss} = \boldsymbol{I}_n$.

2.3.2 Ambiguity of Underdetermined Blind Source Separation

At present, a two-step method is adopted to solve the problem of underdetermined BSS, that is, mixing matrix estimation and source signal recovery. The basic idea of mixing matrix estimating is to cluster observed signals by using the sparsity of source signals. Before estimating the mixing matrix, it is necessary to normalize the observed signal. At present, the common method is 2-norm normalization method. For example, if there are $M = 3$ observation signals, $N = 4$ source signals, and the source signals are sufficiently sparse, then the signal received by the m-th observation channel is:

$$x_m(t) = a_{11}s_1(t) + a_{12}s_2(t) + \cdots + a_{1n}s_n(t) \tag{2.18}$$

If only the i-th source is non-zero signal at time point t_i, then

$$\begin{aligned} x_1(t_i) &= a_{1i}s_i(t_i) \\ x_2(t_i) &= a_{2i}s_i(t_i) \\ x_3(t_i) &= a_{3i}s_i(t_i) \end{aligned} \tag{2.19}$$

normalize $\boldsymbol{x}(t_i) = [x_1(t_i), x_2(t_i), x_3(t_i)]^T$, we have

$$\frac{\boldsymbol{x}(t_i)}{||\boldsymbol{x}(t_i)||_2} = \left[\frac{a_{1i}}{||\boldsymbol{a}_i||_2}, \frac{a_{2i}}{||\boldsymbol{a}_i||_2}, \frac{a_{3i}}{||\boldsymbol{a}_i||_2}\right] \tag{2.20}$$

It can be seen that all the time points when the i-th source signal is non-zero, the normalized observed signals gather at the same spot, so the clustering center estimated by the clustering algorithm corresponds to the column vector of the mixing matrix, but the estimated column vector has a scale stretch to the real column vector. That is to say, the relationship between the estimated mixing matrix and the real mixing matrix is as follows:

$$\hat{\boldsymbol{A}} = \begin{bmatrix} \frac{1}{\|\boldsymbol{a}_1\|_2} & 0 & \cdots & 0 \\ 0 & \frac{1}{\|\boldsymbol{a}_2\|_2} & \cdots & 0 \\ \vdots & \ddots & \ddots & \vdots \\ 0 & \cdots & \cdots & \frac{1}{\|\boldsymbol{a}_N\|_2} \end{bmatrix} \times \boldsymbol{A} \tag{2.21}$$

where $\hat{\boldsymbol{A}}$ is the estimated mixing matrix, $\boldsymbol{A}$ is the real mixing matrix. There is a constant coefficient between the column vector of estimated mixing matrix and the real column vector, so when the estimated mixing matrix is used for source signal recovery, there will be a coefficient between the real source signal and the estimated source signal.

In addition, since the correspondence between the various types and the column vectors of the mixing matrix cannot be determined when the observed signals are clustered, there will be ambiguity in the order of estimated source signals.

Summary

This chapter first analyzes the mathematical models of BSS in detail, namely linear instantaneous mixing model and linear convolution mixing model, and gives the mathematical relationship expressions between transmitted and received signal under different models. The linear instantaneous mixing BSS problem is divided into overdetermined/determined BSS and underdetermined BSS. Secondly, the assumptions required to solve the linear instantaneous mixed BSS are given. The overdetermined/determined BSS and underdetermined BSS are discussed respectively. Finally, the ambiguity inherent in overdetermined/determined and underdetermined BSS is discussed. The content of this chapter is a prelude to the content of the following chapters.

Chapter 3
The Theoretical Basis of Blind Source Separation

To solve the problem of blind source separation, it is necessary to use several kinds related theory of signal analysis. This chapter first introduces the signal analysis theories related to BSS technology, such as Principal Component Analysis (PCA), Independent Component Analysis (ICA), Sparse Component Analysis (SCA), and then gives the common methods and ideas to solve the problem of BSS, and finally explains the performance evaluation index of BSS technology.

3.1 Principal Component Analysis (PCA)

Principal Component Analysis (PCA) [15–16] is a commonly data analysis method. PCA transforms the original data into a set of linearly independent representations of each dimension through linear transformation, which can be used to extract the main feature components of the data, and is often used for dimensionality reduction of high-dimensional data. PCA essentially takes the direction with the largest variance as the main feature, and 'de-correlates' the data in each orthogonal direction, that is, they have no correlation in the different orthogonal directions.

The optimization goal of the dimensionality reduction problem is to reduce a group of N-dimensional vectors to K-dimension ($0 < K < N$) it is to select K unit orthogonal basis, so that the covariance between the vectors obtained after the original data is transformed to this set of bases is 0, and the variance of the vectors is as large as possible.

Assume there are m n-dimension data, which is arranged into a $n \times m$ matrix $\boldsymbol{X}$ in columns after de-meaning, let

$$\boldsymbol{C} = \frac{1}{m}\boldsymbol{X}\boldsymbol{X}^{\mathrm{T}} \tag{3.1}$$

W. Fu, *Principle and Application of Blind Source Separation Technology*,
https://doi.org/10.1007/978-981-96-0022-9_3

then $\boldsymbol{C}$ is a symmetric matrix, the diagonal lines are the variance of each vector, and the i-th row j-th column elements are same with the j-th row i-th column elements, representing the covariance between i-th and j-th vector.

From the above derivation, we find that to achieve the purpose of optimization, it is equivalent to diagonalizing the covariance matrix. That is, all the other elements of the covariance matrix are zero except for the diagonal elements, and arranging the elements diagonally by size from top to bottom, such we achieve the purpose of optimization.

Let the covariance matrix corresponding to the original data matrix $\boldsymbol{X}$ be $\boldsymbol{C}$, and $\boldsymbol{P}$ is a matrix of rows of basis vectors, suppose $\boldsymbol{Y} = \boldsymbol{PX}$, and the covariance matrix of $\boldsymbol{Y}$ is $\boldsymbol{D}$, then

$$\boldsymbol{D} = \frac{1}{m}\boldsymbol{YY}^{\mathrm{T}} = \frac{1}{m}(\boldsymbol{PX})\left(\boldsymbol{PX}^{\mathrm{T}}\right) = \boldsymbol{P}\left(\frac{1}{m}\boldsymbol{XX}^{\mathrm{T}}\right)\boldsymbol{P}^{\mathrm{T}} = \boldsymbol{PCP}^{\mathrm{T}} \tag{3.2}$$

Then $\boldsymbol{P}$ is the matrix that diagonalizes covariance matrix $\boldsymbol{C}$. In other words, the optimization goal becomes to find a matrix $\boldsymbol{P}$, which can make $\boldsymbol{PCP}^{\mathrm{T}}$ is a diagonal matrix, and the diagonal elements are arranged in order from largest to smallest. Then the first K rows of $\boldsymbol{P}$ are the bases to be found, and the matrix composed of the first K rows of $\boldsymbol{P}$ is multiplied by $\boldsymbol{X}$ to reduce $\boldsymbol{X}$ from N-dimension to K-dimension, and meet the above optimization conditions.

From the related knowledge of matrix, we can know that n unit orthogonal eigenvectors can be found from a real symmetric matrix with n rows n columns. Let n eigenvectors be $\boldsymbol{e}_1, \boldsymbol{e}_2, \boldsymbol{e}_3, \ldots, \boldsymbol{e}_n$, we will form a matrix by its columns:

$$\boldsymbol{E} = [\boldsymbol{e}_1, \boldsymbol{e}_2, \boldsymbol{e}_3, \ldots, \boldsymbol{e}_n] \tag{3.3}$$

Then:

$$\boldsymbol{E}^{\mathrm{T}}\boldsymbol{CE} = \boldsymbol{\Lambda} = \begin{pmatrix} \lambda_1 & & & \\ & \lambda_2 & & \\ & & \ddots & \\ & & & \lambda_n \end{pmatrix} \tag{3.4}$$

where $\boldsymbol{\Lambda}$ is the diagonal matrix, and its diagonal elements are the eigenvalues corresponding to each eigenvector (there may be duplications). Therefore, the required matrix $\boldsymbol{P}$ has been found:

$$\boldsymbol{P} = \boldsymbol{E}^{\mathrm{T}} \tag{3.5}$$

$\boldsymbol{P}$ is the matrix obtained by arranging the eigenvectors of the covariance matrix $\boldsymbol{C}$ in rows, each row of $\boldsymbol{P}$ is an eigenvector of $\boldsymbol{C}$.

The whitening preprocessing method in ICA can effectively reduce the complexity of the problem, and the algorithm is simple, and can be completed by traditional PCA.

The whitening preprocessing of the observed signal with PCA makes the original unmixing matrix degenerate into an orthogonal matrix. In addition, when the number of observed signals is greater than the number of source signals, the number of observed signals can be automatically reduced to the same as the source signals after whitening.

3.2 Independent Component Analysis (ICA)

Under the assumption of statistical independence of the source signals, ICA determines an inverse transformation to ensure that the components of the output signal are as independent as possible. BSS aims to separate and reconstruct the source signal regardless of whether the source signal has statistical independence. Therefore, ICA is a subclass of the many ways to implement BSS. Because the source signal is usually transmitted by different physical devices, statistical independence is easily satisfied, ICA is often used to replace BSS. Specifically, ICA takes the non-Gaussian source signals as the research object, and requires there is at most one Gaussian signal in the source signals. Because the linear mixture of multiple Gaussian distributed signals still follows Gaussian distribution and is no longer separable in the sense of independence.

The basic purpose of Independent Component Analysis (ICA) is to determine the linear transformation matrix $\boldsymbol{W}$ so that the output components $\boldsymbol{y}(t) = \boldsymbol{W}\boldsymbol{x}(t)$ after the transformation are as statistically independent as possible. Intuitively, there are two main parts to implement ICA: the objective function and the optimization method.

Generally speaking, different objective functions are obtained by different estimation criteria, and ICA is realized by appropriate optimization methods, that is to find the mixing matrix $\boldsymbol{H}$ and independent component $\boldsymbol{s}$. Most of these optimization methods are gradient-based methods. In order to more clearly describe the method of implementing ICA, suppose that the objective function obtained by different estimation criteria is represented by $J(\boldsymbol{W})$, and the n-th row of $\boldsymbol{W}$ is represented by $\boldsymbol{w}_n$, then the optimization problem is (taking maximization as an example, minimization is equivalent to it) [17–18]:

$$\begin{cases} \max\limits_{\boldsymbol{W}}(J(\boldsymbol{W})) \\ s.t.|\boldsymbol{w}_n| = 1, \quad n = 1, 2, \ldots N \end{cases} \tag{3.6}$$

At present, the main methods of ICA for BSS include non-Gaussian maximization method, mutual information minimization method and maximum likelihood function estimation. What these methods have in common is that they measure the independence of the output signals according to certain criteria. The different measures of independence constructed different objective functions, and different BSS algorithms are derived by optimizing these objective functions with different algorithms. Next, we first give the common independence measurement methods, and then introduce the above three blind separation methods respectively.

In blind source separation, the Kullback–Leibler divergence (also known as relative entropy) is often used as a measure of independence. Kullback–Leibler divergence is defined as the distance between $\boldsymbol{p}_y(y)$ probability density functions and $\boldsymbol{q}(y)$ probability density functions, i.e.

$$D_{pq} = KL[\boldsymbol{p}_y(y), \boldsymbol{q}(y)] = \int \boldsymbol{p}_y(y) \log \frac{\boldsymbol{p}_y(y)}{\boldsymbol{q}(y)} d\boldsymbol{y} \tag{3.7}$$

The Kullback–Leibler distance between probability density function $\boldsymbol{p}_y(y)$ and $\boldsymbol{q}(y)$ satisfy the following inequality:

$$D_{pq} \geq 0 \tag{3.8}$$

The necessary and sufficient condition for equation is that the probability density function $\boldsymbol{p}_y(y)$ and $\boldsymbol{q}(y)$ are equal almost everywhere. Kullback–Leibler divergence is a measure of the degree of deviation between probability density functions.

3.2.1 *Non-Gaussian Maximization Method*

In most classical statistical theories, random variables are assumed to be Gaussian distributions. A classical conclusion in probability theory, the central limit theorem, states that under certain conditions, the sum of independent random variables tends to be Gaussian, and the sum of independent random variables is closer to the Gaussian distribution than any of the original random variables.

Let $y_k = \boldsymbol{w}^{\mathrm{T}}\boldsymbol{x}$, where $\boldsymbol{w}$ is an undetermined column vector, let $\boldsymbol{z} = \boldsymbol{A}^{\mathrm{T}}\boldsymbol{w}$, then

$$y_k = \boldsymbol{w}^{\mathrm{T}}\boldsymbol{x} = \boldsymbol{w}^{\mathrm{T}}\boldsymbol{A}\boldsymbol{s} = \boldsymbol{z}^{\mathrm{T}}\boldsymbol{s} \tag{3.9}$$

So y_k is a linear combination of s_i. Known by the central limit theorem $\boldsymbol{z}^{\mathrm{T}}\boldsymbol{s}$ is closer to the Gaussian distribution than any other s_i, and therefore $\boldsymbol{w}$ can be regarded as a vector of maximizing the non-Gaussianity of $\boldsymbol{w}^{\mathrm{T}}\boldsymbol{x}$. Then y_k is the estimation of one independent components. Therefore, to estimate all the independent components (source signals) $\boldsymbol{s}_i(i = 1, 2, \ldots, N)$, we just need to find $\boldsymbol{N}$ such vectors $\boldsymbol{w}_i(i = 1, 2, \ldots, N)$. Let $\boldsymbol{W} = [\boldsymbol{w}_1, \boldsymbol{w}_2 \ldots, \boldsymbol{w}_N]^{\mathrm{T}}$, the estimation of the source signal can be expressed as:

$$\boldsymbol{y} = \boldsymbol{W}\boldsymbol{x} \tag{3.10}$$

where $\boldsymbol{y} = [y_1, y_2, \ldots, y_N]^{\mathrm{T}}$ is the estimation of N source signals. Therefore, the key of this method is the measurement of non-Gaussianity. The measurement methods of non-Gaussianity include kurtosis method [19] and negative entropy method [20].

Kurtosis, also known as the fourth order cumulative quantity, the kurtosis of y_i is defined as:

$$kurt(y_i) = E\{y_i^4\} - 3E\{y_i^2\}^2, (i = 1, 2, \ldots, N) \quad (3.11)$$

For a Gaussian random variable, its kurtosis is equal to zero; But for most non-Gaussian random variables, its kurtosis is not equal to zero. Kurtosis can be positive or negative, and the positive one is called a super-Gaussian signal and the negative one is called a sub-Gaussian signal. Therefore, the greater the absolute value of the kurtosis of y_i, the stronger its non-Gaussianity, which is a non-Gaussian kurtosis measure.

The non-Gaussian negative entropy measure is introduced based on the concept of entropy in information theory. A basic conclusion of information theory is that the Gaussian variable has the maximal entropy among all random variables with equal variance. This means that entropy can be used as a measure of non-Gaussianity. Negative entropy can be obtained by modifying the definition of entropy, which is defined as follows:

$$J(y) = H(y_{\text{gauss}}) - H(y) \quad (3.12)$$

where $\boldsymbol{y}_{\text{gauss}}$ is a Gaussian random vector with the same covariance of $\boldsymbol{y}$, $H(\boldsymbol{y})$ representing the entropy of $\boldsymbol{y}$. Based on the properties mentioned above, negative entropy is always non-negative, and it is zero if and only if $\boldsymbol{y}$ is a Gaussian distribution. Therefore, the greater the negative entropy of $\boldsymbol{y}$, the better its non-Gaussianity.

3.2.2 Maximum Likelihood (ML)

The so-called maximum likelihood estimation is to find a matrix $\boldsymbol{W}$ so that the distribution of $\boldsymbol{Wx}$ is as close as possible to the distribution of the source signal, it is a common method of ICA it is closely related to information theory, and its essence is the same as minimizing mutual information. Let $\boldsymbol{W} = (\boldsymbol{w}_1, \boldsymbol{w}_2, \ldots, \boldsymbol{w}_N)^{\mathrm{T}}, \boldsymbol{y} = \boldsymbol{Wx}$ is the estimate of the source signal, let $\boldsymbol{q} = [q_1, q_2, \ldots q_N]$ is the probability density function of $\boldsymbol{y}$(assuming it has been obtained), then the probability density of $\boldsymbol{x}$ is:

$$p(\boldsymbol{x}) = |\det \boldsymbol{W}| \cdot \boldsymbol{q}(\boldsymbol{Wx}) = |\det \boldsymbol{W}| \cdot \prod_{i=1}^{N} q_i(\boldsymbol{w}_i^{\mathrm{T}} \boldsymbol{x}) \quad (3.13)$$

If $\boldsymbol{x}(t)(t = 1, 2, \ldots, T)$ is the T samples of $\boldsymbol{x}$, and they are independent of each other, then

$$p(\boldsymbol{x}) = p(\boldsymbol{x}(1)) \times p(\boldsymbol{x}(2)) \cdots \times p(\boldsymbol{x}(N)) \quad (3.14)$$

Taking the logarithm of $p(\boldsymbol{x})$, we can get logarithmic likelihood function as follow.

$$L = \sum_{t=1}^{T}\sum_{i=1}^{N} \log q_i(\boldsymbol{w}_i^{\mathrm{T}} x(t)) + T \log|\det \boldsymbol{W}| \tag{3.15}$$

If the separating matrix $\boldsymbol{W}$ can be found to maximize the likelihood function L, then the individual signal components can be separated by $\boldsymbol{y} = \boldsymbol{W}\boldsymbol{x}$.

3.2.3 Minimizing Mutual Information (MMI)

The mutual information of N random variables $y_i (i = 1, 2, \ldots, N)$ is defined as:

$$I(y_1, y_2, \ldots, y_N) = \sum_{i=1}^{N} H(y_i) - H(\boldsymbol{y}) \tag{3.16}$$

where $H(\boldsymbol{y})$ is the entropy of $\boldsymbol{y}$. Mutual information is a natural measure of correlation, which is always non-negative and is 0 if and only if the variables are statistically independent. An important property of mutual information is that if $\boldsymbol{y} = \boldsymbol{W}\boldsymbol{x}$, then

$$I(y_1, y_2, \ldots, y_N) = \sum_{i=1}^{N} H(y_i) - H(\boldsymbol{x}) - \log|\det \boldsymbol{W}| \tag{3.17}$$

If y_i is uncorrelated and has a unit variance, i.e. $E(\boldsymbol{y}\boldsymbol{y}^T) = \boldsymbol{I}$, then:

$$E(\boldsymbol{y}\boldsymbol{y}^T) = E(\boldsymbol{W}\boldsymbol{x}\boldsymbol{x}^T\boldsymbol{W}^T) = \boldsymbol{W}\boldsymbol{R}_{xx}\boldsymbol{W}^T = \boldsymbol{I} \tag{3.18}$$

Then:

$$\boldsymbol{R}_{xx} = \boldsymbol{W}^{-1}\boldsymbol{W}^{-\mathrm{T}} \tag{3.19}$$

According to Eqs. (3.17) and (3.19),we have:

$$\begin{aligned} I(y_1, y_2, \ldots, y_N) &= \sum_{i=1}^{N} [\mathrm{H}(y_i) - \mathrm{H}(y_{i,\mathrm{gauss}})] + \sum_{i=1}^{N} \mathrm{H}(y_{i,\mathrm{gauss}}) - \mathrm{H}(\boldsymbol{x}) - \log|\det(\boldsymbol{W})| \\ &= -\sum_{i=1}^{N} \mathrm{J}(y_i) + \sum_{i=1}^{N} \mathrm{H}(y_{i,\mathrm{gauss}}) - \mathrm{H}(\boldsymbol{x}) + \frac{1}{2}\log|\det R_{xx}| \end{aligned} \tag{3.20}$$

That is

$$I(y_1, y_2, \ldots, y_N) = c - \sum_{i=1}^{N} \mathrm{J}(y_i) \tag{3.21}$$

where $\mathrm{J}(y_i)$ is the negative entropy of y_i, and c is a constant independent of $\boldsymbol{W}$. Therefore, if we can find an invertible matrix $\boldsymbol{W}$ that minimize the mutual information of N random variables $y_i (i = 1, 2, \ldots, N)$, we have found the maximum direction of negative entropy. According to the non-Gaussian negative entropy metric, the source signal can be estimated.

3.3 Sparse Component Analysis (SCA)

Sparse Component Analysis (SCA) mainly studies the measurement of signal sparsity and sparse representation of signals. At present, a prerequisite to effectively solve the problem of underdetermined BSS is that the source signal has sparsity in time domain or transform domain (such as time–frequency domain). Intuitively, the sparsity of the signal means that the value of the signal is 0 or very small at most times, and only a few times the value is large. The sparsity of source signal can be divided into two cases: sufficient sparsity and insufficient sparsity. A sufficiently sparse source signal means that only one source signal is non-zero at time t, that is $s_i(t) \neq 0$ and $s_j(t) = 0$ for $j \neq i$. In contrast, source signal is not sufficiently sparse means that there are more than one source signal is non-zero at time t.

Since signals with generalized Gaussian distribution exist widely in nature, researchers use the probability density function of generalized Gaussian distribution to study the measure of signal sparsity. The probability of the generalized Gaussian distribution is

$$P(s; \alpha) = \frac{1}{2\Gamma(1 + 1/\alpha)} \exp(-|s|^{\alpha}), \quad \alpha > 0 \tag{3.22}$$

where, $\Gamma(\cdot)$ represents the gamma function,$\Gamma(s) = \int_0^{\infty} v^{s-1} e^{-v} dv > 0$, the shape of the $P(s; \alpha)$ is closely related to the value of the parameter α. As shown in Fig. 3.1 [21], when the parameter α value is small, the shape of the probability density curve is sharper. Parameter α can measure signal sparsity to some extent. In fact, when $\alpha = 2$, $P(s; \alpha)$ is a probability density function of a Gaussian distribution, and when $\alpha = 1$, the source signal follows the Laplacian distribution.

Next, we illustrate how the parameter α measures the sparsity of the signal by the distribution shape of the equal probability density curve (equivalent to the contour line). First, assuming that each component of the source signal vector $\boldsymbol{s} = [s_1, s_2, \ldots, s_N]^{\mathrm{T}}$ is independently identically distributed, and all follow the generalized Gaussian distribution, then the joint probability density function of the source signal vector is

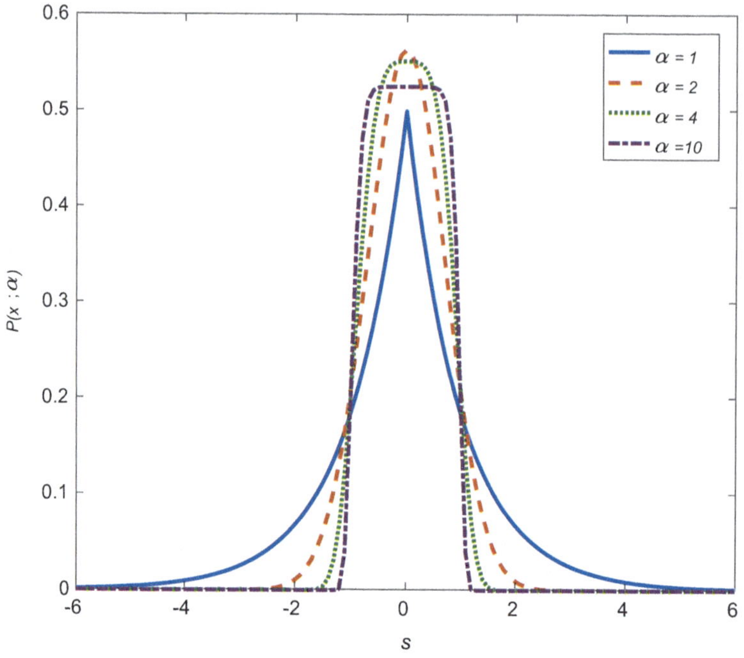

Fig. 3.1 Probability density curve of generalized Gaussian distribution

$$
\begin{aligned}
P(s_1, s_2, \ldots, s_N) &= \prod_{i=1}^{N} \frac{1}{2\Gamma(1+1/\alpha)} \exp(-|s_i|^{\alpha}) \\
&= \left(\frac{1}{2\Gamma(1+1/\alpha)}\right)^N \exp\left(-\sum_{i=1}^{N} |s_i|^{\alpha}\right)
\end{aligned}
\tag{3.23}
$$

In order to find the contour line of the Eq. (3.23), let $P(s_1, s_2, \ldots, s_N)$ be a fixed value $P_c (0 \le P_c \le 1)$, then

$$
\left(\frac{1}{2\Gamma(1+1/\alpha)}\right)^N \exp\left(-\sum_{i=1}^{N} |s_i|^{\alpha}\right) = P_c \tag{3.24}
$$

From Eq. (3.24), we have

$$
\exp\left(-\sum_{i=1}^{N} |s_i|^{\alpha}\right) = P_c[2\Gamma(1+1/\alpha)]^N \tag{3.25}
$$

Take the logarithm base natural number e on both sides of the Eq. (3.25), we have

$$\sum_{i=1}^{N} |s_i|^{\alpha} = -\ln P_c[2\Gamma(1 + 1/\alpha)]^N \tag{3.26}$$

Let $-\ln P_c[2\Gamma(1 + 1/\alpha)]^N = c$, then

$$\sum_{i=1}^{N} |s_i|^{\alpha} = |s_1|^{\alpha} + |s_2|^{\alpha} + \cdots + |s_N|^{\alpha} = c \tag{3.27}$$

Therefore, the contour line distribution shape of Eq. (3.27) is consistent with the distribution of the equiprobability density curve. By changing the value of the constant c in Eq. (3.27), the contour line family can be obtained. Take 2 source signals as an example, that is, the source signal vector $\boldsymbol{s} = [s_1, s_2]^{\mathrm{T}}$. According to Eq. (3.27), contour lines are drawn under different parameter α values to obtain the contour line family as shown in Fig. 3.2 [21], where Fig. 3.2a–f correspond to parameter $\alpha = 2, 1.5, 1, 0.5, 0.1, 0.01$ respectively.

It can be seen from the figure that contour line family is a closed curve family, and with the decrease of the parameter α, the contour line moves closer to the coordinate axis. When $\alpha \to 0^+$, the contour line family is almost condensed into two straight lines. Therefore, the value of α represents the degree of the signal sparsity. when $\alpha = 1$, Eq. (3.27) denotes the ℓ_1 norm

$$\|\boldsymbol{s}\|_1 = \sum_{i=1}^{N} |s_i| \tag{3.28}$$

Now let's think about the case where $\alpha \to 0^+$. Assuming that the set $\{i_1, i_2, \ldots, i_k\} \subset \{1, 2, \ldots, N\}$, k is the number of non-zero elements of the source signal vector $\boldsymbol{s} = [s_1, s_2, \ldots, s_N]^{\mathrm{T}}$ when $i \in \{i_1, i_2, \ldots, i_k\}$, $s_i \neq 0$ but if $i \in \{i_1, i_2, \ldots, i_k\}$ and $i \in \{1, 2, \ldots, N\}$, then $s_i = 0$ so the limit of (3.27) is obtained

$$\lim_{\alpha \to 0^+} \sum_{i=1}^{N} |s_i|^{\alpha} = \lim_{\alpha \to 0^+} \left(\left|s_{i_1}\right|^{\alpha} + \left|s_{i_2}\right|^{\alpha} + \cdots + \left|s_{i_k}\right|^{\alpha}\right) = \sum_{j=1}^{k} \lim_{\alpha \to 0^+} \left|s_{i_j}\right|^{\alpha} \tag{3.29}$$

Due to $0 < \left|s_{i_j}\right| < +\infty$, $\lim_{\alpha \to 0^+} \left|s_{i_j}\right|^{\alpha} = 1$. Thus Eq. (3.29) becomes

$$\lim_{\alpha \to 0^+} \sum_{i=1}^{N} |s_i|^{\alpha} = k \tag{3.30}$$

It is customary to call the left side of Eq. (3.30) the ℓ_0 norm and denote as $\|s\|_0$. Thus, the ℓ_0 norm of a vector represents the number of non-zero elements of the

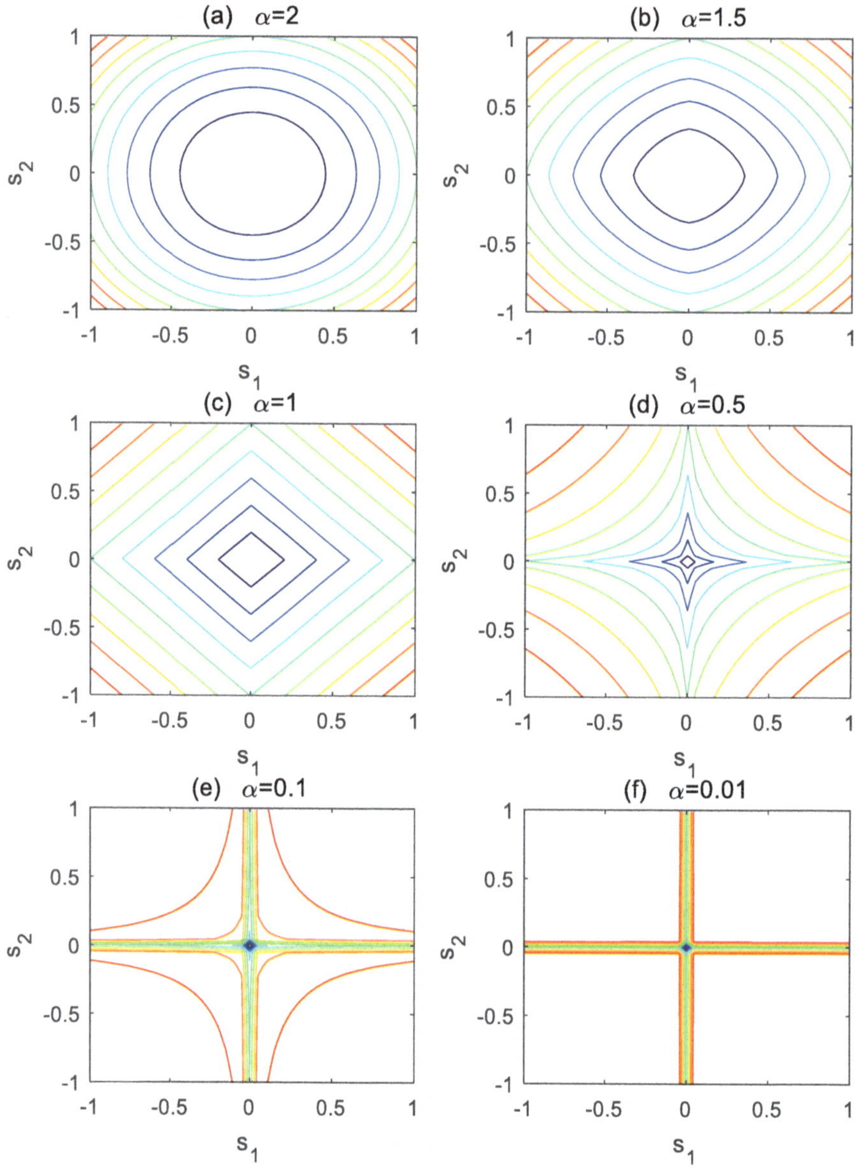

Fig. 3.2 Equiprobability curve of source signal vector

vector. However, the ℓ_0 norm is not a norm in the true sense of the word. From the mathematical definition, the norm must satisfy three conditions:

(1) Non-negativity: when $\boldsymbol{s} \neq \mathbf{0}$ ($\boldsymbol{s}$ is not zero vector), $\|\boldsymbol{s}\| > 0$; When $\boldsymbol{s} = \mathbf{0}$ (the elements of $\boldsymbol{s}$ are 0), $\|\boldsymbol{s}\| = 0$.

(2) Isometry: $\|\lambda s\| = |\lambda| \|s\|$, where λ is a constant.
(3) Additive inequality: $\|s + s'\| \leq \|s\| + \|s'\|$, $\|s\|$, $\|s'\|$ is the norm of any vector.

Obviously, the ℓ_0 norm does not satisfy the isometry. For example, according to Eq. (3.27) $\|s\|_0 = k$, if $\lambda = 2$, then $\|\lambda s\|_0 = \|2s\|_0 = k$, and $|\lambda| \|s\|_0 = 2\|s\|_0 = 2k$. That is $\|\lambda s\|_0 \neq |\lambda| \|s\|_0$. The ℓ_0 norm is therefore not a norm in the mathematical sense, but a customary notation. The ℓ_1 norm is the norm in the true sense because it satisfies the above three conditions.

3.4 Common Methods of Blind Source Separation

3.4.1 Common Methods of Overdetermined/Determined Blind Source Separation

Currently commonly used overdetermined/determined blind source separation algorithms can be divided into the following three categories:

(1) After the signal is transformed, the dependency between different signal components is minimized. This method is called independent component analysis (ICA), was first proposed by Comon in 1994.
(2) A nonlinear transfer function is used to transform signal so that distribution of transformed signal is contained in a finite hypercube, and then the entropy maximization will force signal to be evenly distributed in the hypercube as much as possible. This kind of method is called the entropy maximization method, which was proposed by Bell and Sejnowski in 1995.
(3) Nonlinear principal component analysis is a generalization of linear principal component analysis [22].

It has been proved that entropy maximization method is equivalent to independent component analysis.

3.4.2 Common Methods of Underdetermined Blind Source Separation

At present, the problem of underdetermined BSS is usually solved by two-step method. The 'two-step method' means that the mixing matrix is estimated first, and then the source signals are recovered. When the source signals are sparse, especially when they are sufficiently sparse, the scatterplot of the observed signal shows the characteristics of line clustering, and each clustering center corresponds to each column vector of the mixing matrix. When the source signals are not sufficiently sparse, the observed signal space presents subspace clustering characteristics, and each column vector of the mixing matrix is in the intersection of each subspace. after mixing matrix estimating, the sparse signal reconstruction method is used to recover the source signals.

1. **Mixing Matrix Estimation**

When the source signals are sufficiently sparse, only one source signal is active at any time, for example, only the i'-th source signal is active, then Eq. (2.1) becomes

$$\boldsymbol{x}(t) = \sum_{i=1}^{N} s_i(t)\boldsymbol{a}_i \tag{3.31}$$

Therefore, the observed signal presents the characteristics of linear clustering. As shown in Fig. 3.3, two receiving channels receive four source signals, and the four lines shown in the scatter diagram of observed signal correspond to the four column vectors of the mixing matrix.

After normalizing the observed signals, the scatterplot appears four clustering centers as shown in Fig. 3.4. At this time, the clustering centers are corresponding to normalized four column vectors of the mixing matrix.

Adopting point clustering method, we can estimate each column vector of the mixing matrix by clustering the normalized observed signals. Point clustering methods include k-means clustering, hierarchical clustering [23], etc. These clustering methods require the number of source signals (i.e. the number of clustering

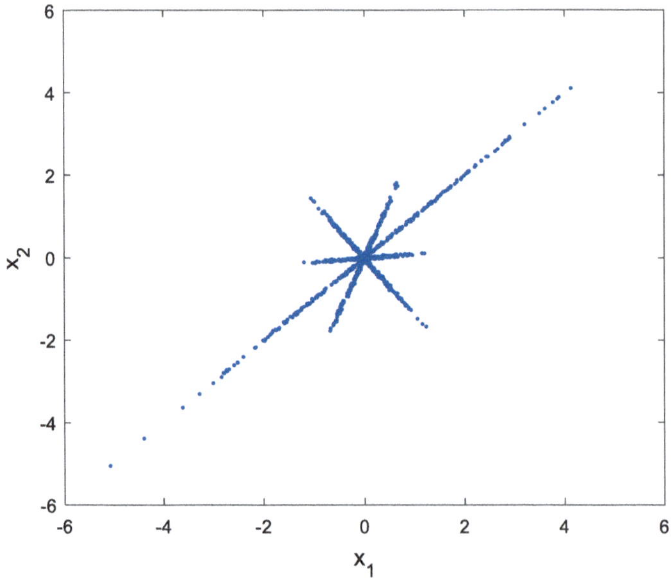

Fig. 3.3 Scatterplot of observed signals in the case of sufficiently sparse source signals

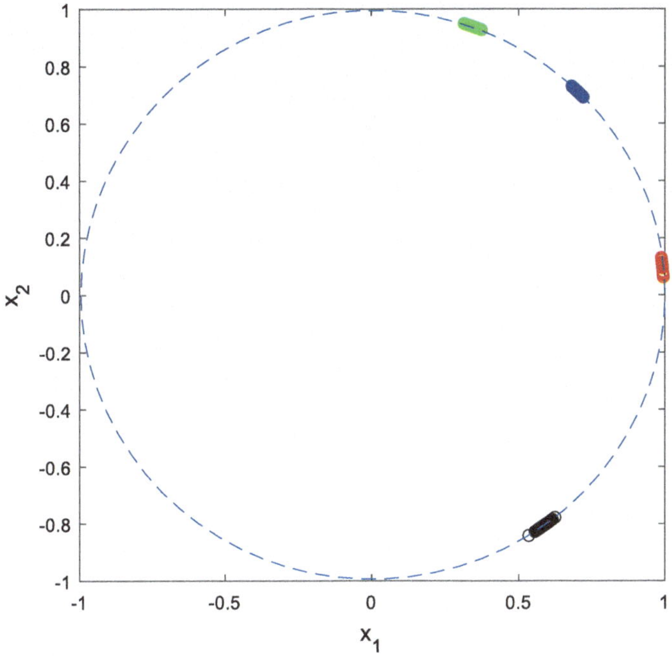

Fig. 3.4 Scatterplot of normalized observed signals under sufficiently sparse source signals

centers) to be known. In view of the unknown number of source signals, an improved k-means clustering method is proposed [24], which is used for point clustering in the following sections. Two definitions as follows are introduced.

Definition 3.1 If at time t, only one source signal is active, that is, the observed signal vector $x(t)$ satisfies the formula (3.31), then $x(t)$ is called a single source point.

Definition 3.2 If several source signals are active at time t, the observed signal vector at time t is.

$$\boldsymbol{x}(t) = s_{i_i}(t)\boldsymbol{a}_{i_1} + s_{i_2}(t)\boldsymbol{a}_{i_2} + \cdots + s_{i_k}(t)\boldsymbol{a}_{i_k} \tag{3.32}$$

where, $\{i_1, i_2, \ldots, i_k\} \subset \{1, 2, \ldots, N\}$, k is the number of source signals that are active at the time t, and $k \geq 2$, then the $\boldsymbol{x}(t)$ in Eq. (3.32) is called the multiple source points.

According to the above definition, in the ideal case where the source signals are sufficiently sparse (no noise and interference points), the observed signal space is composed of single source points. But in practical applications, there will be noise and interference from multiple source points, so how to identify single source points is a problem that needs to be studied.

If there are a large number of multiple source points in the time domain or transformation domain, that is, the source signals are not sufficiently sparse, multiple

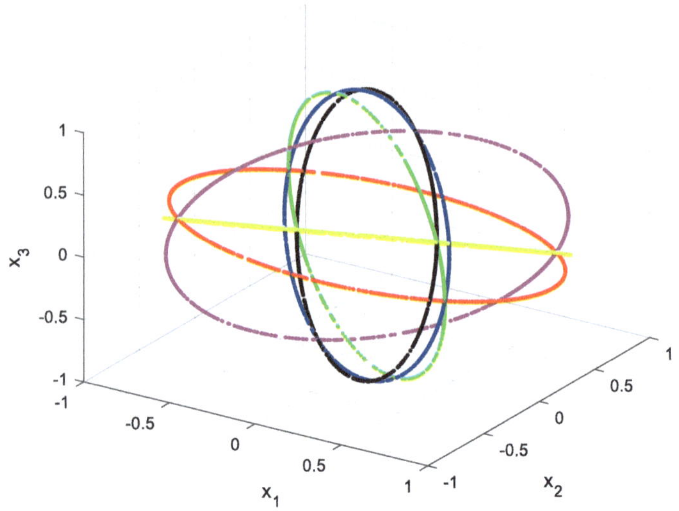

Fig. 3.5 Scatterplot of normalized observed signals in the case of insufficiently sparse source signals

subspaces (hyperplanes) are formed in the observed signal space. For a more intuitive explanation, let $M = 3, N = 4$, there are $k(k = 2)$ source signals active at every moment, as shown in Fig. 3.5. At this time, the scatter plot of normalized observed signals presents the characteristics of plane clustering, with a total of $C_N^k = C_4^2 = 6$ planes.

It is obvious that the four column vectors $\boldsymbol{a}_1, \boldsymbol{a}_2, \boldsymbol{a}_3, \boldsymbol{a}_4$ of the mixing matrix $\boldsymbol{A}$ are hidden in the intersection lines of the plane, and each column vector $\boldsymbol{a}_i (i = 1, 2, 3, 4)$ is in the intersection lines of $C_{N-1}^{k-1} = C_{4-1}^{2-1} = 3$ planes. For the more general case (high dimension), if the number of source signals active at the same time is k and $k \leq M - 1$, then the observed signal $\boldsymbol{x}$ forms multiple subspaces in the space $\boldsymbol{R}^N$, and the individual column vectors of the mixing matrix are at the intersection of the subspaces. Therefore, when the source signals are not sufficient, the mixing matrix estimating is divided into two stages, the first stage is to estimate the subspace (hyperplane), the second stage is to estimate the intersecting line of the column vectors of the mixing matrix.

2. **Source Signal Recovery**

In the absence of noise, if the mixing matrix $\boldsymbol{A}$ is known, the source signal recovery problem can be equivalent to the following optimization problem:

$$\min \sum_{i=1}^{N} \sum_{t=1}^{T} |s_i(t)| \quad \text{subject to } \boldsymbol{X} = \boldsymbol{AS} \tag{3.33}$$

It is not difficult to prove that Eq. (3.33) can be decomposed into T optimization problems

$$\min \sum_{i=1}^{N} |s_i(t)| \quad \text{subject to}\, \boldsymbol{x}(t) = \boldsymbol{A}\boldsymbol{s}(t) \qquad t = 1, 2, \ldots, T \tag{3.34}$$

The objective function in formula (3.34) is the ℓ_1 norm of the source signal vector $\boldsymbol{s}(t)$, and the source signal can be recovered by solving the problem of formula (3.34). In the case of omitting t, write the formula (3.34) as

$$\min \|\boldsymbol{s}\|_1 \quad \text{subject to}\, \boldsymbol{x} = \boldsymbol{A}\boldsymbol{s} \tag{3.35}$$

where, $\|\cdot\|_1$ represents the ℓ_1 norm of the vector. Let's explain the formula (3.35) geometrically. Take $N = 3$ as example, the objective function in formula (3.35) is $F_s = |s_1| + |s_2| + |s_3|$, the corresponding polygonal body of $|s_1| + |s_2| + |s_3| = 1$ is shown in Fig. 3.6a, and the contour line is a closed family of diamond curves (consistent with Fig. 3.2c), so the polygonal body is graphically called a "diamond" ball. Objective function value $F_s \geq 0$.As the value of the F_s increases, the "diamond shape" expands outward continuously, as shown in Fig. 3.6b, and the corresponding contour line also extends outward continuously. When the diamond shape is tangent to the feasible domain $\{s|x = \boldsymbol{A}\boldsymbol{s}\}$ in Eq. (3.35), the optimal solution will be obtained at the vertex of the diamond shape. Obviously, the coordinate vectors of the vertices of the "diamond" ball in Fig. 3.6a are sparse, and the coordinates of the six vertices are $[1\,0\,0;\, 0\,1\,0;\, -1\,0\,0;\, 0\,-1\,0;\, 0\,0\,1;\, 0\,0\,-1]$.

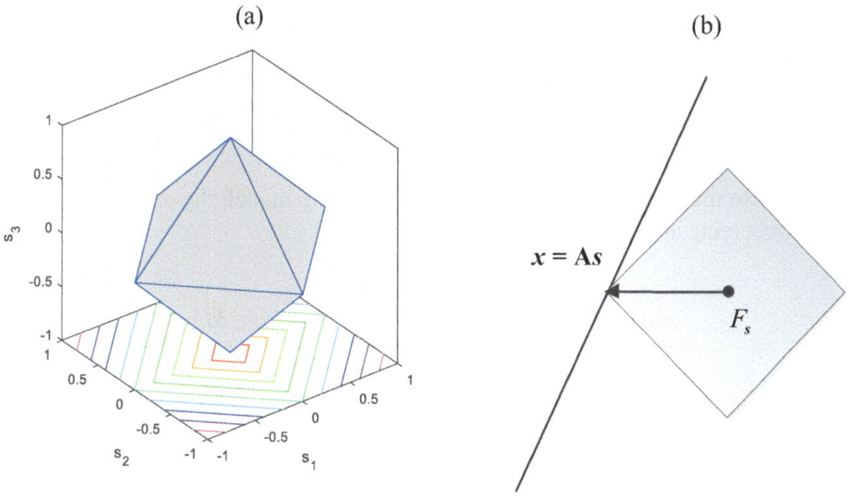

Fig. 3.6 Geometric diagram of optimization ℓ_1 norm

According to the above analysis, the solution of Eq. (3.35) is a sparse solution. In order to find the most sparse solution, the most direct way is to optimize the ℓ_0 norm as follow.

$$\min\|s\|_0 \quad \text{subject to } x = \boldsymbol{A}s \tag{3.36}$$

where, $\|\cdot\|_0$ represents the ℓ_0 norm of the vector. In fact, formulas (3.35) and (3.36) are equivalent under certain conditions. For a matrix $\boldsymbol{A} \in \boldsymbol{R}^{M\times N}$, if there is a minimum constant δ_k that make all the vectors in the set $\{\boldsymbol{s}|\boldsymbol{s} \in \boldsymbol{R}^N, \|\boldsymbol{s}\|_0 = k\}$ satisfy

$$(1-\delta_k)\|\boldsymbol{s}\|_2^2 \le \|\boldsymbol{A}\boldsymbol{s}\|_2^2 \le (1+\delta_k)\|\boldsymbol{s}\|_2^2 \tag{3.37}$$

δ_k is called the isometric constant of the mixing matrix $\boldsymbol{A}$. If the isometric constant $\delta_{2k} < \sqrt{2}-1$, the solution of formula (3.36) is equivalent to the solution of formula (3.35). However, the solution of formula (3.36) belongs to NP-hard problem and is not easy to solve. In order to solve the problem effectively, the approximate ℓ_0 norm is used instead of the ℓ_0 norm.

3.5 Performance Index of Blind Source Separation

In order to quantitatively evaluate the performance of the blind source separation (BSS) algorithm, a variety of performance indexes can be defined, such as interference signal ratio, Cross-talk index, correlation coefficient, etc. In addition, the mixing matrix estimating error can be used to measure the accuracy of the mixing matrix estimation in underdetermined BSS, which is described below one by one [25].

3.5.1 *Interference to Signal Ratio (ISR)*

According to the blind source separation linear mixing model, the output of the i-th sensor (receiving antenna) is:

$$x_i(t) = \sum_{l=1}^{N} a_{il}s_l(t) + v_i(t) \qquad (i = 1, 2, \ldots M) \tag{3.38}$$

Assuming the separating matrix is $\boldsymbol{W}$, the separated signal is:

$$\boldsymbol{y}(t) = \boldsymbol{W}\boldsymbol{x}(t) = \boldsymbol{W}\boldsymbol{A}\boldsymbol{s}(t) + \boldsymbol{W}\boldsymbol{v}(t) \tag{3.39}$$

We define the global matrix $\boldsymbol{G} = \boldsymbol{WA}$, if $\boldsymbol{G}$ is a diagonal matrix, then the j-th separation signal $y_j(t)$ is an estimation of the j-th source signal $s_j(t)$, and

$$y_j(t) = \sum_{i=1}^{M} w_{ji} x_i(t) \tag{3.40}$$

Substituting Eq. (3.38) into Eq. (3.40), we have

$$\begin{aligned} y_j(t) &= \sum_{i=1}^{M} w_{ji} \left\{ \sum_{l=1}^{N} a_{il} s_l(t) + v_i(t) \right\} \\ &= \sum_{i=1}^{M} w_{ji} \sum_{l=1}^{N} a_{il} s_l(t) + \sum_{i=1}^{M} w_{ji} v_i(t) \\ &= \sum_{i=1}^{M} w_{ji} a_{ij} s_j(t) + \sum_{i=1}^{M} \sum_{\substack{l=1 \\ l \neq j}}^{N} w_{ji} a_{il} s_l(t) + \sum_{i=1}^{M} w_{ji} v_i(t) \end{aligned} \tag{3.41}$$

As can be seen from formula (3.41), there are three items in $y_j(t)$, the first is the desired signal, the second is the residual interference, and the third is the output noise. When evaluating the performance of communication system, the second moment or energy of signal is generally used to measure. Then the second moment of the first term in the Eq. (3.41) is the energy of the expected signal, defined as

$$\begin{aligned} S_j &= E\left[\left| \sum_{i=1}^{M} w_{ji} a_{ij} s_j(t) \right|^2 \right] \\ &= E\left[\left\{ \sum_{i=1}^{M} w_{ji} a_{ij} s_j \right\} \left\{ \sum_{k=1}^{M} w_{jk} a_{kj} s_j \right\}^* \right] \\ &= \sum_{i=1}^{M} \sum_{k=1}^{M} w_{ji} a_{ij} w_{jk}^* a_{kj}^* E\left[s_j(t) s_j^*(t) \right] \end{aligned} \tag{3.42}$$

Suppose that the power of the j-th source signal is P_j, i.e.

$$E\left[s_j(t) s_j^*(t) \right] = P_j \tag{3.43}$$

Then the Eq. (3.42) is

$$S_j = P_j \sum_{i=1}^{M} \sum_{k=1}^{M} w_{ji} a_{ij} w_{jk}^* a_{kj}^* \tag{3.44}$$

It can be expressed in vector as

$$S_j = P_j \boldsymbol{W}_{j\cdot} \boldsymbol{A}_{\cdot j} \boldsymbol{A}_{\cdot j}^H \boldsymbol{W}_{j\cdot}^H \tag{3.45}$$

where $\boldsymbol{W}_{j\cdot}$ represents the j-th row of the separating matrix $\boldsymbol{W}$ and the $\boldsymbol{H}_{\cdot j}$ represents the j-th column of the mixing matrix $\boldsymbol{H}$.

The second moment of the second term in formula (3.41) is the residual interference power, which is defined as

$$\begin{aligned} I_j \triangleq E\left[\left|\sum_{i=1}^{M}\sum_{\substack{l=1\\ l\neq j}}^{N} w_{ji}a_{il}s_l\right|^2\right] &= E\left[\left\{\sum_{i=1}^{M}\sum_{\substack{l=1\\ l\neq j}}^{N} w_{ji}a_{il}s_l\right\}\left\{\sum_{k=1}^{M}\sum_{\substack{p=1\\ p\neq j}}^{N} w_{jk}a_{kp}s_p\right\}^*\right] \\ &= \sum_{\substack{l=1\\ l\neq j}}^{N}\sum_{\substack{p=1\\ p\neq j}}^{N} E\left[s_l(t)s_p^*(t)\right]\sum_{i=1}^{M}\sum_{k=1}^{M} w_{ji}a_{il}a_{kp}^*w_{jk}^* \end{aligned} \tag{3.46}$$

According to the hypothesis of statistical independence between source signals:
当 $l \neq p$ 时.
When $l \neq p$, we have

$$E\left[s_l(t)s_p^*(t)\right] = 0 \tag{3.47}$$

When $l = p$, it can be seen from Eq. (3.43) that

$$E\left[s_l(t)s_l^*(t)\right] = P_l \tag{3.48}$$

Then, Eq. (3.48) is substituted into (3.46), and it can be seen from Eq. (3.47) that the residual interference power is simplified as

$$I_j = \sum_{\substack{l=1\\ l\neq j}}^{N} P_l \sum_{i=1}^{M}\sum_{k=1}^{M} w_{ji}a_{il}a_{kl}^*w_{jk}^* \tag{3.49}$$

The Eq. (3.49) can be expressed as follow.

$$I_j = \sum_{\substack{l=1\\ l\neq j}}^{N} P_l \mathbf{W}_{j\cdot} A_{\cdot l} A_{\cdot l}^{\mathrm{H}} \mathbf{W}_{j\cdot}^{\mathrm{H}} \tag{3.50}$$

The second moment of the third term of Eq. (3.41) is the output noise power, defined as

$$N_j = E\left|\sum_{i=1}^{M} w_{ji} v_i(t)\right|^2 = E\left\{\sum_{i=1}^{M} w_{ji} v_i(t)\right\}\left\{\sum_{k=1}^{M} w_{jk} v_k(t)\right\}^* = \sum_{i=1}^{M}\sum_{k=1}^{M} w_{ji} E\left[v_i(t) v_k^*(t)\right] w_{jk}^* \tag{3.51}$$

The above formula can be expressed by vectors as follow.

$$N_j = \boldsymbol{W}_{j\cdot} E\left[\boldsymbol{v}(t)\boldsymbol{v}^{\mathrm{H}}(t)\right]\boldsymbol{W}_{j\cdot}^{\mathrm{H}} \tag{3.52}$$

Suppose the noise covariance matrix is $\boldsymbol{K}_v$, i.e.

$$E\left[\boldsymbol{v}(t)\boldsymbol{v}^{\mathrm{H}}(t)\right] = \boldsymbol{K}_v \tag{3.53}$$

Then the output noise power is

$$N_j = \boldsymbol{W}_{j\cdot}\boldsymbol{K}_v\boldsymbol{W}_{j\cdot}^{\mathrm{H}} \tag{3.54}$$

From the above analysis, we know that residual interference to signal power ratio (interference signal ratio, ISR) can be used to measure the ability of BSS algorithm to suppress interference. The smaller the interference signal ratio, the stronger the ability of the algorithm to suppress interference of other signals, and the better the separating effect of the algorithm.

The interference signal ratio (ISR) of a particular desired signal is defined as:

$$\zeta_j = \frac{I_j}{S_j} \tag{3.55}$$

If Eqs. (3.45) and (3.50) are substituted into formula (3.55), then

$$\zeta_j = \frac{I_j}{S_j} = \frac{\sum\limits_{\substack{l=1 \\ l \neq j}}^{N} P_l \boldsymbol{W}_{j\cdot}\boldsymbol{A}_{\cdot l}\boldsymbol{A}_{\cdot l}^{\mathrm{H}}\boldsymbol{W}_{j\cdot}^{\mathrm{H}}}{P_j \boldsymbol{W}_{j\cdot}\boldsymbol{A}_{\cdot j}\boldsymbol{A}_{\cdot j}^{\mathrm{H}}\boldsymbol{W}_{j\cdot}^{\mathrm{H}}} \tag{3.56}$$

In order to measure the overall effect of the BSS algorithm, it is often measured by the average of the ISR of multiple desired signals. namely

$$ISR_{avg} = \frac{1}{N}\sum_{j=1}^{N} \zeta_j \tag{3.57}$$

Or the maximum ISR

$$ISR_{\max} = \max_{1 \le j \le N} \{\zeta_j\} \tag{3.58}$$

The maximum ISR can ensure that the ISR of all the separated signals will not exceed $ISR_{\max}$.

If the powers of all source signals are equal, that is $P_j = P_l \quad (\forall k, l = 1, 2, \ldots, m)$, then

$$\begin{aligned}
\zeta_j &= \frac{\sum\limits_{\substack{l=1 \\ l \ne j}}^{N} \boldsymbol{W}_{j\cdot}\boldsymbol{A}_{\cdot j}\boldsymbol{A}_{\cdot j}^{\mathrm{H}}\boldsymbol{W}_{j\cdot}^{\mathrm{H}}}{\boldsymbol{W}_{j\cdot}\boldsymbol{A}_{\cdot j}\boldsymbol{A}_{\cdot j}^{\mathrm{H}}\boldsymbol{W}_{j\cdot}^{\mathrm{H}}} \\
&= \frac{\sum\limits_{l=1}^{N} \boldsymbol{W}_{j\cdot}\boldsymbol{A}_{\cdot l}\boldsymbol{A}_{\cdot l}^{\mathrm{H}}\boldsymbol{W}_{j\cdot}^{\mathrm{H}} - \boldsymbol{W}_{j\cdot}\boldsymbol{A}_{\cdot j}\boldsymbol{A}_{\cdot j}^{\mathrm{H}}\boldsymbol{W}_{j\cdot}^{\mathrm{H}}}{\boldsymbol{W}_{j\cdot}\boldsymbol{A}_{\cdot j}\boldsymbol{A}_{\cdot j}^{\mathrm{H}}\boldsymbol{W}_{j\cdot}^{\mathrm{H}}} \\
&= \frac{\sum\limits_{l=1}^{N} |G_{jl}|^2 - |G_{jj}|^2}{|G_{jj}|^2}
\end{aligned} \tag{3.59}$$

where G_{jl} is the (j, l)-th element of the global matrix $\boldsymbol{G} = \boldsymbol{WA}$.

From the Eqs. (3.57) and (3.59), we have

$$ISR_{\text{avg}} = \frac{1}{N}\sum_{j=1}^{N} \zeta_j = \frac{1}{N} \cdot \sum_{j=1}^{N} \left[\frac{\sum\limits_{l=1}^{N} |G_{jl}|^2 - |G_{jj}|^2}{|G_{jj}|^2} \right] \tag{3.60}$$

It can be seen from the above analysis process that the ISR defined by formula (3.60) is given under the assumption that the global matrix is a diagonal matrix or that the j-th separated signal is the estimation of j-th source signal. However, due to the ambiguity of BSS, we know that the order of separated signals and source signals may be different. That is, the j-th separated signal may not be an estimation of the j-th source signal, but an estimation of the i-th source signal. That is, the global matrix $\boldsymbol{G}$ is no longer an approximate diagonal matrix, but a product of an approximate diagonal matrix and a permutation matrix, that is $\boldsymbol{G} = \boldsymbol{\Lambda P}$, where $\boldsymbol{\Lambda}$ is a diagonal matrix, and $\boldsymbol{P}$ is an identity permutation matrix. That is, the global matrix $\boldsymbol{G}$ is approximately a matrix with only one nonzero element in each row and column. At this time, the defined ISR can not be Eqs. (3.59), (3.60), and the ISR of the j-th separated signal is defined as

$$\zeta_j' = \frac{\sum\limits_{l=1}^{N} |G_{jl}|^2 - |G_{\max j}|^2}{|G_{\max j}|^2} \tag{3.61}$$

where $G_{\max j}$ is the element with the largest modular value in j-th row of the global matrix $\boldsymbol{G}$.

The average ISR is defined as

$$ISR_{avg} = \frac{1}{N}\sum_{j=1}^{N}\zeta_j' = \frac{1}{N}\cdot\sum_{j=1}^{N}\left[\frac{\sum_{l=1}^{N}|G_{jl}|^2 - |G_{\max j}|^2}{|G_{\max j}|^2}\right] \tag{3.62}$$

The maximum ISR is

$$ISR_{\max} = \max_{1\le j\le N}\{\zeta_j'\} \tag{3.63}$$

3.5.2 Cross-Talk Index

Cross-talk index is defined as follows:

$$\varepsilon = \frac{1}{N}\cdot\left\{\sum_{i=1}^{N}\left[\sum_{j=1}^{N}\frac{|G_{i,j}|}{\max(|G_{i,j}|)} - 1\right] + \sum_{j=1}^{N}\left[\sum_{i=1}^{N}\frac{|G_{j,i}|}{\max(|G_{j,i}|)} - 1\right]\right\} \tag{3.64}$$

where $G_{i,j}$ is the (i,j)-th element of the global matrix $\boldsymbol{G}$ and N is the number of source signals. $\varepsilon = 0$ indicates that the global matrix $\boldsymbol{G}$ is a permutation matrix with only one nonzero element in each row and column, and there is no interference from other source signals in the separated signal. The closer ε is to 0, the better the separation effect.

3.5.3 Correlation Coefficient

The goal of BSS is to extract the waveforms of each source signal or separate each source signal without prior information. The correlation coefficient can be a good measure of the consistency of waveform changes, if the correlation coefficient of the recovered source signal and the real source signal is 1, their waveform changes are completely consistent, for example, for the voice signal, the voice waveform is consistent at this time, without the amplitude (volume size) is consistent. Therefore, the correlation coefficient is used to measure the recovery accuracy of the source signal, which is defined as

$$R_{coef}(\hat{\boldsymbol{S}}, \boldsymbol{S}) = \frac{1}{N}\sum_{n=1}^{N} \frac{\left|\sum_{t=1}^{T} s(n,t)\hat{s}(n,t)\right|}{\sqrt{\sum_{t=1}^{T} s^2(n,t) \sum_{t=1}^{N} \hat{s}^2(n,t)}} \tag{3.65}$$

where, $\boldsymbol{S}$ is the estimated source signal matrix, $\hat{\boldsymbol{S}}$ is the actual source signal matrix, $s(n, t)$ represents the n-th row t-th column element of $\boldsymbol{S}$, $\hat{s}(n, t)$ represents the n-th row t-th column element of $\hat{\boldsymbol{S}}$. Larger R_{coef} means higher precision, and the maximum value of R_{coef} is 1.

3.5.4 Estimating Error of Mixing Matrix

In underdetermined BSS, the relative error of mixing matrix estimation is defined as

$$\varepsilon = \min\left\{\frac{\|\boldsymbol{A} - \hat{\boldsymbol{A}}\boldsymbol{P}\|_F}{\|\boldsymbol{A}\|_F}\right\} \tag{3.66}$$

where, $\boldsymbol{A}$ is the real mixing matrix, $\hat{\boldsymbol{A}}$ is the estimated mixing matrix, $\boldsymbol{P}$ is the permutation matrix and used to eliminate the order ambiguity, $\|\cdot\|_F$ represents the matrix Frobenius norm. When the value of the estimating error is very small, the contrast curve drawn with ε will be stacked, so the separating performance of the algorithm cannot be visually compared. Therefore, the signal-to-interference ratio (SIR) of mixing matrix estimation is also defined in logarithmic form as follow

$$\text{SIR} = 20 * \log_{10}\left(\min\left\{\frac{\|\boldsymbol{A}\|_F}{\|\boldsymbol{A} - \hat{\boldsymbol{A}}\boldsymbol{P}\|_F}\right\}\right) \tag{3.67}$$

Summary

This chapter mainly introduces the theoretical basis of blind source separation (BSS) technology. Firstly, Independent Component Analysis (ICA) and Principal Component Analysis (PCA), which are widely used in overdetermined/determined BSS, and Sparse Component Analysis (SCA), which are widely used in underdetermined BSS, are introduced. On this basis, the current common methods of BSS are given. Finally, several common performance indexes of BSS algorithms are defined to lay the groundwork for evaluating the performance of BSS algorithms given in the following chapters.

Chapter 4
Preprocessing Technology for Blind Source Separation

Before solving the blind source separation problem, the observed signals need to be preprocessed by centralization and whitening, and the number of source signals needs to be estimated. The centering preprocessing is relatively simple, which makes the observed data subtract its mean. Turning the observed data into zero-mean vectors is one of the necessary conditions for the subsequent whitening operation, and it becomes easier to calculate the covariance matrix of the observed data, which improves the computational efficiency of the algorithm. In this chapter, we will introduce the whitening preprocessing techniques and the estimating techniques of the number of source signals, which are essential in blind source separation.

4.1 Whitening Preprocessing Technology

The purpose of whitening preprocessing is to apply a linear transformation to the observation vector so that its components are uncorrelated and have unit variance, hence whitening is also called Sphered.

The whitening preprocessing [25] is to choose a matrix $\boldsymbol{Q}$ such that the covariance matrix $E\{\boldsymbol{y}\boldsymbol{y}^{\mathrm{H}}\}$ becomes a unit array $\boldsymbol{I}$, where $\boldsymbol{y}(t) = \boldsymbol{Q}\boldsymbol{x}(t)$. The components of the vector $\boldsymbol{y}(t)$ are uncorrelated with each other and they have unit variance. The independence condition is stronger than the uncorrelation condition, which is necessary for statistical independence. The whitening preprocessing is more effective in large-scale signal processing, and sometimes signals are not well separated without whitening. Overdetermined BSS problem is often converted into determined BSS problem by whitening, which on the one hand achieves the purpose of data dimensionality reduction, and on the other hand, it can achieve the purposes of denoising and reducing the correlation between the signals to improve anti-noise ability of the blind separation algorithm, and effectively reduces the complexity of the problem.

W. Fu, *Principle and Application of Blind Source Separation Technology*,
https://doi.org/10.1007/978-981-96-0022-9_4

4.1.1 Standard Batch Whitening Algorithm

A zero-mean random vector $\boldsymbol{y}(t)$ whose covariance matrix is an identity matrix, i.e., $\boldsymbol{R}_{yy} = E\{\boldsymbol{y}\boldsymbol{y}^{\mathrm{H}}\} = \boldsymbol{I}$, is white. In whitening, observed signal $\boldsymbol{x}(t)$ is preprocessed with the following transformations.

$$\boldsymbol{y}(t) = \boldsymbol{Q}\boldsymbol{x}(t) \tag{4.1}$$

where $\boldsymbol{y}(t)$ is the whitened vector. $\boldsymbol{Q}$ is a whitening matrix of $n \times m$. If $m > n$, n is the number of source signals (assumed to be known or estimated), and m is the number of observed signals, then the whitening matrix $\boldsymbol{Q}$ decreases the dimensionality of the observed data vector from m to n. The purpose of the whitening is to select the whitening matrix $\boldsymbol{Q}$ such that the covariance matrix $E\{\boldsymbol{y}\boldsymbol{y}^{\mathrm{H}}\}$ is an identity matrix $\boldsymbol{I}_n$. Thus the components of the whitened vector $\boldsymbol{y}(t)$ are uncorrelated with each other, and have unit variance, i.e.

$$\boldsymbol{R}_{yy} = E\{\boldsymbol{y}\boldsymbol{y}^{\mathrm{H}}\} = E\{\boldsymbol{Q}\boldsymbol{x}\boldsymbol{x}^{\mathrm{H}}\boldsymbol{Q}^{\mathrm{H}}\} = \boldsymbol{Q}\boldsymbol{R}_{xx}\boldsymbol{Q}^{\mathrm{H}} = \boldsymbol{I}_n \tag{4.2}$$

In general, the covariance matrix $\boldsymbol{R}_{xx}$ of the observed signal $\boldsymbol{x}(t)$ is usually symmetric positive definite and it can be eigenvalue decomposed as follow.

$$\boldsymbol{R}_{xx} = \boldsymbol{U}_x\boldsymbol{\Lambda}_x\boldsymbol{U}_x^{\mathrm{H}} = \boldsymbol{U}_x\boldsymbol{\Lambda}_x^{1/2}\boldsymbol{\Lambda}_x^{1/2}\boldsymbol{U}_x^{\mathrm{H}} \tag{4.3}$$

where $\boldsymbol{U}_x$ is an orthogonal matrix and $\boldsymbol{\Lambda}_x = diag\{\lambda_1, \lambda_2, \ldots, \lambda_n, \lambda_{n+1}, \ldots \lambda_m\}$ is a diagonal matrix consisting of the eigenvalues $\lambda_1, \lambda_2, \ldots \lambda_m$ of $\boldsymbol{R}_{xx}$. If there is no noise in the observed signal vector, then

$$\lambda_1 \geq \lambda_2 \geq \cdots \geq \lambda_n > 0, \lambda_{n+1} = \lambda_{n+2} = \cdots = \lambda_m = 0 \tag{4.4}$$

and when there is noise interference in the observed signal, there is

$$\lambda_1 \geq \lambda_2 \geq \cdots \geq \lambda_n > \lambda_{n+1} \approx \lambda_{n+2} \approx \cdots \approx \lambda_m \approx \sigma^2 \tag{4.5}$$

where σ^2 is the power of the noise

Let $\boldsymbol{U}_x = [\boldsymbol{U}_s \ \ \boldsymbol{U}_v]$, $\boldsymbol{\Lambda}_x = \begin{bmatrix} \boldsymbol{\Lambda}_s & \boldsymbol{0} \\ \boldsymbol{O} & \boldsymbol{\Lambda}_v \end{bmatrix}$, and $\boldsymbol{U}_s = [\boldsymbol{u}_1, \boldsymbol{u}_2, \ldots, \boldsymbol{u}_n]$, $\boldsymbol{U}_v = [\boldsymbol{u}_{n+1}, \boldsymbol{u}_{n+2}, \ldots, \boldsymbol{u}_m]$, $\boldsymbol{\Lambda}_s = diag(\lambda_1, \lambda_2, \ldots, \lambda_n)$, $\boldsymbol{\Lambda}_v = diag(\lambda_{n+1}, \lambda_{n+2}, \ldots, \lambda_m)$, $\boldsymbol{O}$ be matrix whose elements are all zero, and $\boldsymbol{u}_i (i = 1, 2, \ldots, m)$ be the eigenvector corresponding to the eigenvalue $\lambda_i (i = 1, 2, \ldots m)$ of $\boldsymbol{R}_{xx}$. Then

$$\boldsymbol{R}_{xx} = [\boldsymbol{U}_s \ \boldsymbol{U}_v]\begin{bmatrix} \boldsymbol{\Lambda}_s & \boldsymbol{O} \\ \boldsymbol{O} & \boldsymbol{\Lambda}_v \end{bmatrix}\begin{bmatrix} \boldsymbol{U}_s^{\mathrm{H}} \\ \boldsymbol{U}_v^{\mathrm{H}} \end{bmatrix} = \boldsymbol{U}_s\boldsymbol{\Lambda}_s\boldsymbol{U}_s^{\mathrm{H}} + \boldsymbol{U}_v\boldsymbol{\Lambda}_v\boldsymbol{U}_v^{\mathrm{H}} \tag{4.6}$$

The subspace tensed by columns vector $\boldsymbol{U}_s$ is generally called the signal subspace, while the subspace tensed by column vectors $\boldsymbol{U}_v$ is referred to as the noise subspace.

Then the whitening matrix is calculated as follows:

$$\boldsymbol{Q} = \boldsymbol{\Lambda}_s^{-1/2}\boldsymbol{U}_s^{\mathrm{H}} = diag\left\{\frac{1}{\sqrt{\lambda_1}}, \frac{1}{\sqrt{\lambda_2}}, \ldots, \frac{1}{\sqrt{\lambda_n}}\right\}\boldsymbol{U}_s^{\mathrm{H}} \tag{4.7}$$

or

$$\boldsymbol{Q} = \boldsymbol{U}\boldsymbol{\Lambda}_s^{-1/2}\boldsymbol{U}_s^{\mathrm{H}} \tag{4.8}$$

where $\boldsymbol{U}$ is an arbitrary orthogonal matrix. It is easy to demonstrate that the observed signal vector $\boldsymbol{y}(t)$ is white when transformed by the matrix $\boldsymbol{Q}$ described in Eq. (4.7) or (4.8). Substituting (4.7) or (4.8) into Eq. (4.2), we have

$$\boldsymbol{R}_{yy} = E\{\boldsymbol{y}\boldsymbol{y}^{\mathrm{H}}\} = \boldsymbol{Q}\boldsymbol{R}_{xx}\boldsymbol{Q}^{\mathrm{H}} = \boldsymbol{U}\boldsymbol{\Lambda}_s^{-1/2}\boldsymbol{U}_s^{\mathrm{H}}\boldsymbol{R}_{xx}\boldsymbol{U}_s\boldsymbol{\Lambda}_s^{-(1/2)\mathrm{H}}\boldsymbol{U}^{\mathrm{H}} \tag{4.9}$$

according to Eq. (4.6), we get

$$\begin{aligned}\boldsymbol{R}_{yy} &= \boldsymbol{U}\boldsymbol{\Lambda}_s^{-1/2}\boldsymbol{U}_s^H[\boldsymbol{U}_s\Lambda_s\boldsymbol{U}_s^H + \boldsymbol{U}_v\Lambda_v\boldsymbol{U}_v^H]\boldsymbol{U}_s\boldsymbol{\Lambda}_s^{-(1/2)H}\boldsymbol{U}^H \\ &= \boldsymbol{U}\boldsymbol{\Lambda}_s^{-1/2}\boldsymbol{U}_s^H\boldsymbol{U}_s\Lambda_s\boldsymbol{U}_s^H\boldsymbol{U}_s\boldsymbol{\Lambda}_s^{-1/2}\boldsymbol{U}^H + \boldsymbol{U}\boldsymbol{\Lambda}_s^{-1/2}\boldsymbol{U}_s^H\boldsymbol{U}_v\Lambda_v\boldsymbol{U}_v^H\boldsymbol{U}_s\boldsymbol{\Lambda}_s^{-(1/2)H}\boldsymbol{U}^H \\ &= \boldsymbol{U}\boldsymbol{\Lambda}_s^{-1/2}\Lambda_s\boldsymbol{\Lambda}_s^{-1/2}\boldsymbol{U}^H \\ &= \boldsymbol{I}\end{aligned} \tag{4.10}$$

The derivation of the above Eq. (4.10) utilizes the orthogonality between $\boldsymbol{U}_s$ and $\boldsymbol{U}_v$. Accordingly, the matrix represented by Eqs. (4.7) and (4.8) satisfies the requirement for whitening matrix, and the above whitening method is known as the standard whitening algorithm for batch processing.

4.1.2 Adaptive Standard Whitening Algorithm

The goal of the adaptive standard whitening algorithm is to find a simple adaptive algorithm for estimating the whitening matrix $\boldsymbol{Q}$ such that the covariance matrix of the output signal is a diagonal matrix, i.e.

$$\boldsymbol{R}_{yy} = E\{\boldsymbol{y}\boldsymbol{y}^{\mathrm{H}}\} = \boldsymbol{\Lambda} \tag{4.11}$$

where $\boldsymbol{\Lambda} = diag\{\lambda_1, \lambda_2, \ldots, \lambda_n\}$ is a diagonal matrix and $\boldsymbol{y}^H$ is the conjugate transpose of $\boldsymbol{y}$. Typically $\boldsymbol{\Lambda} = \boldsymbol{I}$. We know that signal is uncorrelated if all the cross-correlations are zero. That is

$$r_{ij} = E\{y_i y_j^*\} = 0 \quad \forall i \neq j \tag{4.12}$$

and the autocorrelation is non-zero

$$r_{ii} = E\{|y_i|^2\} = \lambda_i > 0 \tag{4.13}$$

Thus the minimization criterion can be written in terms of p-norm as the following objective function

$$J_p(\boldsymbol{W}) = \frac{1}{p}\sum_{i=1}^{n}\sum_{\substack{j=1\\ j\neq i}}^{n}|r_{ij}|^p \tag{4.14}$$

Constraints: $r_{ii} \neq 0, \quad \forall i$, typically $r_{ii} = 1, \quad \forall i$.

The following takes the real-valued signal as an example, we derive the adaptive whitening algorithm when $p = 2$. The objective function is as follow.

$$J_2(\boldsymbol{W}) = \frac{1}{2}\sum_{i=1}^{n}\sum_{\substack{j=1\\ j\neq i}}^{n}|r_{ij}|^2 = \frac{1}{2}\sum_{i=1}^{n}\sum_{j=1}^{n}|E\{y_i y_j\} - \lambda_i\delta_{i,j}|^2 = \frac{1}{2}||E\{\boldsymbol{y}\boldsymbol{y}^{\mathrm{H}}\} - \boldsymbol{\Lambda}||_F^2 \tag{4.15}$$

Here $||\boldsymbol{\Lambda}||_F$ denotes the Frobenius norm of matrix $\boldsymbol{\Lambda}$. Then

$$\boldsymbol{R}_{yy} = E\{\boldsymbol{y}\boldsymbol{y}^{\mathrm{H}}\} = E\{\boldsymbol{Q}\boldsymbol{x}\boldsymbol{x}^{\mathrm{H}}\boldsymbol{Q}^{\mathrm{H}}\} = E\{\boldsymbol{Q}\boldsymbol{H}\boldsymbol{s}\boldsymbol{s}^{\mathrm{H}}(\boldsymbol{Q}\boldsymbol{H})^{\mathrm{H}}\} = \boldsymbol{G}\boldsymbol{R}_{ss}\boldsymbol{G}^{\mathrm{H}} = \boldsymbol{G}\boldsymbol{G}^{\mathrm{H}} \tag{4.16}$$

where $\boldsymbol{G} = \boldsymbol{Q}\boldsymbol{H}$ is the global transformation matrix from $\boldsymbol{s}$ to $\boldsymbol{y}$. The last step of the above equation assumes that the source signal is unit variance. Therefore, the objective function expressed in Eq. (4.15) is

$$J_2(\boldsymbol{W}) = \frac{1}{2}||\boldsymbol{G}\boldsymbol{G}^{\mathrm{H}} - \boldsymbol{\Lambda}||_F^2 = \frac{1}{2}tr\{(\boldsymbol{G}\boldsymbol{G}^{\mathrm{H}} - \boldsymbol{\Lambda})(\boldsymbol{G}\boldsymbol{G}^{\mathrm{H}} - \boldsymbol{\Lambda})\} \tag{4.17}$$

Using standard gradient descent methods and the chain rule, there are

$$\frac{dg_{ij}}{dt} = -\mu\frac{\partial J_2(\boldsymbol{W})}{\partial g_{ij}} = -\mu\sum_{k=1}^{n}\sum_{p=1}^{n}\frac{\partial J_2}{\partial r_{kp}}\frac{\partial r_{kp}}{\partial g_{ij}} \tag{4.18}$$

where r_{kp} is the (k, p) element of the matrix $\boldsymbol{R}_{yy} = \boldsymbol{G}\boldsymbol{G}^T$ and μ is the iteration step, then

$$\frac{dg_{ij}}{dt} = \mu\left[\sum_{k=1}^{n}\lambda_k\frac{\partial r_{kk}}{\partial g_{ij}} - \sum_{k=1}^{n}\sum_{p=1}^{n}r_{kp}\frac{\partial r_{kp}}{\partial g_{ij}}\right] = \mu\left[2\lambda_i g_{ij} - \sum_{k=1}^{n}r_{ik}g_{kj} - \sum_{p=1}^{n}r_{pi}g_{pj}\right] \tag{4.19}$$

Since the covariance matrix $\boldsymbol{R}_{yy}$ is symmetric, i.e., $r_{ij} = r_{ji}$, Eq. (4.19) can be simplified to

$$\frac{dg_{ij}}{dt} = 2\mu\left[\lambda_i g_{ij} - \sum_{k=1}^{n} r_{ik} g_{kj}\right] \quad (i, j = 1, 2, \ldots, n) \tag{4.20}$$

Written in matrix form, then

$$\frac{d\boldsymbol{G}}{dt} = 2\mu(\boldsymbol{\Lambda} - \boldsymbol{R}_{yy})\boldsymbol{G} = 2\mu(\boldsymbol{\Lambda} - \boldsymbol{G}\boldsymbol{G}^{\mathrm{H}})\boldsymbol{G} \tag{4.21}$$

Because $\boldsymbol{G} = \boldsymbol{Q}\boldsymbol{H}$, and we assume that $\boldsymbol{H}$ changes very slowly over time, i.e., $d\boldsymbol{H}/dt \approx \boldsymbol{0}$, we have

$$\frac{d\boldsymbol{Q}}{dt}\boldsymbol{H} = 2\mu(\boldsymbol{\Lambda} - \boldsymbol{R}_{yy})\boldsymbol{Q}\boldsymbol{H} \tag{4.22}$$

i.e.

$$\frac{d\boldsymbol{Q}}{dt} = 2\mu(\boldsymbol{\Lambda} - \boldsymbol{R}_{yy})\boldsymbol{Q} \tag{4.23}$$

The discrete form of the algorithm is

$$\boldsymbol{Q}(l+1) = \boldsymbol{Q}(l) + \eta_l(\boldsymbol{\Lambda} - \boldsymbol{R}_{yy}^{(l)})\boldsymbol{Q}(l) \tag{4.24}$$

Equation (4.24) is the adaptive standard whitening algorithm, where $\boldsymbol{Q}(l)$ is the whitening matrix at the l-th iteration, η_l is the iteration step-size, and $\boldsymbol{R}_{yy}^{(l)}$ is the covariance matrix estimated at the l-th iteration, calculated as follow.

$$\hat{\boldsymbol{R}}_{yy}^{(l)} = \frac{1}{N}\sum_{k=0}^{N-1} \boldsymbol{y}^{(l)}(k)[\boldsymbol{y}^{(l)}(k)]^{\mathrm{H}} \tag{4.25}$$

where $\boldsymbol{y}^{(l)}(k) = \boldsymbol{Q}(l)\boldsymbol{x}(k)$. When $\boldsymbol{\Lambda} = \boldsymbol{I}$, we have

$$\boldsymbol{Q}(l+1) = \boldsymbol{Q}(l) + \eta_l(\boldsymbol{I} - \hat{\boldsymbol{R}}_{yy}^{(l)})\boldsymbol{Q}(l) \tag{4.26}$$

4.1.3 Batch Robust Whitening Algorithm

Although the standard whitening algorithm given in the previous two sections can whiten the observed signal, if there is noise in the observation signal, in order to improve the performance of the BSS algorithm, it is necessary to give a whitening algorithm that can eliminate the effect of noise, the robust whitening algorithm. When the observed signal is contaminated by noise, the whitening matrix $\boldsymbol{Q}$ can be obtained by using the following equation:

$$\boldsymbol{Q} = (\boldsymbol{\Lambda}_s - \sigma^2 \boldsymbol{I}_n)^{-1/2} \times \boldsymbol{U}_s^{\mathrm{H}} \tag{4.27}$$

where $\boldsymbol{\Lambda}_s$, σ^2 and $\boldsymbol{U}_s$ have the same definitions as before. The whitening matrix represented by Eq. (4.27) has the function of denoising. The whitening algorithm consisting of the whitening matrix represented by Eq. (4.27) is what we call a batch robust whitening algorithm.

Theorem 4.1 *In a noisy environment, after whitening using a robust whitening algorithm, the global mixing matrix* $\mathbf{G} = \boldsymbol{QH}$ *from* $\mathbf{s}$ *to* $\mathbf{y}$ *is an orthogonal matrix. In the absence of noise, due to* $\sigma^2 = 0$*, the standard whitening algorithm* (4.7) *or* (4.8) *and the robust whitening algorithm* (4.27) *are equivalent, the global matrices* $\mathbf{G} = \boldsymbol{QH}$ *are orthogonal.*

Proof When there is noise, $\lambda_{n+1} = \lambda_{n+2} = \cdots = \lambda_m = \sigma^2$, that is $\boldsymbol{\Lambda}_v = \sigma^2 \boldsymbol{I}_{m-n+1}$.

$$z = \boldsymbol{Qx} = \boldsymbol{QHs} + \boldsymbol{Qv} = \tilde{z} + \tilde{v} \tag{4.28}$$

where $\tilde{z} = \boldsymbol{QHs} = \boldsymbol{Gs}$, $\tilde{v} = \boldsymbol{Qv}$ are still Gaussian white noise according to (4.27), then

$$E(zz^{\mathrm{H}}) = \boldsymbol{Q}E(\boldsymbol{xx}^{\mathrm{H}})\boldsymbol{Q}^{\mathrm{H}} = (\boldsymbol{\Lambda}_s - \sigma^2 \boldsymbol{I}_n)^{-1/2} \boldsymbol{U}_s^{\mathrm{H}} \boldsymbol{U}_x \boldsymbol{\Lambda}_x \boldsymbol{U}_x^{\mathrm{H}} \boldsymbol{U}_s (\boldsymbol{\Lambda}_s - \sigma^2 \boldsymbol{I}_n)^{-1/2} \tag{4.29}$$

after a simple mathematical derivation, we get

$$E(zz^{\mathrm{H}}) = (\boldsymbol{\Lambda}_s - \sigma^2 \boldsymbol{I}_n)^{-1/2} \boldsymbol{\Lambda}_s (\boldsymbol{\Lambda}_s - \sigma^2 \boldsymbol{I}_n)^{-1/2} \tag{4.30}$$

furthermore

$$\begin{aligned} E[\tilde{z}\tilde{z}^{\mathrm{H}}] &= E[(z - \boldsymbol{Qv})(z - \boldsymbol{Qv})^{\mathrm{H}}] = E[zz^{\mathrm{H}}] - E[\boldsymbol{Qv}z^{\mathrm{H}}] - E[z(\boldsymbol{Qv})^{\mathrm{H}}] + E[\boldsymbol{Qvv}^{\mathrm{H}}\boldsymbol{Q}^{\mathrm{H}}] \\ &= E[zz^{\mathrm{H}}] - E[\boldsymbol{Qv}(\boldsymbol{Gs} + \boldsymbol{Qv})^{\mathrm{H}}] - E[(\boldsymbol{Gs} + \boldsymbol{Qv})(\boldsymbol{Qv})^{\mathrm{H}}] + E[\boldsymbol{Qvv}^{H}\boldsymbol{Q}^{\mathrm{H}}] \\ &= E[zz^{\mathrm{H}}]E[\boldsymbol{vv}^{\mathrm{H}}]\boldsymbol{Q}^{\mathrm{H}} = E[zz^{\mathrm{H}}] - \sigma^2 \boldsymbol{QQ}^{\mathrm{H}} \end{aligned} \tag{4.31}$$

It can be obtained by Eqs. (4.27), (4.30), and (4.31)

$$E[\tilde{z}\tilde{z}^{\mathrm{H}}] = (\boldsymbol{\Lambda}_s - \sigma^2 \boldsymbol{I}_n)^{-1/2} \boldsymbol{\Lambda}_s (\boldsymbol{\Lambda}_s - \sigma^2 \boldsymbol{I}_n)^{-1/2} = -\sigma^2 (\boldsymbol{\Lambda}_s - \sigma^2 \boldsymbol{I}_n)^{-1} \tag{4.32}$$

Since $\boldsymbol{\Lambda}_s$ is a diagonal matrix, we get $E[\tilde{z}\tilde{z}^{\mathrm{H}}] = \boldsymbol{I}$ from Eq. (4.32) and because $\tilde{z} = \boldsymbol{BHs} = \boldsymbol{Gs}$, then $\boldsymbol{G}$ is also an orthogonal matrix.

When there is no noise, $\sigma^2 = 0$ is a special case of noise, and the above conclusion also true, so the Theorem 4.1 is proved.

With the robust whitening algorithm, the global matrix $\boldsymbol{G}$ becomes an orthogonal matrix, but the covariance matrix of the whitened signal $\boldsymbol{z}$ is no longer strictly identity matrix after whitening the observed signal. By using the standard whitening algorithm, the covariance matrix of the whitened signal $\boldsymbol{z}$ is an identity matrix, but the global matrix is no longer strictly orthogonal. Due to its ability to denoise to some extent, the robust whitening algorithm is more commonly used than the standard whitening algorithm.

4.1.4 Adaptive Robust Whitening Algorithm

The adaptive standard whitening algorithm is derived above as follow.

$$\boldsymbol{Q}(l+1) = \boldsymbol{Q}(l) + \eta_l[I - \langle \boldsymbol{y}(k)\boldsymbol{y}^{\mathrm{H}}(k)\rangle]\boldsymbol{Q}(l) \tag{4.33}$$

where $\boldsymbol{y}(k) = \boldsymbol{Q}(l)\boldsymbol{x}(k)$ is whitened signal vector, $\langle\cdot\rangle$ denote statistical average, which are replaced by arithmetic averages in practice. When $\boldsymbol{x}(k)$ contains noise, denoted as $\boldsymbol{x}(k) = \tilde{\boldsymbol{x}}(k) + \boldsymbol{v}(k)$, $\tilde{\boldsymbol{x}}(k)$ and $\tilde{\boldsymbol{y}}(k) = \boldsymbol{Q}(l)\tilde{\boldsymbol{x}}(k)$ are the noise-free estimation of the input signal vector and output signal vector, respectively. It is easy to see that the additive noise $\boldsymbol{v}(k)$ in $\boldsymbol{x}(k)$ introduces bias in the estimated whitening matrix. The covariance matrix of output $\boldsymbol{y}(k)$ can be estimated as

$$\tilde{\boldsymbol{R}}_{yy} = \langle \boldsymbol{y}(k)\boldsymbol{y}^{H}(k)\rangle = \boldsymbol{Q}\tilde{\boldsymbol{R}}_{\tilde{x}\tilde{x}}\boldsymbol{Q}^{H} + \boldsymbol{Q}\tilde{\boldsymbol{R}}_{vv}\boldsymbol{Q}^{H} \tag{4.34}$$

where $\tilde{\boldsymbol{R}}_{\tilde{x}\tilde{x}} = \langle \tilde{x}(k)\tilde{x}^{\mathrm{H}}(k)\rangle = \frac{1}{N}\sum_{k=1}^{N} \tilde{x}(k)\tilde{x}^{\mathrm{H}}(k)$, $\tilde{R}_{vv} = \langle \boldsymbol{v}(k)\boldsymbol{v}^{\mathrm{H}}(k)\rangle$.

Assuming that the power or covariance matrix of the noise is known or has been estimated, e.g., $\tilde{\boldsymbol{R}}_{vv} = \tilde{\sigma}_v^2\boldsymbol{I}$, the iterative formulation of the adaptive robust whitening algorithm is as follow.

$$\boldsymbol{Q}(l+1) = \boldsymbol{Q}(l) + \eta(l)[\boldsymbol{I} - \langle \boldsymbol{y}(k)\boldsymbol{y}^{\mathrm{H}}(k)\rangle + \tilde{\sigma}_v^2\boldsymbol{Q}^{\mathrm{H}}(l)\boldsymbol{Q}(l)]\boldsymbol{Q}(l) \tag{4.35}$$

The whitening algorithm represented by Eq. (4.35) is called the adaptive robust whitening algorithm, which can eliminate the effect of noise.

4.2 Technology for Estimating the Number of Source Signals

At present, most BSS algorithms require the number of source signals to be known, but in the actual communication environment, the number of source signals is often unpredictable, so the number of source signals needs to be reliably estimated.

Existing methods for estimating the number of source signals can generally be classified into the following categories: methods based on eigenvalue decomposition, methods based on information theoretic criteria including Akaika Information Theoretic Criteria (AIC) and Minimum Description Length (MDL) methods, and methods based on the Gerschgorin Criterion (Gerschgorin Disk Estimator, GDE) methods, etc.

4.2.1 Method Based on Eigenvalue Decomposition

The method based on eigenvalue decomposition [26] is the simplest method to estimate the number of source signals. Assuming that M source signals are independent of each other and the number of receiving channels is N, the observed signal $\boldsymbol{x}(t) = [x_1(t), x_2(t), \ldots, x_N(t)]$ at a certain sampling time t can be expressed as:

$$\boldsymbol{x}(t) = \boldsymbol{A}\boldsymbol{s}(t) + \boldsymbol{v}(t) \tag{4.36}$$

where $\boldsymbol{A}$ denotes the channel mixing matrix, $\boldsymbol{s}(t) = [s_1(t), s_2(t), \ldots, s_M(t)]^{\mathrm{T}}$ denotes the source signal vector, and $\boldsymbol{v}(t)$ is zero-mean Gaussian white noise vector. The covariance matrix is computed from the observed signal $\boldsymbol{x}(t)$, with

$$\boldsymbol{R} = E\big[\boldsymbol{x}(t)\boldsymbol{x}(t)^{\mathrm{H}}\big] = \boldsymbol{\Psi} + \sigma^2 I \tag{4.37}$$

where $\boldsymbol{R}$ is a $N \times N$ matrix, $\boldsymbol{\Psi} = \boldsymbol{A}\boldsymbol{R}_s\boldsymbol{A}^{\mathrm{H}}$, $\boldsymbol{R}_s = \mathrm{E}\big[\boldsymbol{s}(t)\boldsymbol{s}(t)^{\mathrm{H}}\big]$, $\boldsymbol{A}$ is full rank matrix, and due to the independence between the individual source signals, $\boldsymbol{R}_s$ is a $M \times M$ full rank matrix and has $rank(\boldsymbol{\Psi}) = rank(\boldsymbol{R}_s) = M$. Thus among the N eigenvalues of the matrix $\boldsymbol{\Psi}$, there are $N - M$ zero eigenvalues. let $\lambda_1 \geq \lambda_2 \geq \cdots \geq \lambda_M$ be the M eigenvalues of the covariance matrix $\boldsymbol{R}$, then N-M smallest eigenvalues of $\boldsymbol{R}$ are equal to σ^2, i.e.

$$\lambda_{M+1} = \lambda_{M+2} = \cdots = \lambda_N = \sigma^2 \tag{4.38}$$

Therefore, when estimating the number of source signals, we can use the characteristics of the source signal eigenvalues to be larger than the noise eigenvalues. Eigenvalue decomposition of the covariance matrix is carried out, and the resulting N eigenvalues are arranged in order from the largest to the smallest, and then the ratios between the adjacent eigenvalues $\gamma_i = \lambda_i/\lambda_{i+1}, i = 1, \ldots, N$ are calculated. The number of source signals can be determined by finding the sequence number i when γ_i is the largest.

Method based on eigenvalue decomposition is very simple and easy to implement, but the disadvantage is that the required data samples are large enough to ensure a very accurate estimation of the covariance, and the noise variance is required to be very small.

If the number of source signals is estimated directly using the received signals, when the number of observed signals is smaller than the number of source signals, the application condition of source signal number estimation algorithm is not satisfied. In addition, even if the number of observed signals is greater than the number of source signals, it is theoretically possible to directly utilize the observed signals to estimate the number of source signals by means of an eigenvalue decomposition of the covariance matrix of the observed signals, but the estimating accuracy is relatively low in practice due to the influence of noise. Therefore, it is necessary to expand the received signals using a channel expansion method to obtain observed signals with a greater number of channel, and then estimate covariance matrix of the expanded observed signals. Then the source number estimating algorithm based on eigenvalue is used.

Here we give a method of interval sampling to extend the channel of each signal in the received signal, the specific process refers to resampling the input signal to obtain a new data vector, which constitutes a multidimensional matrix. Assuming that one observed signal is $x(t)$, it is sampled at an interval T_1 to obtain the discrete form of the observed signal $\tilde{x}(l) = x(lT_1)$. Then the discrete observed signal $\tilde{x}(l)$ is sampled at every B points, and the discretized observed signal $\tilde{x}(l)$ can be rearranged into B virtual multichannel signals $\tilde{x}_1(l), \tilde{x}_2(l), \ldots, \tilde{x}_B(l)$ through interval sampling as shown in Eq. (4.39):

$$\tilde{x}_b(l) = \tilde{x}((l-1)B + b),\ b = 1, 2, \ldots, B \tag{4.39}$$

the multi-channel observed signal matrix $\tilde{\boldsymbol{X}}_B(l)$ formed by the expansion of each signal can be obtained as shown in Eq. (4.40):

$$\tilde{\boldsymbol{X}}_B(l) = \begin{bmatrix} \tilde{x}_1(l) \\ \tilde{x}_2(l) \\ \vdots \\ \tilde{x}_B(l) \end{bmatrix} = \begin{bmatrix} \tilde{x}((l-1)B+1) \\ \tilde{x}((l-1)B+2) \\ \vdots \\ \tilde{x}((l-1)B+B) \end{bmatrix} \tag{4.40}$$

By expanding the received signals using the interval sampling method, they are transformed into B times the number of received signals. Using the expanded multi-channel received signal, it is possible to estimate the number of source signals more accurately than before the expansion.

4.2.2 Methods Based on Information Theoretic Criteria

There are two main approaches based on information theoretic criterion, AIC criterion [27] and MDL criterion [28]. Methods based on information theoretic criteria can be described as follows. Given an observed matrix $\boldsymbol{X}$ with T sampling points and a set of parameterized probability density function models $f(\boldsymbol{X}|\Theta)$, from which the model that best matches the observed data is selected. Where the model when the AIC criterion has minimum value can be defined as:

$$\xi_{AIC} = -2\log f\left(\boldsymbol{X}|\hat{\Theta}\right) + 2\kappa \tag{4.41}$$

$\hat{\Theta}$ is the maximum likelihood estimate of Θ and κ is the degree of freedom in Θ. The first item in Eq. (4.41) is the log-likelihood of the maximum likelihood estimator, and the second item is a correction item such that ξ_{AIC} is an unbiased estimate of the Kullback–Leibler distance between $f(\boldsymbol{X}|\Theta)$ and $f\left(\boldsymbol{X}|\hat{\Theta}\right)$.

The model of the MDL criterion is defined as:

$$\xi_{\mathrm{MDL}} = -\log f\left(\boldsymbol{X}|\hat{\Theta}\right) + \frac{1}{2}\kappa\log(T) \tag{4.42}$$

The MDL criterion and the AIC criterion yield the same conclusions when the sample of observed data is sufficiently large.

The purpose of the number of source signals estimation lies in estimating the rank of the matrix $\boldsymbol{\Psi}$. Before using the MDL criterion or the AIC criterion, a set of models or probability density functions, $f(\boldsymbol{X}|\Theta)$, are developed based on the observed data matrix. According to Eq. (4.37), assuming the rank of the matrix $\boldsymbol{\Psi}$ is k, we have:

$$\boldsymbol{R}^{(k)} = \boldsymbol{\Psi}^{(k)} + \sigma^2\boldsymbol{I} \tag{4.43}$$

where $k \in \{0, 1, \ldots, N-1\}$ traverses all possible numbers of source signals. The algebraic expansion $\boldsymbol{\Psi}^{(k)}$ can be expressed as:

$$\boldsymbol{\Psi}^{(k)} = \sum_{i=1}^{k}\left(\lambda_i - \sigma^2\right)\boldsymbol{v}_i\boldsymbol{v}_i^{\mathrm{H}} + \sigma^2\boldsymbol{I} \tag{4.44}$$

where $\lambda_1, \lambda_2, \ldots, \lambda_k$ and $\boldsymbol{v}_1, \boldsymbol{v}_2, \ldots, \boldsymbol{v}_k$ are the eigenvalues and eigenvectors of $\boldsymbol{R}^{(k)}$, respectively. The vector of parameters in the model of $f(\boldsymbol{X}|\Theta)$ is represented by $\Theta^{(k)}$, i.e.

$$\Theta^{(k)} = \left[\lambda_1, \ldots\lambda_k, \sigma^2, v_1^{\mathrm{T}}, \ldots v_k^{\mathrm{T}}\right] \tag{4.45}$$

Since the observed data sample $\boldsymbol{x}(i)(i = 1, 2, \ldots, T)$ is a statistically independent zero-mean Gaussian random variable, its joint probability density is

$$f\left(\boldsymbol{X}|\Theta^{(k)}\right) = f\left((x(1), \ldots, x(T))|\Theta^{(k)}\right) = \prod_{i=1}^{T}\frac{1}{\pi^N\left|\boldsymbol{R}^{(k)}\right|}\exp\left(-x(i)^{\mathrm{H}}\left[\boldsymbol{R}^{(k)}\right]^{-1}x(i)\right) \tag{4.46}$$

$|\boldsymbol{R}^{(k)}|$ denotes the determinant of $\boldsymbol{R}^{(k)}$.

Taking logarithms on both sides of Eq. (4.46) and ignoring terms that are irrelevant to $\Theta^{(k)}$, we get

$$\log f\left(\boldsymbol{X}|\Theta^{(k)}\right) = -T\log\left|\boldsymbol{R}^{(k)}\right| - \operatorname{tr}\left\{\left[\boldsymbol{R}^{(k)}\right]^{-1}\hat{\boldsymbol{R}}\right\} \tag{4.47}$$

where tr{•} denotes the trace of the matrix, and

$$\hat{\boldsymbol{R}} = \frac{1}{T}\sum_{i=1}^{T}\boldsymbol{x}(i)\boldsymbol{x}(i)^{\mathrm{H}} \tag{4.48}$$

the value of $\Theta^{(k)}$ that maximizes Eq. (4.47) is the maximum likelihood estimate of Θ, and the estimate is:

$$\begin{cases} \hat{\lambda}_i = l_i,\ i = 1, \ldots, k \\ \sigma^2 = \dfrac{1}{N-k}\displaystyle\sum_{i=k+1}^{N} l_i \\ \hat{\boldsymbol{v}}_i = \boldsymbol{h}_i,\ i = 1, \ldots, k \end{cases} \tag{4.49}$$

where $l_1 > l_2 > \cdots > l_N$ and $\boldsymbol{h}_1, \ldots, \boldsymbol{h}_N$ are the eigenvalues and eigenvectors of matrix $\hat{\boldsymbol{R}}$, respectively.

According to Eqs. (4.47) and (4.49) it can be obtained that

$$\begin{cases} \log f\left(\boldsymbol{X}|\Theta^{(k)}\right) = (N-k)T\log\left(\dfrac{\alpha_N(k)}{\beta_N(k)}\right) \\ \alpha_N(k) = \displaystyle\prod_{i=k+1}^{N} l_i^{1/(N-k)} \\ \beta_N(k) = \dfrac{1}{N-k}\displaystyle\sum_{i=k+1}^{N} l_i \end{cases} \tag{4.50}$$

For the number of independent parameters in $\Theta^{(k)}$, it can be found by calculating the degrees of freedom of the space spanned by $\Theta^{(k)}$. $\hat{\lambda}_i$ and $\hat{\boldsymbol{v}}_i$ are real-valued eigenvalues and complex-valued eigenvectors, respectively, so the number of parameters in $\Theta^{(k)}$ is $k + 1 + 2kN$, where the modulus of eigenvectors $\hat{\boldsymbol{v}}_i$ is normalized which can reduce $2k$ degrees of freedom, and the eigenvectors are orthogonal to each other, then $2[k(k-1)/2]$ degrees of freedom is reduced. Thus, the correction term κ is

$$\kappa = k + 1 + 2kN - 2k - k(k-1) = k(2N-k) + 1 \tag{4.51}$$

Substituting Eqs. (4.50) and (4.51) into (4.41) and (4.42), we obtain the formula under the AIC and MDL criteria as follows.

$$\begin{cases} \xi_{\mathrm{AIC}} = -2(N-k)T\log\left(\dfrac{\alpha_N(k)}{\beta_N(k)}\right) + 2[k(2N-k)+1] \\ \xi_{\mathrm{MDL}} = -(N-k)T\log\left(\dfrac{\alpha_N(k)}{\beta_N(k)}\right) + \dfrac{1}{2}[k(2N-k)+1]\log(T) \end{cases} \tag{4.52}$$

The rank k corresponding to minimizing the above criterion when traversing time $k = 0, 1, \ldots N-1$ is the estimate of the number of source signals.

It has been demonstrated that the number of sources obtained by the MDL criterion is an asymptotically consistent estimate of the true value, i.e., when the data samples tend to infinity, the probability of convergence of the estimate to the true value is 1. In contrast, the estimate of the AIC criterion is too large and is not an asymptotically consistent estimate of the true value, which is confirmed by the simulation results. From the above analysis, it can be concluded that the conditions for information-theoretic criterion are: (1) the number of receiving channels N must be greater than the number of sources M; (2) a small amount of noise must be present to ensure that $\lambda_{M+1}, \lambda_{M+2}, \ldots, \lambda_N$ are not equal to 0; and (3) the noise must be white Gaussian noise.

4.2.3 Method Based on Gerschgorin Disk Criterion

1. **Gerschgorin Theorem**

For an $N \times N$ real or complex matrix $\boldsymbol{A}$, let its i-th row and j-th column elements be a_{ij}. The sum of the absolute values of all elements in the i-th row except the i-th column is

$$\xi_i = \sum_{j=1, j\neq i}^{N} |a_{ij}| \tag{4.53}$$

Defining the set of all points in the complex plane with a_{ij} as the center and ξ_i as the radius as the i-th disk O_i, all the eigenvalues of the matrix $\boldsymbol{A}$ will be contained in the concurrent intervals of the disk O_i, where $i = 1, 2 \ldots, N$. This theory is known as the theory of Gerschgorin, and the disk O_i is called Gerschgorin's disk.

According to Eq. (4.37), the covariance matrix for the observed data $\boldsymbol{x}$ is

$$\boldsymbol{R} = E\left[\boldsymbol{x}\boldsymbol{x}^{\mathrm{H}}\right] = \boldsymbol{A}\boldsymbol{R}_s\boldsymbol{A}^{\mathrm{H}} + \sigma^2\boldsymbol{I} \tag{4.54}$$

The eigenvalues of the covariance matrix $\boldsymbol{R}$ will be contained in the concatenation interval of N Gerschgorin disks, and since $\boldsymbol{R}$ is a symmetric matrix with real eigenvalues and Gerschgorin centers, the possible distribution locations of the individual

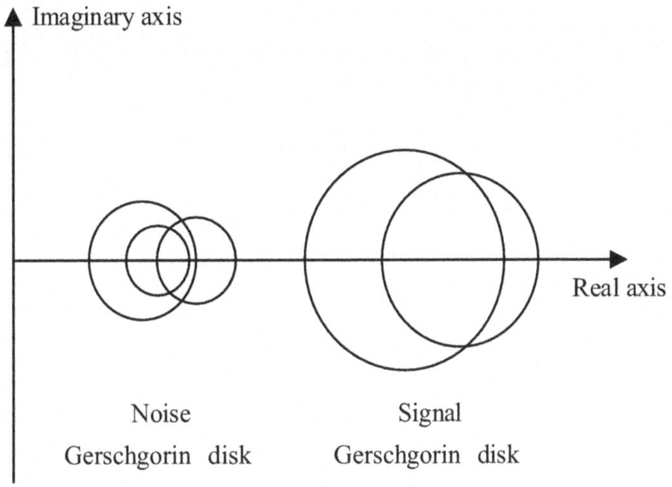

Fig. 4.1 Schematic diagram of a Gerschgorin disk

eigenvalues can be found on the real axis. Ideally, all the Gerschgorin disks can form a union of two disjoint sets as shown in Fig. 4.1, corresponding to the Gerschgorin disk of the source signal and the Gerschgorin disk of the noise, respectively. Then the union set of Gerschgorin disks with larger radius contain the eigenvalues corresponding to M source signals, and the union set of Gerschgorin disks with smaller radius contain the eigenvalues corresponding to the noise.

On the other hand, in practice, the radii of the individual Gerschgorin circles are relatively large and they do not always form two disjoint union sets as shown in Fig. 4.1. Therefore, it is generally necessary to transform the covariance matrix so that the Gerschgorin disk of the noise is as far as possible from the Gerschgorin disc of the source signal and has a smaller radius.

The covariance matrix $\boldsymbol{R}$ of the observed data can be re-expressed as:

$$\boldsymbol{R} = \begin{pmatrix} r_{11} & \cdots & r_{1N} \\ \vdots & \ddots & \vdots \\ r_{N1} & \cdots & r_{NN} \end{pmatrix} = \begin{bmatrix} \boldsymbol{R}_1 & \boldsymbol{r} \\ \boldsymbol{r}^{\mathrm{H}} & r_{NN} \end{bmatrix} \tag{4.55}$$

$\boldsymbol{R}_1$ is the $(N-1)$-th order principal submatrix of $\boldsymbol{R}$. Noting $\boldsymbol{A} = [\boldsymbol{a}_1, \boldsymbol{a}_2, \ldots, \boldsymbol{a}_N]^{\mathrm{T}}$, we have

$$\boldsymbol{r} = \left[r_{1N}, r_{2N}, \ldots r_{(N-1)N}\right]^{\mathrm{H}} = \left[\boldsymbol{a}_1, \boldsymbol{a}_2, \ldots, \boldsymbol{a}_{N-1}\right]^{\mathrm{T}} \boldsymbol{R}_s a_N^* = \boldsymbol{A}_1 \boldsymbol{R}_s a_N^* \tag{4.56}$$

eigenvalue decomposition of $\boldsymbol{R}_1$ yields

$$\boldsymbol{R}_1 = \boldsymbol{U}_1 \boldsymbol{D}_1 \boldsymbol{U}_1^{\mathrm{H}} \tag{4.57}$$

where $\boldsymbol{U}_1 = [\boldsymbol{u}'_1, \boldsymbol{u}'_2, \ldots \boldsymbol{u}'_{N-1}]$, $\boldsymbol{D}_1 = diag(\lambda'_1, \lambda'_2, \ldots, \lambda'_{N-1})$ are the eigenvectors matrix of $\boldsymbol{R}_1$ and the diagonal matrix composed of eigenvalues of $\boldsymbol{R}_1$, respectively. Construct the transforming matrix:

$$\boldsymbol{U} = \begin{bmatrix} \boldsymbol{U}_1 & 0 \\ 0^{\mathrm{T}} & 1 \end{bmatrix} \tag{4.58}$$

then the covariance matrix $\boldsymbol{R}$ is transformed according to the transforming matrix $\boldsymbol{U}$ as follows:

$$\boldsymbol{S} = \boldsymbol{U}^{\mathrm{H}}\boldsymbol{R}\boldsymbol{U} = \begin{bmatrix} \boldsymbol{U}_1^{\mathrm{H}}\boldsymbol{R}_1\boldsymbol{U}_1 & \boldsymbol{U}_1^{\mathrm{H}}\boldsymbol{r} \\ \boldsymbol{r}^{\mathrm{H}}\boldsymbol{U}_1 & r_{NN} \end{bmatrix} = \begin{bmatrix} \boldsymbol{D}_1 & \boldsymbol{\rho} \\ \boldsymbol{\rho}^{\mathrm{H}} & r_{NN} \end{bmatrix} \tag{4.59}$$

where $\boldsymbol{\rho} = [\rho_1, \rho_2, \ldots, \rho_{N-1}]^{\mathrm{T}}$, $\rho_i = \boldsymbol{u}_i'^{\mathrm{H}}\boldsymbol{r} = \boldsymbol{u}_i'^{\mathrm{H}}\boldsymbol{A}_1\boldsymbol{R}_s a_N^*$. When $i = M+1, \ldots N-1$, $\boldsymbol{u}'_i$ are the eigenvectors corresponding to the noise in matrix $\boldsymbol{R}_1$, and $\boldsymbol{R}_1\boldsymbol{u}'_i = \sigma^2\boldsymbol{u}'_i = (\boldsymbol{A}_1\boldsymbol{R}_s\boldsymbol{A}_1^{\mathrm{H}} + \sigma^2\boldsymbol{I})\boldsymbol{u}'_i$, hence $\boldsymbol{A}_1\boldsymbol{R}_s\boldsymbol{A}_1^{\mathrm{H}}\boldsymbol{u}'_i = 0$, and thus $\boldsymbol{A}_1^{\mathrm{H}}\boldsymbol{u}'_i = 0$. Therefore, in matrix $\boldsymbol{S}$, we have $\rho_{M+1} = \cdots = \rho_{N-1} = 0$ and $\rho^*_{M+1} = \cdots = \rho^*_{N-1} = 0$.

$$\boldsymbol{S} = \begin{bmatrix} \lambda'_1 & 0 & \cdots\cdots & 0 & \rho_1 \\ \vdots & \ddots & & & \vdots \\ 0 & \lambda'_M & & & \rho_M \\ 0 & & \sigma^2 & 0 & 0 \\ \vdots & & 0 & \ddots & \vdots \\ \rho_1^* & \cdots \rho_M^* & 0 & \cdots & r_{NN} \end{bmatrix} \tag{4.60}$$

Matrices $\boldsymbol{S}$ and $\boldsymbol{R}$ have the same eigenvalues, and we can estimate the number of sources by analyzing the Gerschgorin circles of matrix $\boldsymbol{S}$. It is known that the first M Gerschgorin circles of matrix $\boldsymbol{S}$ correspond to the source signal, the radius of the circle is $\xi_i = |\rho_i|$. When $i = M+1, \ldots, N-1$, the radius of the Gerschgorin circle corresponding to the noise is 0. Based on this feature, the estimation of the number of sources can be obtained. From the above analysis, the radius of the Gaelic circle is satisfied:

$$\xi_i = |\rho_i| = \left|\boldsymbol{u}_i'^{\mathrm{H}}\boldsymbol{r}\right| = \left|\boldsymbol{u}_i'^{\mathrm{H}}\boldsymbol{A}_1\boldsymbol{R}_s a_N^*\right| = \left|\boldsymbol{u}_i'^{\mathrm{H}}\boldsymbol{A}_1\right|\left|\boldsymbol{R}_s a_N^*\right| = \boldsymbol{\lambda}\left|\boldsymbol{u}_i'^{\mathrm{H}}\boldsymbol{A}_1\right| \tag{4.61}$$

where $\boldsymbol{\lambda}$ is independent of i, and thus only a suitable threshold is needed to separate Gerschgorin circle corresponding to the noise from that corresponding to the source signal and obtain an estimate of the number of source signals.ξ_i is determined by the magnitude of $\boldsymbol{u}_i'^{\mathrm{H}}\boldsymbol{A}_1$, independent of the type and size of the noise, and thus the Gerschgorin-circle-based method is still valid with non-Gaussian noise and small sample data.

The corresponding judgment method can be defined as:

$$\xi_{\mathrm{GDE}}(k) = \xi_k - \frac{D(T)}{N-1}\sum_{i=1}^{N-1}\xi_i \tag{4.62}$$

$D(T)$ is the correction factor, since the radius of Gerschgorin circle of noise decreases with the increase of data samples, $D(T)$ is a non-increasing function of the number of samples T. Therefore, as the number of samples increases, the corresponding sample threshold decreases, and $D(T)$ is generally taken in the range of 0 to 1. k is taken in the range of $1, 2, \ldots, N-2$. When $\xi_{\mathrm{GDE}}(k)$ is negative for the first time, the number of sources is k-1.

2. **Joint Estimating Methods Based on Gerschgorin-Circle and Information-Theoretic Criteria**

Another idea for the estimating algorithm of the number of sources based on Gerschgorin circle approach is to combine Gerschgorin circle theorem and the information-theoretic criterion. The maximum likelihood estimation based on Gerschgorin circle can be obtained by replacing $R^{(k)}$ with $S^{(k)}$ in Eq. (4.47). Where the cost function of Gerschgorin circle likelihood estimation can be expressed as:

$$\log f\left(\boldsymbol{X}|\Theta^{(k)}\right) = -T(N-1-k)\log\left[\frac{\alpha'_N(k)}{\beta'_N(k)}\right] - T\log\left(r_{NN} - \sum_{i=1}^{k}\frac{\xi_i^2}{\lambda'_i}\right) \tag{4.63}$$

where $\alpha'_N(k)$ and $\beta'_N(k)$ have the same form with $\alpha_N(k)$, $\beta_N(k)$ in (4.50). The degrees of freedom of the likelihood function are determined by the k-dimensional submatrices of the matrix $\boldsymbol{S}$ and the corresponding Gerschgorin circles, which have a degree of freedom of $\kappa = k^2 + k$. Estimating of the number of sources is carried out by substituting Eq. (4.63) and κ into the AIC criterion and the MDL criterion of Eqs. (4.41) and (4.42), respectively, to obtain the corresponding Gerschgorin-circle-based AIC criterion (GAIC) and Gerschgorin-circle-based MDL criterion (GMDL). Both GAIC and GMDL are asymptotically consistent estimates of the true number of signals.

The Gerschgorin-circle-based method requires fewer data samples and is suitable for the case of non-Gaussian noise. But again, the number of receiving channels N must be greater than the number of source M and satisfy $N - M > 1$.

4.3 Simulation and Analysis of Source Number Estimation Algorithms

Two QPSK signals and two BPSK signals are used as the source signals $s_1(t)$, $s_2(t)$, $s_3(t)$ and $s_4(t)$ in the simulation, in which the carrier frequency of the first source signal $s_1(t)$ is set to 60 MHz, and the carrier frequencies of the rest of the signals are added sequentially with fixed carrier intervals (the carrier intervals can be adjusted for subsequent experiments). The bit rate is 0.25 Mbit/s, the sampling frequency is set to 5 MHz, and Gaussian white noise is added. This experiment studies both single-channel received signals and multi-channel received signals, and the model of the mixed signal is $\boldsymbol{X}(t) = \sum_{m=1}^{M} \boldsymbol{a}_m s_m(t) + \boldsymbol{v}(t)$.

For the two cases of single-channel received signals and multiple-channel received signals, 1000 Monte Carlo simulation experiments are conducted for each case to count the accuracy of different source signal number estimation methods, and then to compare the performance of source number estimation of the existing methods under different conditions, including methods based on eigenvalue decomposition (EVD), AIC criterion, MDL criterion, and GDE criterion.

4.3.1 Single-Channel Received Signals

For the mixed multiple source signals received by a single sensor, the source signal number estimation method can be applied after the single channel received signal is extended by the interval sampling method.

1. **Influence of Carrier Frequency Interval on Source Number Estimation**

Keeping the actual number of source signals as 3, the single-channel received signals are expanded to 10 channels, the carrier frequency interval (CFI) between the source signals is set to 0.6 MHz and is continuously reduced at fixed intervals of 0.1 MHz, and the estimation accuracies of different methods at different CFI are given in Fig. 4.2 [29].

As can be seen in Fig. 4.2, when the carrier frequency interval between the source signals is gradually reduced, the performance of the various source number estimation methods deteriorates significantly. When carrier frequency interval is 0.6 MHz, the method based on the AIC criterion has only 80% estimation accuracy and is not a consistent estimation even if the SNR is increased to more than 0 dB. The method based on the GDE criterion is able to achieve 100% estimation accuracy when SNR is more than -2 dB and carrier frequency intervals is more than 0.3 MHz. but the method fails when the carrier frequency interval is reduced to 0.2 MHz because the Gerschgorin disks of the signals are close to each other when the carrier frequency intervals are low, and they cannot form a suitable union set of the signals to be distinguished from the noisy Gerschgorin disks. For the method based on eigenvalue decomposition, its estimation accuracy decreases significantly with the reduction of the carrier frequency interval, and when the carrier frequency interval is reduced to 0.2 MHz, the estimation accuracy is 0 even if SNR is increased to 10 dB, because the method based on eigenvalue decomposition requires the covariance to be calculated very accurately, and the signal carrier frequency interval is too small which will bring a great deal of difficulty in calculating the covariance. The method requires a high SNR to accurately estimate the number of source signals. The estimation method based on the MDL criterion also performs well when the signal carrier frequency

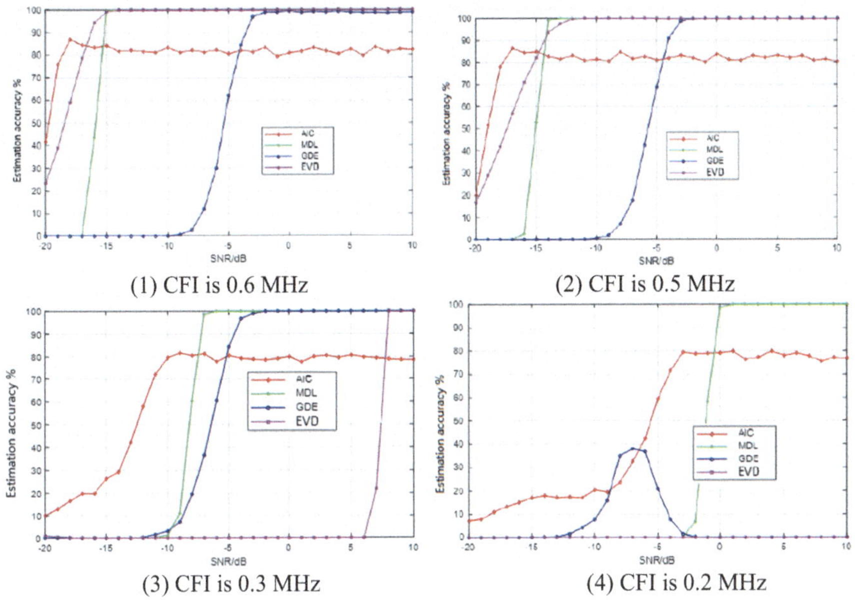

(1) CFI is 0.6 MHz (2) CFI is 0.5 MHz

(3) CFI is 0.3 MHz (4) CFI is 0.2 MHz

Fig. 4.2 Estimation accuracy of different carrier frequency intervals

interval is close, and it is an asymptotically consistent estimation of the true number of source signals, which can achieve 100% estimation accuracy when SNR is more than 1 dB even if the carrier frequency interval is reduced to 0.2 MHz.

Comprehensively, for the source number estimation problem of single-channel received signals, the MDL criteria method and the estimation method based on the eigenvalue decomposition perform better and have significant advantages over other algorithms. Further analysis reveals that the eigenvalue decomposition-based method performs best when the carrier frequency interval is greater than 0.5 MHz, i.e., when the spectrum is not overlapped. When the carrier frequency interval is less than 0.5 MHz, the estimation method based on the MDL criterion performs the best. However, considering that the carrier frequency interval is not known when blind separation is performed in practice, the source number estimation method based on MDL criterion is more practical.

2. **Performance of the Algorithm Under Different Source Numbers**

The analysis of experiment (1) indicates that the eigenvalue decomposition method and MDL criterion method are the most effective under different carrier frequency intervals. Therefore, the subsequent experiments will only compare these two methods. Keeping the carrier frequency interval of the source signals at 0.5 MHz, expanding the single-channel received signals to 10 channels, we change the source numbers, and compare the performances of the eigenvalue decomposition method and the MDL criterion method as shown in Fig. 4.3.

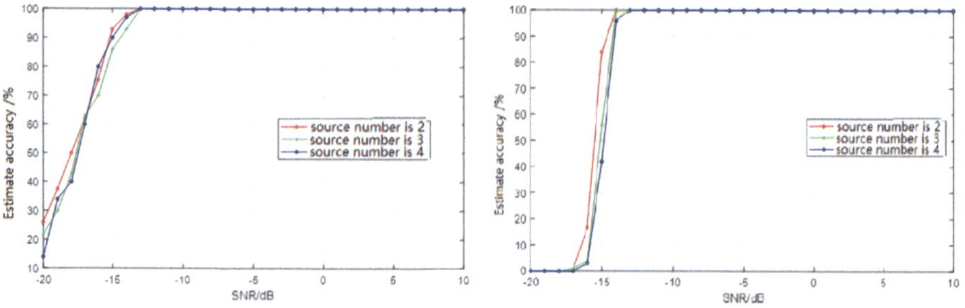

(1) eigenvalue decomposition method (2) MDL criterion method

Fig. 4.3 Estimation accuracy for different number of source signals

As can be seen in Fig. 4.3, in the case of single-channel reception, the performance of the two estimation methods basically shows a trend of decreasing gradually with the increase of the source number, but the loss of estimation performance is not significant. When the number of source signals is increased to 4, the MDL criterion method and eigenvalue decomposition method achieves 100% estimation accuracy at SNR > -13 dB. Therefore, both source number estimation methods have good estimation performance and can be applied in practical situations.

3. **Influence of the Number of Extended Channels on Algorithm Performance**

Keeping the number of source signals as 3 and the carrier frequency interval as 0.5 MHz and varying the number of extended channels (5, 10, and 20), we compare the performance of the eigenvalue decomposition method and the MDL criterion method as shown in Fig. 4.4, where P is the number of extended channels.

Figure 4.4 shows when the number of extended channels is 5, estimation method based on eigenvalue decomposition is out of work, because the reduction in the number of extended channels brings more difficulties in the accurate calculating of the covariance matrix. When the number of extended channels is 10, the number of source signals can be accurately estimated when SNR is more than -15 dB or more. However, when the number of extended channels is 20, there is a certain performance

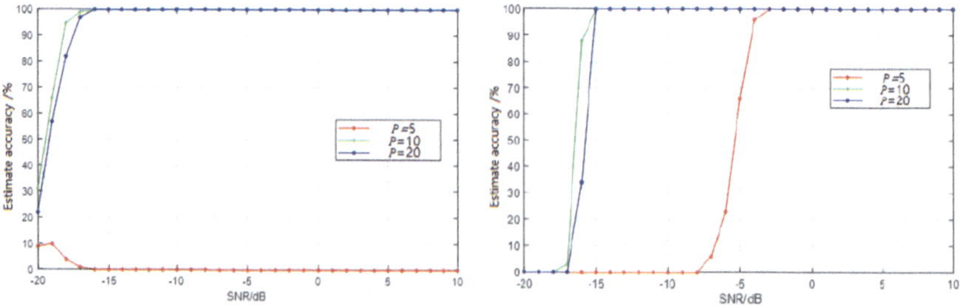

(1) eigenvalue decomposition method (2) MDL criterion method

Fig. 4.4 Estimation accuracy for different number of signal expansion channels

loss because the number of sampling points of each extended signal is reduced, resulting in the retention of information of each source signal is not as complete as the original. For the estimation method based on the MDL criterion, when the number of extended channels is 10, the estimation performance is the best among the three numbers of extended channels. A smaller number of extended channels cannot construct an accurate covariance matrix, while too many extended channels cannot ensure that there are enough sampling points for each extended signal, which leads to the loss of part of the information and a large number of extended channels also leads to an increase in the complexity of the algorithm. Taken together, the estimation method based on eigenvalue decomposition is slightly better than the estimation method based on the MDL criterion when the number of expansion passes is larger, but the estimation method based on the MDL criterion can also estimate the number of source signals when the number of expansion passes is smaller, so it is more practical. In addition, the performance of both source number estimation methods is closely related to the number of expansion channels, and the number of expansion channels for single-channel received signals should not be too few or too much, and in general, it is enough to ensure that the number of expansion channels is 2–4 times of the real number of source signals.

4.3.2 Multi-Channel Received Signals

Receiving mixed multiple signals with multiple sensors (it should be ensured that the number of received signals is larger than the number of source signals), in order to ensure the estimation accuracy of the number of source signals, it is still necessary to use the interval sampling method to extend part or all of the received signals, and then estimate the number of source signals by using the eigenvalue decomposition method and MDL criterion method.

1. **Influence of the Extended Channel Numbers on Source Number Estimation**

Keeping the number of source signals as 3, the carrier frequency interval as 0.5 MHz, using 4 channels to receive the observation signals, changing the number of receiving channels L used to construct the observed signal matrix, and extended the number of channels for each receiving signal to 10 (utilizing the odd number of them for constructing the observed signal matrix), we compare the performance of the two methods for estimating the number of source signals as show in Fig. 4.5.

As can be seen from Fig. 4.5, with the increasing in the number of receiving channels used to construct the observed signal matrix, the estimation accuracy of two methods will gradually improve, because the more receiving channels are utilized to construct the observed signal matrix, the more complete the source signal information can be retained, and the performance of the source signal number estimation methods will be improved. When only one of the receiving channels is used, the eigenvalue decomposition method has poor estimation performance and requires a high SNR to accurately estimate the number of source signals, whereas the MDL criterion method

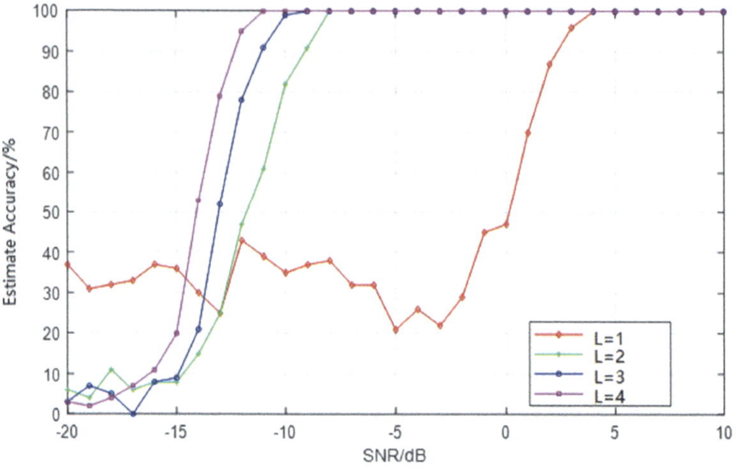

(1) eigenvalue decomposition method

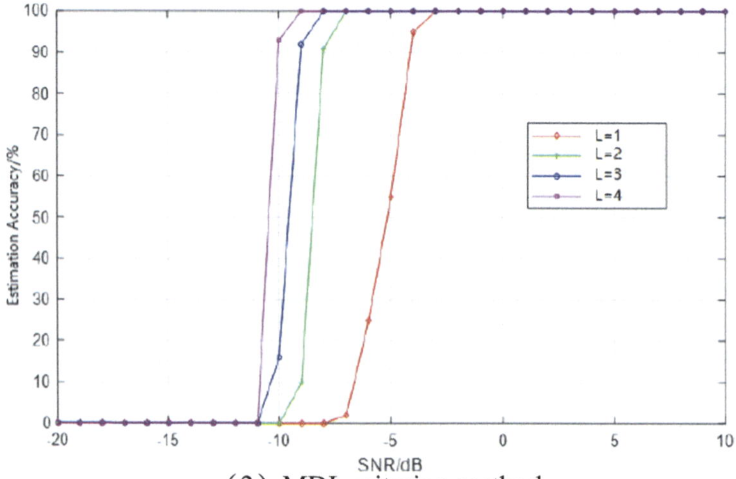

(2) MDL criterion method

Fig. 4.5 Estimation accuracy with different number of receiving channels

can accurately estimate the number of source signals when SNR ≥ -3 dB. When the number of receiving channels for constructing the observed signal matrix is increased to 2, the eigenvalue decomposition method outperforms the MDL criterion method. Therefore, when the majority of the signals received by the sensor are available, the eigenvalue decomposition method is better for estimating the number of source signals.

2. **Influence of the Number of Extended Channels on the Estimation of the Number of Source Signals**

Keeping the number of source signals as 3 and the carrier frequency interval as 0.5 MHz, 4 channels are used to receive the observed signals, and all the received signals are used to construct the observed signal matrix, and the number of extended channels P is changed, and the estimation accuracies of two methods are shown in Fig. 4.6.

As can be seen from Fig. 4.6, the performance of the estimation method based on eigenvalue decomposition gradually improves when the number of extended channels of each observed signal increases, because the covariance matrix can be calculated

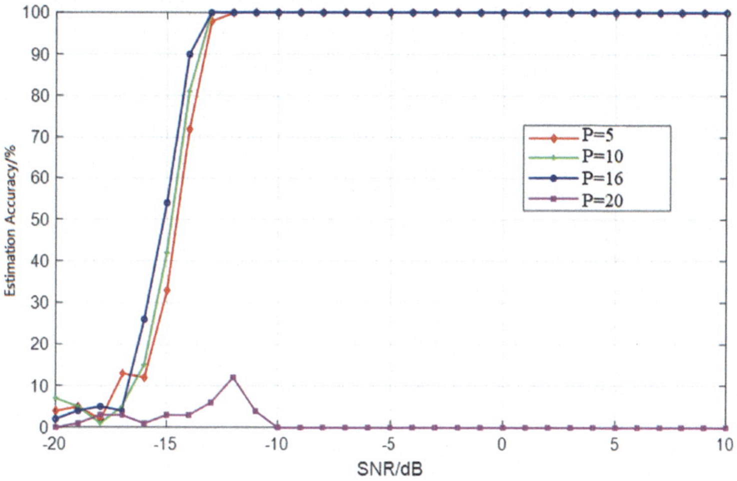

(1) Eigenvalue Decomposition Method

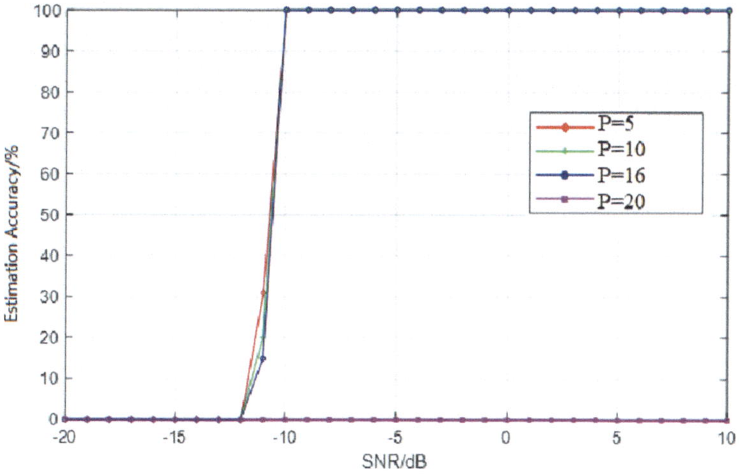

（2）MDL Criterion Method

Fig. 4.6 Estimation accuracy with different number of extended channles

more accurately, slightly improving the performance of the method. The estimation performance of the MDL criterion method does not produce a significant change, which indicates that as long as the extended number is more than the real number of source signals, a good estimation of the number of sources can be achieved. However, when the number of extended channels is 20, both eigenvalue decomposition method and MDL criterion method are out of work. If the number of extended channels is too large, which will result in fewer sampling points for each extended signal, it is not enough to calculate the accurate covariance matrix, and thus lead to the error in the computation of the eigenvalues. In summary, the number of extended channels for each observed signal should not be too few or too many. In general, combined with the consideration of complexity, the number of extended channels for each observed signal should be set at 2–4 times of the real number of source signals.

In summary, eigenvalue decomposition method and MDL criterion method have the best performance, and the selection of these two methods is mainly related to the signal carrier frequency interval. When the carrier frequency interval is very small or even spectral overlapping occurs, the MDL criterion method is suitable, and when the carrier frequency interval is larger, eigenvalue decomposition method is suitable. When the carrier frequency interval is 0.5 MHz, i.e. exactly twice the symbol rate, both methods have better estimation performance.

Summary

This chapter introduces the preprocessing methods for blind source separation in detail. Firstly, various whitening algorithms are investigated, which can reduce the dimensionality of mixed signal matrix and complexity of the problem. Among these algorithms, robust whitening algorithms can eliminate the influence of noise to a certain extent, and therefore they are more commonly used than standard whitening algorithms. Then three types of existing source signal number estimation methods are studied, the algorithm principle and implementation process are derived in detail, and finally, the performance of the source signal number estimation methods under different conditions is evaluated in simulation. From a comprehensive point of view, the methods based on the MDL criterion and the methods based on eigenvalue decomposition have better estimation performance and can estimate the number of source signals more accurately.

Chapter 5
Gradient-Based Overdetermined/Determined BSS Algorithm

There are many ways to solve the overdetermined/determined BSS problem. This chapter focuses on the gradient-based overdetermined/determined BSS algorithm. Firstly, the gradient-based BSS algorithm with a fixed step size is analyzed and derived, followed by the gradient-based BSS algorithm with a variable step size is introduced. Then, the stability of the BSS algorithm as well as the selection of the nonlinear function are analyzed and proved. Finally, the separation performance of the overdetermined/determined BSS algorithm presented in this chapter is simulated and analyzed.

5.1 Gradient-Based Blind Source Separation Algorithm with a Fixed Step Size

According to Chap. 2, the linear instantaneous mixing model for BSS is $\boldsymbol{x}(\mathrm{t}) = \boldsymbol{A}\boldsymbol{s}(\mathrm{t}) + \boldsymbol{v}(t)$, where $\boldsymbol{v}(t)$ is noise. The task of BSS is to find a separation matrix $\boldsymbol{W}$ based on the observed signal vector $\boldsymbol{x}(t)$ such that the components $\boldsymbol{y}(t) = \boldsymbol{W}\boldsymbol{x}(t)$ are independent of each other. To derive a BSS algorithm, one of the key tasks is to give a suitable independence measure function, different objective functions are constructed from different measures of independence, and different blind source separation algorithms are derived by optimizing these objective functions.

According to Chap. 4, the observed signal needs to be whitened before separating. Here we assume that the whitened signal is $\boldsymbol{z} = \boldsymbol{Q}\boldsymbol{x}$, where $\boldsymbol{Q}$ is the whitening matrix, the whitening method can be found in Sect. 4.1, and the later processing is carried out on $\boldsymbol{z}$. Now the global mixing matrix is $\boldsymbol{G} = \boldsymbol{Q} \times \boldsymbol{A}$.

In the following derivation, to simplify the problem, it can be assumed that there is no noise or that the noise is small enough to be ignored.

W. Fu, *Principle and Application of Blind Source Separation Technology*,
https://doi.org/10.1007/978-981-96-0022-9_5

5.1.1 Selection of the Cost Function

In order to obtain an estimate $\boldsymbol{y} = \boldsymbol{W}\boldsymbol{z}$ of the source signal $\boldsymbol{s}$ (here $\boldsymbol{W}$ is referred to as the separating matrix), it is necessary to introduce an objective or loss function $\rho(\boldsymbol{y}, \boldsymbol{W})$ with respect to $\boldsymbol{y}$ and $\boldsymbol{W}$. The expected value of this function is called the risk function (cost function):

$$\boldsymbol{R}(\boldsymbol{W}) = E\{\rho(\boldsymbol{y}, \boldsymbol{W})\} \tag{5.1}$$

the risk function $\boldsymbol{R}(\boldsymbol{W})$ takes its minimum value when the components of $\boldsymbol{y}$ are independent of each other, i.e., when $\boldsymbol{W}$ is the stretching and permutation of $\boldsymbol{A}^{-1}$.

The Kullback–Leibler (K-L) divergence can be used as a measure of independence, let $p_y(\boldsymbol{y}, \boldsymbol{W})$ is the probability density function of the random variable $\boldsymbol{y} = \boldsymbol{W}\boldsymbol{z}$, and $q(\boldsymbol{y})$ is another probability density function $\boldsymbol{y}$ that we chosed. When the components of $\boldsymbol{y}$, y_i are statistically independent, $q(\mathbf{y})$ can be decomposed into the following product form:

$$q(\boldsymbol{y}) = \prod_{i=1}^{m} q_i(y_i) \tag{5.2}$$

This independence probability density function is called the reference function, and the risk function is then:

$$\begin{aligned} \boldsymbol{R}(\boldsymbol{W}) = \mathrm{E}\{\rho(\boldsymbol{y}; \boldsymbol{W})\} = K_{pq}(\boldsymbol{W}) = K[p_y(\boldsymbol{y}; \boldsymbol{W})||q(\boldsymbol{y})] \\ = \int p_y(\boldsymbol{y}, \boldsymbol{W}) \log \frac{p_y(\boldsymbol{y}; \boldsymbol{W})}{q(\boldsymbol{y})} d\boldsymbol{y} \end{aligned} \tag{5.3}$$

The K-L divergence gives a natural measure of the difference between two probability distributions, $K[p_y(\boldsymbol{y}; \boldsymbol{W})||q(\boldsymbol{y})] \geq 0$. If and only if $p_y(\boldsymbol{y}; \boldsymbol{W}) = q(\boldsymbol{y})$, $K[p_y(\boldsymbol{y}; \boldsymbol{W})||q(\boldsymbol{y})] = 0$.

When $q(\mathbf{y})$ is the true distribution p_s of the source signal, if $\boldsymbol{W} = (\boldsymbol{Q}\boldsymbol{A})^{-1}$, then we have $\boldsymbol{y} = \boldsymbol{W}\boldsymbol{z} = \boldsymbol{W}(\boldsymbol{Q}\boldsymbol{A})\boldsymbol{s} = \boldsymbol{s}$ and thus $p_y(\boldsymbol{y}; \boldsymbol{W}) = p_s(\boldsymbol{y})$ and $\boldsymbol{R}(\boldsymbol{W})$ is a minimum of 0, so $K_{pq}(\boldsymbol{W})$ is a suitable cost function. Also, even if $q(\boldsymbol{y})$ is not equal to p_s, $\boldsymbol{W} = (\boldsymbol{Q}\boldsymbol{A})^{-1}$ is still a saddle point for $R(\boldsymbol{W}) = K_{pq}(\boldsymbol{W})$.

Presently, most heuristic algorithms can be explained by the above cost function, the only difference being the choice of the reference function $q(\boldsymbol{y})$. If we choose $q(\boldsymbol{y})$ to be the true distribution of the source signal, it is maximum likelihood methods. However, the true distribution is usually unknown, so for general $q(\boldsymbol{y})$ there is:

$$\boldsymbol{R}(\boldsymbol{W}) = \int p_y(\boldsymbol{y}, \boldsymbol{W}) \log \frac{p_y(\boldsymbol{y}; \boldsymbol{W})}{q(\boldsymbol{y})} d\boldsymbol{y} = -\mathrm{H}(\boldsymbol{y}) - \sum_{i=1}^{n} E\{\log(q_i(y_i))\} \tag{5.4}$$

where $\mathrm{H}(\boldsymbol{y}) = -\int p_y(\boldsymbol{y}, \boldsymbol{W}) \log p_y(\boldsymbol{y}; \boldsymbol{W}) d\boldsymbol{y}$ denotes the differential entropy of $\boldsymbol{y}$.

Since $\boldsymbol{y} = \boldsymbol{W}\boldsymbol{z}$, then $p_y(\boldsymbol{y}) = |\det \boldsymbol{W}|^{-1} p_z(\boldsymbol{z})$, therefore

$$\begin{aligned}
\mathrm{H}(\boldsymbol{y}) &= -\int |\det \boldsymbol{W}|^{-1} p_z(\boldsymbol{z}) \cdot \log \frac{p_z(\boldsymbol{z})}{|\det \boldsymbol{W}|} \cdot |\det \boldsymbol{W}| \cdot d\boldsymbol{z} \\
&= -\int p_z(\boldsymbol{z}) \cdot \log \frac{p_z(\boldsymbol{z})}{|\det \boldsymbol{W}|} d\boldsymbol{z} = \mathrm{H}(\boldsymbol{z}) + \int p_z(\boldsymbol{z}) \log |\det \boldsymbol{W}| \cdot d\boldsymbol{z} \\
&= \mathrm{H}(\boldsymbol{z}) + \log |\det \boldsymbol{W}|
\end{aligned} \tag{5.5}$$

Since $\mathrm{H}(\boldsymbol{z})$ is independent of $\boldsymbol{W}$ and can be omitted, the final cost function is:

$$\boldsymbol{R} = K_{pq} = E\{\rho(\boldsymbol{y}, \boldsymbol{W})\} = -\log|\det(\boldsymbol{W})| - \sum_{i=1}^{m} E\{\log(q_i(y_i))\} \tag{5.6}$$

5.1.2 A Stochastic Gradient Algorithm for Blind Source Separation

From the previous section, we know that in order to correctly separate the source signal, the expected value of the objective function $\rho(\boldsymbol{y}, \boldsymbol{W})$ should be minimized. For this purpose, the stochastic gradient optimization algorithm is first used, then:

$$\Delta \boldsymbol{W}(k) = \boldsymbol{W}(k+1) - \boldsymbol{W}(k) = -\eta \cdot \frac{\partial \rho(\boldsymbol{y}, \boldsymbol{W})}{\partial \boldsymbol{W}} \tag{5.7}$$

where η is called the step size. From Eqs. (5.6)–(5.7), we can see that

$$\Delta \boldsymbol{W}(k) = \eta \cdot \left[\frac{\partial \log|\det \boldsymbol{W}(k)|}{\partial \boldsymbol{W}(k)} + \sum_{i=1}^{m} \left(\frac{\partial \log(q_i(y_i(k)))}{\partial y_i(k)} \cdot \frac{\partial y_i(k)}{\partial \boldsymbol{W}(k)} \right) \right] \tag{5.8}$$

For $\boldsymbol{y}(k) = \boldsymbol{W}(k) \cdot \boldsymbol{z}(k)$, then

$$\Delta \boldsymbol{W}(k) = \eta \cdot \left[\boldsymbol{W}(k)^{-\mathrm{T}} - \sum_{i=1}^{m} (f_i(y_i(k)) \cdot [\tilde{\boldsymbol{X}}]_i) \right] \tag{5.9}$$

where

$$f_i(y_i) = -\frac{d \log q_i(y_i)}{dy_i} = -\frac{dq_i(y_i)/dy_i}{y_i} = -\frac{q_i'(y_i)}{q_i(y_i)} \tag{5.10}$$

$[\tilde{\boldsymbol{X}}]_i$ denotes the matrix with zero elements except for $[z_1, z_2, \ldots, z_m]$ in row i. Further simplification of the above equation can be obtained:

$$\Delta \boldsymbol{W}(k) = \eta \cdot \{\mathbf{W}^{-\mathrm{T}}(k) - \boldsymbol{f}[\mathbf{y}(k)]\mathbf{z}^{\mathrm{T}}(k)\} \tag{5.11}$$

where $\boldsymbol{f}[\mathbf{y}(k)] = [f_1(y_1), f_2(y_2), \ldots, f_m(y_m)]^T$ real-time iteration formula of the separating matrix is:

$$\boldsymbol{W}(k+1) = \boldsymbol{W}(k) + \eta \cdot \{\mathbf{W}^{-\mathrm{T}}(k) - \boldsymbol{f}[\mathbf{y}(k)]\mathbf{z}^{\mathrm{T}}(k)\} \tag{5.12}$$

replacing the instantaneous value with the time average of the signal, we get the batch form of the separating algorithm as follows.

$$\boldsymbol{W}(k+1) = \boldsymbol{W}(k) + \eta \cdot \{\mathbf{W}^{-\mathrm{T}}(k) - \langle \boldsymbol{f}[\mathbf{y}(k)] \cdot \boldsymbol{z}^{\mathrm{T}}(k) \rangle\} \tag{5.13}$$

where $\langle \cdot \rangle$ denotes time average. Equations (5.12) and (5.13) are the iterative formulas of the stochastic gradient BSS algorithm.

5.1.3 Natural Gradient Blind Source Separation Algorithm

To improve the performance of the BSS algorithm, Amari [30] proposed the concept of natural gradient. Because the space in which each parameter exists in independent component analysis (ICA) is not a Euclidean space, but rather a Riemann space. In the Riemann space, the direction of stochastic gradient is no longer the direction in which the objective function descends most steeply but the natural gradient direction is.

The iterative formula for the BSS algorithm based on natural gradient is derived below. The natural gradient of the objective function $\rho(\boldsymbol{y}, \boldsymbol{W})$ is defined as $\frac{\partial \rho(\boldsymbol{y}, \boldsymbol{W})}{\partial \boldsymbol{W}} \boldsymbol{W}^{\mathrm{T}} \boldsymbol{W}$ then:

$$\Delta \boldsymbol{W}(k) = \boldsymbol{W}(k+1) - \boldsymbol{W}(k) = -\eta \cdot \frac{\partial \rho(\boldsymbol{y}, \boldsymbol{W})}{\partial \boldsymbol{W}} \boldsymbol{W}^{\mathrm{T}} \boldsymbol{W} \tag{5.14}$$

according to (5.6), (5.14)

$$\Delta \boldsymbol{W}(k) = \eta \cdot \left[\frac{\partial \log|\det \boldsymbol{W}(k)|}{\partial \boldsymbol{W}(k)} + \sum_{i=1}^{m} \left(\frac{\partial \log(q_i(y_i(k)))}{\partial y_i(k)} \cdot \frac{\partial y_i(k)}{\partial \boldsymbol{W}(k)} \right) \right] \boldsymbol{W}^{\mathrm{T}} \boldsymbol{W} \tag{5.15}$$

and for $\boldsymbol{y}(k) = \boldsymbol{W}(k) \cdot \boldsymbol{z}(k)$, then

$$\Delta \boldsymbol{W}(k) = \eta \cdot [\boldsymbol{W}(k)^{-\mathrm{T}} - \sum_{i=1}^{m} (f_i(y_i(k)) \cdot [\tilde{\boldsymbol{X}}]_i)] \cdot \boldsymbol{W}(k)^{\mathrm{T}} \boldsymbol{W}(k) \tag{5.16}$$

further simplifying the equation above, we yield

$$\Delta \boldsymbol{W}(k) = \eta(\boldsymbol{I} - \boldsymbol{f}[\boldsymbol{y}(k)]\boldsymbol{y}^{\mathrm{T}}(k))\boldsymbol{W}(k) \tag{5.17}$$

thus, the real-time iterative formula for the separating matrix is:

$$\boldsymbol{W}(k+1) = \boldsymbol{W}(k) + \eta\{I - \boldsymbol{f}[\boldsymbol{y}(k)]\boldsymbol{y}^{\mathrm{T}}(k)\}\boldsymbol{W}(k) \tag{5.18}$$

Similarly, the batch form of the separation algorithm can be obtained by replacing the instantaneous value with the time average of the signal.

$$\boldsymbol{W}(k+1) = \boldsymbol{W}(k) + \eta\{I - \langle \boldsymbol{f}[\boldsymbol{y}(k)]\boldsymbol{y}^{\mathrm{T}}(k)\rangle\}\boldsymbol{W}(k) \tag{5.19}$$

Equation (5.19) is the BSS algorithm based on natural gradient, comparing (5.13) and (5.19), it can be seen that the natural gradient BSS algorithm reduces the matrix inverse operation relative to the stochastic gradient algorithm, so the algorithm is more practical relative to the stochastic gradient separation algorithm.

5.1.4 EASI BSS Algorithm

As mentioned above, the observed signal $\boldsymbol{x}$ is whitened by matrix $\boldsymbol{Q}$ prior to BSS, then equivalent mixing matrix $\boldsymbol{G} = \boldsymbol{Q}\boldsymbol{A}$ is an orthogonal matrix.

When the system is free of noise, $\boldsymbol{x} = \boldsymbol{A}\boldsymbol{s}$, and the whitened signal is $z = \boldsymbol{Q}\boldsymbol{x} = \boldsymbol{Q}\boldsymbol{A}\boldsymbol{s} = \boldsymbol{G}\boldsymbol{s}$. since z is white, i.e., $E(zz^{\mathrm{T}}) = \boldsymbol{I}$, then

$$E(zz^{\mathrm{T}}) = E\{\boldsymbol{G}\boldsymbol{s}\boldsymbol{s}^{\mathrm{T}}\boldsymbol{G}^{\mathrm{T}}\} = \boldsymbol{G}E\{\boldsymbol{s}\boldsymbol{s}^{\mathrm{T}}\}\boldsymbol{G}^{\mathrm{T}} = \boldsymbol{I} \tag{5.20}$$

according to the assumptions, $E\{\boldsymbol{s}\boldsymbol{s}^{\mathrm{T}}\} = \boldsymbol{I}$, then

$$\boldsymbol{G}\boldsymbol{G}^{\mathrm{T}} = \boldsymbol{I} \tag{5.21}$$

When there is noise in the system, according to Theorem 4.1 in Chap. 4, there is still $\boldsymbol{G}\boldsymbol{G}^{\mathrm{T}} = \boldsymbol{I}$.

Equation (5.19) represents the iterative formula for the separation matrix when there are no restrictions on $\boldsymbol{W}$. We know that the mixing matrix is an orthogonal matrix after whitening the observed signals, and thus we wish to keep $\boldsymbol{W}$ as orthogonal as possible throughout the iterations. Assuming that $\boldsymbol{W}(k)$ is already orthogonal (which can be satisfied by setting $\boldsymbol{W}(0) = \boldsymbol{I}$), from $\Delta\boldsymbol{W} = \varepsilon \cdot \boldsymbol{W}(k)$, where ε is a very small change matrix, then

$$\boldsymbol{W}(k+1) = \boldsymbol{W}(k) + \varepsilon\boldsymbol{W}(k) \tag{5.22}$$

and

$$\boldsymbol{W}(k+1)\boldsymbol{W}(k+1)^{\mathrm{T}} = \boldsymbol{W}(k)\boldsymbol{W}^{\mathrm{T}}(k) + \boldsymbol{\varepsilon}\boldsymbol{W}(k)\boldsymbol{W}^{\mathrm{T}}(k) + \boldsymbol{W}(k)\boldsymbol{W}^{\mathrm{T}}(k)\boldsymbol{\varepsilon}^{\mathrm{T}} + O(\boldsymbol{\varepsilon})$$

$$= \boldsymbol{I} + \boldsymbol{\varepsilon} + \boldsymbol{\varepsilon}^{\mathrm{T}} + O(\boldsymbol{\varepsilon}) = \boldsymbol{I} \tag{5.23}$$

then

$$\boldsymbol{\varepsilon} = -\boldsymbol{\varepsilon}^{\mathrm{T}} \tag{5.24}$$

Therefore, to maintain the orthogonality of $\boldsymbol{W}$, ε is required to be skew symmetric. Combining with Eq. (5.19), we can take:

$$\begin{aligned} \varepsilon &= \eta \cdot \{[\boldsymbol{I} - \boldsymbol{f}[\boldsymbol{y}(k)]\boldsymbol{y}^{\mathrm{T}}(k)] - [\boldsymbol{I} - \boldsymbol{f}[\boldsymbol{y}(k)]\boldsymbol{y}^{\mathrm{T}}(k)]^{\mathrm{T}}\} \\ &= \eta \cdot \{\boldsymbol{y}(k)\boldsymbol{f}^{\mathrm{T}}[\boldsymbol{y}(k)] - \boldsymbol{f}[\boldsymbol{y}(k)]\boldsymbol{y}^{\mathrm{T}}(k)\} \end{aligned} \tag{5.25}$$

So the iterative formula for the orthogonal separation matrix $\boldsymbol{W}$ is:

$$\begin{cases} \boldsymbol{W}(k+1) = \boldsymbol{W}(k) + \eta \cdot [\boldsymbol{y}(k)\boldsymbol{f}^{\mathrm{T}}[\boldsymbol{y}(k)] - \boldsymbol{f}[\boldsymbol{y}(k)]\boldsymbol{y}^{\mathrm{T}}(k)]\boldsymbol{W}(k) \\ \boldsymbol{y}(k) = \boldsymbol{W}(k) \cdot \boldsymbol{z}(k) \end{cases} \tag{5.26}$$

Similarly, batch form of the separation algorithm is

$$\begin{cases} \boldsymbol{W}(k+1) = \boldsymbol{W}(k) + \eta \cdot [\langle \boldsymbol{y}(k)\boldsymbol{f}^{\mathrm{T}}[\boldsymbol{y}(k)]\rangle - \langle \boldsymbol{f}[\boldsymbol{y}(k)]\boldsymbol{y}^{\mathrm{T}}(k)\rangle]\boldsymbol{W}(k) \\ \boldsymbol{y}(k) = \boldsymbol{W}(k) \cdot \boldsymbol{z}(k) \end{cases} \tag{5.27}$$

Since the signals need to be whitened, according to Eq. (4.26), we konw the iterative formula for the whitening matrix is as follow.

$$\boldsymbol{Q}(k+1) = \boldsymbol{Q}(k) + \eta(I - \langle \boldsymbol{y}\boldsymbol{y}^{\mathrm{T}} \rangle)\boldsymbol{Q}(k) \tag{5.28}$$

If the whitening and separating processes are combined, the total separating matrix is $\tilde{\boldsymbol{W}}(k+1) = \boldsymbol{G}(k+1)\boldsymbol{Q}(k+1)$. After simplification, the iterative formula of the separating matrix is as follows.

$$\begin{cases} \tilde{\boldsymbol{W}}(k+1) = \tilde{\boldsymbol{W}}(k) + \eta \cdot [\boldsymbol{I} - \langle \boldsymbol{y}\boldsymbol{y}^{\mathrm{T}} \rangle + \langle \boldsymbol{y}(k)\boldsymbol{f}^{\mathrm{T}}[\boldsymbol{y}(k)]\rangle - \langle \boldsymbol{f}[\boldsymbol{y}(k)]\boldsymbol{y}^{\mathrm{T}}(k)\rangle]\tilde{\boldsymbol{W}}(k) \\ \boldsymbol{y}(k) = \tilde{\boldsymbol{W}}(k) \cdot \boldsymbol{z}(k) \end{cases} \tag{5.29}$$

Equation (5.29) constitutes the EASI BSS algorithm.

5.2 Gradient BSS Algorithm with Variable Step Sizes

The two BSS algorithms given in the previous section have an fixed iterative step-size. If the step-size is too small, the algorithms converge very slowly, while if the step size is too large, the separating algorithm may not converge. In order to solve the contradiction between the convergence speed and the stability of the algorithm, the above BSS algorithm is improved to some extent. A step-size adaptive BSS algorithm can be obtained, which adaptively adjusts the step-size in order to improve convergence speed without causing the algorithm to diverge.

5.2.1 Natural Gradient BSS Algorithm with Variable Step-Size

As mentioned earlier, the selection of the iterative step-size η has a great impact on the convergence speed of the algorithm as well as the stability of the system. In order to solve this problem, a step-size-adaptive BSS algorithm has been proposed, which has the step-size iteration formula (the algorithm is referred to as "Algorithm 1") [31]

$$\eta(k+1) = \eta(k) + \gamma \times trace(\boldsymbol{A}^{\mathrm{T}}(k) \cdot \boldsymbol{A}(k-1)) \tag{5.30}$$

$$\boldsymbol{A}(k) = \{\boldsymbol{I} - \boldsymbol{f}(\boldsymbol{y}(k))\boldsymbol{y}^{T}(k)\}\boldsymbol{W}(k) \tag{5.31}$$

where γ is a very small constant, also known as the step-size of the step. In order to further improve the convergence speed of the algorithm without affecting the stability of the algorithm, a step-size iterative algorithm can be proposed by using the idea of optimal step-size, which can greatly improve the convergence speed of BSS algorithm and the step-size iterative algorithm based on the idea of optimal step-size converges faster than the algorithms given in (5.30) and (5.31).

Let the step-size of the algorithm at the $(k+1)$-th iteration be

$$\eta(k+1) = \eta(k) + \gamma \times \Delta\eta(k) \tag{5.32}$$

where $\Delta\eta(k)$ is the offset of step-size $\eta(k)$. The offset should be chosen so that step-size $\eta(k+1)$ is optimal, i.e., to minimize the objective function at the next iteration, therefore

$$\Delta\eta(k) = \frac{\partial\rho(\boldsymbol{W}(k+1))}{\partial\eta(k)} = trace\left\{\left[\frac{\partial\rho(\boldsymbol{W}(k+1))}{\partial\boldsymbol{W}(k+1)}\right]^{\mathrm{T}} \cdot \frac{\partial\boldsymbol{W}(k+1)}{\partial\eta(k)}\right\} \tag{5.33}$$

where $trace(\cdot)$ denotes the trace of the matrix. From (5.6), we get

$$\rho(\boldsymbol{W}(k+1)) = -\log|\det(\boldsymbol{W}(k+1))| - \sum_{i=1}^{m} E\{\log(q_i(y_i))\} \tag{5.34}$$

It can be seen that, at the $(k+1)$-th iteration, $\rho(\boldsymbol{W}(k+1))$ is only related to the first term on the right-hand side of Eq. (5.34) (since y_i can be considered fixed when finding the optimal value of the step-size), and thus we have

$$\frac{\partial\rho(\boldsymbol{W}(k+1))}{\partial(\boldsymbol{W}(k+1))} = -\boldsymbol{W}^{-\mathrm{T}}(k+1) \tag{5.35}$$

From (5.13) we know that $\boldsymbol{W}(k+1) = \boldsymbol{W}(k) + \eta(k) \times (\boldsymbol{W}^{-\mathrm{T}}(k) - E(f(\boldsymbol{y})\boldsymbol{z}^{\mathrm{T}}))$, so

$$\frac{\partial \boldsymbol{W}(k+1)}{\partial\eta(k)} = \boldsymbol{W}^{-\mathrm{T}}(k) - E(f(\boldsymbol{y})\boldsymbol{z}^{\mathrm{T}}) \tag{5.36}$$

And from Eqs. (5.33), (5.35), and (5.36), we get

$$\Delta\eta(k) = trace\{-\boldsymbol{W}(k+1)^{-1} \times \{\boldsymbol{W}^{-\mathrm{T}}(k) - E[f(\boldsymbol{y})\boldsymbol{z}^{\mathrm{T}}]\}\} \tag{5.37}$$

Therefore, a variable Step-Size Adaptive Based on Natural Gradient-2 (SABNG-1) algorithm can be obtained by combining Eqs. (5.13), (5.32) and (5.37).

In addition, from Eq. (5.19), we get

$$\frac{\partial \boldsymbol{W}(k+1)}{\partial\eta(k)} = \{I - \langle f[\boldsymbol{y}(k)]\boldsymbol{y}^{\mathrm{T}}(k)\rangle\}\boldsymbol{W}(k) \tag{5.38}$$

Similarly, according to (5.33), (5.35), and (5.38), we have

$$\Delta\eta(k) = trace\{-\boldsymbol{W}(k+1)^{-1} \times [I - \langle f[\boldsymbol{y}(k)]\boldsymbol{y}^{\mathrm{T}}(k)\rangle]\boldsymbol{W}(k)\} \tag{5.39}$$

Then according to Eqs. (5.19), (5.32), and (5.39), another variable step-size natural gradient blind separation algorithm, referred to as SABNG-2, can be obtained.

5.2.2 *EASI BSS Algorithm with Variable Step-Size*

1. **A Step-Size Adaptive EASI Algorithm**

Even though the EASI BSS algorithm is capable of effectively separating individual source signals from linearly mixing observed signals, further improvements to convergence speed can be achieved by utilizing the parameter adaptive optimization idea. The new step-size adaptive EASI algorithm (named 'EASIBSA') can be obtained, which the step-size η is also adaptively iterated.

According to (5.29), the iterative formula for the EASI algorithm is:

$$\boldsymbol{W}(k+1) = \boldsymbol{W}(k) + \eta(k) \cdot \boldsymbol{W}_d(k) \tag{5.40}$$

where

$$\boldsymbol{W}_d(k) = \boldsymbol{I} - \langle \boldsymbol{y} \cdot \boldsymbol{y}^{\mathrm{T}} \rangle - \langle \boldsymbol{f}(\boldsymbol{y}) \cdot \boldsymbol{y}^{\mathrm{T}} \rangle + \langle \boldsymbol{y} \cdot \boldsymbol{f}^{\mathrm{T}}(\boldsymbol{y}) \rangle \tag{5.41}$$

from the previous section, after performing k iterations, the objective function is as follow.

$$\rho(\boldsymbol{W}(k)) = -\log|\det(\boldsymbol{W}(k))| - \sum_{i=1}^{m} \log(q_i(y_i(k))) \tag{5.42}$$

from Eqs. (5.40) (5.42), the derivative of $\rho(\boldsymbol{W}(k))$ with respect to $\eta(k-1)$ is

$$\frac{\partial \rho(\boldsymbol{W}(k))}{\partial \eta(k-1)} = \frac{\partial \rho(\boldsymbol{W}(k))}{\partial \boldsymbol{W}(k)} \otimes \frac{\partial \boldsymbol{W}(k)}{\partial \eta(k-1)} = \boldsymbol{W}_g(k) \otimes \boldsymbol{W}_d(k-1) \tag{5.43}$$

where $\boldsymbol{W}_g(k) = -\boldsymbol{W}(k)^{-\mathrm{T}} + \langle \boldsymbol{f}(\boldsymbol{y}) \cdot \boldsymbol{x}^{\mathrm{T}} \rangle$ and $\otimes$ denotes the inner product between vectors or matrices. Then the iterative formula of step-size is

$$\eta(k) = \eta(k-1) - K(k) \cdot \frac{\partial \rho(\boldsymbol{W}(k))}{\partial \eta(k-1)} = \eta(k-1) - K(k) \cdot \boldsymbol{W}_g(k) \otimes \boldsymbol{W}_d(k-1) \tag{5.44}$$

Here $\eta(k)$ can be either a scalar or a diagonal matrix, and we discuss this in the following cases [32].

(1) $\eta(k)$ **is scalar**

From Eq. (5.44), we get:

$$\begin{aligned} \eta(k) &= \eta(k-1) - K(k) \cdot \boldsymbol{W}_g(k) \otimes \boldsymbol{W}_d(k-1) \\ &= \eta(k-1) - K(k) \cdot \sum_i \sum_j [\boldsymbol{W}_g(k)]_{i,j} \cdot [\boldsymbol{W}_d(k-1)]_{i,j} \end{aligned} \tag{5.45}$$

$K(k)$ can be

$$K(k) = r \cdot \eta(k-1)/v(k) \tag{5.46}$$

Here r is a constant less than 0, and $v(k)$ is the exponential average of $\boldsymbol{W}_g(k) \otimes \boldsymbol{W}_d(k)$, i.e.:

$$v(k) = \gamma \cdot v(k-1) + (1-\gamma) \cdot \boldsymbol{W}_g(k) \otimes \boldsymbol{W}_d(k) \tag{5.47}$$

therefore, it is obtained from the Eqs. (5.45)–(5.46)

$$\eta(k) = \eta(k-1)\left[1 - r\frac{\boldsymbol{W}_g(k) \otimes \boldsymbol{W}_d(k-1)}{v(k)}\right] \tag{5.48}$$

Combination of (5.47), (5.48), and (5.40) constitutes the step-size-adaptive EASI blind separation algorithm, referred to as EASIBSA-1 algorithm.

(2) $\eta(k)$ **is a diagonal matrix**

Let $\eta_i(k)$ be the i-th diagonal element of $\eta(k)$, then its iterative formula is:

$$\begin{aligned}\eta_i(k) &= \eta_i(k-1) - K_i(k) \cdot [\boldsymbol{W}_g(k)]_i \otimes [\boldsymbol{W}_d(k-1)]_i \\ &= \eta_i(k-1) - K_i(k) \cdot \sum_j [\boldsymbol{W}_g(k)]_{ij} \cdot [\boldsymbol{W}_d(k-1)]_{ij}\end{aligned} \tag{5.49}$$

where $[\cdot]_i$ denotes the i-th row of a particular matrix and $[\cdot]_{ij}$ denotes the element in row i and column j of the matrix. Similarly, $K_i(k)$ is taken as:

$$K_i(k) = r \cdot \eta_i(k-1)/v_i(k) \tag{5.50}$$

$$v_i(k) = \gamma \cdot v_i(k-1) + (1-\gamma) \cdot [\boldsymbol{W}_g(k)]_i \otimes [\boldsymbol{W}_d(k)]_i \tag{5.51}$$

From Eqs. (5.49)–(5.50), we have

$$\eta_i(k) = \eta_i(k-1)\left[1 - r\frac{[\boldsymbol{W}_g(k)]_i \otimes [\boldsymbol{W}_d(k-1)]_i}{v_i(k)}\right] \tag{5.52}$$

Combination of (5.51), (5.52), and (5.40) constitutes another step-size adaptive EASI blind separation algorithm, referred to as EASIBSA-2 algorithm.

2. **Step-Size Optimal EASI Algorithm**

Based on the idea of step-size optimization, the step-size in the EASI algorithm can be similarly iterated using Eqs. (5.32) and (5.37), i.e.

$$\eta(k+1) = \eta(k) - \gamma \cdot trace[\boldsymbol{W}^{-1}(k+1) \cdot (\boldsymbol{W}^{-\mathrm{T}}(k) - f(\boldsymbol{y})\boldsymbol{z}^{\mathrm{T}})] \tag{5.53}$$

Combination of Eqs. (5.53) and (5.29) constitutes the step-size-optimized EASI algorithm, referred to as SO-EASI-1.

In addition, since in the EASI algorithm we have

$$\boldsymbol{W}(k+1) = \boldsymbol{W}(k) + \eta(k) \cdot \{\boldsymbol{I} - \langle \boldsymbol{y}\boldsymbol{y}^{\mathrm{T}} \rangle + \langle \boldsymbol{y}(k)\boldsymbol{f}^{\mathrm{T}}[\boldsymbol{y}(k)] \rangle - \langle \boldsymbol{f}[\boldsymbol{y}(k)]\boldsymbol{y}^{\mathrm{T}}(k) \rangle\}\boldsymbol{W}(k) \tag{5.54}$$

Set the iterative formula of the step-size be

$$\eta(k+1) = \eta(k) + \gamma \cdot \Delta\, \eta(k) \tag{5.55}$$

here $\Delta\eta(k)$ is also the offset of the step-size $\eta(k)$. This offset should be chosen so that the step-size $\eta(k+1)$ is optimal, i.e., to minimize the objective function at the next iteration, so that

$$\Delta\eta(k) = \frac{\partial\rho(\mathbf{W}(k+1))}{\partial\eta(k)} = trace\left\{\left[\frac{\partial\rho(\mathbf{W}(k+1))}{\partial\mathbf{W}(k+1)}\right]^{\mathrm{T}} \cdot \frac{\partial\mathbf{W}(k+1)}{\partial\eta(k)}\right\} \tag{5.56}$$

Then:

$$\rho(y, \mathbf{W}(k)) = -\log|\det(\mathbf{W}(k))| - \sum_{i=1}^{m} E\{\log(q_i(y_i(k)))\} \tag{5.57}$$

Therefore,

$$\frac{\partial\rho(\mathbf{W}(k+1))}{\partial\mathbf{W}(k+1)} = -\mathbf{W}^{-\mathrm{T}}(k+1) + E\{f(y(k))z^{\mathrm{T}}\} \tag{5.58}$$

Also from Eq. (5.54), we get

$$\frac{\partial\mathbf{W}(k+1)}{\partial\eta(k)} = [\mathbf{I} - \langle \mathbf{y}\mathbf{y}^{\mathrm{T}}(k)\rangle + \langle y(k)\mathbf{f}^{\mathrm{T}}[y(k)]\rangle - \langle f[y(k)]\mathbf{y}^{\mathrm{T}}(k)\rangle]\mathbf{W}(k) \tag{5.59}$$

According to (5.56) (5.58) (5.59), we get

$$\begin{aligned}\Delta\eta(k) = trace\{[-\mathbf{W}^{-\mathrm{T}}(k+1) + E\{f(y(k))z^{\mathrm{T}}\}]^{\mathrm{T}} \cdot \\ [\mathbf{I} - \langle \mathbf{y}\mathbf{y}^{\mathrm{T}}(k)\rangle + \langle y(k)\mathbf{f}^{\mathrm{T}}[y(k)]\rangle - \langle f[y(k)]\mathbf{y}^{\mathrm{T}}(k)\rangle]\mathbf{W}(k)\}\end{aligned} \tag{5.60}$$

Then Eqs. (5.54) and (5.60) constitute the step-size optimal EASI algorithm 2, referred to as SO-EASI-2 [33].

5.3 Stability Analysis of BSS Algorithms and Selection of Nonlinear Functions

5.3.1 Stability Analysis of Natural Gradient Algorithms

BSS Algorithm is stable means that it converges to a suitable equilibrium point that corresponds to the correct separation of independent sources. Here we discuss local stability conditions for the natural gradient BSS algorithm. According to the iterative Eq. (5.19) for the natural gradient algorithm, we get

$$\Delta \boldsymbol{W}(k) = \eta\{I - \langle \boldsymbol{f}[\boldsymbol{y}(k)]\boldsymbol{y}^{\mathrm{T}}(k)\rangle\}\boldsymbol{W}(k) \tag{5.61}$$

To analyze its behavior, consider the following continuous differential equation

$$\frac{d\boldsymbol{W}}{dt} = \eta\{\boldsymbol{I} - E[\boldsymbol{f}[\boldsymbol{y}(t)]\boldsymbol{y}^{\mathrm{T}}(t)]\}\boldsymbol{W}(t) \tag{5.62}$$

When BSS algorithm reaches the equilibrium point, $\frac{dW}{dt} = 0$, i.e., $\boldsymbol{I} - E[\boldsymbol{f}[\boldsymbol{y}(t)]\boldsymbol{y}^{\mathrm{T}}(t)] = 0$, then

$$E\{f_i(y_i)y_j\} = 0, \quad i \neq j \tag{5.63}$$

$$E\{f_i(y_i)y_i\} = 1, \; i = j \tag{5.64}$$

Equation (5.63) holds when the separated signals y_i, $y_j (i \neq j)$ are independent of each other. Thus the equilibrium of the natural gradient algorithm correctly separates the independent source signals.

The local stability condition for the equilibrium point of the natural gradient algorithm is analyzed below. We know that the natural gradient algorithm is derived on the cost function [34]

$$\rho(\boldsymbol{W}) = -\log|\det(\boldsymbol{W})| - \sum_{i=1}^{m} E\{\log(q_i(y_i))\} \tag{5.65}$$

Therefore, if and only if the expectation of the Hessian matrix of the cost function

$$d^2\rho = \sum \frac{\partial^2 \rho(\boldsymbol{W})}{\partial \omega_{ij} \partial \omega_{kl}} d\omega_{ij} d\omega_{kl} \tag{5.66}$$

is positive definite, the equilibrium of the algorithm is a stable equilibrium.

When the separation matrix changes from $\boldsymbol{W}$ to $\boldsymbol{W} + d\boldsymbol{W}$, the first derivative of the cost function $\rho(\boldsymbol{W})$ is

$$d\rho = \rho(\boldsymbol{W} + d\boldsymbol{W}) - \rho(\boldsymbol{W}) = \sum_{i,j} \frac{\partial \rho(\boldsymbol{W})}{\partial \omega_{i,j}} d\omega \tag{5.67}$$

from Eq. (5.65), the first derivative of the first term of the cost function is

$$\begin{aligned}
-\log|\det(\boldsymbol{W} + d\boldsymbol{W})| + \log|\det(\boldsymbol{W})| &= -\log\left|\frac{\det(\boldsymbol{W} + d\boldsymbol{W})}{\det(\boldsymbol{W})}\right| \\
&= -\log|\det[(\boldsymbol{W} + d\boldsymbol{W})\boldsymbol{W}^{-1}]| = \log|\det(I + d\boldsymbol{W}\boldsymbol{W}^{-1})| \\
&= -\log|1 + trace(d\boldsymbol{W}\boldsymbol{W}^{-1}) + O(d\boldsymbol{W}\boldsymbol{W}^{-1})|
\end{aligned}$$

$$\approx -trace(d\boldsymbol{W}\boldsymbol{W}^{-1}) \tag{5.68}$$

And the first derivative of the second term of the cost function is:

$$\begin{aligned}
&\sum_k \sum_l \frac{\partial \left\{ -\sum_{i=1}^{m} E[\log(q_i(y_i))] \right\}}{\partial \omega_{k,l}} d\omega_{k,l} = -\sum_k \sum_l \sum_{i=1}^{m} \frac{q_i'(y_i)}{q_i(y_i)} \cdot \frac{\partial y_i}{\partial \omega_{k,l}} d\omega_{k,l} \\
&= -\sum_k \sum_l \frac{q_k'(y_k)}{q_k(y_k)} \cdot x_k d\omega_{k,l} = -\sum_k \frac{q_k'(y_k)}{q_k(y_k)} \cdot \sum_l x_k d\omega_{k,l} \\
&= -\sum_k \frac{q_k'(y_k)}{q_k(y_k)} \cdot dy_k = \boldsymbol{f}^T(\boldsymbol{y})d\boldsymbol{y}
\end{aligned} \tag{5.69}$$

Let $d\boldsymbol{C} = d\boldsymbol{W}\boldsymbol{W}^{-1}$, then according to Eqs. (5.65), (5.67) and (5.68), we have

$$d\rho = -trace(d\boldsymbol{C}) + \boldsymbol{f}^{\mathrm{T}}(\boldsymbol{y})d\boldsymbol{y} \tag{5.70}$$

for $\boldsymbol{y} = \boldsymbol{W}\boldsymbol{x}$, then

$$d\boldsymbol{y} = d\boldsymbol{W}\boldsymbol{x} = d\boldsymbol{W}\boldsymbol{W}^{-1}\boldsymbol{y} = d\boldsymbol{C}\boldsymbol{y} \tag{5.71}$$

and

$$d\rho = -trace(dC) + f^{\mathrm{T}}(y)dCy \tag{5.72}$$

from Eq. (5.72), the second derivative of the cost function is

$$\begin{aligned}
d^2\rho &= \boldsymbol{y}^{\mathrm{T}}d\boldsymbol{C}^{\mathrm{T}}\boldsymbol{f}'(\boldsymbol{y})d\boldsymbol{y} + \boldsymbol{f}^{\mathrm{T}}(\boldsymbol{y})d\boldsymbol{C}d\boldsymbol{y} \\
&= \boldsymbol{y}^{\mathrm{T}}d\boldsymbol{C}^{\mathrm{T}}\boldsymbol{f}'(\boldsymbol{y})d\boldsymbol{C}\boldsymbol{y} + \boldsymbol{f}^{\mathrm{T}}(\boldsymbol{y})d\boldsymbol{C}d\boldsymbol{C}\boldsymbol{y}
\end{aligned} \tag{5.73}$$

The expectation of the first term of the above equation is:

$$\begin{aligned}
E\{\boldsymbol{y}^{\mathrm{T}}d\boldsymbol{C}^{\mathrm{T}}\boldsymbol{f}'(\boldsymbol{y})d\boldsymbol{C}\boldsymbol{y}\} &= \sum_{k,j,i} E\{y_i dc_{ji} f_j'(y_j) dc_{j,k} y_k\} \\
&= \sum_{i,j,j\neq i} E\{y_i^2\}E\{f_j'(y_j)\}dc_{ji}^2 + \sum_i E\{y_i^2 f_i'(y_i)\}dc_{ii}^2
\end{aligned} \tag{5.74}$$

Let:

$$\begin{cases}
\alpha_i^2 = E\{y_i^2\} \\
k_i = E\{f_i'(y_i)\} \\
m_i = E\{y_i^2 f_i'(y_i)\}
\end{cases} \tag{5.75}$$

Then

$$E\{\mathbf{y}^T d\mathbf{C}^{\mathrm{T}} \mathbf{f}'(\mathbf{y}) d\mathbf{C}\mathbf{y}\} = \sum_{i,j,i\neq j} \alpha_i^2 k_j (dc_{ji})^2 + \sum_i m_i (dc_{ii})^2 \tag{5.76}$$

Similarly

$$\begin{aligned} E\{\mathbf{f}^{\mathrm{T}}(\mathbf{y}) d\mathbf{C} d\mathbf{C}\mathbf{y}\} &= \sum_{i,j,k} E\{f_i(y_i) dc_{ij} dc_{jk} y_k\} \\ &= \sum_{i,j} E\{f_i(y_i) y_i\} dc_{ij} dc_{ji} \\ &= \sum_{i,j} dc_{ij} dc_{ji} \end{aligned} \tag{5.77}$$

The last step of the above formula is obtained using the normalization condition $E\{f_i(y_i)y_i\} = 1$. It can be seen from Eqs. (5.73), (5.74) and (5.75) that

$$E\{d^2\rho\} = \sum_{i,j,j\neq i} \{\alpha_i^2 k_j (dc_{ji})^2 + dc_{ji} dc_{ij}\} + \sum_i (m_i + 1)(dc_{ii})^2 \tag{5.78}$$

for any pair of (i, j), $i \neq j$, the first term in Eq. (5.78) is rewritten as:

$$k_{ij} = \alpha_i^2 k_j (dc_{ji})^2 + \alpha_j^2 k_i (dc_{ij})^2 + 2 dc_{ji} dc_{ij} \tag{5.79}$$

k_{ij} is a quadratic form of dc_{ij}, dc_{ji} and

$$E\{d^2\rho\} = \sum_{i,j,i>j} k_{ij} + \sum_i (m_i + 1)(dc_{ii})^2 \tag{5.80}$$

from Eq. (5.80), for $E\{d^2\rho\}$ to be positively definite, the following conditions must be satisfied.

$$\begin{cases} (m_i + 1) > 0 \\ k_i > 0 \\ \gamma_{ij} = \alpha_i^2 \alpha_j^2 k_i k_j > 1, \quad \text{for } all \quad 1 \leq i < j \leq m \end{cases} \tag{5.81}$$

That is, the solution of the BSS natural gradient algorithm is the stable equilibrium point when Eq. (5.80) is satisfied. From Eqs. (5.75) and (5.81), the stabilization conditions of the BSS natural gradient algorithm are related to the statistical properties of the source signals and the chosen nonlinear function $f_i(y_i)$.

5.3.2 Selection of Nonlinear Functions in Natural Gradient Algorithms

From the derivation of the natural gradient BSS algorithm, it is clear that $f_i(y_i)$ is only related to the chosen reference probability density function $q(\mathbf{y})$. $q(\mathbf{y})$ is best chosen as the true probability density of the source signal s. However, this is not available, and can generally be approximated by choosing some nonlinear function. For super-Gaussian source signals, $f_i(y_i) = \tanh(\alpha y_i)$, $\alpha \geq 0$, and for sub-Gaussian signals, $f_i(y_i)$ can be taken to be $y_i^3 + \alpha y_i, \quad \alpha \geq 0$. Since digital communication signals are generally sub-Gaussian, $f_i(y_i)$ is

$$f_i(y_i) = y_i^3 + \alpha y_i,\ \alpha \geq 0 \tag{5.82}$$

It is proved below that the nonlinear function of Eq. (5.82) satisfies the stability condition of the algorithm in separating communication signals.

When the algorithm converges, the separated signal y_i is an estimate of a certain source signal s_j. Since $f_i'(y_i) = 3y_i^2 + \alpha$, then:

$$m_i + 1 = E\{y_i^2(3y_i^2 + \alpha)\} + 1 > 0 \tag{5.83}$$

$$k_i = E\{3y_i^2 + \alpha\} > 0 \tag{5.84}$$

$$\alpha_i^2\alpha_j^2 k_i k_j = E(y_i^2)E(y_j^2)E\{3y_i^2 + \alpha\}E\{3y_j^2 + \alpha\} \tag{5.85}$$

according to the normalization condition $E\{f_i(y_i)y_i\} = 1$, we have

$$E\{(y_i^3 + \alpha y_i)y_i\} = 1 \tag{5.86}$$

Then

$$E\{y_i^4\} = 1 - \alpha E\{y_i^2\} \tag{5.87}$$

It is also known from the assumption that the source signal is a sub-Gaussian signal that

$$E\{y_i^4\} - 3E^2\{y_i^2\} < 0 \tag{5.88}$$

According to (5.87) and (5.88), we have

$$E\{y_i^2\} > \frac{\sqrt{\alpha^2 + 12} - \alpha}{6} \tag{5.89}$$

When $\alpha = 0$, $E\{y_i^2\} > \frac{1}{\sqrt{3}}$, and

$$\alpha_i^2 \alpha_j^2 k_i k_j = E(y_i^2) E(y_j^2) E\{3y_i^2\} E\{3y_j^2\} > 1 \tag{5.90}$$

when $\alpha > 0$, it can be shown that the above equation also holds.

Thus, when separating sub-Gaussian communication signals, the nonlinear function $f_i(y_i) = y_i^3 + \alpha y_i$, $\alpha \geq 0$ satisfies the stability condition of the natural gradient algorithm.

5.3.3 Stability Analysis of the EASI Algorithm

From the derivation process of the EASI algorithm, it can be seen that algorithm is obtained by using the orthogonality constraint of separation matrix W on the basis of natural gradient algorithm. The separation matrix $\boldsymbol{W}$ is orthogonal, i.e., $\boldsymbol{W}^T = \boldsymbol{W}^{-1}$, then

$$\boldsymbol{W}\boldsymbol{W}^{\mathrm{T}} = \boldsymbol{I} \Rightarrow d\boldsymbol{W}\boldsymbol{W}^{\mathrm{T}} + \boldsymbol{W}d\boldsymbol{W}^{\mathrm{T}} = 0 \tag{5.91}$$

according to previous definition $d\boldsymbol{C} = d\boldsymbol{W}\boldsymbol{W}^{-1}$, we get

$$d\boldsymbol{C} = d\boldsymbol{W}\boldsymbol{W}^{\mathrm{T}} \tag{5.92}$$

from Eqs. (5.91) and (5.92), it can be seen that

$$d\boldsymbol{C} = -d\boldsymbol{C}^{\mathrm{T}} \tag{5.93}$$

i.e., $d\boldsymbol{C}$ is skew symmetric, then $trace(d\boldsymbol{C}) = 0$. Thus, according to Eq. (5.73), we get

$$d\rho = f^{\mathrm{T}}(y)dy = f^{\mathrm{T}}(y)d\boldsymbol{C}y \tag{5.94}$$

Similarly, for the EASI algorithm to be locally stable, the Hessian matrix $d^2\rho$ is required to be positive definite.

First, let's see if the equilibrium point of EASI algorithm is the correct solution [35], that is, whether it can correctly separate independent source signals. From Eq. (5.29), we have

$$\Delta \boldsymbol{W}(k) = \eta \cdot [\boldsymbol{I} - \langle \boldsymbol{y}\boldsymbol{y}^{\mathrm{T}} \rangle + \langle \boldsymbol{y}(k)\boldsymbol{f}^{\mathrm{T}}[\boldsymbol{y}(k)] \rangle - \langle \boldsymbol{f}[\boldsymbol{y}(k)]\boldsymbol{y}^{\mathrm{T}}(k) \rangle] \tag{5.95}$$

In the same manner, we consider the following continuous differential equation:

$$\frac{d\boldsymbol{W}}{dt} = \eta \cdot [\boldsymbol{I} - E\{\boldsymbol{y}(t)\boldsymbol{y}^{\mathrm{T}}(t)\} + E\{\boldsymbol{y}(t)\boldsymbol{f}^{\mathrm{T}}[\boldsymbol{y}(t)]\} - E\{\boldsymbol{f}[\boldsymbol{y}(t)]\boldsymbol{y}^{\mathrm{T}}(t)\}] \tag{5.96}$$

when the separating system reaches equilibrium, there is $\frac{dW}{dt} = 0$, i.e.

$$E\{y_i^2\} = 1 \tag{5.97}$$

and

$$-E\{y_i y_j\} + E\{y_i f(y_j)\} - E\{y_j f(y_i)\} = 0 \text{ , } i \neq j \tag{5.98}$$

Equation (5.97) determines the amplitude or power of the separated signals. When the separated signals y_i, y_j are independent of each other, there are

$$\begin{aligned} E\{y_i y_j\} &= 0, i \neq j \\ E\{y_i f(y_j)\} &= 0, i \neq j \end{aligned} \tag{5.99}$$

That is, Eq. (5.98) holds. Thus, the equilibrium point of the separated system can separate the independent source signals.

The following section discusses whether the equilibrium point of the system is locally stable. From Eq. (5.94),we have

$$d\rho = f^{\mathrm{T}}(y) d\boldsymbol{C}\boldsymbol{y} = \sum_{i,j} f_i(y_i) dc_{ij} y_j = \sum_{i,j,i>j} [f_i(y_i) y_j - f_j(y_j) y_i] dc_{ij} \tag{5.100}$$

The last step of the above equation is derived utilizing the skew symmetry condition for $d\boldsymbol{C}$, i.e., $dc_{ii} = 0$, $dc_{ij} = -dc_{ji}$.

$$\begin{aligned} d^2\rho = &\sum_{i,j,i>j} [f_i'(y_i) y_j \sum_k dc_{ik} y_k - f_j'(y_j) y_i \sum_k dc_{jk} y_k \\ &+ f_i(y_i) \sum_k dc_{jk} y_k - f_j(y_j) \sum_k dc_{ik} y_k] dc_{ij} \\ = &\sum_{i,j,i>j} \sum_k [f_i'(y_i) y_j y_k dc_{ik} - f_j(y_j) y_k dc_{ik} - f_j'(y_j) y_i y_k dc_{jk} + f_i(y_i) y_k dc_{jk}] dc_{ij} \end{aligned} \tag{5.101}$$

From the normalization constraints shown in Eq. (5.97) and the skew-symmetry condition for $d\boldsymbol{C}$, the Hessian matrix $d^2\rho$ at the desired solution $\boldsymbol{W}$ is

$$\begin{aligned} E\{d^2\rho\} &= \sum_{i,j,i>j} [E\{f_i'(y_i) y_j^2\} dc_{ij} - E\{f_j(y_j) y_j\} dc_{ij} - E\{f_j'(y_j) y_i^2\} dc_{ji} + E\{f_i(y_i) y_i\} dc_{ji}] dc_{ij} \\ &= \sum_{i,j,i>j} [E\{f_i'(y_i) y_j^2\} - E\{f_j(y_j) y_j\} + E\{f_j'(y_j) y_i^2\} - E\{f_i(y_i) y_i\}] (dc_{ij})^2 \end{aligned} \tag{5.102}$$

The derivation of the above equation utilizes the condition $E(y_i y_j) = 0$, because the separated signals y_i, y_j are independent of each other at the desired solution $\boldsymbol{W}$. It also follows from $E\{y_i^2\} = 1$ that

$$E\{d^2\rho\} = \sum_{i,j,i>j} [E\{f_i'(y_i)\} - E\{f_j(y_j)y_j\} + E\{f_j'(y_j)\} - E\{f_i(y_i)y_i\}](dc_{ij})^2 \quad (5.103)$$

Let $\chi_i = E\{f_i'(y_i)\} - E\{f_i(y_i)y_i\}$, then $E\{d^2\rho\}$ is positive definite when $\chi_i + \chi_j > 0$. Thus, the stability condition for the EASI algorithm is

$$\chi_i + \chi_j > 0 \quad (5.104)$$

For each y_i, the condition

$$\chi_i > 0 \quad (5.105)$$

is a sufficient condition for the stability of the EASI algorithm.

5.3.4 The Selection of Nonlinear Functions in EASI

For BSS EASI algorithm for communication signals, we also choose the nonlinear function

$$f_i(y_i) = y_i^3 + \alpha y_i, \quad \alpha \geq 0 \quad (5.106)$$

It is shown below that the nonlinear function satisfies the stability condition (5.105).

Due to

$$f_i'(y_i) = 3y_i^2 + \alpha \quad (5.107)$$

so

$$E\{f_i'(y_i)\} = 3E\{y_i^2\} + \alpha \quad (5.108)$$

as well as

$$E\{f_i(y_i)y_i\} = E\{(y_i^3 + \alpha y_i)y_i\} = E\{y_i^4\} + \alpha E\{y_i^2\} \quad (5.109)$$

Then it is known from $E\{y_i^2\} = 1$ that

$$\chi_i = E\{f_i'(y_i)\} - E\{f_i(y_i)y_i\} = 3 - E\{y_i^4\} \quad (5.110)$$

Assuming that y_i is an estimate of the source signal s_j, it is known from the sub-Gaussian character of the communication signal that

$$E\{y_i^4\} - 3E^2\{y_i^2\} < 0 \tag{5.111}$$

That is, $E\{y_i^4\} - 3 < 0$, hence $\chi_i > 0$. Thus, the function satisfies the stability condition.

It should be noted that the various algorithms given above are derived under the assumption that the signal is a real valued signal, and if the signal is a complex valued signal, the transpose operation in the above algorithm needs to be changed to a conjugate transpose, and the nonlinear function is chosen to be $f_i(y_i) = |y_i|^2 \cdot y_i + \alpha y_i, \quad \alpha \geq 0$. It also can be proved that the nonlinear function satisfies the stability condition.

5.4 Simulation and Analysis of Algorithm Performance

5.4.1 Simulation of Natural Gradient BSS Algorithm

1. **Simulation 1**

Assume that three source signals are a 2ASK signal with carrier frequency of 610 kHz, a 2FSK signal with carrier frequency of 650 kHz, and a 2PSK signal with carrier frequency of 710 kHz. The symbol rate is 70,000 baud/s and the sampling frequency is 560 kHz. Assume that there are four observed signals and the mixing matrix is a randomly generated matrix:

$$\boldsymbol{H} = \begin{bmatrix} 0.3545 & 0.2621 & 0.7297 \\ 0.9493 & 0.6866 & 0.9682 \\ 0.4358 & 0.1027 & 0.5110 \\ 0.6496 & 0.6580 & 0.9287 \end{bmatrix} \tag{5.112}$$

Assuming that there is no noise in the observed signals, the observed signals are processed using the whitening algorithm in Chap. 2 and separated by "Algorithm 1", the SABNG-1 algorithm as well as the SABNG-2 algorithm, and the Fixed-Step Natural Gradient algorithm (referred to as the "FSNG" algorithm). Figure 5.1 shows the relationship between the number of iterations and the ISR of the four natural gradient algorithms. The initial value of the step-size is set to 0.08 for all simulations, and set $\gamma = 0.01$.

From Fig. 5.1, it can be seen that two variable step-size algorithms given in this chapter have a great improvement in convergence speed, and when ISR reaches −30 dB, the 'SABNG-1' and 'SABNG-2' algorithms need only about 440 iterations, while 'Algorithm 1' needs 950 iterations, and 'FSNG' needs 1,820 iterations to reach −30 dB. Therefore, the convergence speed of the proposed 'FSNG' algorithm is twice faster than that of 'algorithm 1', and more than 3 times faster than that of 'FSNG' algorithm [34].

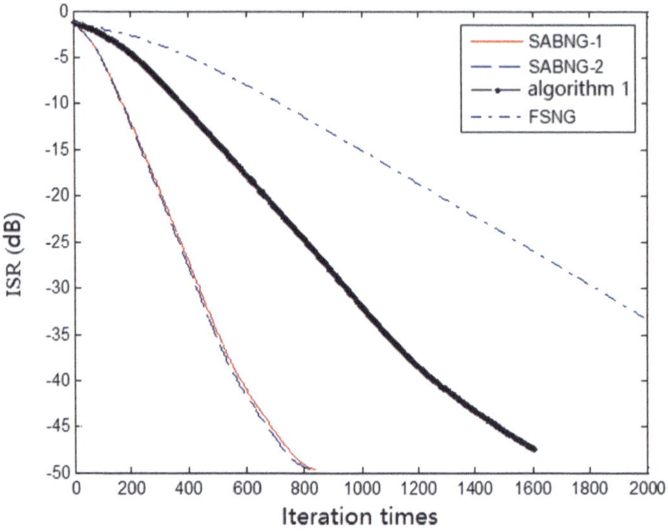

Fig. 5.1 Performance comparison of natural gradient algorithms (without noise)

2. **Simulation 2**

The source signals are the same as simulation 1, and the mixing matrix is randomly generated as

$$\boldsymbol{H} = \begin{bmatrix} 0.0461 & 0.2707 & 0.2710 \\ 0.4787 & 0.8321 & 0.5437 \\ 0.3919 & 0.8258 & 0.4173 \\ 0.5669 & 0.0494 & 0.2615 \end{bmatrix} \tag{5.113}$$

The noise is Gaussian white noise with SNR = 20 dB, and the observed signals are also whitened by algorithm in Chap. 4, and then separated by using "Algorithm 1", the SABNG-1 algorithm, the SABNG-2 algorithm, and the Fixed-Step Natural Gradient algorithm (referred to as "FSNG"). The other conditions of the simulation are the same as in Simulation 1. Figure 5.2 shows the relationship between the number of iterations and the ISR of the four natural gradient algorithms. Figure 5.2 shows that three adaptive algorithms converge faster than the fixed-step algorithm, the two algorithms proposed in this chapter converge faster than "Algorithm 1", and the separation effect after convergence is not much worse.

3. **Simulation 3**

The previous two simulations were performed with random mixing matrix. In communication reconnaissance, multiple signals are often received through antenna arrays, so here we give a comparison of the performance when an antenna array is used for reception.

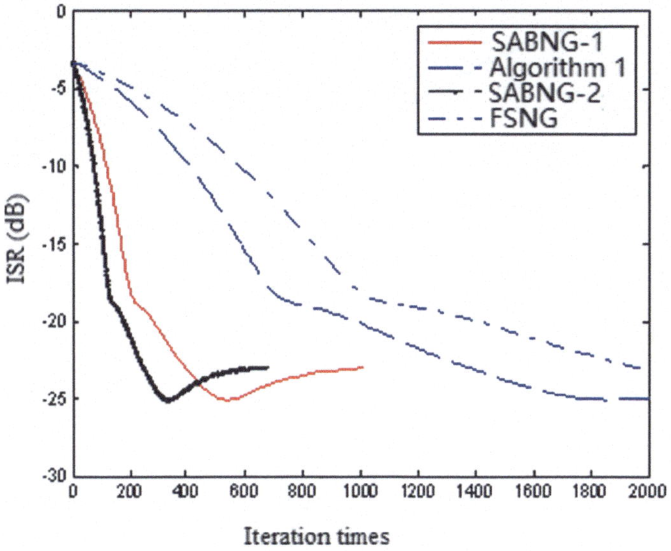

Fig. 5.2 Performance comparison of natural gradient algorithms (SNR = 20 dB)

It is assumed that the source signal is an 8QAM signal with carrier frequency of 630 kHz, a 2FSK signal with carrier frequency of 650 kHz, a 2PSK signal with carrier frequency of 690 kHz, and a QPSK signal with a carrier frequency of 710 kHz. A uniform circular array of 6 array elements is used for reception, the radius of the array elements is half of the wavelength, and the incident angles are 20°, 50°, 80°, and 100°, respectively, so the mixing matrix is

$$\boldsymbol{H} = \begin{bmatrix} 0.4050 - 0.9143i & -0.9471 - 0.3208i & 0.9421 + 0.3352i & -0.9234 - 0.3837i \\ -0.9463 - 0.3233i & 0.7353 - 0.6777i & -0.9538 + 0.3005i & 0.9997 + 0.0235i \\ -0.0876 - 0.9962i & -0.4790 + 0.8778i & -0.7979 + 0.6029i & -0.9322 + 0.3620i \\ 0.4050 + 0.9143i & -0.9471 + 0.3208i & 0.9421 - 0.3352i & -0.9234 + 0.3837i \\ -0.9463 + 0.3233i & 0.7353 + 0.6777i & -0.9538 - 0.3005i & 0.9997 - 0.0235i \\ -0.0876 + 0.9962i & -0.4790 - 0.8778i & -0.7979 - 0.6029i & -0.9322 - 0.3620i \end{bmatrix} \tag{5.114}$$

It can be seen that the signal is a complex valued signal, and the various algorithms given earlier can separate the complex-valued signal as long as the transpose operation in them is changed to a conjugate transpose operation. Figure 5.3 gives the performance curves of several algorithms when SNR = 20 dB. The simulations are all done with an initial value of 0.08 for the step-size, $\gamma = 0.01$, and a nonlinear function $f(y) = |y|^2 \cdot y + 2y$.

Figure 5.3 show that the two improved variable-step natural gradient algorithms (SABNG-1 and SABNG-2) have much-improved convergence speeds compared to the fixed-step natural gradient (FSNG) algorithm when the array antenna is used for

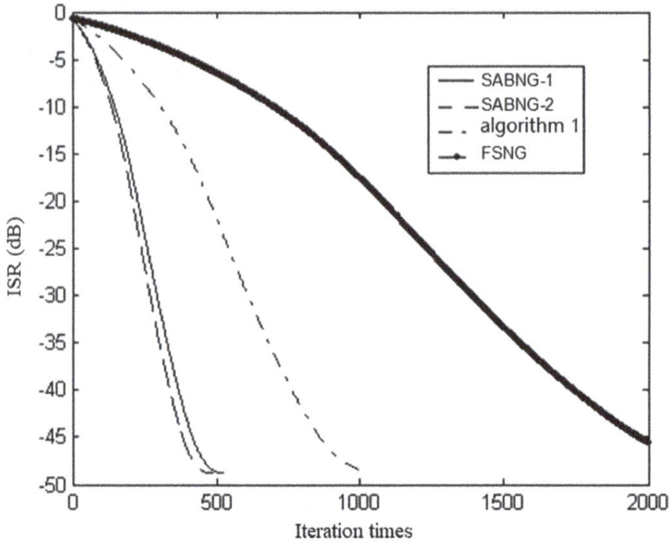

Fig. 5.3 Separation performance when receiving with array antenna (SNR = 20 dB)

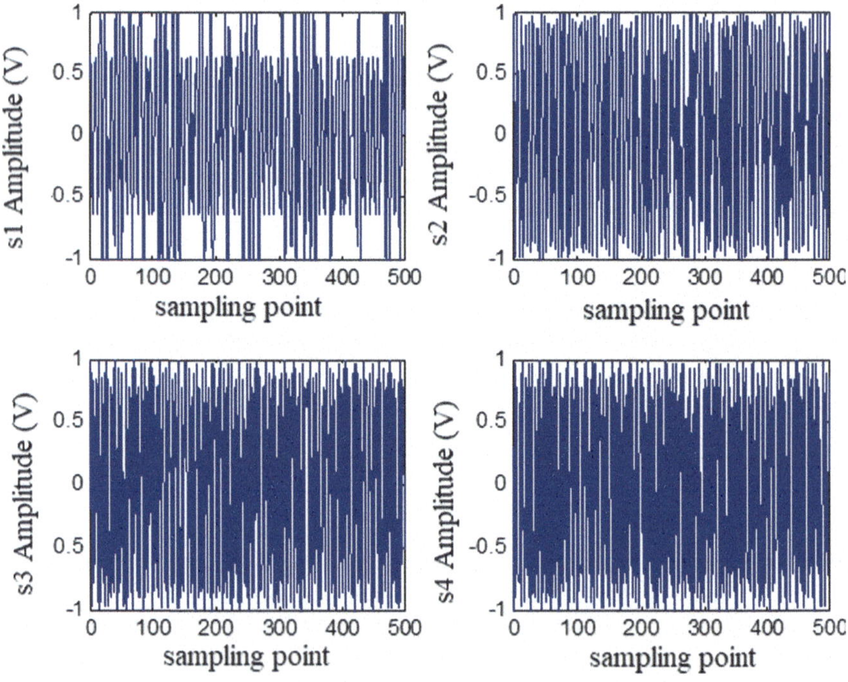

Fig. 5.4 Time-domain waveforms of Source signals (first 500 sample points)

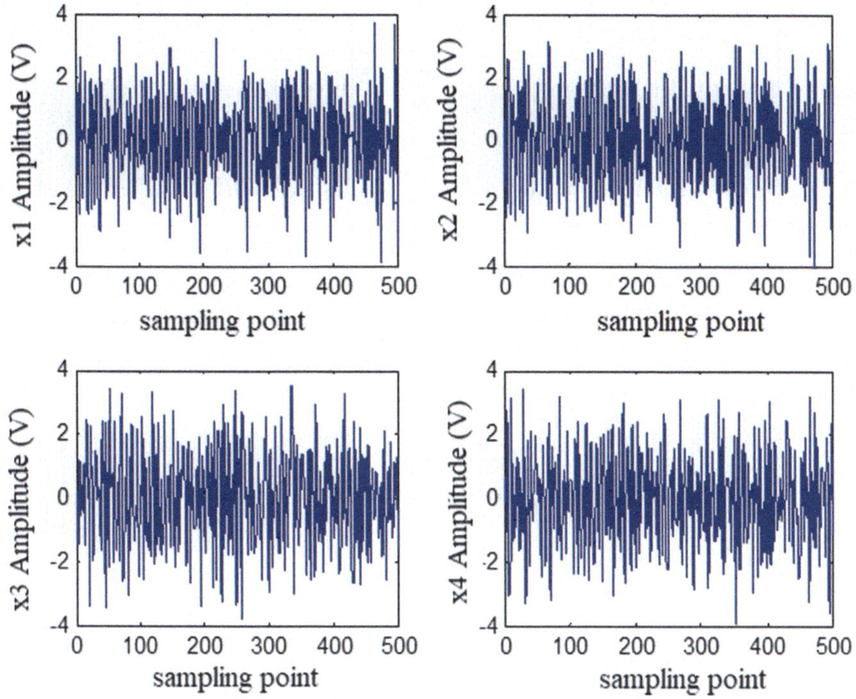

Fig. 5.5 Time-domain waveforms of the first four observed signals (first 500 samples)

reception. when SNR = 20 dB, the two algorithms given in this paper only need about 400 iterations to achieve the ISR of −45 dB, while "Algorithm 1" needs about 800 iterations, and FSNG algorithm needs more iterations, nearly 2000 iterations.

In order to see the effect of the algorithm in this chapter on separating communication signals more directly, the time-domain waveforms of the source signals, the first four observed signals and the separated signals when the algorithm SABNG-2 converges are respectively shown in Figs. 5.4, 5.5 and 5.6.

In addition, when the algorithm converges, the global matrix $\boldsymbol{G} = \boldsymbol{W} \times \boldsymbol{Q} \times \boldsymbol{H}$ is shown below:

$$\boldsymbol{G} = \begin{bmatrix} -0.0002 - 0.0002i & -0.0010 + 0.0014i & -0.0005 - 0.0008i & 0.4830 + 0.0409i \\ 0.0003 + 0.0003i & -0.0022 + 0.0018i & -0.2667 - 0.4056i & 0.0006 + 0.0003i \\ -0.0049 - 0.4754i & -0.0001 + 0.0003i & 0.0001 + 0.0001i & -0.0000 - 0.0002i \\ -0.0004 + 0.0002i & 0.3089 + 0.3733i & -0.0030 + 0.0019i & 0.0002 - 0.0020i \end{bmatrix} \tag{5.115}$$

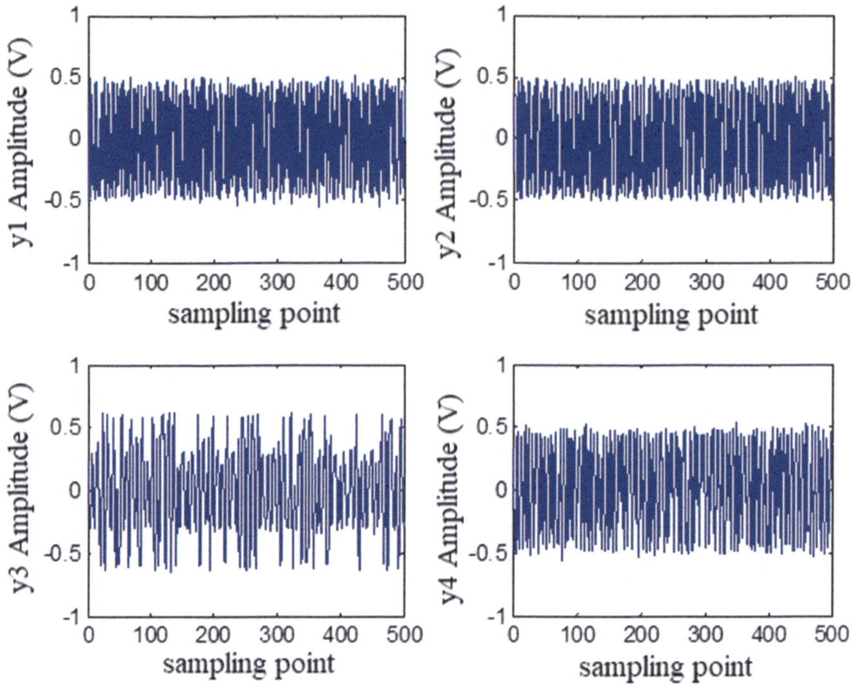

Fig. 5.6 Time-domain waveforms of separated signals (first 500 samples)

$$|\boldsymbol{G}| = \begin{bmatrix} 0.0003 & 0.0017 & 0.0009 & 0.4847 \\ 0.0004 & 0.0028 & 0.4854 & 0.0007 \\ 0.4754 & 0.0003 & 0.0001 & 0.0002 \\ 0.0004 & 0.4845 & 0.0036 & 0.0020 \end{bmatrix} \tag{5.116}$$

After separation, ISR can be calculated as −48.7 dB. Figure 5.6 and the value of ISR show that SABNG-2 algorithm recovers the waveforms of source signals very well.

5.4.2 Simulation of EASI Algorithm

1. **Performance Simulation with Arbitrary Mixing Matrices**

Assume that the four source signals are a 2FSK signal with carrier frequency of 780 kHz, a QPSK signal with carrier frequency of 740 kHz, a 4ASK signal with carrier frequency of 730 kHz, and an 8QAM signal with carrier frequency of 760 kHz. Assuming there are six observed signals, the mixing matrix is randomly generated as:

$$
\boldsymbol{H} = \begin{bmatrix} 0.6942 & 0.1538 & 0.8144 & 0.4794 \\ 0.3740 & 0.8648 & 0.0535 & 0.7178 \\ 0.3072 & 0.8085 & 0.7549 & 0.4065 \\ 0.4540 & 0.6350 & 0.0617 & 0.1907 \\ 0.1686 & 0.6132 & 0.8917 & 0.5881 \\ 0.6012 & 0.8324 & 0.7381 & 0.0050 \end{bmatrix} \tag{5.117}
$$

The noise is Gaussian white noise with SNR=20 dB. After whitening the observed signal, the fixed-step EASI algorithm (referred to as "FS-EASI"), the step-size adaptive EASI separation algorithms(EASIBSA-1 and EASIBSA-2) given in Sect. 5.2.2, step-size optimal EASI algorithms (SO-EASI-1 and SO-EASI-2) are used for separating, respectively. For the convenience of comparison, the initial value of the step-size is 0.02 for all simulations, and the parameter $\gamma = 0.001$ in the two step-size optimal EASI algorithms, while the parameter in the step-size-adaptive EASI algorithms is chosen to be $\gamma = 0.9, r = 0.01$. The nonlinear function is $f(y) = y^3 + 2y$.

Figure 5.7 compares the performance of the two step-size optimal EASI algorithms with the FS-EASI algorithm.

Figure 5.7 shows that the SO-EASI-1 algorithm has the fastest convergence speed, requiring only 30 iterations, while the SO-EASI-2 algorithm requires 130 iterations to reach convergence, and the FS-EASI algorithm converges the slowest, requiring up to 250 iterations to converge. After the convergence of the three algorithms, the

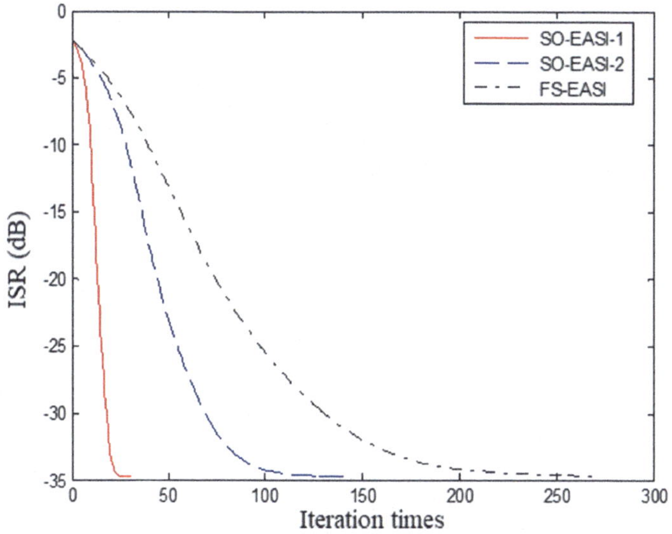

Fig. 5.7 Performance comparison of the step-size optimized EASI algorithms with the FS-EASI algorithm

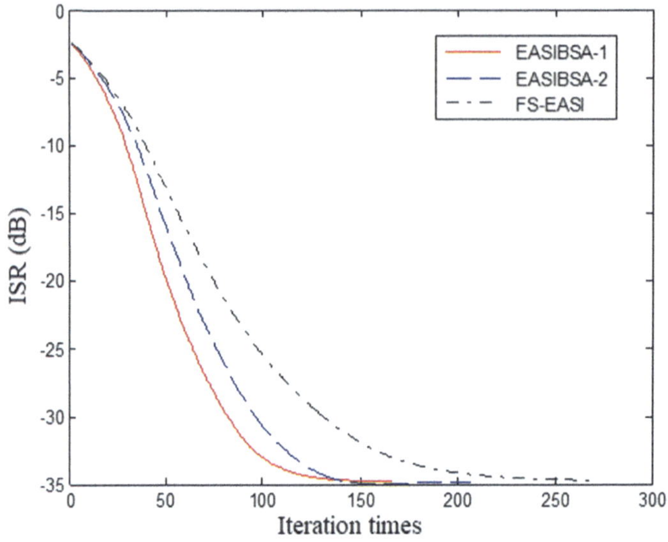

Fig. 5.8 Performance comparison of the step-size adaptive EASI algorithm with the fixed step-size EASI algorithm

ISR is almost the same, so SO-EASI algorithm can improve the convergence speed by more than 7 times without worsening the separation effect of the algorithm.

The performance comparison of two step-size adaptive EASI algorithms with the fixed step-size EASI algorithm is given in Fig. 5.8. It can be seen that EASIBSA-1/2 algorithm can converge after about 150 iterations. Compared with fixed step-size EASI algorithm, the convergence speed is improved by about 70%, while the separation performance is not deteriorated after convergence.

2. **Performance Simulation of Array Antenna Receiving**

The four source signals are an 8QAM signal with carrier frequency of 610 kHz, a 2FSK signal with carrier frequency of 630 kHz, a QPSK signal with carrier frequency of 750 kHz, and a linear frequency hopping signal with starting frequency of 710 kHz and rate of frequency change of 560 kHz/s, whose incidence angles are 30°, 50°, 100°, 70° respectively a 6-array uniform circular array is for reception. The noise is Gaussian white noise with SNR = 20 dB.

After whitening the observed signal, the fixed-step EASI algorithm (referred to as "FS-EASI"), the step-size adaptive EASI separation algorithms (EASIBSA-1 and EASIBSA-2), step-size optimal EASI algorithms (SO-EASI-1 and SO-EASI-2) are used for separating, respectively. For the convenience of comparison, the initial value of the step-size is 0.02 for all simulations, and the parameter $\gamma = 0.001$ in the two step-size optimal EASI algorithms, while the parameter in the step-size-adaptive EASI algorithms is chosen to be $\gamma = 0.9, r = 0.01$. The nonlinear function is $f(y) = |y|^2 \cdot y + 2y$.

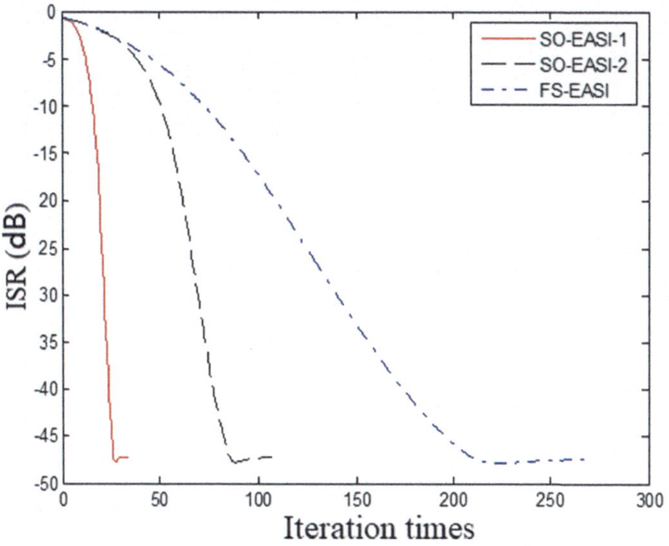

Fig. 5.9 Performance comparison of the step-size optimized EASI algorithm and FS-EASI algorithm for array antenna reception

Figure 5.9 compares the performance of the two step-size optimal EASI algorithm with the FS-EASI algorithm.

Figure 5.9 shows that SO-EASI-1 algorithm has the fastest convergence speed, which needs only 35 iterations to converge, SO-EASI-2 algorithm requires 100 iterations to converge, and the FS-EASI has the slowest convergence speed, which needs up to 240 iterations to converge. Three algorithms converge with almost the same ISR, so the step-size optimal EASI algorithm can increase the convergence speed by more than seven times without worsening the separation effect of the algorithm.

The performance comparison of two step-size adaptive EASI algorithms with FS-EASI algorithm is given in Fig. 5.10. It can be seen that EASIBSA-1/2 algorithm converges after 120 iterations. Compared with fixed step-size EASI algorithm, the convergence speed is about double, but the separation performance is not worsened after convergence.

From the above two simulations, it can be seen that step-size optimal EASI (SO-EASI-1) algorithm converges much faster than the step-size adaptive EASI algorithm.

Summary

In this chapter, the gradient-based overdetermined/determined BSS algorithms are investigated. First, the principles of fixed step-size stochastic gradient BSS algorithms, natural gradient BSS algorithms, and EASI BSS algorithms are introduced in detail. On this basis, variable step-size BSS algorithms are investigated, including variable step-size natural gradient BSS algorithms and variable step-size EASI algorithms. and then, the stability conditions of the above algorithms as well as the

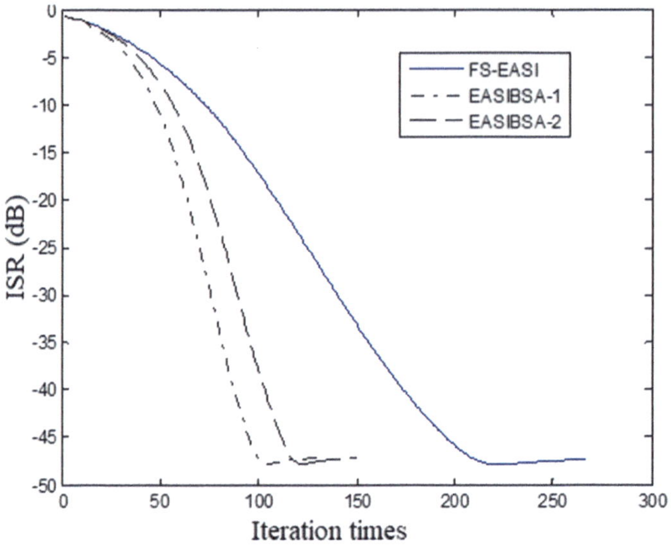

Fig. 5.10 Performance comparison of step-size adaptive EASI algorithms with FS-EASI algorithm for array antenna reception

selection of the nonlinear functions are derived in detail. Finally, the performance of algorithms presented in the chapter are analyzed by simulation.

Chapter 6
Fast-ICA Algorithm for Overdetermined/Determined BSS

Fixed-point iterative algorithm, proposed by Hyvarinen et al., University of Helsinki, Finland, is a fast optimization iterative algorithm. Different from ordinary neural network algorithms, this algorithm adopts batch processing, that is, a large number of sample data are involved in each iteration. Compared with other independent component analysis (ICA) algorithms, fixed-point iterative algorithm has a faster convergence speed, so it is also called Fast-ICA algorithm. Fast-ICA algorithm is an efficient batch processing algorithm. It studies the structure of the data by projecting the data with higher dimensions into the space with lower dimensions. It removes the projected results from the original data, and then extracts the remaining data again by projection. By repeating the above steps, multiple independent components in the mixed signal can be estimated in turn.

In this chapter, the principle of fast fixed-point algorithm based on kurtosis is analyzed first, and the iterative formula of algorithm is given. Secondly, the principle and iterative formula of fast fixed-point algorithm based on negative entropy are analyzed. Above two fast fixed-point algorithms can only separate one source signal at a time. In order to separate multiple source signals, the multivariate fast fixed-point algorithm is introduced in Sect. 6.3. Aiming at blind source separation of complex-valued signals, a fast fixed-point algorithm for complex-valued signals is introduced. Finally, the performance of the algorithm is simulated and analyzed.

6.1 Fast Fixed-Point Algorithm Based on Kurtosis

6.1.1 Definition and Characteristics of Kurtosis

Kurtosis is a statistic that measures the characteristics of a signal distribution, which is a fourth-order cumulant. Kurtosis also reflects the distance between a random signal and a random Gaussian signal with the same variance. Kurtosis can be used

W. Fu, *Principle and Application of Blind Source Separation Technology*,
https://doi.org/10.1007/978-981-96-0022-9_6

to measure the non-Gaussianity of a random signal. For a random variable y, its kurtosis, denoted by kurt(y), is defined as:

$$kurt(y) = E\{y^4\} - 3(E\{y^2\})^2 \tag{6.1}$$

After preprocessed by algorithm described in Chap. 4, all the original signals can be assumed random variables with zero-mean, unit variance. Therefore, $E\{y^2\} = 1$, then the right side of Eq. (6.1) can be simplified to $E\{y^4\} - 3$, which shows that kurtosis is the normalized value of the fourth moment $E\{y^4\}$ of random variable. For a Gaussian random variable y, the fourth moment is $3(E\{y^2\})^2$, so the Gaussian random variable has a kurtosis of 0, and most (but not all) non-Gaussian random variables have a kurtosis of non-zero.

The kurtosis of random variables can be positive or negative. A random variable with negative kurtosis is called a sub-gaussian signal, while a random variable with positive kurtosis is called a super-gaussian signal. In statistics, this is also called a sharp peak or wide peak signal. Super-gaussian random variables have a typical sharp peak probability density function, that is, the probability density is large at the zero value of the variable and small at the larger value of the variable. A typical example is the laplacian distribution, whose probability density function is:

$$p(y) = \frac{1}{\sqrt{2}} \exp(-\sqrt{2}|y|) \tag{6.2}$$

Sub-gaussian random variables, on the other hand, have a flat probability density function, a typical example is a uniform distribution, whose probability density function is

$$p(y) = \begin{cases} \frac{1}{2\sqrt{3}}, & \text{if } |y| \leq \sqrt{3} \\ 0, & \text{others} \end{cases} \tag{6.3}$$

Figure 6.1 shows the probability density distributions for three variables. It can also be seen from Fig. 6.1 that the difference of the probability density distribution of super-gaussian, sub-gaussian and gaussian random variables.

In BSS technology, kurtosis is used to measure the non-gaussianity of signals, mainly because of its simplicity, both in mathematical calculation and in theory. Mathematically, we only need to estimate the fourth moment of the sampled signal to estimate the kurtosis of the signal. Because the kurtosis has the following characteristics: if $x1$, $x2$ are two independent random variables, then

$$kurt(x1 + x2) = kurt(x1) + kurt(x2) \tag{6.4}$$

$$kurt(\alpha \cdot x1) = \alpha^4 \cdot kurt(x1) \tag{6.5}$$

Here α is a constant, $kurt(\cdot)$ represents the kurtosis of the variable.

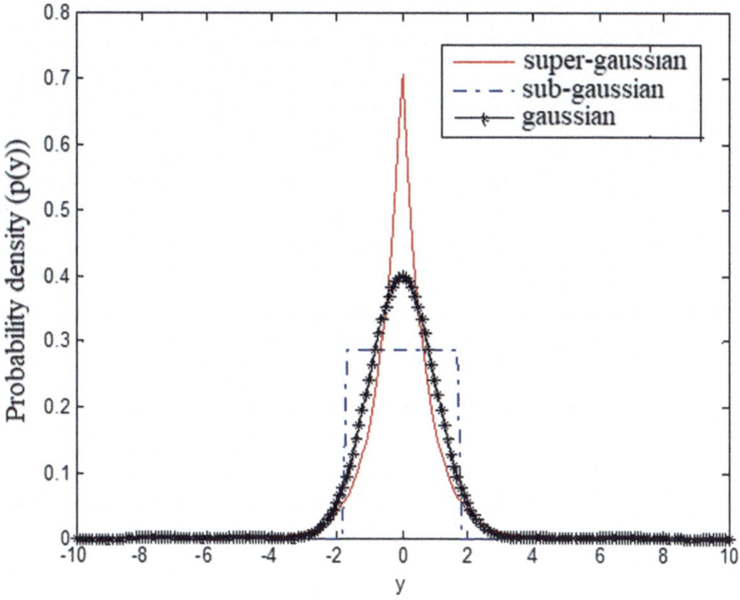

Fig. 6.1 Probability density distribution of super Gaussian, sub-Gaussian and Gaussian random variables

6.1.2 Gradient Algorithm Using Kurtosis

In fact, to maximize the absolute value of the signal kurtosis, we first need to start with some vector w, use the sample $z(1), z(2), \ldots z(T)$ of the mixed signal vector z after whitening to calculate the fastest direction in which the absolute kurtosis of the signal $\boldsymbol{y} = \boldsymbol{w}^T z$ is growing, and then move the vector $\boldsymbol{w}$ along that direction. The gradient of the absolute kurtosis of the signal $\boldsymbol{y} = \boldsymbol{w}^{\mathrm{T}} \boldsymbol{z}$ is

$$\frac{\partial |kurt(\boldsymbol{w}^{\mathrm{T}} \boldsymbol{z})|}{\partial \boldsymbol{w}} = 4\mathrm{sign}(kurt(\boldsymbol{w}^{\mathrm{T}} \boldsymbol{z}))[E\{\boldsymbol{z}(\boldsymbol{w}^{\mathrm{T}} \boldsymbol{z})^3\} - 3\boldsymbol{w}||\boldsymbol{w}||^2] \tag{6.6}$$

For the whitened observed signal, we have $E\{(\boldsymbol{w}^{\mathrm{T}} z)^2\} = ||\boldsymbol{w}||^2$. According to the constraints condition of the original signals have unit variance, we have $E\{(\boldsymbol{w}^{\mathrm{T}} z)^2\} = ||\boldsymbol{w}||^2 = 1$.

Therefore, in order to ensure that $||\boldsymbol{w}||^2 = 1$, in the iterative process of the gradient algorithm, $\boldsymbol{w}$ must be projected onto the unit sphere at each iterative step, and this process can be completed only by dividing $\boldsymbol{w}$ by its norm.

To further simplify the algorithm, we note that $3\boldsymbol{w}||\boldsymbol{w}||^2$ in Eq. (6.6) only changes the norm of the vector $\boldsymbol{w}$ without changing its direction, and we require that $||\boldsymbol{w}||^2 = 1$, so this term can be deleted, thus obtaining the following gradient algorithm:

$$\Delta \boldsymbol{w}(k) \propto \mathrm{sign}(kurt(\boldsymbol{w}(k)^{\mathrm{T}} \boldsymbol{z}))\mathrm{E}\{\boldsymbol{z}(\boldsymbol{w}(k)^{\mathrm{T}} \boldsymbol{z})^3\} \tag{6.7}$$

$$\boldsymbol{w}(k+1) = \boldsymbol{w}(k) + \eta\Delta\boldsymbol{w}(k) \tag{6.8}$$

$$\boldsymbol{w}(k+1) := \boldsymbol{w}(k+1)/||\boldsymbol{w}(k+1)|| \tag{6.9}$$

where, $\boldsymbol{w}(k)$ is separating matrix vector in k-th iteration, η is the iterative step-size, z is the observed signal after whitening, and $E(\cdot)$ represents the expectation. When the above iterative process converges, the obtained $\boldsymbol{w}(k)$ is the final separating vector $\boldsymbol{w}^*$, and the separated signal is $\boldsymbol{y} = (\boldsymbol{w}^*)^{\mathrm{T}}z$. The above algorithm is batch processing, and the online iterative version of the algorithm is that replaces the formula (6.7) with the following formula (6.10) and the rest remains.

$$\Delta\boldsymbol{w}(k) \propto \mathrm{sign}(kurt(\boldsymbol{w}(k)^{\mathrm{T}}z)) \cdot \{z(\boldsymbol{w}(k)^{\mathrm{T}}z(k))^3\} \tag{6.10}$$

The separated signal in the online iterative version is $\boldsymbol{y}(k) = (\boldsymbol{w}(k))^{\mathrm{T}}z(k)$, where $\boldsymbol{y}(k)$ is the k-th sampling point of the separated signal, $z(k)$ is the k-th sampling point of the observed signal after whitening.

6.1.3 Fast Fixed-Point Algorithm Using Kurtosis

In the previous section, a gradient BSS algorithm for non-Gaussian maximization, which uses kurtosis to measure the non-Gaussianity of signal, is derived. The advantage of this algorithm is that it is closely connected with neural network learning algorithm and can quickly reach convergence in non-stationary environments. But at the same time, the convergence speed is very slow, and the convergence speed also depends on the choice of learning rate. Poor choice of learning rate will destroy the convergence of the algorithm. Therefore, it is necessary to design a reliable and fast convergence algorithm, and fixed-point iterative algorithm is one such algorithm.

To derive a more efficient fixed-point iterative algorithm, we note that at the equilibrium point of the gradient algorithm, the gradient must point in the direction of $\boldsymbol{w}$. That is, the gradient must be equal to $\boldsymbol{w}$ times a constant so that adding the gradient to $\boldsymbol{w}$ does not change its direction and the algorithm can converge. This means that after the normalization of $\boldsymbol{w}$, its value will not change, except that it may change the sign of $\boldsymbol{w}$.

$$\boldsymbol{w} \propto E\{z(\boldsymbol{w}^{\mathrm{T}}z)^3\} - 3\boldsymbol{w}||\boldsymbol{w}||^2 \tag{6.11}$$

This formula gives a fixed-point algorithm, that is

$$\boldsymbol{w}(k+1) = E\{z(\boldsymbol{w}(k)^{\mathrm{T}}z)^3\} - 3\boldsymbol{w}(k) \tag{6.12}$$

After each fixed-point iteration, divide $\boldsymbol{w}$ by its norm to maintain $||\boldsymbol{w}||^2 = 1$, that is

$$w(k+1) = w(k+1)/||w(k+1)|| \tag{6.13}$$

This is why $\|w\|^2$ is omitted in formula (6.12). An independent component of the original signal can be estimated from the obtained vector w as follows.

$$y = w^{\mathrm{T}} z \tag{6.14}$$

Equations (6.12), (6.13), and (6.14) are iterative formulas for fast fixed-point algorithm based on kurtosis, also called Kurtosis FastICA. The convergence of the Kurtosis FastICA algorithm means that the values of the old and new w point to the same direction, i.e. their inner product is almost equal to 1.

In summary, the steps of FastICA algorithm based on kurtosis are as follows:

Step 1: The observed signal x is de-averaged and whitened to obtain signal z;
Step 2: Let $k = 0$, select the initial vector $w(0) \neq 0$;
Step 3: Iterate according to formula (6.12);
Step 4: Normalize $w(k+1)$ according to formula (6.13);
Step 5: If it is not convergent, i.e. $w(k+1)^{\mathrm{T}} w(k)|$ is not close to 1, then $k := k + 1$ and return to Step 3; Otherwise, go to Step 6.
Step 6: If $w(k+1)$ meets the requirements, stop the iteration and extract the independent component $y = w(k+1)^{\mathrm{T}} z$.

6.2 Fast Fixed-Point Algorithm Based on Negative Entropy

A fast fixed-point algorithm based on kurtosis is presented above, but when the kurtosis is estimated by the sample of data, it has some defects in practical application. The main problem is that kurtosis is very sensitive to outliers. For example, if a random variable with zero mean and unit variance contains a value equal to 10 in its 1000 sample points, then the kurtosis of the random variable is at least: $10^4/1000 - 3 = 7$. This means that one variate can make kurtosis large, which means that kurtosis is not a robust measure of non-Gaussianity.

Therefore, under certain conditions, other measuring methods of non-gaussianity may be better than kurtosis method. Next, we discuss another important measuring method of non-gaussianity—negative entropy method. Its properties are the opposite of kurtosis in many ways. The negative entropy method is robust but computationally complex. In order to solve the problem of computation complexity, the simple approximate calculation method of negative entropy is introduced, so that the advantages of two measuring methods can be more or less combined.

Negative entropy is always greater than or equal to zero. It is equal to zero only when the signal is a Gaussian signal, and for signals with the same variance, the entropy of the signal following the Gaussian distribution is the largest. The independent components of the observed signal can be separated by iterative search based on the direction of the maximum negative entropy. This algorithm uses the traditional projection pursuit algorithm. Because this algorithm is a fixed-point iterative algorithm, the iterative results are relatively stable, and batch approach of Fast ICA makes iterative process require fewer iterations than other algorithms. These advantages make this algorithm become the research hotspot of blind source separation.

6.2.1 Negative Entropy Measurement of Non-Gaussianity

In addition to kurtosis, entropy and negative entropy are also a measure of the non-Gaussianity properties of random signals. Entropy is a measure of the state of matter or the degree to which it is in that state. The entropy of a random variable is related to the information provided by the observations of the random variable, and the greater the randomness, that is, the random variable is more unpredictable or unconstructed, the greater the entropy. The entropy H($\boldsymbol{y}$) of a random variable $\boldsymbol{y}$ with probability density is $p_y(\eta)$ is defined as follows:

$$\mathrm{H}(\boldsymbol{y}) = -\int p_y(\eta) \log p_y(\eta) \mathrm{d}\eta \tag{6.15}$$

A basic conclusion of information theory is that of all random variables with equal variance, Gaussian random variables have the highest entropy, which means that the above Eq. (6.15) can be used as a measure of non-Gaussianity. This also shows that the Gaussian distribution is the most random of all distributions. For distributions that are obviously concentrated on certain values, entropy is very low.

In order to obtain a non-Gaussian measurement that is 0 for a Gaussian random variable and is always non-negative, people often use the normalized value of entropy, also known as negative entropy. Negative entropy J is defined as follows:

$$J(\boldsymbol{y}) = H(\boldsymbol{y}_{\text{gauss}}) - H(\boldsymbol{y}) \tag{6.16}$$

Here $\boldsymbol{y}_{\text{gauss}}$ is a Gaussian random variable with the same covariance matrix $\sum$ as $\mathbf{y}$. Based on above properties, negative entropy is always non-negative and is 0 if and only if y is a Gaussian distribution. According to the definition of entropy, we have

$$H(\boldsymbol{y}_{\text{gauss}}) = \frac{1}{2} \log |\det \sum| + \frac{n}{2}(1 + \log 2\pi) \tag{6.17}$$

where n is the dimension of $\boldsymbol{y}$.

Negative entropy is invariant for invertible linear transformations. For $\boldsymbol{y} = \boldsymbol{M}\boldsymbol{x}$, it has

$$\mathrm{E}\{\boldsymbol{y}\boldsymbol{y}^{\mathrm{T}}\} = \boldsymbol{M} \sum \boldsymbol{M}^{\mathrm{T}} \tag{6.18}$$

Then:

$$\begin{aligned}
J(\boldsymbol{Mx}) &= \frac{1}{2}\log|\det(\boldsymbol{M}\sum\boldsymbol{M}^{T})| + \frac{n}{2}[1+\log 2\pi] - (H(\boldsymbol{x}) + \log|\det\boldsymbol{M}|) \\
&= \frac{1}{2}\log|\det\sum| + 2\times\frac{1}{2}\log|\det\boldsymbol{M}| + \frac{n}{2}[1+\log 2\pi] - H(\boldsymbol{x}) - \log|\det\boldsymbol{M}| \\
&= \frac{1}{2}\log|\det\sum| + \frac{n}{2}[1+\log 2\pi] - H(\boldsymbol{x}) \\
&= H(\boldsymbol{x}_{\text{gauss}}) - H(\boldsymbol{x}) \\
&= J(\boldsymbol{x})
\end{aligned} \tag{6.19}$$

The advantage of negative entropy as a non-Gaussian measurement is that it can be well justified by statistical theory. In fact, in a sense, negative entropy is the best estimate of non-Gaussianity. However, the disadvantage of using negative entropy to measure non-gaussianity is that the calculation is very complicated, and using the definition to calculate negative entropy will need to estimate the probability density of the signals, so the simple approximate calculation of negative entropy is very useful. The approximate calculation method of negative entropy is given below.

6.2.2 *Approximate Calculation of Negative Entropy*

The typical method of negative entropy approximation is obtained by using higher order cumulants and polynomial extension of probability density as follows

$$J(y) \approx \frac{1}{2}E\{y^3\}^2 + \frac{1}{48}kurt(y)^2 \tag{6.20}$$

The random variable y is assumed to have zero mean and unit variance. When y is a zero-mean random variable with symmetric distribution, the first term on the right of Eq. (6.20) is 0, then this equation is equivalent to the square of kurtosis, Maximizing the square of kurtosis is equivalent to maximizing the absolute value of kurtosis. This approximation will lead to poor robustness of the algorithm, just like the problem faced by the kurtosis method above, so it is necessary to propose a more mature approximate calculation method of negative entropy.

In general, we'll use other function G_i $(i \geq 2)$ to replace $\boldsymbol{y}^3$, $\boldsymbol{y}^4$. As one of the simplest special cases, you can take the G_1 and G_2 as any non-quadratic function to make G_1 is the odd function, and G_2 is the even function, then

$$J(y) \approx k_1(E\{G_1(y)\})^2 + k_2(E\{G_2(y)\} - E\{G_2(v)\})^2 \tag{6.21}$$

where k_1, k_2 is positive constant, $\boldsymbol{v}$ is a Gaussian random variable with zero mean and unit variance, and the variable $\boldsymbol{y}$ is assumed to have zero mean and unit variance.

We should note that even under these conditions, this approximation is not the most accurate. The formula (6.21) can be used to construct a non-Gaussian measurement that is consistent in the sense that the values are always non-negative and is 0 only if $\boldsymbol{y}$ is a gaussian distribution. When we use only one non-quadratic function G, the negative entropy can be approximated as

$$\mathrm{J}(\boldsymbol{y}) \propto [\mathrm{E}\{\mathrm{G}(\boldsymbol{y})\} - \mathrm{E}\{\mathrm{G}(\boldsymbol{v})\}]^2 \tag{6.22}$$

Formula (6.22) is a generalization of formula (6.20) when $\boldsymbol{y}$ has a symmetric distribution, because the first term in formula (6.20) is 0. In this case let $G = y^4$, the approximation based on kurtosis is obtained.

The key point here is that a better approximation of the negative entropy relative to (6.20) can be obtained by selecting G widely. By choosing a function G that does not grow too fast, we can obtain a more robust estimator. The following options have been proved useful [37].

$$\mathrm{G}_1(\boldsymbol{y}) = \frac{1}{a_1}\log\cosh(a_1\boldsymbol{y}) \tag{6.23}$$

$$\mathrm{G}_2 = -\mathrm{e}^{(-y^2/2)} \tag{6.24}$$

where $1 \le a_1 \le 2$ is the appropriately chosen constant, generally taken as 1.

6.2.3 Gradient Algorithm Based on Negative Entropy

When the signal $\boldsymbol{y}$ in formula (6.22) is a separated signal, then $\boldsymbol{y} = \boldsymbol{w}^{\mathrm{T}}\boldsymbol{z}$. At this time, the gradient of the negative entropy shown in Eq. (6.22) is approximated, and the iterative formula of the separating vector $\boldsymbol{w}$ is obtained by using the assumed condition $E\{(\boldsymbol{w}^T\boldsymbol{z})\} = ||\boldsymbol{w}||^2 = 1$ of the source signal as follows

$$\boldsymbol{w}(k+1) = \boldsymbol{w}(k) + \Delta\boldsymbol{w}(k) = \boldsymbol{w}(k) + \gamma E\{\boldsymbol{z}g(\boldsymbol{w}(k)^{\mathrm{T}}\boldsymbol{z})\} \tag{6.25}$$

$$\boldsymbol{w}(k+1) \leftarrow \boldsymbol{w}(k+1)/||\boldsymbol{w}(k+1)|| \tag{6.26}$$

$$\gamma = E\{G(\boldsymbol{w}(k)^T\boldsymbol{z})\} - E\{G(\boldsymbol{v})\} \tag{6.27}$$

where $\boldsymbol{w}(k)$ is the column vector of the separating matrix at the k-th iteration, $\boldsymbol{v}$ is the standard (zero mean, unit variance) Gaussian random variable, and $\boldsymbol{z}$ is the observed signal after whitening.

The normalization shown in Eq. (6.26) is to project $\boldsymbol{w}$ onto the unit sphere, thus keeping the variance of $\boldsymbol{w}^T\boldsymbol{z}$ constant. Function $g(\cdot)$ is the derivative of function $G(\cdot)$ used to approximate negative entropy, and the expectation symbol can be omitted in an online (adaptive) stochastic gradient algorithm. Equations (6.25), (6.26) and (6.27) are the iterative formulas of the negative entropy gradient algorithm.

For the function $g(\cdot)$, it can be obtained by taking the derivative of the negative entropy approximation function shown in (6.23)–(6.24), as well as by taking the derivative of the fourth power function representing kurtosis, that is

$$g_1(y) = \tanh(a_1 y) \tag{6.28}$$

$$g_2(y) = y \cdot \exp(-y^2/2) \tag{6.29}$$

$$g_3(y) = y^3 \tag{6.30}$$

where $1 \le a_1 \le 2$ is the appropriately chosen constant, generally taken as 1.

6.2.4 Fast Fixed-Point Algorithm Based on Negative Entropy

Like the fast fixed-point algorithm based on kurtosis, we can use the fixed-point algorithm to obtain a maximum negative entropy algorithm that is faster than the gradient algorithm. The resulting FastICA algorithm is to find a direction, that is, the unit vector $\boldsymbol{w}$, which maximizes the non-Gaussianity of the projection $\boldsymbol{w}^{\mathrm{T}}\boldsymbol{z}$. Here the non-Gaussianity is measured by the approximate negative entropy $J(\boldsymbol{w}^{\mathrm{T}}\boldsymbol{z})$ of Eq. (6.22). The variance of the $\boldsymbol{w}^{\mathrm{T}}\boldsymbol{z}$ must be limited to 1, which is equivalent to the norm of $\boldsymbol{w}$ being limited to 1 for whitened data.

The FastICA algorithm is based on the fixed-point iteration mechanism to find the vector $\boldsymbol{w}$ that maximizes the non-Gaussiality of $\boldsymbol{w}^T\boldsymbol{z}$. It can be derived using the approximate Newton method. The FastICA algorithm based on negative entropy combines the superior properties of fixed-point iteration mechanism with the excellent statistical properties of negative entropy. The following is the algorithm derivative process.

According to the gradient method in Eq. (6.25), the following fixed-point iterative algorithm can be obtained

$$\boldsymbol{w}(k+1) = E\{\boldsymbol{z}g(\boldsymbol{w}(k)^{\mathrm{T}}\boldsymbol{z})\} \tag{6.31}$$

$$\boldsymbol{w}(k+1) \leftarrow \boldsymbol{w}(k+1)/||\boldsymbol{w}(k+1)|| \tag{6.32}$$

where $\boldsymbol{w}(k)$ represents the column vector of the separating matrix at the k-th iteration, γ can be omitted because the normalization of formula (6.32) can cancel it out. Of course, the iterative algorithm of formula (6.31) does not have the convergence properties as good as the kurtosis FastICA algorithm, because the non-polynomial moments do not have the algebraic properties as good as the real cumulants like kurtosis. Therefore, it is necessary to modify the iterative algorithm of formula (6.31), an iterative approximate Newton algorithm has been proposed.

To derive the approximate Newton algorithm, we first notice that the approximate negative entropy of $w^T z$ has a maximum value at the extreme point of $\mathrm{E}\{G(w^T z)\}$. According to the Kuhn-Tucker condition, the extremum point of $\mathrm{E}\{G(w^T z)\}$ under the restriction of $\mathrm{E}\{(w^T z)^2\} = ||w||^2 = 1$ is obtained at the point satisfying the following formula (6.33).

$$\mathrm{E}\{zg(w^T z)\} + \beta w = 0 \tag{6.33}$$

Here $\beta = \mathrm{E}\{w^{*T} zg(w^{*T} z)\}$, w^* is the optimized separation matrix column vector, the function $g(\cdot)$ is the derivative of $G(\cdot)$. We now use Newton method to solve the Eq. (6.33), which is equivalent to using Newton method to find the optimal point of Lagrange. Define the left expression of Eq. (6.33) as F, then its gradient is:

$$\frac{\partial F}{\partial w} = \mathrm{E}\{zz^T g'(w^T z)\} + \beta \boldsymbol{I} \tag{6.34}$$

In order to simplify the matrix inversion operation, the first term on the right of Eq. (6.34) can be approximated as follows.

$$\mathrm{E}\{zz^T g'(w^T z)\} = \mathrm{E}\{zz^T\} E\{g'(w^T z)\} = \mathrm{E}\{g'(w^T z)\} \boldsymbol{I} \tag{6.35}$$

Then the gradient represented by Eq. (6.34) becomes a diagonal matrix, which is easy to inverse. Thus, we obtain the following approximate Newton iteration formula:

$$w(k+1) = w(k) - [\mathrm{E}\{zg(w(k)^T z)\} + \beta w(k)]/[\mathrm{E}\{g'(w(k)^T z)\} + \beta] \tag{6.36}$$

In order to further simplify the algorithm, both sides of the above formula are multiplied by $E\{g'(w(k)^T z)\} + \beta$, then

$$\begin{aligned} w(k+1) &= [E\{g'(w(k)^T z)\} + \beta] w(k) - [E\{zg(w(k)^T z)\} + \beta w(k)] \\ &= E\{g'(w(k)^T z)\} w(k) - E\{zg(w(k)^T z)\} \end{aligned} \tag{6.37}$$

In order to limit the separating vector w as the unit norm, a normalization process should be carried out every iteration, so the coefficient $E\{g'(w_k^T z)\} + \beta]$ in the first term on the left of formula (6.37) can be omitted. Due to the inherent ambiguity of blind source separation, the positive and negative sign of w will not affect the separation effect of the algorithm, so

$$w(k+1) = E\{zg(w(k)^T z)\} - E\{g'(w(k)^T z)\} w(k) \tag{6.38}$$

Equations (6.38) and (6.32) are the iterative formulas of FastICA algorithm based on negative entropy. where $g(\cdot)$ is a nonlinear function, $g'(\cdot)$ is a derivative function of $g(\cdot)$, and there are generally three choices of $g(\cdot)$, as shown in Eqs. (6.28)–(6.30).

In summary, the steps of FastICA algorithm based on negative entropy are as follows.

Step 1: First, the observed signal matrix $\boldsymbol{x}(t) = [\boldsymbol{x}_1(t), \boldsymbol{x}_2(t), \ldots, \boldsymbol{x}_M(t)]^{\mathrm{T}}$ is de-averaged to obtain the observed signal with zero mean $\overline{\boldsymbol{x}}(t) = [\overline{\boldsymbol{x}}_1(t), \overline{\boldsymbol{x}}_2(t), \ldots, \overline{\boldsymbol{x}}_M(t)]^{\mathrm{T}}$, where $\overline{x}_m(t) = x_m(t) - E(x_m(t))$, $(m = 1, 2, \ldots, M)$.
Step 2: Then use the whitening algorithm introduced in Chap. 4 to obtain the observed signal vector $z(t)$ after whitening.
Step 3: Let $k = 0$, select an initial unit vector $\boldsymbol{w}(0)$.
Step 4: Update: $\boldsymbol{w}(k+1) = E\{zg(\boldsymbol{w}(k)^{\mathrm{T}}z(t))\} - E\{g'(\boldsymbol{w}(k)^{\mathrm{T}}z(t))\}\boldsymbol{w}(k)$.
Step 5: Normalize: $\boldsymbol{w}(k+1) \leftarrow \boldsymbol{w}(k+1)/||\boldsymbol{w}(k+1)||$.
Step 6: If it does not converge, namely $||\boldsymbol{w}(k+1) - \boldsymbol{w}(k)|| > \varepsilon$, return to Step 4. If it converges, the separated signal $y(t) = \boldsymbol{w}(k)^{\mathrm{T}}z(t)$ is obtained.

6.3 Multivariate Fast Fixed-Point Algorithm

The fast fixed-point algorithm derived in the above section can only separate one independent component, which is called the unary algorithm. In order to extract multiple independent components, the above unary algorithm can be repeated for many times. At the same time, in order to avoid that the independent components extracted at the n-th time are the same as the independent components extracted at the first n-1 times, the vector $\boldsymbol{w}_1, \boldsymbol{w}_2, \ldots \boldsymbol{w}_n$ needs to be orthogonalized after each iteration operation. Two orthogonalization methods are given below.

6.3.1 Gram-Schmidt Orthogonalization

Theorem *Let $\boldsymbol{\alpha}_1, \boldsymbol{\alpha}_2, \ldots, \boldsymbol{\alpha}_m$, $m \le n$ is a linearly independent set of vectors in R^m, if*

$$\begin{gathered}
\boldsymbol{\beta}_1 = \boldsymbol{\alpha}_1 \\
\boldsymbol{\beta}_2 = \boldsymbol{\alpha}_2 - \frac{\langle \boldsymbol{\alpha}_2, \boldsymbol{\beta}_1 \rangle}{\langle \boldsymbol{\beta}_1, \boldsymbol{\beta}_1 \rangle}\boldsymbol{\beta}_1 \\
\cdots \\
\boldsymbol{\beta}_m = \boldsymbol{\alpha}_m - \frac{\langle \boldsymbol{\alpha}_m, \boldsymbol{\beta}_1 \rangle}{\langle \boldsymbol{\beta}_1, \boldsymbol{\beta}_1 \rangle}\boldsymbol{\beta}_1 - \frac{\langle \boldsymbol{\alpha}_m, \boldsymbol{\beta}_2 \rangle}{\langle \boldsymbol{\beta}_2, \boldsymbol{\beta}_2 \rangle}\boldsymbol{\beta}_2 - \cdots - \frac{\langle \boldsymbol{\alpha}_m, \boldsymbol{\beta}_{m-1} \rangle}{\langle \boldsymbol{\beta}_{m-1}, \boldsymbol{\beta}_{m-1} \rangle}\boldsymbol{\beta}_{m-1}
\end{gathered} \tag{6.39}$$

Then $\boldsymbol{\beta}_1, \boldsymbol{\beta}_2, \ldots, \boldsymbol{\beta}_m$ is a set of orthogonal vectors, and let

$$\boldsymbol{e}_i = \frac{\boldsymbol{\beta}_i}{\|\boldsymbol{\beta}_i\|}(i = 1, 2, m) \tag{6.40}$$

You can get a set of standard orthogonal vectors $\boldsymbol{e}_1, \boldsymbol{e}_2, \ldots, \boldsymbol{e}_m$*, and this set of vectors is equivalent to* $\boldsymbol{\alpha}_1, \boldsymbol{\alpha}_2, \ldots, \boldsymbol{\alpha}_m$.

The above is the method of constructing set of orthonormal vectors by using set of linearly independent vectors, which is Schmidt orthogonalization method.

The following takes $m = 3$ as an example to prove the correctness of the above theorem.

Assuming vectors $\boldsymbol{\alpha}_1, \boldsymbol{\alpha}_2, \boldsymbol{\alpha}_3$ are linearly independent, we first construct the set of orthogonal vectors $\boldsymbol{\beta}_1, \boldsymbol{\beta}_2, \boldsymbol{\beta}_3$ and make them equivalent to $\boldsymbol{\alpha}_1, \boldsymbol{\alpha}_2, \ldots, \boldsymbol{\alpha}_r (r = 1, 2, \ldots, m)$. According to the conditions required, $\boldsymbol{\beta}_1$ is the linear combination of $\boldsymbol{\alpha}_1$, $\boldsymbol{\beta}_2$ is linear combination of $\boldsymbol{\alpha}_1, \boldsymbol{\alpha}_2$. for convenience, let

$$\boldsymbol{\beta}_1 = \boldsymbol{\alpha}_1, \boldsymbol{\beta}_2 = \boldsymbol{\alpha}_2 - k\boldsymbol{\beta}_1 \tag{6.41}$$

where the value of k should meet that $\boldsymbol{\beta}_1$ and $\boldsymbol{\beta}_2$ are vertical, that is

$$\langle \boldsymbol{\beta}_2, \boldsymbol{\beta}_1 \rangle = \langle \boldsymbol{\alpha}_2, \boldsymbol{\beta}_1 \rangle - k \langle \boldsymbol{\beta}_1, \boldsymbol{\beta}_1 \rangle = 0 \tag{6.42}$$

Notice that $\langle \boldsymbol{\beta}_1, \boldsymbol{\beta}_1 \rangle > 0$, then $k = \frac{\langle \boldsymbol{\alpha}_2, \boldsymbol{\beta}_1 \rangle}{\langle \boldsymbol{\beta}_1, \boldsymbol{\beta}_1 \rangle}$, and

$$\boldsymbol{\beta}_1 = \boldsymbol{\alpha}_1, \boldsymbol{\beta}_2 = \boldsymbol{\alpha}_2 - \frac{\langle \boldsymbol{\alpha}_2, \boldsymbol{\beta}_1 \rangle}{\langle \boldsymbol{\beta}_1, \boldsymbol{\beta}_1 \rangle} \boldsymbol{\beta}_1 \tag{6.43}$$

according to $\boldsymbol{\beta}_1$ and $\boldsymbol{\beta}_2$, we construct vector $\boldsymbol{\beta}_3$, in order to meet the requirements, let

$$\boldsymbol{\beta}_3 = \boldsymbol{\alpha}_3 - k_1 \boldsymbol{\beta}_1 - k_2 \boldsymbol{\beta}_2 \tag{6.44}$$

where, the selection of k_1, k_2 should make $\boldsymbol{\beta}_3$ vertical to vector $\boldsymbol{\beta}_1$ and $\boldsymbol{\beta}_2$, that is

$$\langle \boldsymbol{\alpha}_3, \boldsymbol{\beta}_1 \rangle - k_1 \langle \boldsymbol{\beta}_1, \boldsymbol{\beta}_1 \rangle = 0, \langle \boldsymbol{\alpha}_3, \boldsymbol{\beta}_2 \rangle - k_2 \langle \boldsymbol{\beta}_2, \boldsymbol{\beta}_2 \rangle = 0 \tag{6.45}$$

Thus,

$$k_1 = \frac{\langle \boldsymbol{\alpha}_3, \boldsymbol{\beta}_1 \rangle}{\langle \boldsymbol{\beta}_1, \boldsymbol{\beta}_1 \rangle}, k_2 = \frac{\langle \boldsymbol{\alpha}_3, \boldsymbol{\beta}_2 \rangle}{\langle \boldsymbol{\beta}_2, \boldsymbol{\beta}_2 \rangle} \tag{6.46}$$

then

$$\boldsymbol{\beta}_3 = \boldsymbol{\alpha}_3 - \frac{\langle \boldsymbol{\alpha}_3, \boldsymbol{\beta}_1 \rangle}{\langle \boldsymbol{\beta}_1, \boldsymbol{\beta}_1 \rangle} \boldsymbol{\beta}_1 - \frac{\langle \boldsymbol{\alpha}_3, \boldsymbol{\beta}_2 \rangle}{\langle \boldsymbol{\beta}_2, \boldsymbol{\beta}_2 \rangle} \boldsymbol{\beta}_2 \tag{6.47}$$

It is easy to verify that the set of vectors $\boldsymbol{\beta}_1, \boldsymbol{\beta}_2, \ldots, \boldsymbol{\beta}_r$ is equivalent of the vectors $\boldsymbol{\alpha}_1, \boldsymbol{\alpha}_2, \ldots, \boldsymbol{\alpha}_r$. Normalize $\boldsymbol{\beta}_1, \boldsymbol{\beta}_2, \boldsymbol{\beta}_3$, that is let

$$e_i = \frac{\boldsymbol{\beta}_i}{\|\boldsymbol{\beta}_i\|}(i = 1, 2, 3) \tag{6.48}$$

Thus e_1, e_2, e_3 is a set of orthonormal vectors meet the requirements.

Based on the above theorems, we get a multivariate fast fixed-point algorithm based on Gram-Schmidt orthogonalization. When we have estimated p independent components or p vectors $w_1, w_2, \dots w_p$, we run any unary algorithm to w_{p+1}, and every iteration, extract the projection $(w_{p+1}^T w_j)w_j, \quad j = 1, 2, \dots p$ of the previously estimated p vectors from w_{p+1}, and then re-normalize, the steps of Gram-Schmidt orthogonalization algorithm are as follows.

Step 1: Choose the number m of independent components to estimate, and let $p = 1$.
Step 2: Initialization w_p(for example, random vector).
Step 3: Do one unary algorithm iteration on w_p.
Step 4: Perform the following orthogonalization process:

$$w_p = w_p - \sum_{j=1}^{p-1} (w_p^T w_j)w_j \tag{6.49}$$

Step 5: Normalize w_p: $w_p = w_p/||w_p||$.
Step 6: If w_p is not convergent, go back to step 3; Otherwise go to step 7.
Step 7: Set $p = p + 1$. If $p \leq m$, go back to step 2; Otherwise, the algorithm ends.

6.3.2 Symmetric Orthogonalization

The Gram-Schmidt orthogonalization method has the disadvantage of cumulative error, and in some applications, it is necessary to use the method of symmetric decorrelation, which makes the vectors w_i are not estimated one by one, but are estimated simultaneously, and the symmetric orthogonalization method can compute multiple independent components in parallel. The symmetric orthogonalization method first performs unary iterations on every vector w_i in parallel, and then orthogonalizes all of the w_i by a special symmetric orthogonalization method. the steps of symmetric orthogonalization are as follows.

Step 1: Select the number M of independent components to estimate.
Step 2: Initialization w_i, $i = 1, 2, \dots, M$ (e.g. random vector).
Step 3: Perform one unary iteration on all w_i simultaneously.
Step 4: Symmetric orthogonalize the matrix $W = (w_1, w_2, \dots, w_M)^{\mathrm{T}}$.
Step 5: If it does not converge, go back to step 3.

Symmetric orthogonalization can be accomplished by the following formula:

$$W \leftarrow (WW^{\mathrm{T}})^{-1/2}W \tag{6.50}$$

$(\boldsymbol{WW}^{\mathrm{T}})^{-1/2}$ can be obtained according to the eigenvalue decomposition of $\boldsymbol{WW}^{\mathrm{T}}$, that is $\boldsymbol{WW}^{\mathrm{T}} = \boldsymbol{E}\,diag(d_1, d_2, \ldots, d_M)\boldsymbol{E}^{\mathrm{T}}$, therefore

$$(\boldsymbol{WW}^{\mathrm{T}})^{-1/2} = \boldsymbol{E}\,diag(d_1^{-1/2}, d_2^{-1/2}, \ldots, d_m^{-1/2})\boldsymbol{E}^{\mathrm{T}} \tag{6.51}$$

where $d_1, d_2, \ldots, d_M$ is the eigenvalue of $\boldsymbol{WW}^{\mathrm{T}}$, then $\boldsymbol{E}$ is the eigenvector matrix of $\boldsymbol{WW}^{\mathrm{T}}$, $\boldsymbol{E}$ is the orthogonal matrix.

6.3.3 Multivariate Fast Fixed-Point Algorithm

If you want to separate multiple source signals simultaneously by using the fast fixed-point algorithm, you can run the unary fast fixed-point algorithm several times. In order to avoid recovering the same signal repeatedly, when separating the p-th signal, the Gram-Schmidt algorithm is used to orthogonalize once for each iteration. For the p-th separated signal, the iterative formula is as follows.

$$\boldsymbol{w}_p := \boldsymbol{w}_p - \sum_{j=1}^{p-1} (\boldsymbol{w}_p^{\mathrm{T}}\boldsymbol{w}_j)\boldsymbol{w}_j \tag{6.52}$$

$$\boldsymbol{w}_p \Leftarrow \boldsymbol{w}_p/||\boldsymbol{w}_p|| \tag{6.53}$$

where $\boldsymbol{w}_1$, $\boldsymbol{w}_2, \ldots, \boldsymbol{w}_{p-1}$ are the first $p-1$ separating vectors that have been obtained. Using (6.38), (6.52), and (6.53), all source signals can be separated one by one.

On the other hand, the separating matrix $\boldsymbol{W}$ composed of multiple separating vectors can also be iterated, so as to separate all simultaneously. But in the iteration, it is also necessary to carry out orthogonalization, the iterative equations are as follows.

$$\boldsymbol{W}_{k+1} = E\{g(\boldsymbol{W}_k\boldsymbol{z})\boldsymbol{z}^{\mathrm{T}}\} - diag\{E[g'(\boldsymbol{W}_k\boldsymbol{z})]\}\boldsymbol{W}_k \tag{6.54}$$

$$\boldsymbol{W}_{k+1} \leftarrow (\boldsymbol{W}_{k+1}\boldsymbol{W}_{k+1}^{\mathrm{T}})^{-1/2}\boldsymbol{W}_{k+1} \tag{6.55}$$

Repeat the execution (6.54) and (6.55) until convergence, and multiple source signals can be separated simultaneously, and $\boldsymbol{y} = \boldsymbol{W}_k\boldsymbol{z}$ is the estimation of the source signal vector.

6.4 Fast Fixed-Point Algorithm for Complex Signals

The fast fixed-point algorithm derived above can only be applied to the case of real signals, but in BSS problem of communication signals, the system mixing model is often a complex mixing matrix, and the source signals are also complex. A FastICA algorithm for complex signals is presented here. Assume that both the independent components and the mixing matrix are complex, and assume that the number of independent components is equal to the number of observed signals (even if they are not equal, as long as the number of observed signals is greater than the number of independent components, we can also use whitening algorithm to reduce the dimension of observed signals, so that the number of observed signals after whitening is equal to the number of independent components), and the mixing matrix $\boldsymbol{A}$ column is full rank. It is also assumed that the source signals are independent of each other, and that the real and imaginary parts of each source signal are not correlated and the variance is the same.

6.4.1 Basic Concepts of Complex Random Variables

A complex random variable y can be expressed as $y = u + i \cdot v$, where u, v is a real-valued random variable. The probability density of y is $f(y) = f(u, v)$, and the expectation of y is $E\{y\} = E\{u\} + i \cdot E\{v\}$. If two random variables y_1, y_2 satisfy $E\{y_1 y_2^*\} = E\{y_1\}E\{y_2^*\}$, then they are not related, where $y^* = u - i \cdot v$ represents the complex conjugation of y. The covariance matrix of a zero-mean complex random vector $\boldsymbol{y} = (y_1, y_2, \ldots, y_n)$ is

$$E\{\boldsymbol{y}\boldsymbol{y}^{\mathrm{H}}\} = \begin{bmatrix} C_{11}, C_{12}, \ldots, C_{1n} \\ \vdots \quad \ddots \quad \ldots \quad \ddots \\ C_{n1}, C_{n2}, \ldots, C_{nn} \end{bmatrix} \tag{6.56}$$

where $C_{jk} = E\{y_j y_k^*\}$, and $\boldsymbol{y}^{\mathrm{H}}$ represents the conjugate transpose matrix of $\boldsymbol{y}$.

In the independent component analysis model of complex signal, all independent components s_i are required to have zero mean and unit variance, and the real and imaginary parts are not correlated and have the same variance, that is $E\{\boldsymbol{s}\boldsymbol{s}^{\mathrm{H}}\} = \boldsymbol{I}$ and $E\{\boldsymbol{s}\boldsymbol{s}^{\mathrm{T}}\} = \boldsymbol{0}$, these assumptions mean that the source signal s_i is strictly complex and its imaginary part cannot be 0.

The definition of kurtosis is easy to generalize. For example, for a zero-mean complex random variable y, kurtosis can be defined as

$$kurt(y) = E\{|y|^4\} - E\{yy^*\} \cdot E\{yy^*\} - E\{yy\}E\{y^*y^*\} - E\{yy^*\} \cdot E\{y^*y\} \tag{6.57}$$

Of course, the definition varies with the position of the conjugate symbol (*), that is, there are 16 definition methods, here we choose the following definition

$$kurt(y) = E\{|y|^4\} - 2E\{|y|^2\}^2 - |E\{y^2\}|^2 = E\{|y|^4\} - 2 \tag{6.58}$$

When y is white, i.e. $E\{y^2\} = 0$, $E\{|y|^2\} = 1$, the last equal sign in the above equation holds.

6.4.2 Selection of Non-Gaussian Measurement

In the complex-valued case, the distribution of the complex random variable is often spherically symmetric, so only its module value is of interest. In this way, we can make use of the non-Gaussian measurement based on module values. On basis of non-Gaussian measurement in Eq. (6.22), the following measurement function can be defined.

$$J_G(\boldsymbol{w}) = E\{G(|\boldsymbol{w}^{\mathrm{H}}\boldsymbol{z}|^2)\} \tag{6.59}$$

where $G(\cdot)$ is a smooth even function, $\boldsymbol{w}$ is an n-dimensional complex vector and $E\{|\boldsymbol{w}^{\mathrm{H}}\boldsymbol{z}|^2\} = ||\boldsymbol{w}||^2 = 1$, $\boldsymbol{z}$ is the whitened observed data. Compared with the kurtosis of formula (6.58), if $G(y) = y^2$, then

$$J_G(\boldsymbol{w}) = E\{|\boldsymbol{w}^{\mathrm{H}}\boldsymbol{z}|^4\} \tag{6.60}$$

Now $J_G(\boldsymbol{w})$ is actually measuring the kurtosis of $\boldsymbol{w}^{\mathrm{H}}\boldsymbol{z}$.

Maximize $J_G(\boldsymbol{w})$ can estimate one independent component. Multiple independent components can be estimated by maximizing the sum of n non-Gaussian measurements and adding the constraints of orthogonalization. So the following optimization problem is obtained.

For all of $\boldsymbol{w}_j$, $(j = 1, 2, \ldots, n)$, under constraints $E\{\boldsymbol{w}_k^{\mathrm{H}}\boldsymbol{w}_j\} = \delta_{j,k}$, we maximize

$$\sum_{j=1}^{n} J_G(\boldsymbol{w}_j) \tag{6.61}$$

where

$$\delta_{j,k} = \begin{cases} 1, & j = k \\ 0, & j \neq k \end{cases} \tag{6.62}$$

The slower the function $G(\cdot)$ grows with the increase of input parameters, the stronger the robustness of the algorithm to external interference. Three options of $G(\cdot)$ and their derived functions $g(\cdot)$ are given as follows

$$G_1(y) = \sqrt{a_1 + y} \quad , \quad g_1(y) = \frac{1}{2\sqrt{a_1 + y}} \tag{6.63}$$

$$G_2(y) = \log(a_2 + y), \quad g_2(y) = \frac{1}{a_2 + y} \tag{6.64}$$

$$G_3(y) = \frac{1}{2}y^2, \quad g_3(y) = y \tag{6.65}$$

where a_1, a_2 are arbitrary constants (such as $a_1 \approx 0.1, a_2 \approx 0.1$ is a suitable choice), G_1, G_2 grow more slowly than G_3, so they give more robust estimating algorithms for independent components. The selection of G_3 is derived from the kurtosis of formula (6.58).

6.4.3 Derivation of Fixed-Point Algorithm

We first derive a fixed-point algorithm for one variable. Let $w = w_r + i \cdot w_i, x = x_r + i \cdot x_i$. To simplify the derivation, the algorithm updates the real and imaginary parts of $\boldsymbol{w}$ separately. According to the assumptions of the source signal, we have $E\{\boldsymbol{s}\boldsymbol{s}^H\} = \boldsymbol{I}, E\{\boldsymbol{s}\boldsymbol{s}^T\} = \boldsymbol{0}$ and the observed signal vector after whitening has $E\{zz^{\mathrm{H}}\} = \boldsymbol{I}$.

According to the Kuhn-Tucker condition, under the constraint condition $E\{|\boldsymbol{w}^{\mathrm{H}}z|^2\} = ||\boldsymbol{w}||^2 = 1$, extreme point of $E\{G(|\boldsymbol{w}^{\mathrm{H}}z|^2)\}$ is at

$$\nabla E\{G(|\mathrm{w}^{\mathrm{H}}z|^2)\} - \beta \nabla E\{|\mathrm{w}^{\mathrm{H}}z|^2\} = 0 \tag{6.66}$$

where β, is the real number, and the gradient is calculated separately with respect to the real and imaginary parts of $\boldsymbol{w}$, then the first term of the formula (6.66) is

$$\nabla E\{G(|\boldsymbol{w}^{\mathrm{H}}\boldsymbol{z}|^2)\} = \begin{bmatrix} \frac{\partial}{\partial w_{1r}} \\ \frac{\partial}{\partial w_{1i}} \\ \vdots \\ \frac{\partial}{\partial w_{nr}} \\ \frac{\partial}{\partial w_{ni}} \end{bmatrix} \tag{6.67}$$

$$E\{G(|\boldsymbol{w}^{\mathrm{H}}\boldsymbol{x}|^2)\} = 2\begin{pmatrix} E\{\mathrm{Re}\{z_1(\boldsymbol{w}^{\mathrm{H}}\boldsymbol{z})^*\}g(|\boldsymbol{w}^{\mathrm{H}}\boldsymbol{z}|^2)\} \\ E\{\mathrm{Im}\{z_1(\boldsymbol{w}^{\mathrm{H}}\boldsymbol{z})^*\}g(|\boldsymbol{w}^{\mathrm{H}}\boldsymbol{z}|^2)\} \\ \vdots \\ E\{\mathrm{Re}\{z_n(\boldsymbol{w}^{\mathrm{H}}\boldsymbol{z})^*\}g(|\boldsymbol{w}^{\mathrm{H}}\boldsymbol{z}|^2)\} \\ E\{\mathrm{Im}\{z_n(\boldsymbol{w}^{\mathrm{H}}\boldsymbol{z})^*\}g(|\boldsymbol{w}^{\mathrm{H}}\boldsymbol{z}|^2)\} \end{pmatrix}$$

The second item in the formula (6.66) is

$$\nabla E\{|\boldsymbol{w}^{\mathrm{H}}\boldsymbol{z}|^2\} = 2\begin{pmatrix} \mathrm{Re}(w_1) \\ \mathrm{Im}(w_1) \\ \vdots \\ \mathrm{Re}(w_n) \\ \mathrm{Im}(w_n) \end{pmatrix} = 2\boldsymbol{w} \tag{6.68}$$

$E\{zz^{\mathrm{H}}\} = \boldsymbol{I}$ is used here.

Newton method can be used to solve Eq. (6.66). The Jacobian matrix of $\nabla E\{G(|\boldsymbol{w}^{\mathrm{H}}\boldsymbol{z}|^2)\}$ in (6.67) can be approximated as follows.

$$\begin{aligned} \nabla^2 E\{G(|\boldsymbol{w}^{\mathrm{H}}\boldsymbol{z}|^2)\} &= E\{\nabla^2|\boldsymbol{w}^{\mathrm{H}}\boldsymbol{z}|^2 g(|\boldsymbol{w}^{\mathrm{H}}\boldsymbol{z}|^2) \\ &\quad + (\nabla|\boldsymbol{w}^{\mathrm{H}}\boldsymbol{z}|^2)(\nabla|\boldsymbol{w}^{\mathrm{H}}\boldsymbol{z}|^2)^T g'(|\boldsymbol{w}^{\mathrm{H}}\boldsymbol{z}|^2)\} \\ &\approx 2E\{g(|\boldsymbol{w}^{\mathrm{H}}\boldsymbol{z}|^2) + |\boldsymbol{w}^{\mathrm{H}}\boldsymbol{z}|^2 g'(|\boldsymbol{w}^{\mathrm{H}}\boldsymbol{z}|^2)\}\boldsymbol{I} \end{aligned} \tag{6.69}$$

where the above approximation is obtained by finding the expectation separately, and using $E\{zz^{\mathrm{T}}\} = \boldsymbol{0}$ (it can be obtained by $E\{\boldsymbol{ss}^{\mathrm{T}}\} = \boldsymbol{0}$ directly). In addition, according to (6.68), the Jacobian matrix of $\beta\nabla E\{|\boldsymbol{w}^H\boldsymbol{z}|^2\}$ is as follows.

$$\beta\nabla^2 E\{|\boldsymbol{w}^{\mathrm{H}}\boldsymbol{z}|^2\} = 2\beta\boldsymbol{I} \tag{6.70}$$

The total Jacobian matrix of formula (6.66) is:

$$J = 2(E\{g(|\boldsymbol{w}^{\mathrm{H}}z|^2) + |\boldsymbol{w}^{\mathrm{H}}z|^2 g'(|\boldsymbol{w}^{\mathrm{H}}z|^2)\} - \beta)\boldsymbol{I} \tag{6.71}$$

It's a diagonal matrix so it's easy to invert. Therefore, the approximate Newtonian iterative formula of BSS algorithm is

$$\boldsymbol{w}^{+} = \boldsymbol{w} - \frac{E\{z(\boldsymbol{w}^{\mathrm{H}}z)^{*} g(|\boldsymbol{w}^{\mathrm{H}}z|^2)\} - \beta \boldsymbol{I}}{E\{g(|\boldsymbol{w}^{\mathrm{H}}z|^2) + |\boldsymbol{w}^{\mathrm{H}}z|^2 g'(|\boldsymbol{w}^{\mathrm{H}}z|^2)\} - \beta} \tag{6.72}$$

$$\boldsymbol{w}_{new} = \frac{\boldsymbol{w}^{+}}{||\boldsymbol{w}^{+}||} \tag{6.73}$$

Multiply both sides of Eq. (6.72) by $\beta - E\{g(|\boldsymbol{w}^{\mathrm{H}}z|^2) + |\boldsymbol{w}^{\mathrm{H}}z|^2 g'(|\boldsymbol{w}^{\mathrm{H}}z|^2)\}$ to obtain a fixed-point algorithm as follows

$$\boldsymbol{w}^{+} = E\{z(\boldsymbol{w}^{\mathrm{H}}z)^{*} g(|\boldsymbol{w}^{\mathrm{H}}z|^2)\} - E\{g(|\boldsymbol{w}^{\mathrm{H}}z|^2) + |\boldsymbol{w}^{\mathrm{H}}z|^2 g'(|\boldsymbol{w}^{\mathrm{H}}z|^2)\}\boldsymbol{w} \tag{6.74}$$

$$\boldsymbol{w}_{new} = \frac{\boldsymbol{w}^{+}}{||\boldsymbol{w}^{+}||} \tag{6.75}$$

Equations (6.74)–(6.75) constitutes a fast fixed-point algorithm applicable to complex signals.

6.5 Performance Simulation and Analysis of Algorithms

6.5.1 Fast Fixed-Point Algorithm Simulation of Real Signal

Parameter setting: Four voice signals are used as source signals and mixed by a randomly generated real matrix, with the number of receiving channels P is 4, 6 or 8. The separating performance of the fast fixed-point algorithm based on kurtosis and that based on negative entropy under different number of receiving channels is simulated respectively. A total of 200 Monte Carlo experiments were performed to calculate the average output Interference to Signal Ratio (ISR).

Figure 6.2 shows the changes of ISR with SNR of the fast fixed-point algorithm based on kurtosis. we can see that the more the number of receiving channels P, the lower the ISR after separating. When the number of receiving channels is 8, the minimum ISR can reach −32 dB.

Figures 6.3 and 6.4 show the changes of ISR with SNR of the fast fixed-point algorithm based on negative entropy we can see that the performance variation trend of the fast fixed-point algorithm based on negative entropy is similar to that of the fast fixed-point algorithm based on kurtosis. When the number of receiving antennas is 8, the minimum ISR of the separated signal is also about −32 dB. Therefore, the two fast fixed-point algorithms can separate multiple speech signals well.

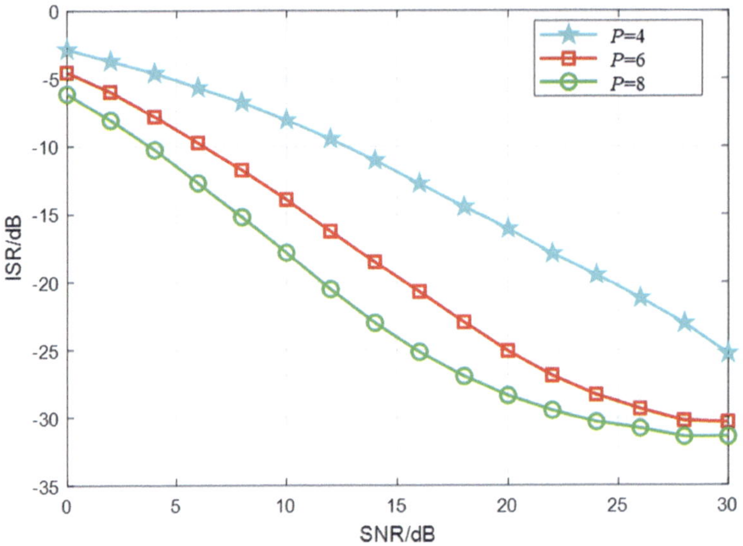

Fig. 6.2 Separating performance of fast fixed-point algorithm based on kurtosis

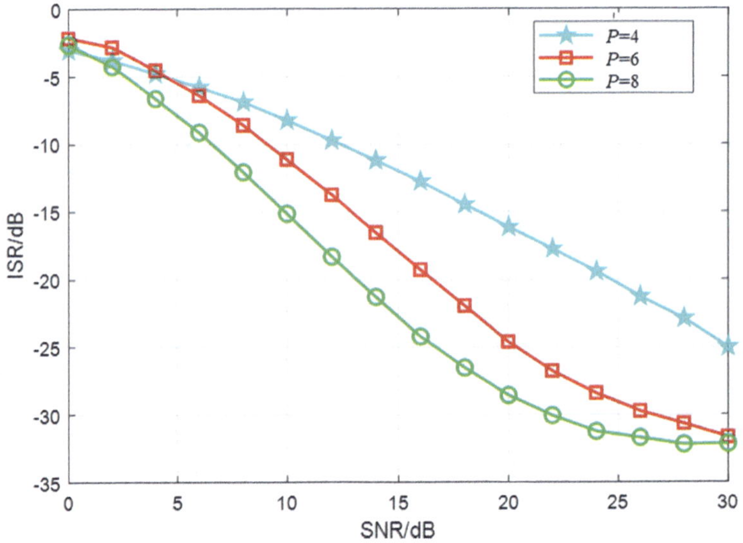

Fig. 6.3 Separating performance of fast fixed-point algorithm based on negative entropy (nonlinear function $g(y) = y \cdot \exp(-y^2/2)$)

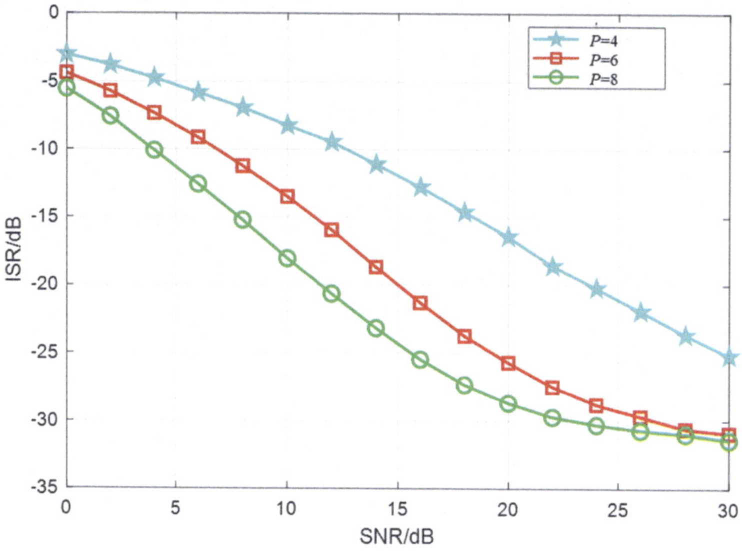

Fig. 6.4 Separating performance of fast fixed-point algorithm based on negative entropy (nonlinear function $g(y) = y^3$)

6.5.2 *Fast Fixed-Point Algorithm Simulation of Complex Signal*

Assume the four source signals (these four signals are located in the 20 MHz bandwidth of intermediate frequency 70 MHz) are Two QPSK signals with a carrier frequency of 65 MHz, an 8PSK signal with a carrier frequency of 65 MHz, and a 16QAM signal with a carrier frequency of 75 MHz. Incident angles are 40, 60, 80, 100 degrees respectively, receiving antenna is a uniform linear array with array elements of 4,6,8, respectively, and the array element interval is half of the intermediate frequency carrier wavelength. The fast fixed-point algorithm for complex signals can choose three kinds of nonlinear functions, including nonlinear function $G_1(y) = \frac{1}{2}y^2$, $g_1(y) = y$, $G_2(y) = \sqrt{a_1 + y}$, $g_2(y) = \frac{1}{2\sqrt{a_1+y}}$ and $G_3(y) = \log(a_2 + y)$, $g_3(y) = \frac{1}{a_2+y}$. A total of 200 Monte Carlo experiments were conducted to compare the separating performance of the algorithm with different nonlinear functions, and their average ISR under different number of receiving array elements were shown in Fig. 6.5a–c.

As can be seen from Fig. 6.5, the complex signal fast fixed-point algorithm can well separate the four source signals, which can not only separate the communication signals of different modulation modes, but also the communication signals of the same modulation mode. Signals with different carrier frequencies can be separated, and signals with the same carrier frequency can also be separated. When the number of receiving array elements increases, the separating performance of the algorithm will be improved. When ISR of the algorithm converges, the SNR required will be

Fig. 6.5 Performance curve of complex signal fast fixed point algorithm

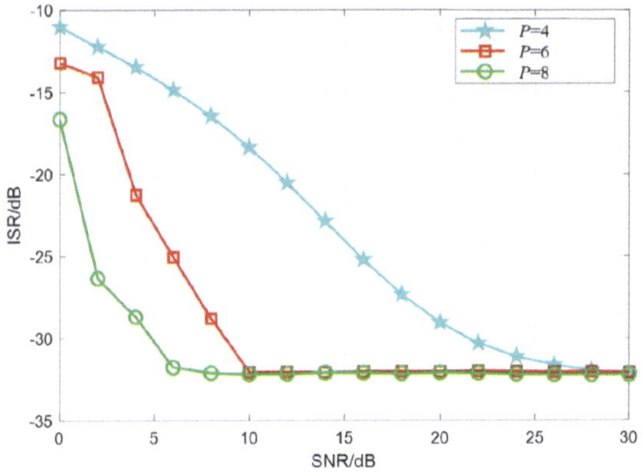

(a) Non-function $G_1(y)=\frac{1}{2}y^2$, $g_1(y)=y$

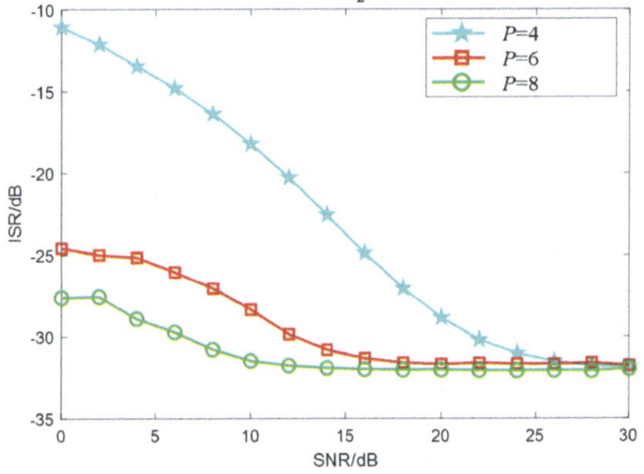

(b) Nonlinear function $G_2(y)=\sqrt{a_1+y}$, $g_2(y)=\frac{1}{2\sqrt{a_1+y}}$

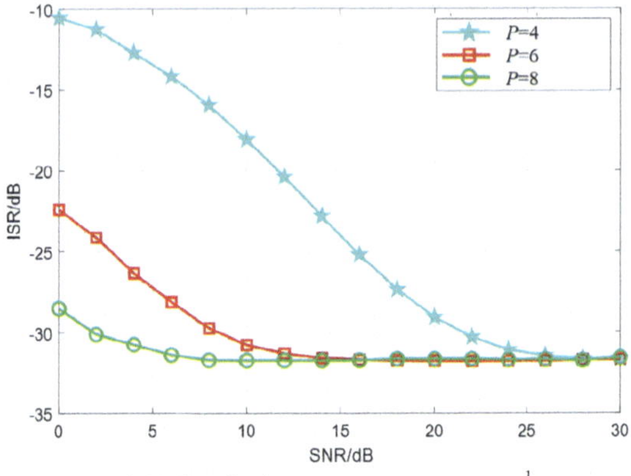

(c) Nonlinear function $G_3(y)=\log(a_2+y)$, $g_3(y)=\frac{1}{a_2+y}$

reduced. In addition, when the algorithm adopts three different nonlinear functions, the ISR after convergence is about −32 dB.

Summary

This chapter introduces the fast fixed-point algorithms in overdetermined blind source separation in detail, including the fast fixed-point algorithm based on kurtosis and the fast fixed-point algorithm based on negative entropy. The algorithm principle and implementation flow were deduced in detail, and then the multivariate fast fixed-point algorithm and fast fixed-point algorithm for complex signal were introduced. Finally, for speech signal and communication signal, the performances of the fast fixed-point algorithm suitable for real signal and complex signal are simulated and analyzed respectively. The experimental results show that both the fast fixed-point algorithms based on kurtosis and negative entropy have better separation performance.

Chapter 7
Overdetermined/Determined BSS Algorithm Based on Joint Diagonalization

The initial research of BSS algorithm based on joint diagonalization mainly focuses on the orthogonal joint diagonalization method. The main feature of this algorithm is that mixing matrix to be estimated is a (unitary) orthogonal matrix. It is necessary to whiten the observed signal first, then estimate the unitary orthogonal diagonalization factor matrix from the whitened observed signal, and realize the separating of the source signal. Whitening is the key technique of BSS algorithm based on orthogonal joint diagonalization. The whitening process can not only reduce the dimension of mixing matrix to be estimated, but also enhance the stability of the algorithm. The principles and methods of whitening algorithm are described in Chap. 4. This chapter mainly introduces the joint diagonalization algorithm based on second-order cumulant and fourth-order cumulant.

7.1 Joint Diagonalization BSS Principle Based on Second-Order Cumulants

The earliest BSS method based on second-order statistics is AMUSE (Algorithm for Multiple Unknown Source), proposed by L. Tong. This algorithm uses eigenvalue decomposition to obtain source signals. Then, another algorithm called SOBI (Second Order Blind Identification) appeared, which used the joint diagonalization of the delay correlation matrix to find the separating matrix and then get the source signals.

Second-order statistics include second-order cumulants and second-order moments, which are equal when the mean is zero. The second moment expression of the signal $\boldsymbol{x}$ is $E(\boldsymbol{x}\boldsymbol{x}^{\mathrm{T}})$, i.e. the covariance matrix, including the time delay covariance matrix. By linear transformation, it becomes a diagonal matrix. The inverse matrix of the transformation matrix $\boldsymbol{W}$ can be regarded as a mixing matrix $\boldsymbol{A}$, and $\boldsymbol{W}$ is the de-mixing matrix. There are many methods to obtain diagonal matrix, AMUSE method

W. Fu, *Principle and Application of Blind Source Separation Technology*,
https://doi.org/10.1007/978-981-96-0022-9_7

uses eigenvalue decomposition (EVD) method, but the results of this algorithm are not stable. A better way is to adopt the idea of joint diagonalization, which combines all the time-delayed covariance matrices into a set of matrices and diagonalizes the individual covariance matrices. Of course, it is impossible to fully diagonalize all the delay matrices, and generally only approximate diagonalization can be achieved. Set a threshold, as long as the sum of the squares of the non-diagonal elements is less than this threshold, it can be considered that the diagonalization is realized. Simulations prove that the algorithm is stable and can obtain accurate source signals.

In the blind source separation algorithm based on second-order statistics, the source signal is required to be uncorrelated both spatially and temporally, and the noise signal is uncorrelated with the source signal, the noise variance is much smaller than the variance of the source signal and the noise variance is equal, and the source signal is a second-order stationary signal.

Suppose the observed signal $\boldsymbol{x}(t)$ is:

$$\boldsymbol{x}(t) = \boldsymbol{A}\boldsymbol{s}(t) + \boldsymbol{v}(t) \tag{7.1}$$

First, the observed signal is preprocessed by whitening algorithm. The whitening matrix is $\boldsymbol{B}$, and the equivalent mixing matrix $\boldsymbol{U} = \boldsymbol{B}\boldsymbol{A}$ is an orthogonal matrix. The signal after whitening is:

$$\boldsymbol{z}(t) = \boldsymbol{B}\boldsymbol{A}\boldsymbol{s}(t) + \widetilde{\boldsymbol{v}}(t) = \boldsymbol{U}\boldsymbol{s}(t) + \widetilde{\boldsymbol{v}}(t) \tag{7.2}$$

Correlation matrices with a delay of $\tau \neq 0$ for $\boldsymbol{z}(t)$ are:

$$\begin{aligned} \boldsymbol{R}_{zz}(\tau) &= E\{\boldsymbol{z}(t)\boldsymbol{z}^{\mathrm{H}}(t+\tau)\} = E\{\boldsymbol{U}\boldsymbol{s}(t)\boldsymbol{s}^{\mathrm{H}}(t+\tau)\boldsymbol{U}^{\mathrm{H}}\} + E\{\widetilde{\boldsymbol{v}}(t)\widetilde{\boldsymbol{v}}^{\mathrm{H}}(t+\tau)\} \\ &= \boldsymbol{U}\boldsymbol{R}_{ss}(\tau)\boldsymbol{U}^{\mathrm{H}} + \boldsymbol{R}_{\widetilde{v}\widetilde{v}}(\tau) \end{aligned} \tag{7.3}$$

When the observed signal is a real signal, $\boldsymbol{R}_{zz}(\tau)$ is a set of real symmetric matrices. When the observation signal is complex signal, $\boldsymbol{R}_{zz}(\tau)$ is a set of complex conjugated symmetric matrices, also called Hermitian matrix, and a real symmetric matrix is a special case of Hermitian matrix. Since the noise is assumed to be independent Gaussian white noise, then $\boldsymbol{R}_{\widetilde{v}\widetilde{v}}(\tau) = \boldsymbol{0}$ when $\tau \neq 0$. And because the source signals are assumed to be independent of each other, $\boldsymbol{R}_{ss}(\tau)$ is a diagonal matrix for any $\tau \neq 0$, so it can be obtained from Eq. (7.3) that

$$\boldsymbol{R}_{ss}(\tau) = \boldsymbol{U}^{\mathrm{H}}\boldsymbol{R}_{zz}(\tau)\boldsymbol{U} \tag{7.4}$$

According to Eq. (7.4), $\boldsymbol{U}$ is an orthogonal matrix diagonalizing a set of delay correlation matrices $\boldsymbol{R}_{zz}(\tau_i)$ $(i = 1, 2, \ldots, L)$ of the whitened signal $\boldsymbol{z}(t)$. Therefore, if we can find such an orthogonal matrix $\hat{\boldsymbol{U}}$, it is an estimate of $\boldsymbol{U}$, and this process is called joint diagonalization. After the orthogonal matrix $\boldsymbol{U}$ is estimated, the de-mixing matrix (also known as the separating matrix) is

$$W = U^{\mathrm{H}} B \tag{7.5}$$

We know that for one or two Hermitian matrices, exact diagonalization can be achieved, but it is impossible to diagonalize multiple matrices exactly simultaneously, so the joint diagonalization mentioned in this chapter generally refers to approximate joint diagonalization.

7.2 Joint Diagonalization BSS Principle Based on Fourth-Order Cumulants

The fourth order cumulant of a signal vector z is defined as:

$$Q_z(i,j,k,l) = cum(z_i, z_j^*, z_k, z_l^*),\ \ i,j = 1,2,\ldots,n \tag{7.6}$$

where

$$\begin{aligned} cum(z_i, z_j^*, z_k, z_l^*) &= E(z_i z_j^* z_k z_l^*) - E(z_i z_j^*)E(z_k z_l^*) \\ &\quad - E(z_i z_k)E(z_j^* z_l^*) - E(z_i z_l^*)E(z_j^* z_k) \end{aligned} \tag{7.7}$$

In addition, for any $n \times n$ matrix M, define a cumulant matrix C

$$C = Q_Z(M) \tag{7.8}$$

The entries in row i and column j of the matrix C are as follows

$$c_{ij} = \sum_{k,l=1}^{n} Q_z(i,j,k,l) m_{lk}$$

If z is an observed signal processed by a robust whitening algorithm, it can be seen from (7.2) that:

$$z = \tilde{z} + \tilde{v} = Us + \tilde{v} \tag{7.9}$$

where U is the orthogonal matrix, $\tilde{v}$ is Gaussian white noise, then

$$\begin{aligned} cum(z_i, z_j^*, z_k, z_l^*) = cum\Bigg(&\sum_{p=1}^{n} U_{i,p} s_p + \tilde{v}_i, \sum_{p=1}^{n} U_{j,p}^* s_p^* \\ &+ \tilde{v}_j^*, \sum_{p=1}^{n} U_{k,p} s_p + \tilde{v}_k, \sum_{p=1}^{n} U_{l,p}^* s_p^* + \tilde{v}_l^* \Bigg) \end{aligned} \tag{7.10}$$

According to the property of cumulant and the characteristic that the fourth-order cumulant of Gaussian white noise is 0, we have

$$\begin{aligned} cum(z_i, z_j^*, z_k, z_l^*) &= \sum_{p=1}^{n} cum(\boldsymbol{U}_{i,p}\boldsymbol{s}_p, \boldsymbol{U}_{j,p}^*\boldsymbol{s}_p^*, \boldsymbol{U}_{k,p}\boldsymbol{s}_p, \boldsymbol{U}_{l,p}^*\boldsymbol{s}_p^*) + cum(\tilde{v}_i, \tilde{v}_j^*, \tilde{v}_k, \tilde{v}_l^*) \\ &= \sum_{p=1}^{n} \boldsymbol{U}_{ip}\boldsymbol{U}_{j,p}^*\boldsymbol{U}_{k,p}\boldsymbol{U}_{l,p}^* \cdot cum(\boldsymbol{s}_p, \boldsymbol{s}_p^*, \boldsymbol{s}_p, \boldsymbol{s}_p^*) \end{aligned} \tag{7.11}$$

Let $k_p = cum(\boldsymbol{s}_p, \boldsymbol{s}_p^*, \boldsymbol{s}_p, \boldsymbol{s}_p^*)$, according to Eq. (7.11), we get

$$c_{ij} = \sum_{k,l=1}^{n} \left\{ \sum_{p=1}^{n} k_p \boldsymbol{U}_{ip}\boldsymbol{U}_{j,p}^*\boldsymbol{U}_{k,p}\boldsymbol{U}_{l,p}^* \right\} m_{lk} = \sum_{p=1}^{n} \left\{ k_p \cdot \sum_{k,l=1}^{n} [\boldsymbol{U}_{k,p}\boldsymbol{U}_{l,p}^* m_{lk}] \cdot \boldsymbol{U}_{ip}\boldsymbol{U}_{j,p}^* \right\} \tag{7.12}$$

So

$$Q_Z(\boldsymbol{M}) = \sum_{p=1}^{n} k_p(\boldsymbol{u}_p^{\mathrm{H}}\boldsymbol{M}\boldsymbol{u}_p)\cdot\boldsymbol{u}_p\boldsymbol{u}_p^{\mathrm{H}} = \boldsymbol{U}\boldsymbol{\Lambda}_{\boldsymbol{M}}\boldsymbol{U}^{\mathrm{H}}, \quad \forall \boldsymbol{M} \tag{7.13}$$

where $\boldsymbol{u}_p$ is the p-th column of the orthogonal matrix $\boldsymbol{U}$, $\boldsymbol{\Lambda}_{\boldsymbol{M}} = diag(k_1\boldsymbol{u}_1^{\mathrm{H}}\boldsymbol{M}\boldsymbol{u}_1, k_2\boldsymbol{u}_2^{\mathrm{H}}\boldsymbol{M}\boldsymbol{u}_2, \ldots, k_n\boldsymbol{u}_n^{\mathrm{H}}\boldsymbol{M}\boldsymbol{u}_n)$. It can be seen from Eq. (7.13) that as long as an orthogonal matrix $\widehat{\boldsymbol{U}}$ is found, such that the matrix $\widehat{\boldsymbol{U}}^H Q_Z(\boldsymbol{M})\widehat{\boldsymbol{U}}$ is diagonal for any matrix $\boldsymbol{M}$, then $\widehat{\boldsymbol{U}}$ is the estimation of $\boldsymbol{U}$. The problem of estimating orthogonal matrices $\boldsymbol{U}$ is transformed into the problem of joint diagonalization of the set of matrices $Q_Z(\boldsymbol{M}_i)$ $(i = 1, 2, \ldots)$, where $\boldsymbol{M}_i (i = 1, 2, \ldots)$ are the multiple matrices selected. In the same way, it is an overdetermined problem to diagonalize multiple matrices exactly at the same time, so the approximate joint diagonalization method is generally used to deal with it.

Theorem *For an n dimensional complex random vector with fourth-order cumulants, there are* n^2 *real numbers* $\lambda_1, \lambda_2, \ldots, \lambda_{n^2}$ *and* n^2 *matrices* $\boldsymbol{M}_1, \boldsymbol{M}_2, \ldots \boldsymbol{M}_{n^2}$, *satisfy* $\boldsymbol{Q}_z(\boldsymbol{M}_i) = \lambda_i \boldsymbol{M}_i$.

Proof Set

$$(\tilde{\boldsymbol{Q}})_{ab} = cum(z_i, z_j^*, z_k, z_l^*) \tag{7.14}$$

where $a = i + (j-1) \cdot n$,$b = l + (k-1) \cdot n$, and define two operators $vec(\cdot)$ and $vec^{-1}(\cdot)$. $\tilde{\boldsymbol{m}}_i = vec(\boldsymbol{M}_i)$ represents the column vector of $n^2 \times 1$ that arrange the matrix $\boldsymbol{M}_i \in C^{n \times n}$ in columns, and $vec^{-1}(\cdot)$ represents the inverse of $vec(\cdot)$.

From the above definition and the formula (7.8) of the cumulant matrix, it can be seen that if $\tilde{\boldsymbol{q}}_z(\boldsymbol{M}_i) = vec[\boldsymbol{Q}_z(\boldsymbol{M}_i)]$, $\tilde{\boldsymbol{m}}_i = vec(\boldsymbol{M}_i)$, then $\tilde{\boldsymbol{q}}_z(\boldsymbol{M}_i) = \tilde{\boldsymbol{Q}}\tilde{\boldsymbol{m}}_i$. It is not difficult to see from formula (7.13) that $\tilde{\boldsymbol{Q}}$ is $n^2 \times n^2$ Hermitian matrix, therefore, the eigenvalue decomposition of $\tilde{\boldsymbol{Q}}$ is $\tilde{\boldsymbol{Q}} = \tilde{\boldsymbol{U}}\tilde{\boldsymbol{D}}\tilde{\boldsymbol{U}}^H$, Where $\tilde{\boldsymbol{U}} = [\tilde{\boldsymbol{u}}_1, \tilde{\boldsymbol{u}}_2, \ldots, \tilde{\boldsymbol{u}}_{n^2}]$ is a matrix composed of eigenvectors of $\tilde{\boldsymbol{Q}}$, and $\tilde{\boldsymbol{D}} = diag(\lambda_1, \lambda_2, \ldots, \lambda_{n^2})$ is a real diagonal matrix composed of its eigenvalues. Then

$$\tilde{\boldsymbol{Q}}\tilde{\boldsymbol{u}}_i = \lambda_i \tilde{\boldsymbol{u}}_i, \ (i = 1, 2, \ldots, n^2) \tag{7.15}$$

If $\tilde{\boldsymbol{m}}_i = \tilde{\boldsymbol{u}}_i$, that is $\boldsymbol{M}_i = vec^{-1}(\tilde{\boldsymbol{m}}_i) = vec^{-1}(\tilde{\boldsymbol{u}}_i)$, then $\tilde{\boldsymbol{q}}_z(\boldsymbol{M}_i) = \tilde{\boldsymbol{Q}}\tilde{\boldsymbol{m}}_i = \lambda_i \tilde{\boldsymbol{u}}_i$, so

$$\boldsymbol{Q}_z(\boldsymbol{M}_i) = vec^{-1}(\tilde{\boldsymbol{q}}_z(\boldsymbol{M}_i)) = vec^{-1}(\lambda_i \tilde{\boldsymbol{u}}_i) = \lambda_i vec^{-1}(\tilde{\boldsymbol{u}}_i) = \lambda_i \boldsymbol{M}_i \tag{7.16}$$

The theorem is proved [38].

According to the theorem, matrix $vec^{-1}(\tilde{\boldsymbol{u}}_i)$ composed of n eigenvectors corresponding to n largest eigenvalues of matrix $\tilde{\boldsymbol{Q}}$ can be chosen as $\boldsymbol{M}_i$. Then the estimate $\widehat{\boldsymbol{U}}$ of $\boldsymbol{U}$ can be obtained by approximating the joint diagonalization of n matrices $\lambda_i vec^{-1}(\tilde{\boldsymbol{u}}_i)$, $(i = 1, 2, \ldots, n)$.

7.3 Approximate Joint Diagonalization Method

Consider a set of target matrices $\{\boldsymbol{R}_k | \boldsymbol{R}_k \in \boldsymbol{C}^{N \times N}, k = 1, \ldots, K\}$ composed of K matrices. The purpose of joint diagonalization is to estimate a non-singular matrix $\boldsymbol{A} \in \boldsymbol{C}^{N \times N}$, $\boldsymbol{B} \in \boldsymbol{C}^{M \times N}$ such that one of the following two equations is true:

$$\boldsymbol{B}\boldsymbol{R}_k\boldsymbol{B}^{\mathrm{H}} = \boldsymbol{\Lambda}_k, k = 1, \ldots, K \tag{7.17}$$

$$\boldsymbol{R}_k = \boldsymbol{A}\boldsymbol{D}_k\boldsymbol{A}^{\mathrm{H}}, k = 1, \ldots, K \tag{7.18}$$

where $\boldsymbol{D}_k \in \boldsymbol{C}^{N \times N}$ and $\boldsymbol{\Lambda}_k \in \boldsymbol{C}^{N \times N}$ are diagonal matrix, $\boldsymbol{A}$ is called the mixing matrix and $\boldsymbol{B}$ is called the de-mixing matrix. When the matrix $\boldsymbol{A}$ and the de-mixing matrix $\boldsymbol{B}$ are orthogonal (unitary) matrices, the joint diagonalization BSS algorithm is called orthogonal joint diagonalization. When $\boldsymbol{A}$ and $\boldsymbol{B}$ are not orthogonal constraints, the joint diagonalization BSS algorithm is called non-orthogonal joint diagonalization.

The joint diagonalization [39] of multiple matrices is not exactly complete diagonalization, but as far as possible under certain criteria, and the so-called "as far as possible" refers to the case that the objective function reaches the minimum, which is regarded as diagonalization.

For the above two Eqs. (7.17) and (7.18), the corresponding objective function has the following two kinds.

1. **Objective Function Based on Least Square Fitting**

According to Eq. (7.16), in the precise joint diagonalization algorithm, both the mixing matrix $\boldsymbol{A}$ and the diagonal matrix $\boldsymbol{D}_k$ corresponding to each target matrix need to be estimated. In this case, the objective function based on least square fitting is as follows.

$$J_2(\boldsymbol{A}, \boldsymbol{D}_{\mathrm{k}}) = \sum_{k=1}^{K} \left\| \boldsymbol{R}_k - \boldsymbol{A}\boldsymbol{D}_k\boldsymbol{A}^{\mathrm{H}} \right\|_F^2 \tag{7.19}$$

It should be noted that the minimization step is performed not only on matrix $\boldsymbol{A}$, but also on a set of diagonal matrices $\boldsymbol{D}_i$, because they are also unknown. Therefore, this problem can be solved using the Alternating Least Squares (ALS) technique, which alternately minimizes one set of components and keeps the other set fixed. In particular, assume that at i-th iteration, $\boldsymbol{A}$ is obtained, the next step is to minimize $J_2(\boldsymbol{A}, \boldsymbol{D}_k)$ on $\boldsymbol{D}_k$.

2. **Objective Function Based on Non-diagonal Error Criterion**

From Eq. (7.17), an objective function based on the non-diagonal error criterion can be proposed, which is the most common objective function in matrix joint diagonalization. The specific description is as follows.

$$J_1(\boldsymbol{B}) = \sum_{k=1}^{K} \mathit{off}\,(\boldsymbol{B}\boldsymbol{R}_k\boldsymbol{B}^{\mathrm{H}}) \tag{7.20}$$

$$\mathit{off}\,(\boldsymbol{B}\boldsymbol{R}_k\boldsymbol{B}^{\mathrm{H}}) = \sum_{i=1}^{N} \sum_{\substack{j=1 \\ i \neq j}}^{M} (\boldsymbol{B}\boldsymbol{R}_k\boldsymbol{B}^{\mathrm{H}})_{ij}^2 \tag{7.21}$$

It is obviously $\boldsymbol{B} = \boldsymbol{0}$ is a minimal solution of it, but it is not the desired solution, so the above criteria need to be constrained to avoid falling into the situation of degenerate solution.

JADE and SOBI algorithm constrain the de-mixing matrix $\boldsymbol{B}$ to orthogonal matrix in order to avoid degenerate solution. They first pre-whiten the observed signals, then set the de-mixing matrix to be the product of Givens matrices or Jacobi matrices, and find the orthogonal matrix $\boldsymbol{B}$ by Givens rotation or Jacobi rotation. In this case let $\boldsymbol{U} = \boldsymbol{B}$, the problem of estimating orthogonal matrices $\boldsymbol{U}$ can be converted to the minimization problem of the following cost functions:

$$\tilde{J}(\boldsymbol{U}) = \sum_{k=1}^{K} \mathit{off}\,\{\boldsymbol{U}^{\mathrm{T}}\boldsymbol{R}_k\boldsymbol{U}\} \tag{7.22}$$

where $off\{\boldsymbol{M}\} = \sum_{i \neq j} |m_{ij}|^2$.

The following describes a common method of Givens rotation.

Givens rotation is the basis of Jacobi algorithm, it is a method to achieve orthogonal normalization of a random vector through iterative steps in accordance with certain optimization requirements, through repeating a series of coordinate plane rotation to achieve the purpose of orthogonal normalization. The following describes the joint diagonalization algorithm using continuous Givens rotations.

Consider joint diagonalize K 2×2 submatrices,

$$\boldsymbol{H}_k = \begin{pmatrix} a_k & b_k \\ c_k & d_k \end{pmatrix}, k = 1, \ldots, K \tag{7.23}$$

our aim is to find a unitary matrix $\boldsymbol{U}$ such that $\boldsymbol{H}'_k = \boldsymbol{U}^{\mathrm{T}}\boldsymbol{H}_k\boldsymbol{U}, k = 1, \ldots, K$ can minimize the cost function defined by Eq. (7.22).

The unitary transformation $\boldsymbol{U}$ is parameterized with complex Givens rotation as follows.

$$\boldsymbol{U} = \begin{pmatrix} \cos(\theta) & e^{j\phi}\sin(\theta) \\ -e^{-j\phi}\sin(\theta) & \cos(\theta) \end{pmatrix} \tag{7.24}$$

If the coefficients of the matrix $\boldsymbol{H}'_k$ are denoted as a'_k, b'_k, c'_k, d'_k, then the minimization of Eq. (7.22) is to maximize $\sum_{k=1}^{K} |a'_k|^2 + |d'_k|^2$ by finding θ and Φ. Note $2\left(|a'_k|^2 + |d'_k|^2\right) = |a'_k - d'_k|^2 + |a'_k + d'_k|^2$, and the trace $a'_k + d'_k$ is invariant in the unitary transformation, so the minimization of the cost function of Eq. (7.22) is equivalent to taking the maximum value of Q for each Givens rotation.

$$Q = \sum_{k=1}^{K} |a'_k - d'_k|^2 \tag{7.25}$$

Easy to verify that

$$\begin{aligned} a'_k - d'_k = {} & (a_k - d_k)\cos(2\theta) - (b_k + c_k)\sin(2\theta)\cos(\Phi) \\ & - j(c_k - b_k)\sin(2\theta)\sin(\Phi), \quad k = 1, \ldots, K \end{aligned} \tag{7.26}$$

If defined

$$\boldsymbol{v}^T = [a'_1 - d'_1, \ldots, a'_K - d'_K] \tag{7.27}$$

$$\boldsymbol{u}^T = [\cos(2\theta), -\sin(2\theta)\cos(\phi), -\sin(2\theta)\sin(\phi)] \tag{7.28}$$

$$g_k^T = [a_k - d_k, b_k + c_k, j(c_K - b_K)] \tag{7.29}$$

Then the K equations in Eq. (7.26) can be written as $\boldsymbol{v} = \boldsymbol{Gu}$, where

$$\boldsymbol{G}^T = [\boldsymbol{g}_1, \ldots, \boldsymbol{g}_K] \tag{7.30}$$

Therefore, the formula (7.25) can be rewritten as

$$Q = \boldsymbol{v}^{\mathrm{H}}\boldsymbol{v} = \boldsymbol{u}^{\mathrm{T}}\boldsymbol{G}^{\mathrm{H}}\boldsymbol{Gu} = \boldsymbol{u}^{\mathrm{T}}Re(\boldsymbol{G}^{\mathrm{H}}\boldsymbol{G})\boldsymbol{u} \tag{7.31}$$

The following fact is used: $\boldsymbol{G}^{\mathrm{H}}\boldsymbol{G}$ is Hermitian matrix, whose imaginary part is antisymmetric, and it has no contribution to the above quadratic form.

It is easy to see from the structure of Eq. (7.28) that $\boldsymbol{u}^{\mathrm{T}}\boldsymbol{u} = 1$. So maximize Q of Eq. (7.31) is to take $\boldsymbol{u}$ as the eigenvector corresponding to the largest eigenvalue of the symmetric matrix $Re(\boldsymbol{G}^H\boldsymbol{G})$. Since $Re(\boldsymbol{G}^H\boldsymbol{G})$ is a real 3×3 symmetric matrix, the analytical representation of the Givens rotation parameter can be derived directly from the coordinates of the eigenvector.

Suppose that the eigenvector corresponding to the maximum eigenvalue of the symmetric matrix $Re(\boldsymbol{G}^{\mathrm{H}}\boldsymbol{G})$ is $\boldsymbol{u}$, then according to Eq. (7.27), we get

$$\cos(\theta) = \sqrt{\frac{1 + \boldsymbol{u}(1)}{2}},\ e^{j\phi}\sin(\theta) = 0.5 \times \frac{-\boldsymbol{u}(2) - j \cdot \boldsymbol{u}(3)}{\cos(\theta)} \tag{7.32}$$

The joint diagonalization method of 2×2 matrix is analyzed above. When the dimensionality of the matrix to be diagonalized is greater than 2, it can be iterated multiple times. Take out a 2×2 submatrix each time, calculate the corresponding Givens rotation matrix according to the above method, rotate the matrix that needs to be diagonalized, get a new matrix group to be diagonalized, and then take out another 2×2 submatrix. Repeat the above process.

The selection principle of the submatrix is as follows. First calculate the sum of all the non-diagonal elements of the matrix group $\boldsymbol{R}_k(k = 1, 2, \ldots, K)$ to be diagonalized, and then find the row i and column j corresponding to the largest non-diagonal element, then the 2×2 submatrixs $\boldsymbol{H}_k$ are as follows.

$$\boldsymbol{H}_k = \begin{bmatrix} \boldsymbol{R}_k(i, i) & \boldsymbol{R}_k(i, j) \\ \boldsymbol{R}_k(j, i) & \boldsymbol{R}_k(j, j) \end{bmatrix}, k = 1, 2, \ldots, K \tag{7.33}$$

where $\boldsymbol{R}_k(i, j)$ represents the value of the element in row i and column j of the matrix $\boldsymbol{R}_k$.

After obtaining the submatrix group $\boldsymbol{H}_k$, $Re(\boldsymbol{G}^{\mathrm{H}}\boldsymbol{G})$ is calculated according to (7.23), (7.29) and (7.30), then the eigenvalue decomposition of $Re(\boldsymbol{G}^H\boldsymbol{G})$ is performed to find the eigenvector $\boldsymbol{u}$ corresponding to the maximum eigenvalue of $Re(\boldsymbol{G}^{\mathrm{H}}\boldsymbol{G})$, and then $\cos(\theta)$ and $e^{j\phi}\sin(\theta)$ are calculated according to (7.32), and 2×2 Givens rotation matrix $\boldsymbol{U}$ is obtained according to (7.24).

If the rotation matrix corresponding to the first iteration is $\boldsymbol{P}_{(1)}$, then

$$\boldsymbol{P}_{(1)} = \begin{bmatrix} 1 & 0 & \cdots & \cdots\cdots\cdots\cdots & \cdots & \cdots & 0 \\ 0 & 1 & 0 & \cdots\cdots\cdots\cdots & \cdots & \cdots & 0 \\ \vdots & \vdots & 1 & \cdots \vdots \cdots\cdots & \cdots & \cdots & 0 \\ 0 & \cdots & \cos(\theta) & 0 \cdots 1 \cdots & e^{j\phi}\sin(\theta) & \cdots & 0 \\ \vdots & \cdots & \vdots & \vdots \cdots \vdots \cdots & \vdots & \cdots & \vdots \\ 0 & \cdots & e^{j\phi}\sin(\theta) & 0 \cdots 1 \cdots & \cos(\theta) & \cdots & 0 \\ \vdots & \cdots & \vdots & \vdots \cdots \vdots \cdots & \vdots & \cdots & \vdots \\ 0 & \cdots & \vdots & \vdots \cdots \vdots \cdots & \vdots & \cdots & 1 \end{bmatrix} \tag{7.34}$$

$\boldsymbol{P}_{(1)}$ is the matrix obtained by replacing the (i, i), (i, j), (j, i), (i, j) elements of an $\boldsymbol{M}$-dimensional identity matrix $\boldsymbol{I}$ with four elements of $\boldsymbol{U}$.

The matrix group $\boldsymbol{R}_k(k = 1, 2, \ldots, K)$ is rotated by the rotation matrix $\boldsymbol{P}_{(1)}$ to obtain a new matrix group $\boldsymbol{R}'_k(k = 1, 2, \ldots, K)$.

$$\boldsymbol{R}'_k = \boldsymbol{P}^{\mathrm{T}}_{(1)}\boldsymbol{R}_k\boldsymbol{P}_{(1)}, \ (k = 1, 2, \ldots, K) \tag{7.35}$$

The above process is repeated until the sum of the non-diagonal elements of the matrix group is less than a certain threshold value, then the iteration stops, and the matrix group obtained is the matrix group after joint approximation diagonalization. The obtained joint diagonalized orthogonal matrix $\widehat{\boldsymbol{G}}$ is as follows.

$$\widehat{\boldsymbol{G}} = \boldsymbol{P}_{(1)}\boldsymbol{P}_{(2)}\ldots\boldsymbol{P}_{(L)} \tag{7.36}$$

where $\boldsymbol{P}_{(l)}, l = 1, 2, \ldots, L$ is the rotation matrix at l-th iteration.

7.4 Steps of BSS Algorithm Based on Joint Diagonalization

1. **Steps of Joint Diagonalization BSS Algorithm Based on Second-order Cumulant Matrix**

The joint diagonalization BSS algorithm based on the second-order cumulative matrix is summarized as follows:

Step 1: Whiten the observed signal $\boldsymbol{x}$ according to the method introduced in Chap. 4, the whitening matrix is $\boldsymbol{B}$, the signal after whitening is $\boldsymbol{z} = \boldsymbol{Bx}$.

Step 2: Find a set of delay correlation matrices $\boldsymbol{R}_{zz}(\tau_i)$ $(i = 1, 2, \ldots, L)$ for $\boldsymbol{z}$.

Step 3: Joint diagonalization of L delayed correlation matrices is carried out using Givens rotation method given in section 7.3 to obtain the orthogonal matrix $\overset{\frown}{\boldsymbol{G}}$.

Step 4: Estimate the source signal $\hat{\boldsymbol{s}} = \overset{\frown}{\boldsymbol{G}}^{-1}\boldsymbol{z} = \overset{\frown}{\boldsymbol{G}}^{\mathrm{H}}\boldsymbol{Bx}$, i.e., the separating matrix $\boldsymbol{W} = \overset{\frown}{\boldsymbol{G}}^{\mathrm{H}}\boldsymbol{B}$.

2. **Steps of Joint Diagonalization BSS Algorithm Based on Fourth-order Cumulant Matrix**

In summary, the joint diagonalization BSS algorithm based on the fourth-order cumulative matrix is summarized as follows.

Step 1: Whiten the observed signal $\boldsymbol{x}$ according to the method introduced in Chap. 4, the whitening matrix is $\boldsymbol{B}$, the signal after whitening is $\boldsymbol{z} = \boldsymbol{Bx}$.

Step 2: Find the fourth order cumulant for $\boldsymbol{z}$, and map it to the $n^2 \times n^2$ matrix $\tilde{\boldsymbol{Q}}$ according to the rule of Eq. (7.14).

Step 3: Perform eigenvalue decomposition on $\tilde{\boldsymbol{Q}}$, obtain n largest eigenvalues $\tilde{d}_i$ of $\tilde{\boldsymbol{Q}}$ and corresponding eigenvector $\tilde{\boldsymbol{u}}_i$, $i = 1, 2, \ldots, n$.

Step 4: According to Eq. (7.16), obtain n matrices $\boldsymbol{Q}_z(\boldsymbol{M}_i)$, $(i = 1, 2, \ldots, n)$, and then use the Givens rotation method given in Sect. 7.3 to joint diagonalize these n matrices and obtain the orthogonal matrix $\overset{\frown}{\boldsymbol{G}}$.

Step 5: Estimate the source signal $\hat{\boldsymbol{s}} = \overset{\frown}{\boldsymbol{G}}^{-1}\boldsymbol{z} = \overset{\frown}{\boldsymbol{G}}^{\mathrm{H}}\boldsymbol{Bx}$, i.e., the separating matrix $\boldsymbol{W} = \overset{\frown}{\boldsymbol{G}}^{\mathrm{H}}\boldsymbol{B}$.

7.5 Computer Simulation Results

7.5.1 Simulation of Second-order Cumulant Joint Diagonalization Algorithm

Suppose there are four source signals, namely a BPSK signal with carrier frequency of 650 kHz, a QPSK signal with carrier frequency of 710 kHz, an 8PSK signal with carrier frequency of 770 kHz, and a 2ASK signal with carrier frequency of 610 kHz. The incident angles are 40°, 60°, 80°, 100°. The receiving antenna array is a uniform circular array of 6 elements, the radius of which is half of wavelength. Assume that the noise is Gaussian white noise, and SNR = 20 dB. After whitening of 6 observed signals, a group of delay correlation matrices of the whitened signals is obtained, and then Givens rotation method is used to approximate joint diagonalization of the matrix, so as to estimate the separation matrix. Assume the global matrix is $\boldsymbol{G} = \boldsymbol{UBA} = \boldsymbol{WA}$, and $\boldsymbol{W} = \boldsymbol{UB}$ is the total separating matrix.

Here we take the number of matrices $L = 20$, delay $\tau_i = i\,(i = 1, 2, \ldots, L)$, the unit is sampling point. The global matrix obtained is

$$G = \begin{bmatrix} -0.0016 + 0.0050i & 0.8740 + 0.4842i & 0.0009 + 0.0014i & -0.0003 - 0.0015i \\ 0.0063 + 0.0034i & 0.0021 + 0.0045i & -0.4573 + 0.8903i & -0.0014 + 0.0012i \\ -0.9987 + 0.0335i & -0.0019 - 0.0042i & -0.0010 + 0.0042i & 0.0142 + 0.0002i \\ -0.0086 - 0.0096i & 0.0012 - 0.0072i & 0.0010 - 0.0009i & -0.4914 + 0.8442i \end{bmatrix}$$

the module of G is

$$|G| = \begin{bmatrix} 0.0053 & 0.9992 & 0.0017 & 0.0015 \\ 0.0072 & 0.0050 & 1.0009 & 0.0018 \\ 0.9993 & 0.0046 & 0.0043 & 0.0142 \\ 0.0129 & 0.0073 & 0.0013 & 0.9768 \end{bmatrix}$$

Figure 7.1 shows the relationship between the number of iterations of SOBI algorithm and ISR. We can see that ISR can reach −38 dB after convergence. In addition, it can also be seen from G that the global matrix is very close to the quasi identity matrix with only one non-zero element in each row and column, so the source signals can be separated well.

In order to visually see the separation effect, waveform of source signals, separated signals and observed signals are shown in Figs. 7.2, 7.3 and 7.4 respectively, from which the effectiveness of the algorithm can be visually seen.

Figure 7.5 shows the convergence effect of the algorithm under different SNR conditions. As can be seen from the figure, when SNR is 6 dB, ISR reaches –30 dB.

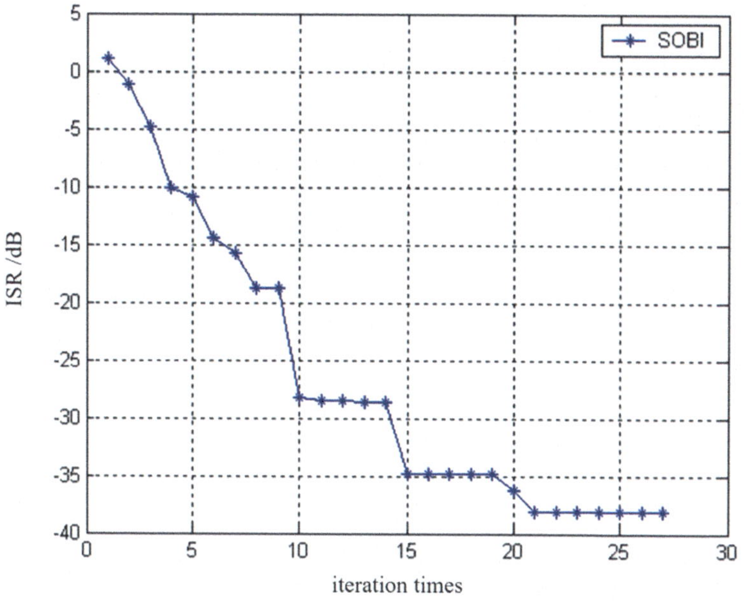

Fig. 7.1 The relationship between the number of iterations of SOBI separation algorithm and ISR (Number of matrices L = 20)

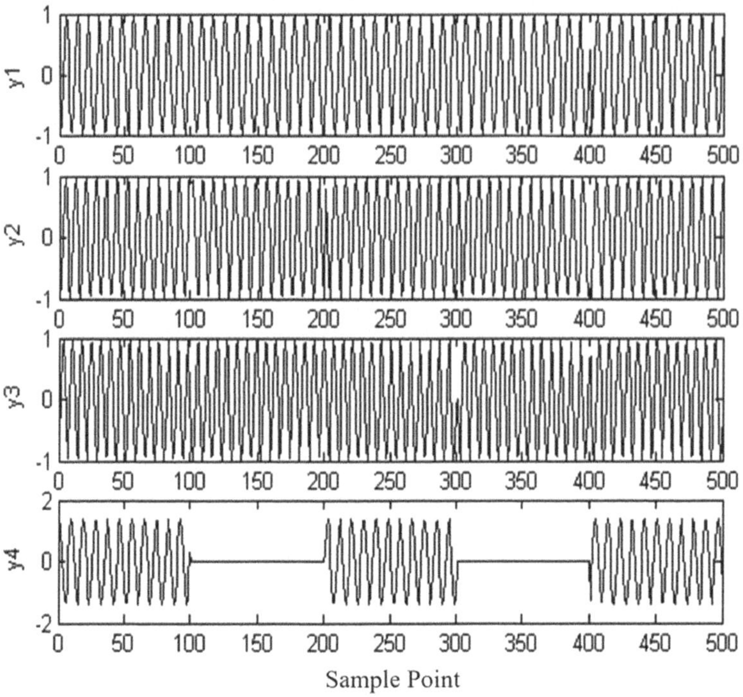

Fig. 7.2 Waveform diagram of source signal

It should be noted that SOBI algorithm can only use quasi-orthogonal whitening algorithm (or robust whitening algorithm) when whitening the observed signal, otherwise if the standard whitening algorithm is used, the whitening matrix does not meet the orthogonal conditions in a noisy environment, and the performance of the algorithm will deteriorate.

Figure 7.6 shows the separating performance of SOBI algorithm with two whitening algorithms. When the robust whitening algorithm is adopted, its separating performance is very good, and ISR reaches −25 dB when the SNR is 4 dB. When the standard whitening algorithm is used, the separating effect is much worse.

7.5.2 Simulation of Fourth-Order Cumulant Joint Diagonalization Algorithm

Assume that the four source signals are (these four signals are located in a 200 kHz bandwidth with an intermediate frequency of 70 MHz): A square 4QAM signal with carrier frequency of 65 MHz, a QPSK signal with carrier frequency of 61 MHz, an 8PSK signal with carrier frequency of 67 MHz, and a square 16QAM signal with

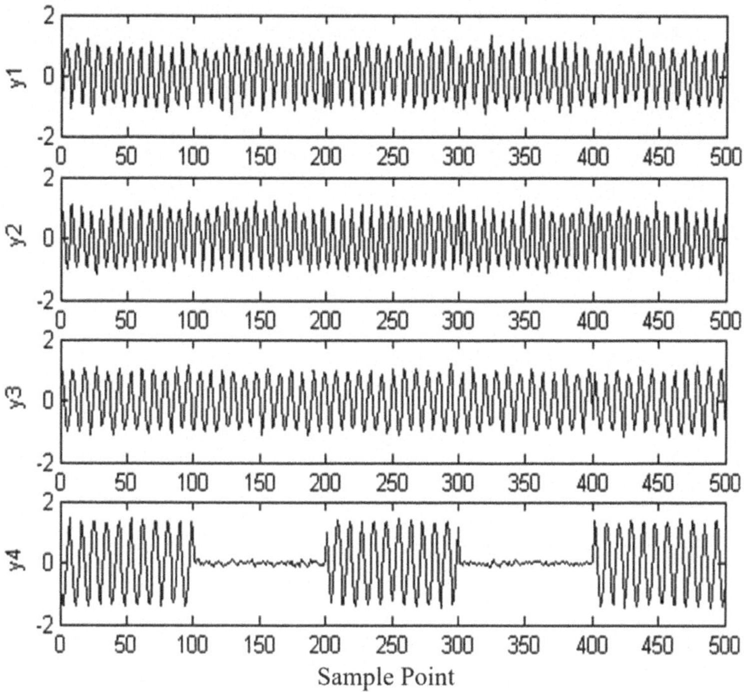

Fig. 7.3 Waveforms of separated signals

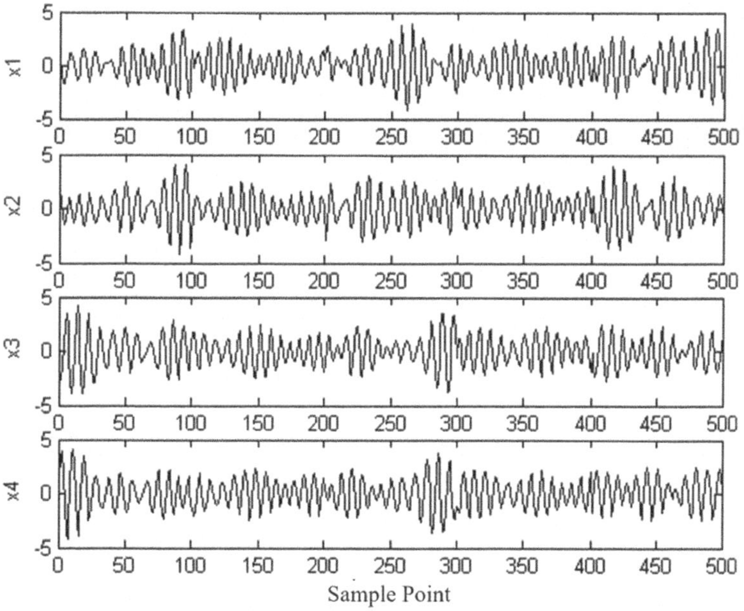

Fig. 7.4 Waveforms of the first four observed signals

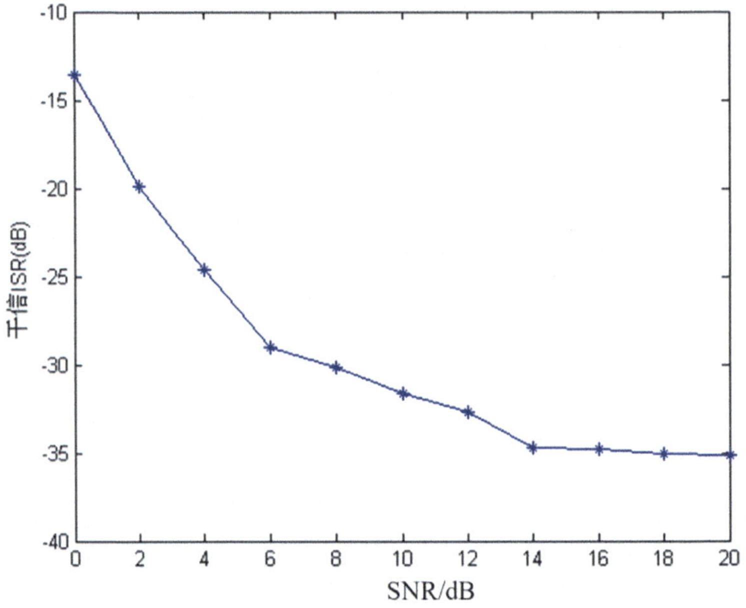

Fig. 7.5 Relationship between the ISR and SNR after SOBI algorithm convergence

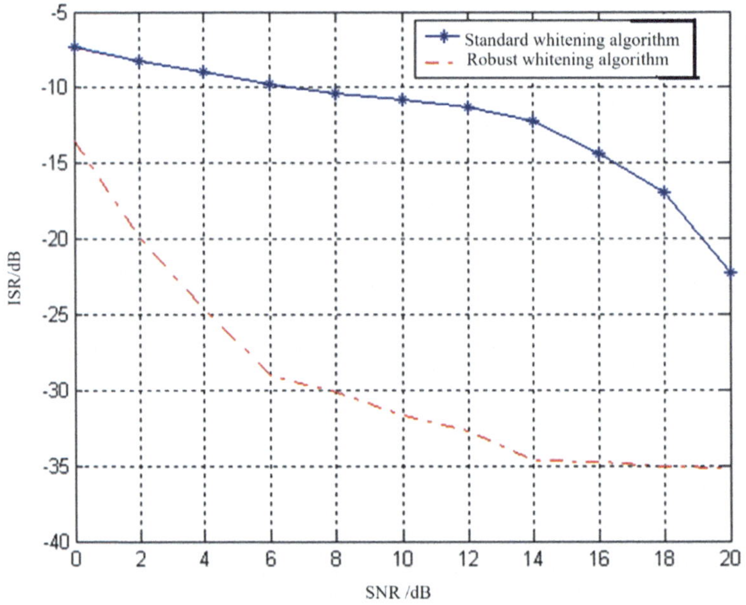

Fig. 7.6 Separating performance of SOBI algorithm with different whitening algorithms

carrier frequency of 69 MHz are received by a uniform circular array of 6 elements with an array elements interval of half wavelength. The noise is Gaussian white noise, and the SNR is 25 dB. The fourth order cumulant joint diagonalization algorithm introduced in this paper is used to separate the observed signals. After the algorithm converges, the ISR is −41 dB, and the global mixing matrix is:

$$\boldsymbol{G} = \begin{bmatrix} -0.0000 + 0.0005i & 0.0048 - 0.0025i & -0.0067 - 0.0005i & -0.5320 + 0.8410i \\ 0.0128 - 0.0126i & 0.8181 - 0.5777i & 0.0000 + 0.0002i & 0.0012 - 0.0031i \\ -0.7405 + 0.6728i & 0.0150 - 0.0105i & -0.0009 - 0.0021i & 0.0000 + 0.0006i \\ 0.0007 - 0.0009i & 0.0005 - 0.0010i & -0.6399 + 0.7691i & -0.0009 - 0.0011i \end{bmatrix}$$

It can be seen from the global matrix $\boldsymbol{G}$ that the algorithm can effectively separate the source signal. The time domain waveform of the first four observed signals, the original signals and the four separated signals are shown in Figs. 7.7, 7.8 and 7.9 respectively. It can be seen more intuitively that the algorithm effectively separates the four digital communication signals.

Figure 7.10 shows the baseband constellation of the separated signal, from which it can also be seen that the constellation of the separated signal has been restored.

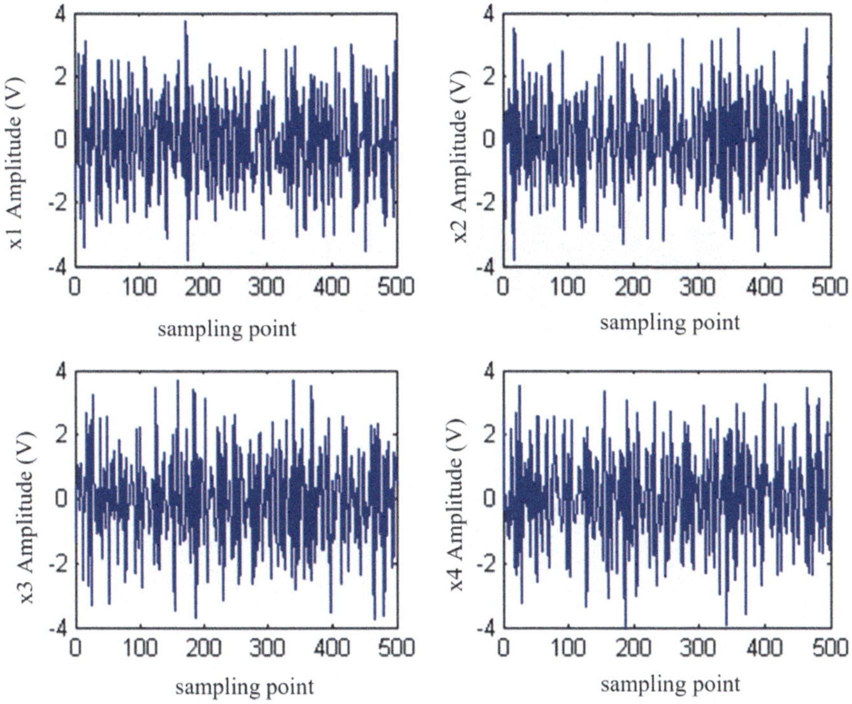

Fig. 7.7 Time domain waveform of the first four observed signals (first 500 sampling points)

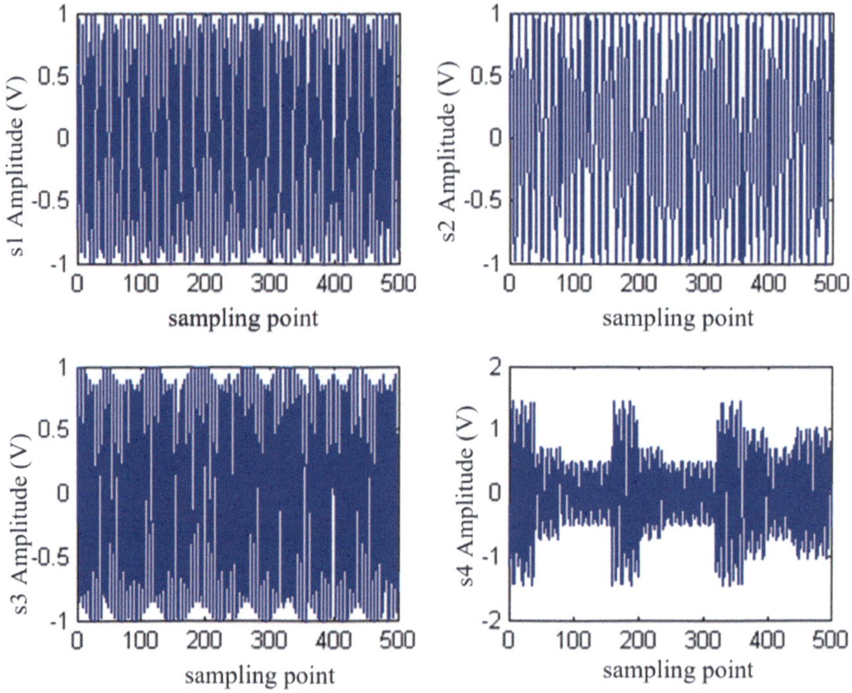

Fig. 7.8 Time domain waveform of four source signals (first 500 sampling points)

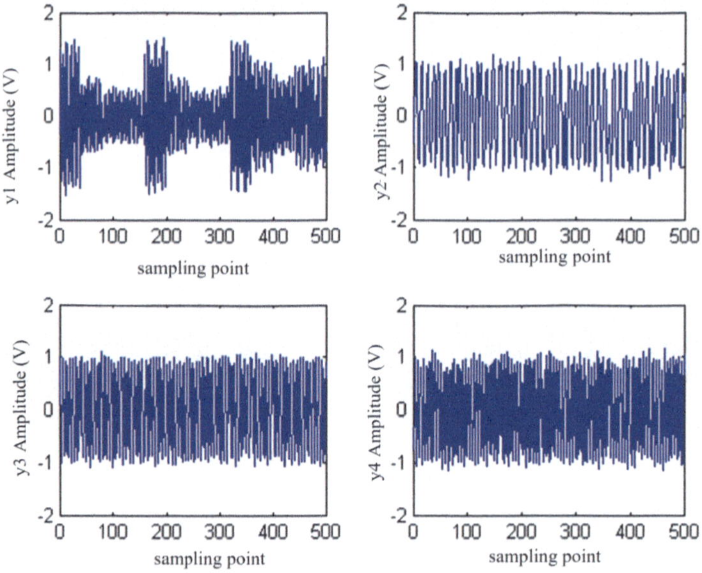

Fig. 7.9 Time domain waveform of separated signals (first 500 sampling points)

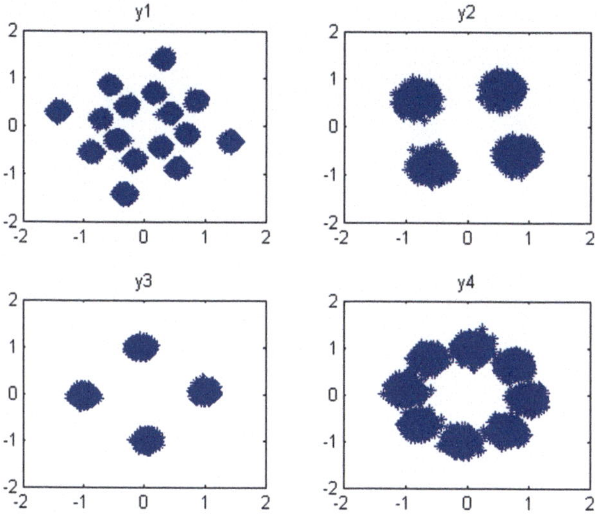

Fig. 7.10 Constellation of separated signals

Summary

This chapter focuses on the BSS algorithm based on joint diagonalization. First, the principle of joint diagonalization BSS algorithm based on second-order cumulant and fourth-order cumulant is introduced, and then the specific joint diagonalization method is given. Finally, the steps of joint diagonalization BSS algorithm based on second-order cumulant and fourth-order cumulant are summarized and the simulation results of two algorithms are given.

Chapter 8
Overdetermined Convolutional Blind Source Separation Techniques

The process of convolutional mixing blind source separation (BSS) is similar to that of the instantaneous BSS and the difference is that the convolutional mixing model needs to obtain a separation filter matrix, while the instantaneous mixing model only needs to recover the unmixing matrix and then obtain the separated signal.

At the present stage, the methods of convolutional mixing blind source separation can be divided into time-domain method and frequency-domain method.

The basic idea of the time-domain method is to compute and estimate a separation filter in the time domain with the same or larger order of the mixing filter. When the mixing filter order is small, the order of corresponding separation filter is also small and the computation amount is relatively small. However, in the real environment, the order of the mixing filter is often very large. For example, to simulate the convolutional mixing of speech signals in a small space, the order of the mixing filter has reached several hundred orders. At this point, if you want to get a better separating effect, the separating filter should also be in a larger order, which makes the coefficients involved in the calculation too much, the computational cost is heavy, and the coefficients of the separating filter are also difficult to converge.

Compared to the time-domain method, the frequency-domain method transforms the received signals to the frequency-domain and then processes them, and its basic idea is to convert the convolutional mixing model in the time-domain to the linear instantaneous mixing model in the frequency-domain. The advantage of the frequency-domain approach is that the mixed signals converted to the frequency-domain can be separated by using several mature instantaneous BSS algorithms. Also, the computational amount of the frequency-domain BSS algorithms is greatly reduced compared to the time-domain algorithms in the case of larger filter orders.

In this chapter, the time-domain convolutional BSS algorithm and the frequency-domain convolutional BSS algorithm are introduced respectively.

W. Fu, *Principle and Application of Blind Source Separation Technology*,
https://doi.org/10.1007/978-981-96-0022-9_8

8.1 Time-Domain Convolutional Blind Source Separation Algorithm

There are two time-domain algorithms for convolutional mixing blind source separation. One is to separate the received signal directly without model transformation. The main idea is that the convolutional mixed signal is deconvolved by a FIR filter, and then the separated signal is obtained. The mathematical model of this separating system is as follows:

$$\boldsymbol{Y}(t) = \sum_{p=0}^{P} \boldsymbol{W}_p(p)\boldsymbol{X}(t-p) \quad (P > L) \tag{8.1}$$

where each element of matrix $\boldsymbol{W}_p$ is the impulse response of the filter and P is the length of demixing filter. If $P = 0$, then the observed signal at each moment t is only related to the source signal at the same moment, and then the convolutional mixing BSS problem degenerates into an instantaneous mixing BSS problem. This type of algorithm is more difficult to model the signal as the channel order increases, the algorithm becomes unstable, and the separation effect deteriorates.

Another important convolutional mixing time-domain BSS method uses sliding window W to transform the source signal vector, the observed signal vector, and the noise signal vector in the time domain as follows:

$$\tilde{\boldsymbol{X}}(t) = [\tilde{\boldsymbol{x}}_1(t), \tilde{\boldsymbol{x}}_2(t), \ldots, \tilde{\boldsymbol{x}}_M(t)]^{\mathrm{T}} \tag{8.2}$$

$$\tilde{\boldsymbol{V}}(t) = [\tilde{\boldsymbol{v}}_1(t), \tilde{\boldsymbol{v}}_2(t), \ldots, \tilde{\boldsymbol{v}}_M(t)]^{\mathrm{T}} \tag{8.3}$$

$$\tilde{\boldsymbol{S}}(t) = [\tilde{\boldsymbol{s}}_1(t), \tilde{\boldsymbol{s}}_2(t), \ldots, \tilde{\boldsymbol{s}}_N(t)]^{\mathrm{T}} \tag{8.4}$$

where

$$\tilde{\boldsymbol{x}}_j(t) = [\boldsymbol{x}_j(t), \boldsymbol{x}_j(t-1), \ldots, \boldsymbol{x}_j(t-W+1)]^{\mathrm{T}},\ j = 1, 2, \ldots, M \tag{8.5}$$

$$\tilde{\boldsymbol{v}}_j(t) = [\boldsymbol{v}_j(t), \boldsymbol{v}_j(t-1), \ldots, \boldsymbol{v}_j(t-W+1)]^{\mathrm{T}},\ j = 1, 2, \ldots, M \tag{8.6}$$

$$\tilde{\boldsymbol{s}}_i(t) = [\boldsymbol{s}_i(t), \boldsymbol{s}_i(t-1), \ldots, \boldsymbol{s}_i(t-(L+W)+1)]^{\mathrm{T}},\ i = 1, 2, \ldots, N \tag{8.7}$$

the convolutional mixing model is transformed into an instantaneous mixing model

$$\tilde{\boldsymbol{X}}(t) = \tilde{\boldsymbol{A}}\tilde{\boldsymbol{S}}(t) \tag{8.8}$$

where $\tilde{\boldsymbol{A}} \in \mathbb{R}^{MW \times N(L+W)}$ is the equivalent mixing matrix, which is defined as

$$\tilde{A} = \begin{bmatrix} A_{11} & \cdots & A_{1N} \\ \vdots & \ddots & \vdots \\ A_{M1} & \cdots & A_{MN} \end{bmatrix}, A_{ji} = \begin{bmatrix} a_{ji}(0) & \cdots & a_{ji}(L) & \cdots & 0 \\ \vdots & \ddots & \ddots & \ddots & \vdots \\ 0 & \cdots & a_{ji}(0) & \cdots & a_{ji}(L) \end{bmatrix} \tag{8.9}$$

In order to realize the time-domain algorithm for convolutional BSS, it is also necessary to satisfy the following assumptions: (1) the number of observed signals is greater than or equal to the number of source signals, which means, for the convolutional mixing model, the equivalent mixing matrix $\tilde{A}$ is required to be columns full rank and satisfy $MW \geq N(L + W)$; (2) the channel impulse response is a causal finite-length sequence, which ensures that the convolutional system can be recognized.

A common convolutional mixing BSS time-domain algorithm is the joint block diagonalization algorithm, the basic principle of which is to change the time-delay autocorrelation matrix of the observed signals into the block diagonal element dominatine matrix.

Assuming that the extended delay autocorrelation matrix of the source signal vector $\tilde{S}(t)$ at time-delay τ can be written as

$$\begin{aligned} R_{\tilde{S}}(t,\tau) &= \mathrm{E}\left\{\tilde{S}(t)\tilde{S}^{\mathrm{H}}(t+\tau)\right\} \\ &= \mathrm{E}\left\{[\tilde{s}_1(t), \tilde{s}_2(t), \ldots, \tilde{s}_N(t)]^{\mathrm{H}}[\tilde{s}_1(t+\tau), \tilde{s}_2(t+\tau), \ldots, \tilde{s}_N(t+\tau)]\right\} \\ &= \begin{bmatrix} R_{\tilde{s}_{11}}(t,\tau) & R_{\tilde{s}_{12}}(t,\tau) & \cdots & R_{\tilde{s}_{1N}}(t,\tau) \\ R_{\tilde{s}_{21}}(t,\tau) & R_{\tilde{s}_{22}}(t,\tau) & \cdots & R_{\tilde{s}_{2N}}(t,\tau) \\ \vdots & \vdots & \ddots & \vdots \\ R_{\tilde{s}_{N1}}(t,\tau) & R_{\tilde{s}_{N2}}(t,\tau) & \cdots & R_{\tilde{s}_{NN}}(t,\tau) \end{bmatrix} \end{aligned} \tag{8.10}$$

where $R_{\tilde{s}_{ij}}(t,\tau)$ denotes the delay correlation matrix between $\tilde{s}_i$ and $\tilde{s}_j$. Assuming that the source signals are statistically independent of each other, we have

$$R_{\tilde{s}_{ij}}(t,\tau)\begin{cases} \neq \mathbf{0}, & i = j \\ = \mathbf{0}, & i \neq j \end{cases} \tag{8.11}$$

According to Eq. (8.11), $R_{\tilde{S}}(t,\tau)$ is a block diagonal matrix.

The delay autocorrelation matrix is calculated according to the observed signal $\tilde{X}(t)$ to obtain $R_{\tilde{X}}(t,\tau)$. Neglecting the effect of noise, the relationship between the observed signal autocorrelation matrix and the source signal autocorrelation matrix is shown below

$$R_{\tilde{X}}(t,\tau) = \mathrm{E}\left\{\tilde{X}(t)\tilde{X}^{\mathrm{H}}(t+\tau)\right\} = \tilde{A}R_{\tilde{S}}(t,\tau)\tilde{A}^{\mathrm{H}} \tag{8.12}$$

Taking k different time delay $\tau_1, \tau_2, \ldots, \tau_k$, a set of delay autocorrelation matrices of observed signal is obtained $\left\{R_{\tilde{X}_1}, R_{\tilde{X}_2}, \ldots, R_{\tilde{X}_k}\right\}$, where $R_{\tilde{X}_k} = R_{\tilde{X}}(t,\tau_k)$. The idea of the joint block diagonalization algorithm is to find a separating matrix $\boldsymbol{B} \in \mathbb{R}^{MW \times N(L+W)}$ such that, for any time delay τ, the following expression is satisfied

$$\boldsymbol{B}^H \boldsymbol{R}_{\tilde{\boldsymbol{X}}_k} \boldsymbol{B} = \boldsymbol{D}_k \tag{8.13}$$

Then the signal is separated successfully. Where $\boldsymbol{D}_k$ represents the block diagonal matrix containing N diagonal blocks, the diagonal block dimensions are $(L+W) \times (L+W)$, and the non-diagonal block elements are zero or near-zero numbers, and Eq. (8.13) can also be expressed as $\boldsymbol{B}^{\mathrm{H}} \boldsymbol{R}_{\tilde{\boldsymbol{X}}_k}(t,\tau)\boldsymbol{B} = \boldsymbol{R}_{\tilde{\boldsymbol{S}}}(t,\tau)$. Due to the existence of ambiguities in the order of arrangement between the separated signals and the source signals, $\boldsymbol{B}^H$ is the inverse of $\boldsymbol{P}\tilde{\boldsymbol{A}}$, and $\boldsymbol{P}$ is the identity permutation matrix. Therefore, the source signals can be obtained by the following equation

$$\hat{\boldsymbol{S}}(t) = \boldsymbol{B}^{\mathrm{H}} \tilde{\boldsymbol{X}}(t) \tag{8.14}$$

8.1.1 JBDun Algorithm

The key of convolutional BSS mentioned above is to find separating matrix, which can transform the autocorrelation matrices $\boldsymbol{R}_{\tilde{\boldsymbol{X}}_k}$ (hereafter abbreviated as $\boldsymbol{R}_k$) into block diagonal matrices. Ghennioui proposes a non-diagonal block error cost function for joint block diagonalization based on the least mean square criterion:

$$\min_{\boldsymbol{B}} \mathcal{J}_1(\boldsymbol{B}) = \sum_{k=1}^{K} \left\| \mathrm{offB}(\boldsymbol{B}^{\mathrm{H}} \boldsymbol{R}_k \boldsymbol{B}) \right\|_{\mathrm{F}}^2 \tag{8.15}$$

where $\boldsymbol{R}_k (k = 1, 2, \ldots, K)$ is the $MW \times MW$ autocorrelation matrix of observed signal that needs to be diagonalized, and the symbol $\mathrm{offB}(\cdot)$ is a generalization of the $\mathrm{off}(\cdot)$, which means setting the non-diagonal block elements to zero. By partitioning the column vectors of matrix $\boldsymbol{B}$ into N groups, i.e. $\boldsymbol{B} = [\boldsymbol{B}_1\ \boldsymbol{B}_2\ \ldots\ \boldsymbol{B}_N]$, where $\boldsymbol{B}_i = \left[\boldsymbol{b}_i^1, \boldsymbol{b}_i^2, \ldots, \boldsymbol{b}_i^{n_i}\right] (i = 1, 2, \ldots, N)$ denotes N block matrices of $MW \times (L+W)$, the cost function is:

$$\begin{aligned}
\mathcal{J}_1(\boldsymbol{B}) &= \sum_{k=1}^{K} \sum_{i=1}^{N} \sum_{j=1, j\neq i}^{N} \left\| \boldsymbol{B}_i^{\mathrm{H}} \boldsymbol{R}_k \boldsymbol{B}_j \right\|_{\mathrm{F}}^2 \\
&= \sum_{k=1}^{K} \sum_{i=1}^{N} \sum_{j=1, j\neq i}^{N} \mathrm{tr}\left(\boldsymbol{B}_i^{\mathrm{H}} \boldsymbol{R}_k \boldsymbol{B}_j \boldsymbol{B}_j^{\mathrm{H}} \boldsymbol{R}_k^{\mathrm{H}} \boldsymbol{B}_i \right) \\
&= \sum_{i=1}^{N} \mathrm{tr}\left\{ \boldsymbol{B}_i^H \left[\sum_{k=1}^{K} \boldsymbol{R}_k \left(\sum_{j=1, j\neq i}^{N} \boldsymbol{B}_j \boldsymbol{B}_j^{\mathrm{H}} \right) \boldsymbol{R}_k^{\mathrm{H}} \right] \boldsymbol{B}_i \right\}
\end{aligned} \tag{8.16}$$

Let $\boldsymbol{B}_{(i)} = [\boldsymbol{B}_1, \ldots, \boldsymbol{B}_{i-1}, \boldsymbol{B}_{i+1}, \ldots \boldsymbol{B}_N]$ be the $MW \times (N-1)(L+W)$ dimensional matrix obtained by removing the i-th group from matrix $\boldsymbol{B}$. Thus we have

$$\sum_{j=1, j\neq i}^{N} \boldsymbol{B}_j \boldsymbol{B}_j^{\mathrm{H}} = \boldsymbol{B}\boldsymbol{B}^{\mathrm{H}} - \boldsymbol{B}_i \boldsymbol{B}_i^{\mathrm{H}} = \boldsymbol{B}_{(i)} \boldsymbol{B}_{(i)}^{\mathrm{H}} \tag{8.17}$$

Substituting the formula (8.17) into (8.16) , we have

$$\mathcal{J}_1(\boldsymbol{B}) = \sum_{s=1}^{n_i} \sum_{i=1}^{N} \boldsymbol{b}_i^{s\mathrm{H}} \left[\sum_{k=1}^{K} \boldsymbol{R}_k \boldsymbol{B}_{(i)} \boldsymbol{B}_{(i)}^{\mathrm{H}} \boldsymbol{R}_k^{\mathrm{H}} \right] \boldsymbol{b}_i^s = \sum_{s=1}^{n_i} \sum_{i=1}^{N} \boldsymbol{b}_i^{s\mathrm{H}} \boldsymbol{Q}_i \boldsymbol{b}_i^s \tag{8.18}$$

where $n_i = L + W$ is the number of columns of matrix $\boldsymbol{B}_i$.

$$\boldsymbol{Q}_i = \sum_{k=1}^{K} \boldsymbol{R}_k \boldsymbol{B}_{(i)} \boldsymbol{B}_{(i)}^{\mathrm{H}} \boldsymbol{R}_k^{\mathrm{H}} \tag{8.19}$$

is a semipositive definite Hermitian matrix.

The JBDun algorithm directly selects the eigenvectors corresponding to the n_i smallest eigenvalues of matrix $\boldsymbol{Q}_i$ as the optimal solution for the grouping matrix $\boldsymbol{B}_i$. Then, it iteratively updates the column vectors in each group to minimize the cost function. Finally, Eq. (8.14) is used to separate the signals.

In summary, the steps of the JBDun algorithm are as follows [39]:

Step 1: For any $i = \{1, 2, \ldots, N\}$, initialize $\boldsymbol{B}_i = \boldsymbol{I} \in \mathbb{R}^{(L+W)\times MW}$ and $\boldsymbol{I}$ is the identity matrix.

Step 2: Set $l = 1$.

Step 3: Calculate $\boldsymbol{Q}_i, i = 1, 2, \ldots, N$ based on Eqs. (8.17) and (8.19).

Step 4: Obtain the n_i smallest eigenvalues $\lambda_i^m(l)(m = 1, 2, \ldots n_i)$ and their corresponding eigenvectors $\boldsymbol{b}_i^m(l)$ of matrix $\boldsymbol{Q}_i$. The matrix formed by $\boldsymbol{b}_i^m(l)$ is the new $\boldsymbol{B}_i, i = 1, 2, \ldots, N$.

Step 5: Update $l \leftarrow l + 1$ and if $\lambda_i^{m(l)} - \lambda_i^{m(l-1)} < \varepsilon$, stop the iteration and go to step 6; otherwise, go back to step 3.

Step 6: Obtain the separating matrix $\boldsymbol{B}$, apply Eq. (8.14) to separate the signals and obtain the separated signals.

8.1.2 F-JBDun Algorithm

The cost function of the classical JBDun algorithm does not satisfy the full-rank property of $\boldsymbol{B}$, does not guarantee that the column vectors of $\boldsymbol{B}$ are uncorrelated with each other, and is easy to converge to a singular solution. The F-JBDun algorithm [40] avoids singular solutions by adding a constraint term to the nondiagonal error cost function, which is as follows:

$$\min_{\boldsymbol{B}} J(\boldsymbol{B}) = \mathcal{J}_1(\boldsymbol{B}) - \mathcal{J}_2(\boldsymbol{B}) = \sum_{k=1}^{K} \left\| \mathrm{offB}(\boldsymbol{B}^{\mathrm{H}}\boldsymbol{R}_k\boldsymbol{B}) \right\|_{\mathrm{F}}^2 - \log|\det(\boldsymbol{B})| \tag{8.20}$$

The first part of Eq. (8.20) ensures that $\boldsymbol{B}^H\boldsymbol{R}_k\boldsymbol{B}$ are diagonalized, and the second part can make all column vectors of $\boldsymbol{B}$ uncorrelated. The $|\cdot|$ means to take absolute values.

According to the method of cyclic minimization, a specified grouping of matrix $\boldsymbol{B}$, say $\boldsymbol{B}_l$, is updated and the values of the other groupings are kept constant. Equation (8.16) is wirten a function of $\boldsymbol{B}_l$ as

$$\begin{aligned}
\mathcal{J}_1(\boldsymbol{B}_l) &= \sum_{j=1,j\neq l}^{N} \sum_{k=1}^{K} \|\boldsymbol{B}_l^{\mathrm{H}}\boldsymbol{R}_k\boldsymbol{B}_j\|_F^2 + \sum_{i=1,i\neq l}^{N} \sum_{k=1}^{K} \|\boldsymbol{B}_i^{\mathrm{H}}\boldsymbol{R}_k\boldsymbol{B}_l\|_F^2 \\
&= \sum_{j=1,j\neq l}^{N} \sum_{k=1}^{K} \mathrm{tr}\left(\mathbf{B}_l^H\mathbf{R}_k\mathbf{B}_j\mathbf{B}_j^{\mathrm{H}}\mathbf{R}_k^{\mathrm{H}}\mathbf{B}_l\right) + \sum_{i=1,i\neq l}^{N} \sum_{k=1}^{K} \mathrm{tr}(\mathbf{B}_l^{\mathrm{H}}\mathbf{R}_k^{\mathrm{H}}\mathbf{B}_i\mathbf{B}_i^{\mathrm{H}}\mathbf{R}_k\mathbf{B}_l) \\
&= \mathrm{tr}\left\{\boldsymbol{B}_l^{\mathrm{H}}\left[\sum_{k=1}^{K} \boldsymbol{R}_k\boldsymbol{B}_{(l)}\boldsymbol{B}_{(l)}^{\mathrm{H}}\boldsymbol{R}_k^{\mathrm{H}} + \boldsymbol{R}_k^{\mathrm{H}}\boldsymbol{B}_{(l)}\boldsymbol{B}_{(l)}^{\mathrm{H}}\boldsymbol{R}_k\right]\boldsymbol{B}_l\right\} \\
&= \sum_{i=1}^{n_i} \boldsymbol{b}_l^{i\mathrm{H}}\boldsymbol{P}_l\boldsymbol{b}_l^i
\end{aligned} \tag{8.21}$$

$$\boldsymbol{P}_l = \sum_{k=1}^{K} \boldsymbol{R}_k\boldsymbol{B}_{(l)}\boldsymbol{B}_{(l)}^{\mathrm{H}}\boldsymbol{R}_k^{\mathrm{H}} + \boldsymbol{R}_k^{\mathrm{H}}\boldsymbol{B}_{(l)}\boldsymbol{B}_{(l)}^{\mathrm{H}}\boldsymbol{R}_k \tag{8.22}$$

where $\mathrm{tr}(\cdot)$ denotes the trace of the matrix, $\boldsymbol{b}_l^i$ denotes the i-th column of matrix $\boldsymbol{B}_l$, and $\boldsymbol{B}_{(l)}$ denotes the matrix obtained by removing the l-th group from matrix $\boldsymbol{B}$. Expanding the constraint terms using the determinant, the s-th column of matrix $\boldsymbol{B}_l$ is denoted as $\boldsymbol{b}_l^s = [b_l^s(1), \ldots, b_l^s(M)]^{\mathrm{T}}$, then we have

$$\det(\boldsymbol{B}) = \boldsymbol{w}_l^{s\mathrm{T}}\boldsymbol{b}_l^s \tag{8.23}$$

where $\boldsymbol{w}_l^s = [w_l^s(1), \ldots, w_l^s(M)]^{\mathrm{T}}$ and

$$\boldsymbol{w}_l^s(m) = (-1)^p \det(\boldsymbol{B}_{lsm}) \tag{8.24}$$

where $p = \sum_{q=1}^{l-1} n_q + s + m$, $\boldsymbol{B}_{lsm}$ are $(MW-1) \times (MW-1)$ dimensional submatrices obtained by deleting m-th row and $\left(\sum_{q=1}^{l-1} n_q + s\right)$-th column of matrix $\boldsymbol{B}$, and $\boldsymbol{w}_l^s$ and $\boldsymbol{b}_l^s$ are uncorrelated. Thus, the constraint term is obtained as follows.

$$\mathcal{J}_2(\boldsymbol{B}_l) = \log\left|\boldsymbol{w}_l^{s\mathrm{H}}\boldsymbol{b}_l^s\right| = \frac{1}{2}\log\left(\boldsymbol{b}_l^{s\mathrm{H}}\boldsymbol{w}_l^s\boldsymbol{w}_l^{s\mathrm{H}}\boldsymbol{b}_l^s\right) \tag{8.25}$$

Combining Eqs. (8.21) and (8.25), we can get

$$\mathcal{J}(\boldsymbol{B}_l) = \sum_{i=1}^{n_i}\boldsymbol{b}_l^{i\mathrm{H}}\boldsymbol{P}_l\boldsymbol{b}_l^i - \frac{1}{2}\log\left(\boldsymbol{b}_l^{s\mathrm{H}}\boldsymbol{w}_l^s\boldsymbol{w}_l^{s\mathrm{H}}\boldsymbol{b}_l^s\right) \tag{8.26}$$

In order to minimize the above equation, we use the loop minimization technique again, which keeps the other column vectors in $\boldsymbol{B}_l$ unchanged, and adjusts the column vector $\boldsymbol{b}_l^s$ to minimize the Eq. (8.26), and then uses the same method to update all the columns of the matrix $\boldsymbol{B}_l$.

Write Eq. (8.26) as a function of vector $\boldsymbol{b}_l^s$:

$$\mathcal{J}(\boldsymbol{b}_l^s) = \boldsymbol{b}_l^{s\mathrm{T}}\boldsymbol{P}_l\boldsymbol{b}_l^s - \frac{1}{2}\log\left(\boldsymbol{b}_l^{s\mathrm{T}}\boldsymbol{w}_l^s\boldsymbol{w}_l^{s\mathrm{T}}\boldsymbol{b}_l^s\right) \tag{8.27}$$

Calculate the conjugate gradient of $\mathcal{J}\left(\boldsymbol{b}_l^s\right)$ about $\boldsymbol{b}_l^s$ as follows.

$$\frac{\partial\mathcal{J}\left(\boldsymbol{b}_l^s\right)}{\partial\boldsymbol{b}_l^s} = \boldsymbol{P}_l\boldsymbol{b}_l^s - \frac{\boldsymbol{w}_l^s}{2\boldsymbol{b}_l^{sH}\boldsymbol{w}_l^s} \tag{8.28}$$

Since the corresponding *Hessian* matrix is $\frac{\partial\mathcal{J}(\boldsymbol{b}_l^s)}{\partial\boldsymbol{b}_l^{sT}\partial\boldsymbol{b}_l^s} = \boldsymbol{P}_l$, which is always semipositive definite, letting the conjugate gradient be zero, we get that the optimal solution satisfies the condition

$$\boldsymbol{P}_l\boldsymbol{b}_l^s = \frac{\boldsymbol{w}_l^s}{2\boldsymbol{b}_l^{s\mathrm{H}}\boldsymbol{w}_l^s} \tag{8.29}$$

Let $\beta = 2\boldsymbol{b}_l^{s\mathrm{H}}\boldsymbol{w}_l^s$, then the optimal solution is

$$\boldsymbol{b}_l^s = \frac{1}{2\beta}\boldsymbol{P}_l^{-1}\boldsymbol{w}_l^s \tag{8.30}$$

Substituting Eq. (8.30) into the defining equation of β and considering that $\boldsymbol{P}_l$ is a *Hermitian* matrix, then β is expressed as $\beta = \sqrt{\boldsymbol{w}_l^{s\mathrm{H}}\boldsymbol{P}_l^{-1}\boldsymbol{w}_l^s/2}$, and the optimal solution is

$$\boldsymbol{b}_l^s = \frac{\boldsymbol{P}_l^{-1}\boldsymbol{w}_l^s}{\sqrt{2\boldsymbol{w}_l^{s\mathrm{H}}\boldsymbol{P}_l^{-1}\boldsymbol{w}_l^s}}, \quad s \in \mathcal{I}_{nl} \tag{8.31}$$

In summary, the steps of the F-JBDun algorithm are as follows.

Input: Set of matrices $\{\boldsymbol{R}_1, \ldots, \boldsymbol{R}_K\}$ to be diagonalized, number of source signals N, dimension of each matrix n_i.

Output: Separating matrix $\boldsymbol{B}$, separated signals $\boldsymbol{Y}$.

Step 1: Initialize matrix $\boldsymbol{B}(0) \leftarrow \boldsymbol{I}, \quad \boldsymbol{B}(0) = [\boldsymbol{B}_1(0), \ldots, \boldsymbol{B}_N(0)]. \quad t \leftarrow 1.$

Step 2: For $l = 1 : N$.

(1) Calculate $\boldsymbol{P}_l$ based on Eq. (8.22).

(2) Update matrix $\boldsymbol{B}_l$ according to the following steps:
for $s = 1 : n_i$
1. Calculate $\boldsymbol{w}_l^s$ based on Eq. (8.24).
2. Calculate $\boldsymbol{b}_l^s$ based on Eq. (8.31).
end

Step 3: Obtain the separating matrix $\boldsymbol{B}$ and get the separated signal $\boldsymbol{Y}$ according to Eq. (8.14).

The F-JBDun algorithm consists of two loops. The inner loop updates all the columns of a particular column-grouping matrix sequentially, and the outer loop updates all the groups sequentially and finally a whole iteration of the algorithm is completed.

8.1.3 JBDns Algorithm

The F-JBDun algorithm improves the problem of low success rate of the classical JBDun algorithm. But the constraint term takes the determinant of the separating matrix $\boldsymbol{B}$, which makes the algorithm limited to application scenarios where the equivalent mixing matrix is a square matrix. To eliminate the limitation of square matrix, the JBDns algorithm [41–42] improves the cost function by using multi-criteria optimization to represent the cost function as two parts:

$$\begin{aligned} &\min_{\boldsymbol{B}} \mathcal{J}_1(\boldsymbol{B}), \quad \max_{\boldsymbol{B}} \det(\boldsymbol{B}^{\mathrm{H}}\boldsymbol{B}) \\ &\text{s.t. } \boldsymbol{b}_i^{\mathrm{H}}\boldsymbol{b}_i = 1\,, i = 1, 2, \ldots, N \end{aligned} \tag{8.32}$$

The first part is derived using a loop optimization algorithm as described in Sect. 8.1.2, where the basic idea is to first divide the parameters to be optimized into groups and then optimize a particular group of parameters with the other parameters fixed. The overall optimization is achieved by alternately optimizing all the grouped parameters. For the l-th group $\boldsymbol{B}_l$, after some matrix operations, we have

$$\mathcal{J}_1(\boldsymbol{B}_l) = \operatorname{tr}\left\{\boldsymbol{B}_l^{\mathrm{H}}\boldsymbol{P}_l\boldsymbol{B}_l\right\} = \sum_{i=1}^{n_i} \boldsymbol{b}_l^{i\mathrm{H}}\boldsymbol{P}_l\boldsymbol{b}_l^i \tag{8.33}$$

where

$$P_l = \sum_{k=1}^{K} R_k B_{(l)} B_{(l)}^{\mathrm{H}} R_k^{\mathrm{H}} + R_k^{\mathrm{H}} B_{(l)} B_{(l)}^{\mathrm{H}} R_k \tag{8.34}$$

Let $\overline{B}_l = [b_{lm}, B_{(lm)}]$, where b_{lm} denotes the l-th column of B_l and $B_{(lm)}$ denotes the matrix obtained by removing b_{lm} from B, then the second part of the formula is as follows:

$$\det(B^{\mathrm{H}}B) = \det(\overline{B}^{\mathrm{H}}\overline{B}) = \det\begin{pmatrix} b_{lm}^{\mathrm{H}} b_{lm} & b_{lm}^{\mathrm{H}} B_{(lm)} \\ B_{(lm)}^{\mathrm{H}} b_{lm} & B_{(lm)}^{\mathrm{H}} B_{(lm)} \end{pmatrix} = \det(B_{(lm)}^{\mathrm{H}} B_{(lm)}) b_{lm}^{\mathrm{H}} P_{lm}^{\perp} b_{lm} \tag{8.35}$$

$$P_{lm}^{\perp} = I - B_{(lm)} [B_{(lm)}^{\mathrm{T}} B_{(lm)}]^{-1} B_{(lm)}^{\mathrm{T}} \tag{8.36}$$

Since b_{lm} is irrelevant to $B_{(lm)}$, the new cost function is obtained as follows.

$$\begin{aligned} &\min_{b_{lm}} b_{lm}^{\mathrm{H}} P_l b_{lm}, \ \max_{b_{lm}} b_{lm}^{\mathrm{H}} P_{lm}^{\perp} b_{lm} \\ &\text{s.t. } b_{lm}^{\mathrm{H}} b_{lm} = 1,\ l = 1, 2, \ldots, N,\ m = 1, 2, \ldots, n_l \end{aligned} \tag{8.37}$$

The optimization problem of Eq. (8.32) can be carried out in two stages. In the initialization phase, the matrix P_l is invertible and is taken as $b_{lm} = v_{\max}$, where $v_{\max}$ denotes the generalized eigenvector corresponding to the largest generalized eigenvalue of $\left(P_{lm}^{\perp}, P_l\right)$. In the second stage, when the algorithm converges gradually and matrix P_l is a singular matrix of rank $N - n_l$ (n_l is the number of columns of B_l), the algorithm becomes unstable. U_0 is used to represent the matrix composed of the eigenvectors corresponding to the $M - N + n_l$ smallest eigenvalues of P_l. Let

$$b_{lm} = U_0 w \tag{8.38}$$

where w denotes the $M - N + n_l$ weight coefficients. Equation (8.38) is substituted into the second optimization criterion in Eq. (8.37), then the optimal vector is

$$w_{opt} = \arg\max_{w} w^{\mathrm{T}} U_0^{\mathrm{T}} P_{lm}^{\perp} U_0 w \tag{8.39}$$

In summary, the steps of the JBDns algorithm are as follows.

Step 1: Given the number of diagonal blocks r, initialize matrix $B = I \in \mathbb{R}^{N(L+W) \times MW}$.

Step 2: For $l = 1 : r$.

(1) Calculate P_l based on Eq. (8.34).

(2) Update B_l according to the following steps:
for $m = 1 : n_l$
(2.1) Calculate $P_{lm}^{\perp}$ based on Eq. (8.36).

(2.2) If $\boldsymbol{P}_l$ is invertible, set $\boldsymbol{b}_{lm} = \boldsymbol{v}_{\max}$; otherwise, solve for $\boldsymbol{b}_{lm}$ using Eqs. (8.38) and (8.39).
(2.3) Update m: $m \rightarrow m + 1$.
end
(3) Update l: $l \rightarrow l + 1$.

end

Step 3: Obtain the separating matrix $\boldsymbol{B}$ and get the separated signals according to Eq. (8.14).

Because the time-domain method requires a lot of computation, and the algorithm will become more and more unstable with the increase of convolutional channel length. The separating effect of time-domain method is not ideal, so we often use the frequency-domain method to solve the problem of convolutional mixing BSS.

8.2 Frequency-Domain Convolution BSS Algorithm

8.2.1 The Principle of Frequency-Domain Methods

First, the mixed signal is transformed to the time-frequency domain by Short-Time Fourier Transform (STFT).The mathematical expression of STFT is:

$$X(\tau, f) = \sum_{n=0}^{L-1} \text{win}(n) x(n + \tau l) \exp(-j2\pi fn/L) \tag{8.40}$$

where τ represents the time index in the time-frequency domain, $f = 0, 1, \ldots, L-1$ is the frequency index, L and l denote the frame length (nfft) and the sliding step-size of the Fourier transform, respectively, n represents the time variable of the signal after sampling, and $\text{win}(n)$ is the window function. Perform the STFT on both sides of the Eq. (2.6) and get the mixing model of the signal in the time-frequency domain:

$$\boldsymbol{X}(\tau, f) = \boldsymbol{A}(f)\boldsymbol{S}(\tau, f) \tag{8.41}$$

where $\boldsymbol{S}(\tau, f) = [S_1(\tau, f), S_2(\tau, f), \ldots S_n(\tau, f)]^{\mathrm{T}}$ denotes the time-frequency domain form of the source signal and $\boldsymbol{X}(\tau, f) = [X_1(\tau, f), X_2(\tau, f), \ldots X_n(\tau, f)]^{\mathrm{T}}$ denotes the time-frequency domain form of the mixed signal. If the short-time Fourier frame length L is much larger than the order P of the convolutional mixing filter, it can be approximated that the time-domain convolutional mixing model expressed in Eq. (2.6) has the instantaneous mixing model expressed in Eq. (8.41) at each frequency point f. At this point, the convolutional mixing model in the time domain is converted into instantaneous mixing model at each frequency point in the frequency domain.

The instantaneous BSS algorithm can then be used for separation at each frequency point. However, since the separation process at each frequency point is independent of each other and the BSSalgorithm itself has two ambiguities of amplitude and ordering, this will result in the amplitude and order of the separated signals at each frequency point not being unified, and thus not being able to correctly recover the time-domain signals. Therefore, before the Inverse Short-Time Fourier Transform (ISTFT) is performed on the frequence-domain separated signal, these two ambiguities must be eliminated to obtain the correct estimated signal.

Next, take the blind separation system of two-input and two-output speech signals as an example to summarize the flow of frequency-domain method. The convolutional mixing model in frequency domain is as follows.

$$X_1(\tau,f) = A_{11}(f)S_1(\tau,f) + A_{12}(f)S_2(\tau,f) \tag{8.42}$$

$$X_2(\tau,f) = A_{21}(f)S_1(\tau,f) + A_{22}(f)S_2(\tau,f) \tag{8.43}$$

where $S_i(\tau,f)$ denotes a time-frequency point coefficient of the source speech signal $S_i(\tau,f)$ at τ-th data frame and the frequency point f, $A_{ji}(f)$ represents the frequency-domain transmission coefficient of the mixing filter h_{ji} from the source speech signal $s_i(t)$ to the microphone $x_j(t)$, and $X_j(\tau,f)$ is a frequency domain representation of the received signal $x_j(t)$.

Diagram of the speech signal frequency domain convolutional mixing BSS system is shown in Fig. 8.1 [43]. The source speech signal is convolved to obtain a mixed signal, the received convolutional mixed speech signal is converted to the time-frequency domain by STFT, and at each frequency point the convolution mixing form becomes an instantaneous mixing form, and then the instantaneous mixing BSS algorithm is used to obtain the blind separating matrix $\boldsymbol{W}(f)$ in the frequency domain and the separated signals $Y_i(\tau,f)$. Next, we need to further solve the problem of amplitude and permutation ambiguities. Finally, ISTFT is used to convert the separated frequency domain signals into the time domain to obtain the estimation of source signal.

8.2.2 Methods for Resolving Amplitude Ambiguity

After performing STFT on a convolutional mixed signal, the mixing process at a single frequency point in the time-frequency domain can be expressed as $\boldsymbol{X}(\tau,f) = \boldsymbol{A}(f)\boldsymbol{S}(\tau,f)$. The mathematical process of separation using an instantaneous BSS algorithm at a particular frequency point can be expressed as:

$$\boldsymbol{Y}(\tau,f) = \boldsymbol{W}(f)\boldsymbol{X}(\tau,f) = \boldsymbol{W}(f)\boldsymbol{A}(f)\boldsymbol{S}(\tau,f) \tag{8.44}$$

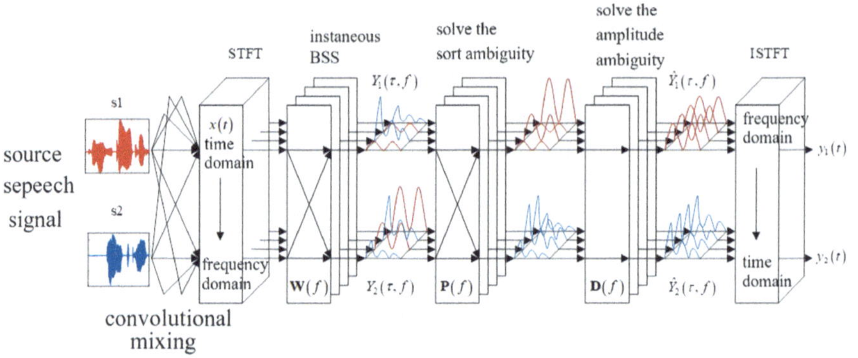

Fig. 8.1 Speech signal convolution mixing frequency domain blind separation system diagram

However, when we use the instantaneous mixing BSS algorithm to obtain the demixing matrix and separated signal at each frequency point, the time-domain separated signal can not be obtained directly by ISTFT transformation. This is because of the lack of priori information, and there will be amplitude ambiguity and permutation ambiguity, which makes it not a simple inverse process between the demixing matrix and the mixing matrix at each frequency point, and the separating matrix $\boldsymbol{W}(f)$ satisfies the following equation:

$$\boldsymbol{W}(f) = \boldsymbol{D}(f)\boldsymbol{P}(f)\boldsymbol{A}^{-1}(f) \tag{8.45}$$

where $\boldsymbol{D}(f)$ is the diagonal matrix, reflecting magnitude ambiguity, and $\boldsymbol{P}(f)$ is the permutation matrix, reflecting permutation ambiguity. Then the separated signal can be expressed as

$$\boldsymbol{Y}(\tau, f) = \boldsymbol{D}(f)\boldsymbol{P}(f)\boldsymbol{S}(\tau, f) \tag{8.46}$$

The above equation shows that the output signals at each frequency point have amplitude ambiguity $\boldsymbol{D}(f)$ and permutation ambiguity $\boldsymbol{P}(f)$. Compared with the permutation ambiguity problem, the effect of amplitude fuzziness is relatively small, only the signal will be distorted, and it does not affect the correspondence between the separated signal and the source signal at each frequency point. The algorithm based on Minimal Distortion Principle (MDP) [44] proposed by K. Matsuoka can solve the amplitude ambiguity problem better, and its main idea is to normalize the demixing matrix at each frequency point.

The basic idea to solve the problem of amplitude ambiguity is to make the amplitude of the separated signal at each frequency unified with the source signal, or its transformation scale is the same relative to the source signal. When the separated signal at each frequency meets the condition, it is considered that the MDP criterion is met. It may be assumed that the permutation ambiguity problem has been solved (i.e., $\boldsymbol{P}(f) = \boldsymbol{I}$ at all frequency points), and according to the description of the MDP, although it is difficult to guarantee $Y_i(\tau, f) = S_i(\tau, f), i = 1, 2 \ldots N$ at each frequency point, it is possible to allow $Y_i(\tau, f) = H_{ii}(f) S_i(\tau, f)$, i.e., the product of the separating and mixing matrices is a diagonal array, which means that $\boldsymbol{W}(f)\boldsymbol{A}(f) = \boldsymbol{D}(f)$ is satisfied. Then it is obtained from equation (8.45).

$$\boldsymbol{W}(f) = \boldsymbol{D}(f)\boldsymbol{A}^{-1}(f) = \operatorname{diag}(\boldsymbol{A}(f))\boldsymbol{A}^{-1}(f) \tag{8.47}$$

where diag(·) represents matrix diagonalization. It is known that $\boldsymbol{A}(f) = \boldsymbol{D}(f)\boldsymbol{W}^{-1}(f)$. Then separating matrix $\boldsymbol{W}_{\mathrm{MDP}}(f)$ that satisfies the MDP criterion is deduced as follows.

$$\begin{aligned} \boldsymbol{W}_{\mathrm{MDP}}(f) &= \operatorname{diag}(\boldsymbol{A}(f))\boldsymbol{A}^{-1}(f) \\ &= \operatorname{diag}\left(\boldsymbol{W}^{-1}(f)\boldsymbol{D}(f)\right)\boldsymbol{D}^{-1}(f)\boldsymbol{W}(f) \\ &= \operatorname{diag}\left(\boldsymbol{W}^{-1}(f)\right)\boldsymbol{W}(f) \end{aligned} \tag{8.48}$$

Therefore, the separating matrix obtained by instantaneous BSS algorithm at each frequency point is converted into the separating matrix satisfying the MDP criterion by the operation process of formula (8.48), which means that the amplitude ambiguity problem is solved.

8.2.3 Methods for Resolving Permutation Ambiguity

For the amplitude ambiguity problem, it can be better solved by using the MDP criterion, and the current difficulty and research focus in the field of convolutional BSS in the frequency domain lies in how to effectively eliminate the permutation ambiguity of the signal.

In the current frequency-domain algorithms for convolutional BSS, the traditional methods to solve the permutation ambiguity can be roughly divided into two categories. One is the mutual parameter method, which is based on the consistency of the intrinsic characteristics of the source signal, such as the amplitude correlation sorting method. The other category is geometric methods, the most representative of which is the DOA sorting method, which sorts by extracting the orientation information in the separating matrix. These two algorithms have different ideas and have their own advantages and disadvantages. Besides, there is also an algorithm derived from independent component analysis—independent vector analysis algorithm, which can avoid the appearance of permutation ambiguity in principle.

1. **DOA Sorting Algorithm**

The Direction-of-Array (DOA) [45] sorting method solves the permutation ambiguity problem by estimating DOA information of the source signal. According to the beamforming theory, a beamforming expression can be constructed by taking each row of the separating matrix $\boldsymbol{W}(f)$ as the weight vector at each frequency point, and the beam pattern drawn by the expression will form a zero in the direction of the interference signal and obtain a larger value in the direction of the target signal. When the orientation information of each separated signal is estimated in this way, the order of the separated signals at each frequency point can be adjusted to eliminate the permutation ambiguity. The mathematical description of the algorithm is as follows.

According to the beamforming theory, the frequency-domain form of the impulse response $a_{jk}(t)$ of the source signal s_k to the array element x_j can be approximated as

$$A_{jk}(f) = e^{j2\pi fc^{-1}d_j \cos\theta_k} \tag{8.49}$$

where c refers to the signal propagation speed (if the source signal is a speech signal, here c represents the speed of sound; if the source signal is a communication signal, here c represents the speed of light), d_j is the relative position of the array element x_j, and θ_k is the direction angle of the source signal s_k. In this approximate formula, assuming that the arrival wave is a plane wave and there is no reverberation effect, the overall frequency response function of the entire mixing and separating process is $\boldsymbol{G}(f) = \boldsymbol{W}(f)\boldsymbol{A}(f)$, and the frequency-domain form of the impulse response function from the source signal s_k to the output signal y_i is:

$$G_{ik}(f) = \sum_{j=1}^{M} W_{ij}(f)A_{jk}(f) = \sum_{j=1}^{M} W_{ij}(f)e^{j2\pi fc^{-1}d_j \cos\theta_k} \tag{8.50}$$

If θ_k is replaced by a variable θ, the above equation becomes

$$G_i(f,\theta) = \sum_{j=1}^{M} W_{ij}(f)e^{j2\pi fc^{-1}d_j \cos\theta} \tag{8.51}$$

From the above equation, it can be seen that the frequency response will change with the change of azimuth angle θ. Equation (8.51) can be regarded as a conventional beam-forming equation constituted with the elements of each row of the separation matrix as the weight vector. The beam pattern drawn according to Eq. (8.51) is shown in Fig. 8.2.

Figure 8.2 illustrates the beam gain $C_i(f,\theta) = |G_i(f,\theta)|$ of two source signals (135° and 45° directions, respectively) in an indoor convolution mixing scene. The upper half of the figure represents the beam gain at a frequency of 3 kHz, and it can be seen that $C_1(3000,\theta)$ has zero gain at 45° and $C_2(3000,\theta)$ has zero gain at 135°. According to the angle information corresponding to the zero gain, the source signal's DOA information contained in the separating matrix can be obtained, and this DOA can be used for sorting.

Taking a two-input, two-output system as an example, the basic steps of the DOA sorting method are as follows.

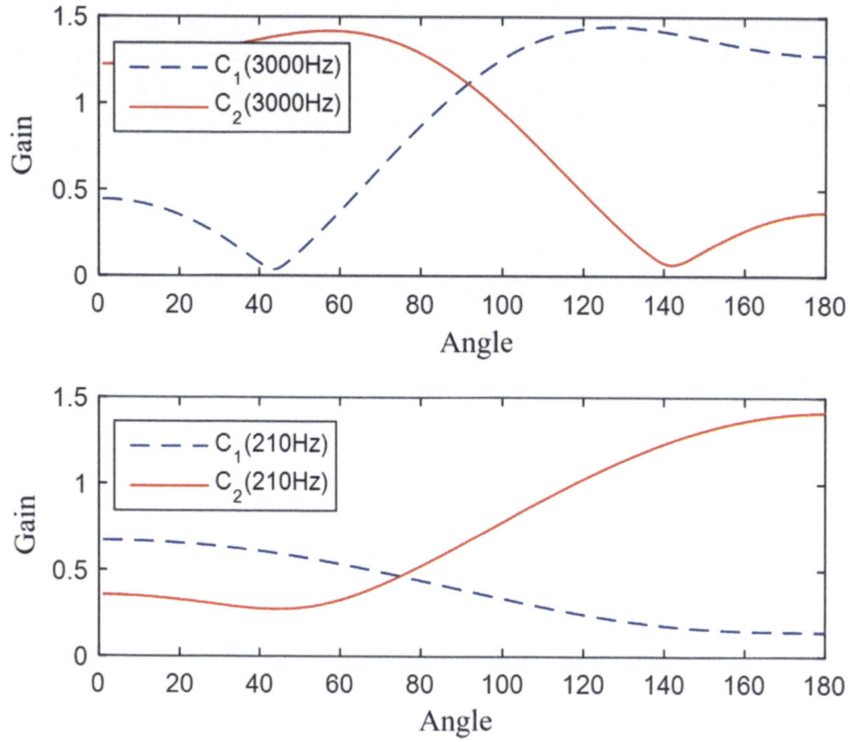

Fig. 8.2 Beam pattern of two source signals

(1) Find the minimum of $C_1(f, \theta)$ and $C_2(f, \theta)$ at each frequency point to obtain the zero directions $\theta_1(f)$ and $\theta_1(f)$, respectively;
(2) The DOA of the two sources $\hat{\theta}_1$ and $\hat{\theta}_2$ are estimated by Eq. (8.52):

$$\hat{\theta}_1 = \frac{1}{F}\sum_{i=1}^{F} \min\left[\theta_1(f_i), \theta_2(f_i)\right]$$
$$\hat{\theta}_2 = \frac{1}{F}\sum_{i=1}^{F} \min\left[\theta_1(f_i), \theta_2(f_i)\right] \tag{8.52}$$

where F is equal to the frame length of the short-time Fourier transform, representing the total number of frequency points.
(3) Determine whether the sorting is correct or not: if $\theta_1(f)$ is near $\hat{\theta}_1$ and $\theta_2(f)$ is near $\hat{\theta}_2$, i.e.$|\theta_1(f)-\hat{\theta}_1|+|\theta_2(f)-\hat{\theta}_2| < |\theta_1(f)-\hat{\theta}_2|+|\theta_2(f)-\hat{\theta}_1|$, it is considered that the separated signals are sorted correctly at frequency f. Conversely, if $|\theta_1(f)-\hat{\theta}_1|+|\theta_2(f)-\hat{\theta}_2| > |\theta_1(f)-\hat{\theta}_2|+|\theta_2(f)-\hat{\theta}_1|$, it is considered that the separated signals are sorted incorrectly at frequency f, i.e., the order of the separated signals is exchanged to solve the problem of permutation ambiguity.

It is worth noting that the lower half of Fig. 8.2 shows a low-frequency beam pattern at a frequency of 210 Hz. It can be seen that at 210 Hz frequency, the beam pattern does not show direction information, the zero gain direction of $C_2(210, \theta)$ is very vague, while $C_1(210, \theta)$ basically does not appear zero gain, and the DOA information of the signal at this frequency cannot be obtained. Therefore, the DOA method tends to be ineffective at low frequencies, which is a drawback of the method. In addition to this, there are more limitations in using the DOA sorting method. One of the limitations is that the distance between the receiving array elements needs to be less than the half wavelength of source signals. When the distance between the array elements is greater than half wavelength, there will be more than one zero in the beam gain pattern. Another limitation is that when there are more than two source signals, there will also be multiple zeros in the beam gain pattern, which will also result in ineffective acquisition of the DOA information of the source signals.

2. **Amplitude Correlation Sorting Algorithm**

(1) Anemuller Algorithm

Scholar Anemuller pointed out that the signals emitted from the same acoustic source have a certain degree of continuity and the modulation mode is also similar, so the amplitude correlation between adjacent frequency points of the same signal is much larger than that of the signal of a different source, which can be clearly observed through the simulation diagram. Taking the two mixed signals as an example, the amplitude envelope curves of the two separated signals obtained by the instantaneous BSS algorithm at different frequency points are shown in Fig. 8.3. Observing the amplitude envelope curves of the separated signals at 1500 and 1531 Hz, it is not difficult to find that the separated signals belonging to the same source have very high similarity in the signal envelope curves at adjacent frequency points, i.e., the amplitude envelopes of $|Y_1(1500\,\text{Hz})|$ and $|Y_1(1531\,\text{Hz})|$, $|Y_2(1500\,\text{Hz})|$ and $|Y_2(1531\,\text{Hz})|$ are basically similar, while the separated signals from different source signals do not have similar signal envelopes at adjacent frequency points. Meanwhile, by observing the signal envelopes at 1500 and 3000 Hz or 1531 and 3000 Hz we can find that if the two frequency points are far apart, then the amplitude envelopes are not similar regardless of whether they come from the same source signal or not.

Anemuller amplitude correlation sorting algorithm (Anemuller Algorithm for short) is an intuitive and convenient sorting algorithm based on the above characteristics. The principle of the algorithm is as follows. When sorting signals on the current frequency point, refer to the signal order on the previous frequency point. By calculating and comparing the magnitude of signal amplitude correlation coefficient, we believe that the two signals with large amplitude correlation coefficient have strong correlation and come from the same signal source. By adjusting the signal order at the current frequency point, the two signals with strong correlation are located in the same way. The mathematical description of the method is as follows.

The correlation coefficient ρ_{ij} of the amplitude of the separated signal is first calculated and used to provide a reasonable measure of the correlation of the signal. As an example, the amplitude envelopes $|Y_i(f)|$ and $\left|Y_j(f)\right|$ of the separated signals

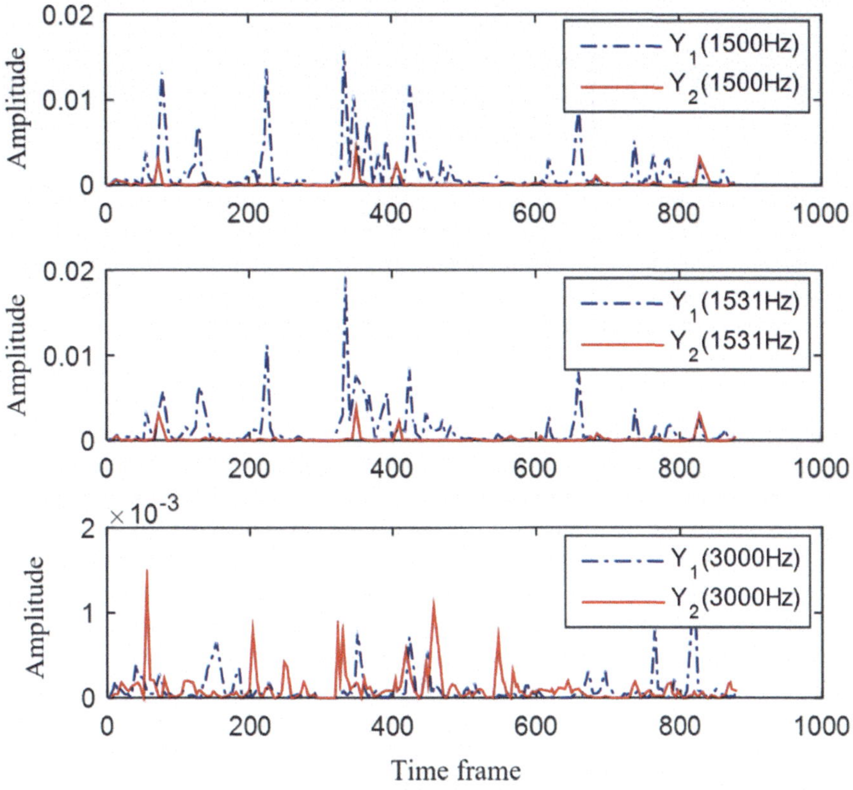

Fig. 8.3 Envelope plot of separated signals at different frequency points

$Y(f)$ at frequency points i and j are calculated as follows:

$$\rho_{ij} = \mathrm{cor}\left(|Y_i(f)|, \left|Y_j(f)\right|\right) = \frac{C\left(|Y_i(f)|, \left|Y_j(f)\right|\right)}{\sqrt{C(|Y_i(f)|, |Y_i(f)|) \cdot C\left(\left|Y_j(f)\right|, \left|Y_j(f)\right|\right)}} \tag{8.53}$$

where $C(\cdot)$ represents the covariance function.

$$C\left(|Y_i(f)|, \left|Y_j(f)\right|\right) = \sum_{\tau} \left\{\left[|Y_i(\tau, f)| - E(|\mathbf{Y}_i(f)|)\right] \times \left[\left|Y_j(\tau, f)\right| - E\left(\left|\mathbf{Y}_j(f)\right|\right)\right]\right\} \tag{8.54}$$

The following takes two source signals as an example, the following correlation coefficients need to be calculated to solve the permutation ambiguity problem by using the Anemuller algorithm.

$$\begin{cases} \rho_{11} = \text{cor}(|Y_1(f)|, |Y_1(f+1)|) \\ \rho_{12} = \text{cor}(|Y_1(f)|, |Y_2(f+1)|) \\ \rho_{21} = \text{cor}(|Y_2(f)|, |Y_1(f+1)|) \\ \rho_{22} = \text{cor}(|Y_2(f)|, |Y_2(f+1)|) \end{cases} \tag{8.55}$$

Next, the magnitude of the correlation coefficient can be used to correct the signal order, i.e., if $\rho_{11}+\rho_{22} < \rho_{12}+\rho_{21}$, it is considered that there is a problem with the ordering of the signals at the current frequency point f, and it is necessary to switch the order of the two separated signals, or else keep them unchanged. When the above operations are done for all frequencies, all sorting is completed.

(2) Murata Algorithm

When using the Anemuller algorithm for sorting, if an error occurs in the sorting of the separated signals at a certain frequency point, it will inevitably result in an error in the sorting of the next frequency point, and then the sorting of all the subsequent frequency points will be wrong, just like the collapse of "dominoes".

In order to avoid the error transmission phenomenon caused by too strong dependence on a single frequency, the algorithm proposed by scholar Murata (referred to as Murata algorithm) adjusts the sequence of separated signals by calculating the sum of amplitude correlation coefficients between frequency points to be sorted and frequency points in a certain neighborhood, so as to eliminate the permutation ambiguity. In other words, when sorting the signals at frequency point f, we take the signals in a certain neighborhood ($|g-f| \leq L$) as a reference, take the sum of amplitude correlation coefficients as the basis for sorting, select an arrangement that can make the value reach the maximum, and this arrangement is the best arrangement at the frequency point f.

Knowing that the arrangement of the separated signals at the sorted frequency point g is Π_g, the arrangement Π_f at the frequency point f can be obtained by the following equation:

$$\Pi_f = \arg\max_{\Pi} \sum_{|g-f|\leq L} \sum_{i=1}^{N} cor\left(\left|\mathbf{Y}_i^{\Pi}(f)\right|, \left|\mathbf{Y}_i^{\Pi_g}(g)\right|\right) \tag{8.56}$$

where $\mathbf{Y}_i^{\Pi}(f) = [\mathbf{Y}_i^{\Pi}(f, 1), \ldots, \mathbf{Y}_i^{\Pi}(f, \tau), \ldots, \mathbf{Y}_i^{\Pi}(f, T)]$ is the i-th separated signal at frequency point f, Π is an arrangement on frequency point f, τ is the time index, T is the time length, $\mathbf{Y}_i^{\Pi_g}(g) = [\mathbf{Y}_i^{\Pi_g}(g, 1), \ldots, \mathbf{Y}_i^{\Pi_g}(g, \tau), \ldots, \mathbf{Y}_i^{\Pi_g}(g, T)]$ is the i-th separated signal at the sorted frequency point g, $cor(\cdot)$ represents the correlation coefficient, the specific formula is shown in (8.53), and $|\cdot|$ represents the amplitude of the signals.

The permutaion ambiguity is considered resolved when the above operation is completed for all frequency points.

3. **Energy Correlation Sorting Algorithm Based on Frequency Correction**

Although the amplitude correlation sorting algorithm has the advantages of small computation and low complexity, when dealing with the permutation ambiguity, the algorithm may occur "one step is wrong, every step is wrong" due to the strong correlation of the sorting on the successive frequency points, therefore, how to improve the robustness of the algorithm is an urgent problem to be solved. On this basis, an energy correlation sorting algorithm based on frequency correction is proposed.

The signals on the neighboring frequency points have strong correlation, i.e., the correlation of the signals decreases with the increase of the distance between the frequency points. Figure 8.4 shows the amplitude correlation plot between frequency points of the same signal, the lighter color in the plot represents the larger value of the amplitude correlation coefficient of the signal between the frequency points. The amplitude correlation coefficient can well reflect that the correlation of the same source signal at the neighboring frequency points is significantly larger than the correlation between the distant frequency points. However, within a certain neighborhood, the magnitude correlation coefficients are kept at a relatively high level with almost no difference. In this range, the amplitude correlation plot can not reflect relationship between correlation and distance well, which will inevitably affect the performance of the algorithm using signal correlation for sorting.

The energy correlation plot between frequency points of the same signal is shown in Fig. 8.5. By comparing Fig. 8.4 with Fig. 8.5, we can see that utilizing the energy correlation coefficient of a signal can effectively compensate for the above shortcoming. It can more clearly reflect the trend that the correlation of a signal between neighboring frequency points decreases with increasing distance. Because the larger

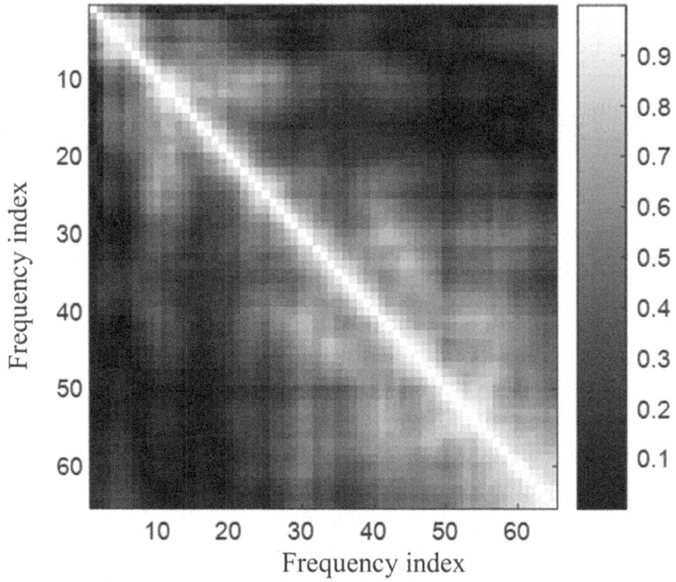

Fig. 8.4 Amplitude correlation plot between frequency points for the same signal

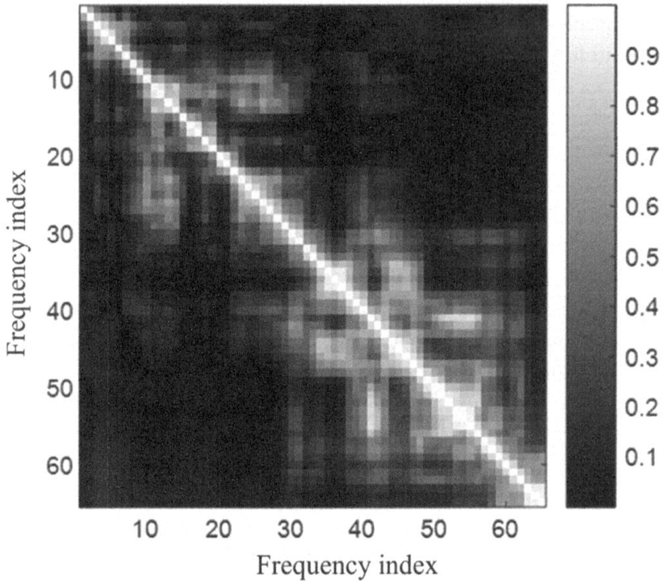

Fig. 8.5 Energy correlation plot between adjacent frequency points for the same signal

the distance between frequency points, the worse their reliability, using the energy correlation of the signal can reduce the impact of these less reliable frequency points on the performance of the algorithm to a certain extent. Therefore, if the energy correlation of signals between neighboring frequency points is used for sorting, the neighboring frequency points with different reliability are reasonably assigned different reference weights, so that more accurate sorting results can be obtained. Assuming that the alignment of the separated signals at the sorted frequency point g is Π_g, the sorting Π_f at the frequency point f can be obtained from equation (8.57).

$$\Pi_f = \arg\max_{\Pi} \sum_{|g-f|\le L} \sum_{i=1}^{N} cor\left(\left|\boldsymbol{Y}_i^{\Pi}(f)\right|^2, \left|\boldsymbol{Y}_i^{\Pi_g}(g)\right|^2\right) \tag{8.57}$$

Due to the presence of noise and the lack of accuracy of the blind separation algorithm in the frequency domain, which can affect the energy correlation of the signals between neighboring frequency points, errors can still occur in the sorting process. Since the ordering of the signals at each frequency point serves as a reference for the ordering at its subsequent frequency points, if there is an error in the order of the signals at a certain frequency point, this will inevitably affect the accuracy of the ordering at subsequent frequency points. In order to effectively prevent the error from continuing backward, we introduce a correction step, i.e., to judge whether the ordering of the frequency points is reliable, and correct the unreliable frequency points in time. Specifically, we need to calculate the average energy correlation coefficient between a frequency point and the frequency points in the reference neighborhood range, and compare it with the set threshold value U_{th}. If the average energy correlation coefficient is greater than or equal to the threshold, the ordering at the frequency point is considered reliable, and the ordering is postponed to the next frequency point; if the average energy correlation coefficient is less than the threshold, the ordering at the frequency point is judged unreliable, and the signal at the current frequency is replaced by the signal at the previous ordering frequency point.

The selection of the threshold value plays a crucial role in judging whether the sorting is correct or not, when the average energy correlation coefficient is greater than the threshold value of U_{th}, it can be assumed that the signal at the frequency point is sorted correctly, or else the signal at the frequency point should be corrected. We can choose a combination of constants and adaptive variables to determine the threshold in order to make a more reliable judgment, i.e., $U_{th} = \min(u_1, u_2)$, where u_1 is a constant chosen between 0 and 1, u_2 is the percentage of the maximum value of the average energy correlation coefficient in the sorted frequency point, notation $u_2 = \beta\% * \rho_{\max}$, $\beta \in [0, 100]$. It is known through simulation experiments that the algorithm gets a better performance when the threshold value is chosen as $\mathrm{U_{th}} = \min(0.6, 70\% * \rho_{\max})$.

In summary, the steps of the energy correlation sorting algorithm based on frequency correction are as follows:

Step 1: Do STFT for the observed signal $x_1(t), \ldots, x_N(t)$, respectively, to obtain an instantaneous mixed signal $\mathrm{X}_1(m,f), \ldots, \mathrm{X}_N(m,f)$ in the frequency domain;

Step 2: Instantaneous blind source separation is performed on the frequency-domain mixed signal $\mathrm{X}_1(m,f), \ldots, \mathrm{X}_N(m,f)$ to obtain separated signals $\mathrm{Y}_1(m,f), \ldots, \mathrm{Y}_N(m,f)$ at each frequency point, and amplitude ambiguity are removed at each frequency point using the minimum distortion criterion described in Sect. 8.2.2;

Step 3: Given the most suitable neighborhood length $L_{\max}$ (through simulation experiments, it can be seen that the neighborhood length of 6 is generally a better balance between computational complexity and performance), based on the signal order at the first frequency point, signal order on the subsequent frequency points is adjusted accordingly. The second frequency point is taking as the starting position for sorting operation, and let the frequency point to be sorted be f_k. If $|f_k - f_1| < L_{\max}$, the neighborhood length $L = |f_k - f_1|$, else $L = L_{\max}$. Then the frequency points are sorted according to the Eq. (8.50).

Step 4: Calculate the energy correlation coefficient ρ_{ij} ($\rho_{ij} = cor(|Y_i(f)|^2, |Y_j(g)|^2)$) between the signal at the current frequency and the signal at each frequency in the neighborhood. Average ρ_{ij} to obtain the average energy correlation coefficient C and compare it with the threshold value U_{th}. If the average energy correlation coefficient C is smaller than the set threshold value U_{th}, then the signal on the current frequency is replaced by the signal on the previous frequency obtained by the energy correlation sorting algorithm. If the average energy correlation coefficient C is greater than or equal to the threshold value U_{th}, then the sorting continues for the next frequency point.

Step 5: Repeating the above process until the last frequency point, the signal is restored to the time domain through ISTFT to obtain the final estimated signal $y_1(t), \ldots, y_N(t)$.

In subsequent simulations, the algorithm presented in this section will be referred to as the Rectify-E-Murata algorithm.

The algorithm flowchart is shown in Fig. 8.6.

4. **Independent Vector Analysis Algorithm**

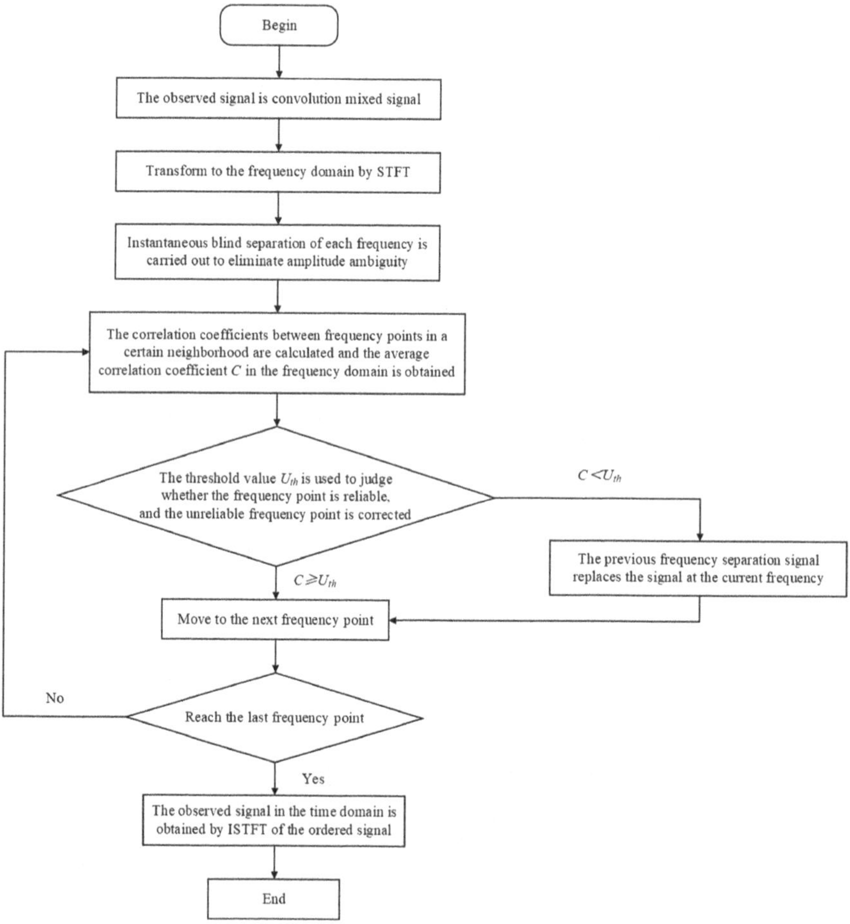

Fig. 8.6 Process flowchart of energy-correlation sorting algorithm based on frequency correction

Independent vector analysis (IVA) is a generalization algorithm of independent component analysis (ICA). The idea of traditional frequency domain method is to use ICA to separate the signal with uncertain ordering at each frequency point, while IVA algorithm uses the internal relationship between the elements in the multidimensional vector component to separate the independent components of all frequency points from the whole frequency domain, thus avoiding the problem of permutation ambiguity. This makes IVA algorithm one of the most effective methods to solve the permutation ambiguity problem. The model and mathematical principles of IVA are described below.

The model of IVA consists of several ICA models in frequency domain. Since the source signal components at different frequency points are mappings of the same signal in the frequency domain, there is an intrinsic relationship between the signals at each frequency point of the same source. The IVA algorithm utilizes this connection to summarize all the frequency components into a multidimensional frequency domain vector. This multidimensional frequency domain vector is separated as a whole during separation, and each component of the vector represents the separated signal at each frequency point.

Figure 8.7 represents a two-input two-output IVA mixing model. Each layer in this model represents a single frequency instantaneous mixing model represented by Eq. (8.34) (the time index is omitted here to facilitate the derivation of the formulas later). Thus, the mixing model of IVA matches the mixing model of convolutional blind separation in the time-frequency domain. Because the separation process of each frequency point is not independent but unified in IVA algorithm, the algorithm avoids the appearance of permutation ambiguity theoretically.

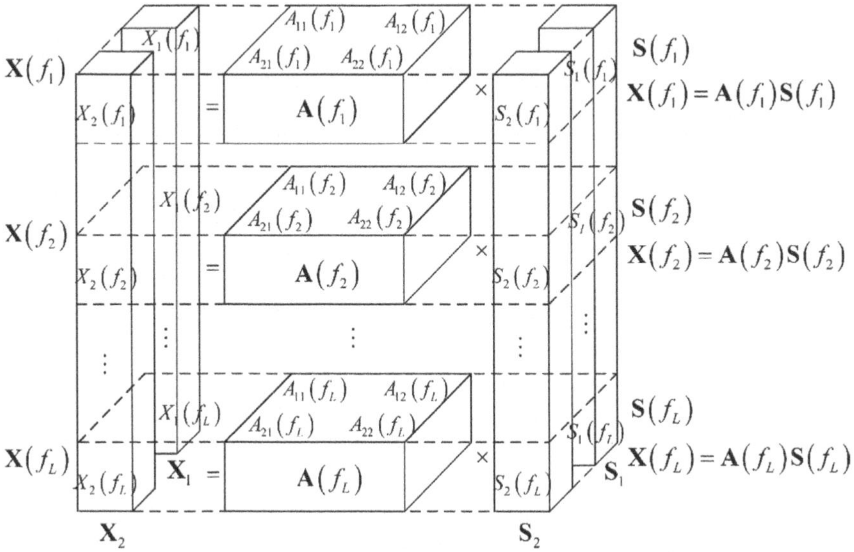

Fig. 8.7 Mixing model of the independent vector analysis algorithm

The IVA algorithm and the traditional ICA algorithm are similar in principle, and also need to select the objective function and the optimization process. The selection of objective function is related to the measure of independence, which can be measured by KL divergence, as shown in Eq. (8.58).

$$\Phi(\boldsymbol{Y}) = \mathrm{KL}\left(p(\boldsymbol{Y}), \prod_{i=1}^{N} q(\boldsymbol{Y}_i)\right) = \int p(\boldsymbol{Y})\ln\frac{p(\boldsymbol{Y})}{\prod_{i=1}^{N} q(\boldsymbol{Y}_i)}\mathrm{d}\boldsymbol{Y}$$
$$= \int p(\boldsymbol{Y})\ln p(\boldsymbol{Y})\mathrm{d}\boldsymbol{Y} - \sum_{i=1}^{N}\int p(\boldsymbol{Y})\ln q(\boldsymbol{Y}_i)\mathrm{d}\boldsymbol{Y} \tag{8.58}$$

where $\boldsymbol{Y}_i = \left[Y_i(f_1), Y_i(f_2), \ldots, Y_i(f_L)\right]^T$, $p(\boldsymbol{Y}) = p(\boldsymbol{Y}_1, \boldsymbol{Y}_2, \ldots, \boldsymbol{Y}_i, \ldots, \boldsymbol{Y}_N)$ is the joint probability density functions, and $\prod_{i=1}^{N} q(\boldsymbol{Y}_i)$ denotes the product of the individual signal probability density functions.

The first term on the right-hand side of the equal sign of Eq. (8.58) has the form of differential entropy $H(\cdot)$, then

$$\Phi(\boldsymbol{Y}) = -H(\boldsymbol{Y}) - \sum_{i=1}^{N}\int p(\boldsymbol{Y})\ln q(\boldsymbol{Y}_i)\mathrm{d}\boldsymbol{Y} \tag{8.59}$$

Also because $\boldsymbol{Y}(f) = \boldsymbol{W}(f)\boldsymbol{X}(f)$, the differential entropy can be introduced in the following form:

$$H(\boldsymbol{Y}) = H(\boldsymbol{X}) + \sum_{f} \ln|\det(\boldsymbol{W}(f))| \tag{8.60}$$

Since the signal has been whitened and de-averaged, $H(\boldsymbol{X})$ in Eq. (8.53) is necessarily constant.

For the second term on the right-hand side of the equality sign of Eq. (8.58), there is:

$$\int p(\boldsymbol{Y})\ln q(\boldsymbol{Y}_i)\mathrm{d}\boldsymbol{Y} = \int \ln q(\boldsymbol{Y}_i)\int p(\boldsymbol{Y})\mathrm{d}\overset{\smile i}{\boldsymbol{Y}}\,\mathrm{d}\boldsymbol{Y}_i$$
$$= \int p(\boldsymbol{Y}_i)\ln q(\boldsymbol{Y}_i)\mathrm{d}\boldsymbol{Y}_i$$
$$= E(\ln q(\boldsymbol{Y}_i)) \tag{8.61}$$

where $\overset{\smile i}{\boldsymbol{Y}}$ represents the set of vectors after removing $\boldsymbol{Y}_i$ from $\boldsymbol{Y}$.

Thus, the objective function can be written in the following form:

$$\Phi(\boldsymbol{W}, \boldsymbol{Y}) = -\sum_{f} \ln|\det(\boldsymbol{W}(f))| - \sum_{i=1}^{N} E \ln q(\boldsymbol{Y}_i) \tag{8.62}$$

where $q(\boldsymbol{Y}_i)$ is the probability density function, which is selected with respect to the type of source signal. When the source signal is a communication signal, $q(\boldsymbol{Y}_i)$ can be

$$q(\boldsymbol{Y}_i) = e^{-\|\boldsymbol{Y}_i\|_2} \tag{8.63}$$

Thus the final form of the objective function is

$$\Phi(\boldsymbol{W}) = -\sum_{f} \ln|\det(\boldsymbol{W}(f))| + \sum_{i} E\left[G\left(\sum_{f} |Y_i(f)|^2\right)\right] \tag{8.64}$$

where $G(z) = \sqrt{z}$ denotes the nonlinear function corresponding to $-\ln q(\boldsymbol{Y}_i)$.

The natural gradient method is selected to optimize the objective function, and finally the iterative formula of the separation matrix can be derived as

$$\boldsymbol{w}_{i,k+1}(f) = \boldsymbol{w}_{i,k}(f) + \eta_k \cdot \left\{\boldsymbol{I} - E\left[G\prime\left(\sum_{f} \left|Y_{i,k}(f)\right|^2\right) Y_{i,k}(f)^{\mathrm{T}}\right]\right\} \boldsymbol{w}_{i,k}(f) \tag{8.65}$$

where k denotes the number of iterations, $\boldsymbol{w}_{i,k}(f)$ represents the i-th row of the separation matrix $\boldsymbol{W}(f)$, η_k denotes the step-size, and $G'(\cdot)$ denotes the first order derivative of $G(\cdot)$.

In fact, the separation principle of IVA is similar to that of ICA, and separation matrix is still solved separately at each frequency point, but its iteration and optimization process is carried out at all frequency points, and the resulting separation matrix at each frequency point is interrelated, and the sorting information they contain is uniform.

5. Step-Size Adaptive IVA Convolutional Blind Separation Algorithm

The IVA algorithm is one of the most effective methods for solving the permutation ambiguity problem, but the algorithm still has shortcomings: (1) Since each iteration of IVA algorithm needs to calculate the objective function value of the separation matrix of all frequency points, it inevitably leads to a large amount of computation for each optimization of the algorithm. In addition, the initialization of the separation matrix is not targeted, resulting in many iterations and long operation time. In a few cases, the mismatch of initial values leads to local convergence of the algorithm and affects the separation effect. (2) The optimization process of the algorithm mainly adopts the gradient class method, which requires the selection of the step-size. The step size increases, the convergence speed is accelerated, but at the same time this may cause the instability of the system, so that the convergence accuracy decreases;

on the contrary, the step size decreases, the stability of the system is enhanced, but the convergence speed is also relatively reduced, so that the operation time is further increased.

In view of the above shortcomings, an IVA convolutional blind separation algorithm based on step-size adaptation is proposed. The algorithm uses the JADE algorithm to initialize the separation matrix, so that the separation matrix can obtain a reasonable initial value for the mixed signals, which effectively avoids the emergence of local convergence; and the algorithm optimizes the step-size during iteration. Compared with the traditional IVA algorithm with fixed step size, it can significantly improve the convergence speed while maintaining the convergence accuracy.

Among the instantaneous blind separation algorithms, the JADE algorithm is efficient and robust, and can better separate the linear mixed signals in the complex domain, so it is widely used in the blind separation of signals in the frequency domain. The initial value generated by the JADE algorithm is closer to the final result, which can reduce the number of iterations and shorten the operation time. When the convolution order becomes larger, the IVA algorithm needs to operate on the separation matrix of more frequency points, and the solution process will be more complicated. In this case, it is often unreasonable to use the identity matrix as the initial value, which makes the algorithm easy to fall into local convergence. It is more reasonable to take the separation matrix obtained by JADE algorithm as the initial value, which can avoid the occurrence of local convergence and obtain the global optimal value.

In the IVA algorithm, the optimization speed of separation matrix $\boldsymbol{W}$ is determined by the step size η. Usually, it is difficult for the algorithm to balance the relationship between convergence speed and convergence stability by using a fixed step size. In order to accelerate the convergence and maintain the accuracy at convergence, the step-size adaptive IVA algorithm can be obtained by using the idea of step-size optimization.

Suppose that on the $(k+1)$-th iteration, the step size is

$$\eta_{k+1} = \eta_k + \gamma \times \Delta\eta_k \tag{8.66}$$

where $\Delta\eta_k$ denotes the update amount for each iteration of step η_k and γ is a smaller constant. The significance of the update amount is to make the step size η_{k+1} converge to the optimal value, i.e., to make the objective function Φ can be minimized at the next iteration. Then

$$\Delta\eta_k = \frac{\partial \Phi_{k+1}}{\partial \eta_k} \tag{8.67}$$

For the step size at each frequency, there is

$$\Delta\eta_{k,f} = \frac{\partial \Phi(\mathbf{W}_{k+1}(f))}{\partial \eta_k} = \left\langle \frac{\partial \Phi(\mathbf{W}_{k+1}(f))}{\partial \mathbf{W}_{k+1}(f)} \cdot \frac{\partial \mathbf{W}_{k+1}(f)}{\partial \eta_k} \right\rangle$$

$$= \text{trace}\left\{\left[\frac{\partial \Phi(\boldsymbol{W}_{k+1}(f))}{\partial \boldsymbol{W}_{k+1}(f)}\right]^{\mathrm{T}} \cdot \frac{\partial \boldsymbol{W}_{k+1}(f)}{\partial \eta_k}\right\} \tag{8.68}$$

where $\langle\cdot\rangle$ is to find the inner product of the matrix and trace$(\cdot)$ is to find the trace of the matrix.

On the one hand, according to the iterative formula (8.64) of the IVA algorithm, we have

$$\frac{\partial \boldsymbol{W}_{k+1}(f)}{\partial \eta_k} = \left[\boldsymbol{I} - f(\boldsymbol{Y}_k) \cdot \boldsymbol{Y}_k^{\mathrm{T}}(f)\right]\boldsymbol{W}_k(\omega) \triangleq \Delta \boldsymbol{W}_k(f) \tag{8.69}$$

where the nonlinear function $f(\boldsymbol{Y}_k) = G\prime\left(\sum_f |\boldsymbol{Y}_k(f)|^2\right)$. On the other hand, $\Phi(\boldsymbol{W})$ is an instantaneous estimate of the objective function based on the natural gradient, therefore

$$\frac{\partial \Phi(\boldsymbol{W}_{k+1}(f))}{\partial \boldsymbol{W}_{k+1}(f)} = \left[\boldsymbol{I} - f(\boldsymbol{Y}_{k+1}) \cdot \boldsymbol{Y}_{k+1}^{\mathrm{T}}(f)\right]\boldsymbol{W}_{k+1}(f) \triangleq \Delta \boldsymbol{W}_{k+1}(f) \tag{8.70}$$

Substituting Eqs. (8.69) and (8.70) into Eq. (8.68), we have

$$\Delta \eta_{k,f} = \text{trace}\left(\Delta \boldsymbol{W}_{k+1}^{\mathrm{T}}(f) \cdot \Delta \boldsymbol{W}_k(f)\right) \tag{8.71}$$

Finally, the update amount of the iterative step-size of all frequency points is unified, and the following is obtained.

$$\Delta \eta_k = \frac{1}{N}\sum_{f=1}^{N} \Delta \eta_{k,f} \tag{8.72}$$

In summary, the final step-size update formula can be obtained as:

$$\eta_{k+1} = \eta_k + \gamma \cdot \frac{1}{N}\sum_{f=1}^{N} \text{trace}\left(\Delta \boldsymbol{W}_{k+1}^{\mathrm{T}}(f) \cdot \Delta \boldsymbol{W}_k(f)\right) \tag{8.73}$$

In addition, an upper limit and a small positive lower limit should be set for the value of η to avoid extreme values of the step-size.

In summary, the steps of IVA convolutional blind separation algorithm based on step-size adaptation are as follows [46]:

Step 1: Convert the convolutional mixed signal to instantaneous mixed signal at each frequency point using STFT;

Step 2: Pre-whiten the mixed signal at each frequency point;

Step 3: Initialize the separation matrix $\boldsymbol{W}_0(f)$ for each frequency point by JADE algorithm and set the initial iterative step-size η_0;

Step 4: Separate the signal at each frequency point based on $\boldsymbol{Y}(f) = \boldsymbol{W}(f)\boldsymbol{X}(f)$;
Step 5: According to separated signal, find the objective function value and the nonlinear function value, and judge whether it converges. If convergence, perform step 9;
Step 6: Update the separation matrix for each frequency point based on Eq. (8.65);
Step 7: Update the iteration step-size based on Eq. (8.73);
Step 8: If the maximum number of iterations is reached, perform step 9, otherwise return to step 4;
Step 9: The estimated source signals are obtained by converting the separated signals at each frequency point to time domain separated signals using ISTFT.

In subsequent simulations, the algorithm introduced in this section will be referred to as the JS-IVA algorithm (JADE initialization and Step-size adaptive IVA).

8.2.4 Algorithm Performance Simulation Analysis

1. **Performance Comparison of Amplitude Correlation and Energy Correlation Sorting Algorithms**

Two speech signals are selected as the source signals, and the performance of the improved algorithm is tested under the filter convolutional mixing model of different orders and the simulated indoor convolutional mixing model, in which the simulated indoor environment has a room size of $4\text{ m} \times 4\text{ m} \times 2.5\text{ m}$, and the impulse response function of the generated mixing filter has an order of more than 400 orders. Meanwhile, Gaussian white noise with different signal-to-noise ratio (SNR) are mixed in the blind separation process to test the performance of the algorithm at different SNRs.

100 experiments are carried out under the 10-order and 20-order convolutional mixing model and 500 experiments are carried under the simulated indoor mixing model, and the average of the signal-to-interference ratios(SIR) of all the separated signals is calculated and will be used as the final result.

Figures 8.8, 8.9, and 8.10 show the comparative results of the Rectify-E-Murata algorithm, the Murata algorithm, and the Anemuller algorithm in terms of their performance in separating speech signals in the convolutional mixing of 10-order and 20-order order filters and in the simulated indoor convolutional mixing environments, respectively.

Comparing the signal-to-interference ratio (SIR) values in the above simulation results, we can see that the Rectify-E-Murata sorting method has significantly better separation performance compared to the Murata algorithm and the Anemuller algorithm. In the 10-order convolutional mixing model, the performance of Rectify-E-Murata sorting algorithm is about 3 dB higher than Murata algorithm and 4 dB higher than Anemuller algorithm.In the 20-order convolutional mixing model, the SIR of Rectify-E-Murata sorting algorithm is about 2 dB higher than Murata algorithm

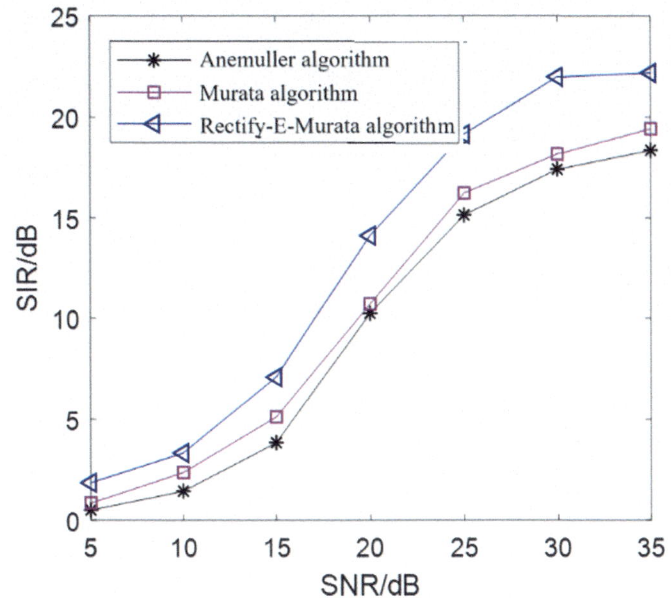

Fig. 8.8 Performance comparison of algorithms under a 10-order convolution mixing model

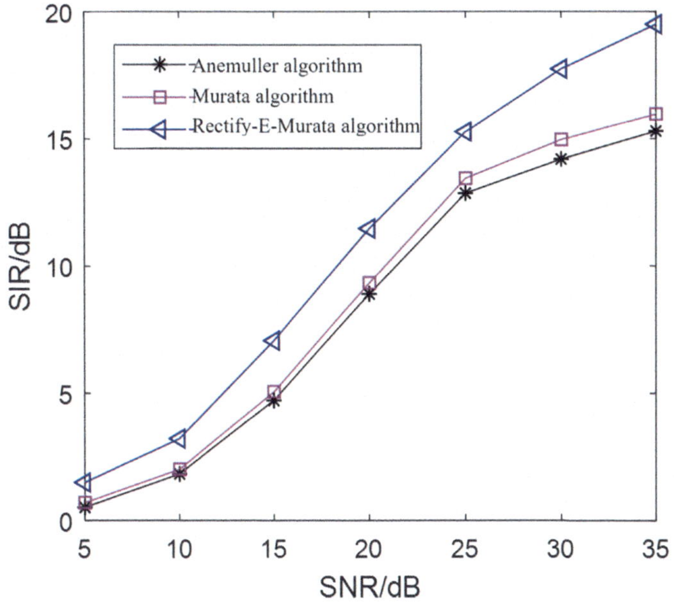

Fig. 8.9 Performance comparison of algorithms under a 20-order convolution mixture model

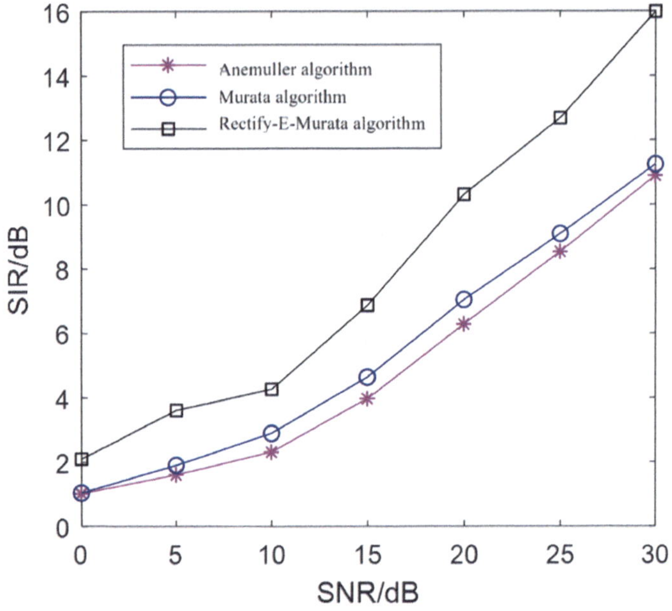

Fig. 8.10 Performance comparison of algorithms under simulated indoor convolution mixing model

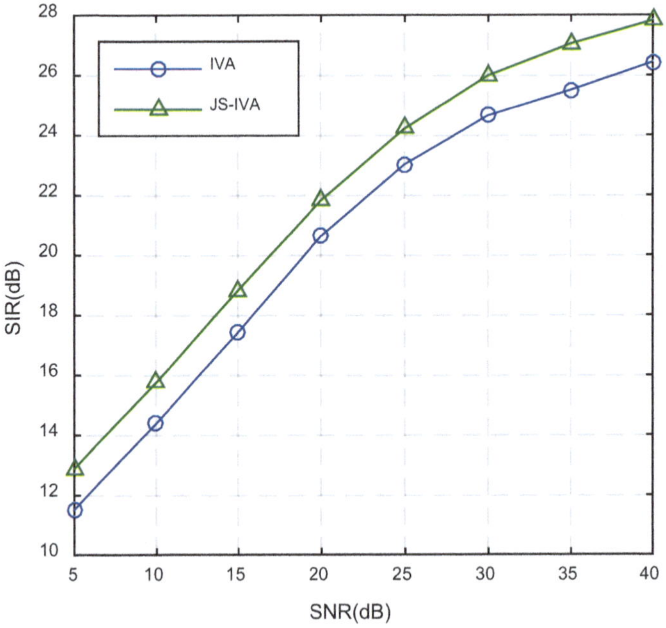

Fig. 8.11 Performance comparison of separation (5-order convolution mixing model, nfft = 128)

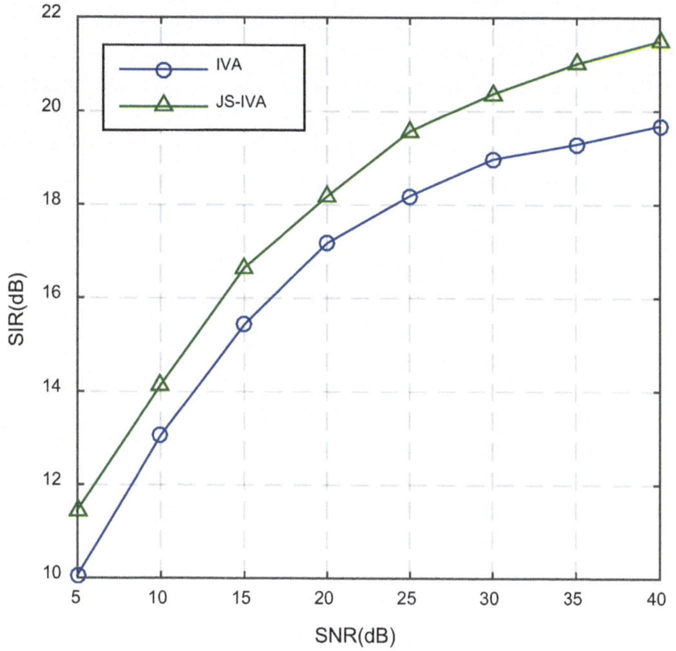

Fig. 8.12 Performance comparison of separation (20-order convolution mixing model, nfft = 256)

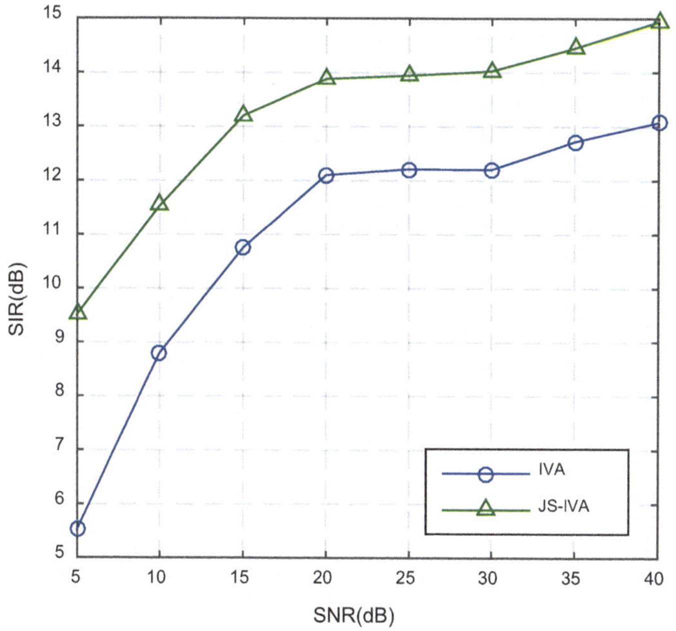

Fig. 8.13 Performance comparison of separation (simulated indoor mixing model, nfft = 2048)

and 3 dB higher than Anemuller algorithm. In the simulated indoor convolutional mixing environment, the Rectify-E-Murata sorting algorithm has a more obvious performance advantage over the other two algorithms, which also indicates that the performance of the other two algorithms is more obviously affected in the high-order convolutional mixing model. As can be seen from the figure, the Rectify-E-Murata sorting algorithm outperforms the Anemuller sorting algorithm by about 2–5 dB and outperforms the Murata sorting algorithm by about 2–4 dB.

2. **Performance Comparison between IVA Algorithm and JS-IVA Algorithm**

Two speech signals are used as source signals for convolution mixing. The signal-to-interference ratio (SIR) is used as a performance measure. The convolutional mixing environment is set up as a 5-order filter, a 20-order filter and a filter for simulated indoor mixing. The parameters of the simulated indoor mixing are a room of 5 m $\times$ 5 m $\times$ 5 m with a microphone spacing of 20 cm, and the generated mixing filter order is about 450 or so. The experiments are simulated by adding Gaussian white noise to the observed signal with SNR as the variable. The parameters of two algorithms are as follows.

IVA algorithm: the step-size is fixed to $\eta = 0.1$, the initial of the separation matrix is an identity matrix, and the maximum number of iterations is 1000.

JS-IVA algorithm: step-size initial value $\eta_0 = 0.1$, step-size update parameter $\gamma = 0.001$, JADE algorithm is used to initialize the separation matrix of each frequency point, and the maximum number of iterations is 1000.

The average SIRs are shown in Figs. 8.11, 8.12 and 8.13.

As can be seen from Figs. 8.11, 8.12 and 8.13, the performance of the JS-IVA algorithm is improved by 1–2 dB compared with the IVA algorithm in the case of convolutional mixing model of lower order (such as 5-order and 20-order); in the case of simulated indoor mixing model with of 400-order, the SIR of the separated signals of the JS-IVA algorithm improves more significantly, by about 2–4 dB.

In the single iteration operation of the IVA algorithm, the calculation process of the objective function needs to perform matrix operations on the separated signals at all frequency points, which is time-consuming, so the key to improve the efficiency is to reduce the number of iterations. The JS-IVA algorithm adds the JADE initialization process before iteration, and the step-size updating process is added to the process of each iteration, which will unavoidably increase the computation amounts of initialization and single iteration, but the number of iterations will also be significantly reduced, so the time consuming of the JS-IVA algorithm will be significantly reduced.

Table 8.1 shows the total time values of the algorithm running in the simulation experiments. As seen in Table 8.1, the JS-IVA algorithm runs 48.4% more efficiently than the IVA algorithm for the same amount of total simulation.

For further comparison, Table 8.2 lists the average time values for a single iteration of the two algorithms in each set of simulation experiments; and Table 8.3 lists the average number of iterations and the average runtime of the two algorithms for a single blind separation process in each set of simulation experiments.

Table 8.1 Simulation runtime of different algorithms

Algorithm	The number of iterations	Total running time(s)
IVA	1840	23803.8
JS-IVA	1840	12283.4

Table 8.2 Single iteration time (ms)

Algorithm	5-order	20-order	Indoor
IVA	27.81	34.73	51.18
JS-IVA	28.70	38.31	63.10

Table 8.3 Number of iterations and running time (numbers/s)

Algorithm	5-order	20-order	Indoor
IVA	114/3.17	167/5.80	658/33.68
JS-IVA	62/1.78	89/3.41	174/10.98

As can be seen from Tables 8.2 and 8.3, the average time of a single iteration of the JS-IVA algorithm is slightly higher than that of the traditional IVA algorithm due to the addition of the step size updating process. However, the JS-IVA algorithm has a significantly lower number of iterations compared to the traditional algorithm due to the optimization effect brought by the step-size update and JADE initialization, so overall the operational efficiency of the JS-IVA algorithm is significantly improved compared to the traditional algorithm. Under the conditions of 5-order convolutional mixing model and 20-order convolutional mixing model, the JS-IVA algorithm reduces the number of iterations by more than 45% and the running time by more than 40% compared with the IVA algorithm. Under complex simulated indoor convolution mixing model, the number of iterations is reduced by more than 73% and the running time is reduced by more than 65%.

Summary

In this chapter, the convolutional mixing blind source separation problem in multipath channel environment is investigated. Firstly, the mathematical model of the convolutional mixing blind source separation problem is analyzed, and it is pointed out that there are two basic ideas for solving this problem, namely, the time-domain method and the frequency-domain method. On this basis, the convolutional blind separation algorithm in the time domain and the convolutional blind separation algorithm in the frequency domain are introduced respectively. For the time-domain methods, JBDun algorithm, F-JBDun algorithm and JBDus algorithm are introduced respectively, the principles of the algorithms are analyzed in detail and the algorithm steps are summarized. For the frequency-domain convolutional blind separation algorithm, the basic idea of the algorithm is introduced firstly. After analysis, it is found that the most critical problem of the frequency-domain blind separation algorithm is the resolution of the permutation ambiguity problem, so this chapter focuses on several sorting algorithms, including the amplitude correlation sorting algorithm, energy correlation sorting algorithm, and independent vector analysis algorithm, etc. Finally, the separation performance of different algorithms is simulated and analyzed.

Chapter 9
Sufficiently Sparse Underdetermined Mixing Matrix Estimation Technique

In solving the underdetermined blind source separation (UBSS) problem by using the idea of sparse component analysis (SCA), the "two-step method" is the most commonly used method, i.e., the mixing matrix is estimated in the first step, and then the individual source signals are recovered in the second step. Therefore, the estimation accuracy of the mixing matrix will directly affect the recovery effect of the source signal in the second step. The estimation method of the mixing matrix is related to whether the source signal is sufficiently sparsee. A sufficiently sparse source signal means that there is at most one non-zero source signal (or a source signal with a large value) at any time.

This chapter focuses on underdetermined mixing matrix estimation techniques under the condition of sufficiently sparse source signal, including sufficiently sparse underdetermined mixing matrix estimation technique based on clustering, similarity detection, and bullseye retrieval.

9.1 Sufficiently Sparse Underdetermined Mixing Matrix Estimation Technique Based on Clustering

9.1.1 Sufficiently Sparse Underdetermined Mixing Matrix Estimation Technique Based on k-*Means Clustering*

Cluster analysis is an important method in data analysis, a major task in knowledge discovery and data mining, and an important branch of unsupervised (uninstructed) pattern classification in pattern recognition. The essence of clustering is the process of grouping or classifying a dataset that does not mark any category. In this process, the dataset is divided into several subsets only on the basis of a certain degree of dissimilarity (or similarity) between the data, so that the similarity between the data

W. Fu, *Principle and Application of Blind Source Separation Technology*,
https://doi.org/10.1007/978-981-96-0022-9_9

of the same subset is as large as possible, and the similarity between the data of different subsets is as small as possible.

Suppose there are T observations $\boldsymbol{X} = [\boldsymbol{x}_1, \boldsymbol{x}_2, \ldots, \boldsymbol{x}_T]$, $\boldsymbol{x}_i$ is m-dimensional column vectors. The Euclidean distance is used as an evaluation criterion to measure whether two observations belong to the same class. That is, the smaller the Euclidean distance between two observations, the greater the possibility that the two observations belong to the same class. The calculation rule of Euclidean distance is shown in Eq. (9.1).

$$d(\boldsymbol{x}_i, \boldsymbol{x}_j) = \sqrt{(\boldsymbol{x}_i - \boldsymbol{x}_j)^{\mathrm{T}}(\boldsymbol{x}_i - \boldsymbol{x}_j)} \tag{9.1}$$

Let the observation data can be divided into k groups. Among all the observations, k observations are randomly selected as the initial clustering centers, and the Euclidean distance between each observation and k clustering centers can be obtained according to Eq. (9.1). If the distance from an observation to a certain clustering center is less than the set threshold, the observation is classified into that class. Then the new clustering centers are calculated by means of the mean value method in each category, and the classification is updated according to the Euclidean distance principle until the cluster centers do not change.

For UBSS, if the source signals are sufficiently sparse, then the observed data presents the characteristics of line clustering, and the observed data is distributed in the straight line direction determined by the column vectors of the mixing matrix $\boldsymbol{A}$. Figure 9.1 shows the observed signal in a noiseless background with source number $n = 3$, and observed signal number $m = 2$. If the observed signals are normalized, the observed signals are clustered at points determined by the column vectors of the normalized mixing matrix, as shown in Fig. 9.2 [47].

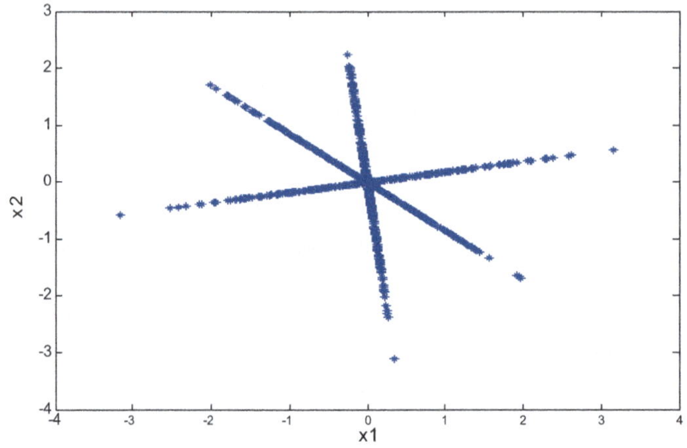

Fig. 9.1 Observed signals in the case of $n = 3$, $m = 2$

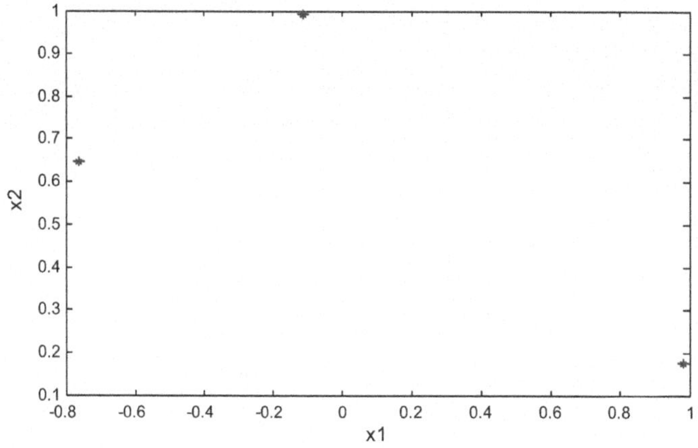

Fig. 9.2 Observed signals in the case of $n = 3$, $m = 2$

This clustering property of the observed signals in UBSS problem can be exactly solved using the k-means approach. As can be seen in Fig. 9.2, the normalized observed signals are clearly divided into 3 classes. The observed signals in the same class have very small Euclidean distances between them, much smaller than the Euclidean distances between them and the observed signals in the other classes. In the absence of noise, the Euclidean distance between observed signals in the same class is zero.

The steps of mixing matrix estimation based on k-means clustering are as follows:

Step 1: Normalize the observed signal.
Step 2: Identify k clustering centers.
Step 3: Calculate the distance from each observed signal to k clustering centers separately according to Eq. (9.1).
Step 4: Each observed signal is categorized into its nearest class and the observed signals in each class are averaged to get the new clustering center.
Step 5: If the clustering center obtained by two successive iterations is the same, the clustering center is the estimation of the mixed matrix; otherwise, go back to step (3).

9.1.2 Sufficiently Sparse Underdetermined Mixing Matrix Estimation Technique Based on Fuzzy C-Means (FCM) Clustering

Fuzzy C-means (FCM) clustering method is one of the most perfect and widely used algorithms in the theory of fuzzy clustering algorithm based on objective function. In order to solve the clustering problem by means of the objective function method, the nonlinear programming function with constraints is constructed by means of the mean square approximation theory.

The FCM divides the T observation vectors $\boldsymbol{x}(j)$, $(j = 1, 2, \ldots T)$ into C fuzzy groups and finds the clustering center of each group to minimize the cost function of the non-similarity index. The generalized form of the cost function (or objective function) of FCM is

$$J(\boldsymbol{U}, \boldsymbol{Z}) = \sum_{i=1}^{C} J_i = \sum_{i=1}^{C} \sum_{j=1}^{T} u_{ij}^{p} d^2(\boldsymbol{x}(j), z(i)) \tag{9.2}$$

where T represents the number of samples and C represents the number of clusters. $u_{ij} \subseteq \boldsymbol{U}$ is the membership function of vector $\boldsymbol{x}(j)$ belonging to class i, satisfies $u_{ij} \in [0, 1]$, and

$$\sum_{i=1}^{C} u_{ij} = 1, \forall j = 1, \ldots, T \tag{9.3}$$

$\boldsymbol{U}$ is a $C \times T$ membership matrix. The dimension of the observation vector $\boldsymbol{x}(t)$ is m, i.e.

$$\boldsymbol{x}(j) = [x_1(j), x_2(j), \ldots, x_m(j)]^{\mathrm{T}} \tag{9.4}$$

then

$$\boldsymbol{X} = [\boldsymbol{x}(1), \boldsymbol{x}(2), \ldots, \boldsymbol{x}(T)] \tag{9.5}$$

is an $m \times T$ matrix. The clustering center

$$\boldsymbol{Z} = [z(1), z(2), \ldots, z(C)] \tag{9.6}$$

is a matrix of $m \times C$.

$$d^2(\boldsymbol{x}(j), z(i)) = \|\boldsymbol{x}(j) - z(i)\|^2 \tag{9.7}$$

is the Euclidean distance between the i-th clustering center and the j-th data point, and $p \in [1, \infty)$ is a weighted index (p usually takes a value between 2 and 5).

Construct the following new objective function

$$\begin{aligned} \overline{J}(\boldsymbol{U}, \boldsymbol{Z}, \lambda_1, \ldots, \lambda_T) &= J(\boldsymbol{U}, \boldsymbol{Z}) + \sum\nolimits_{j=1}^{T} \lambda_j \left(\sum_{i=1}^{C} u_{ij} - 1 \right) \\ &= \sum_{i=1}^{C} \sum_{j}^{T} u_{ij}^{p} d^2(\boldsymbol{x}(j), z(i)) + \sum_{j=1}^{T} \lambda_j \left(\sum_{i=1}^{C} u_{ij} - 1 \right) \end{aligned} \tag{9.8}$$

where $\lambda_j \in [0, 1]$ is the Lagrange multiplier of the following T constraints

$$\sum_{i=1}^{C} u_{ij} = 1, \forall j = 1, \ldots, T \tag{9.9}$$

The necessary conditions to minimize Eq. (9.8) are

$$z(i) = \frac{\sum\limits_{j=1}^{T} u_{ij}^{p} \boldsymbol{x}(j)}{\sum\limits_{j=1}^{T} u_{ij}^{p}} \tag{9.10}$$

and

$$u_{ij} = \frac{1}{\sum\limits_{k=1}^{C} \left(\frac{d(\boldsymbol{x}(j), z(i))}{d(\boldsymbol{x}(j), z(k))} \right)^{2/(p-1)}} \tag{9.11}$$

According to Eqs. (9.10) and (9.11), cluster analysis is carried out on the observed data. After several iterations, when the algorithm converges, a set of clustering centers as well as the corresponding membership matrices $\boldsymbol{U}$ can be estimated. And the mean information entropy (ME) of $\boldsymbol{U}$ can be further derived. Firstly, the calculation formula of ME index is defined as

$$\mathrm{I}(k') = -\sum_{i=1}^{k'} \sum_{t=1}^{T} \left\{ \left[u_{it} \times \log 2(u_{it}) + (1 - u_{it}) \times \log 2(1 - u_{it}) \right] / T \right\} \tag{9.12}$$

where k' denotes the number of clusters and $\mathrm{I}(k')$ denotes the value of the ME index when the number of clusters is k'. The more reasonable the classification of clusters is, i.e., the the greater probability of observed signal sampling points belonging to a cluster, the smaller the ME index value of the cluster. Therefore, of the number of clusters k ($k \in [C_{\min}, C_{\max}]$) is given, where $C_{\min}$ and $C_{\max}$ denote the minimum and maximum number of clusters, respectively. Using the FCM and the given number of clusters, the observed signal sampling points are clustered. The membership matrix obtained under each cluster number is used to calculate the ME index value, which is used as the basis for determining the final cluster number of observation data, so as to obtain the most appropriate cluster number and the corresponding clustering center $\boldsymbol{Z}$, which will be used as the estimation of the mixed matrix.

In summary, the steps of the mixing matrix estimation method based on FCM clustering are as follows:

Step 1: Set the number of cycles $N_{k'}$, $N_{k'} = C_{\max} - C_{\min} + 1$ and set the initial values $k' = C_{\min}$. Where $C_{\min}$ and $C_{\max}$ denote the hypothesized minimum number of clusters and maximum number of clusters, respectively.

Step 2: Randomly generate a set of initial clustering centers $\tilde{\mathbf{Z}}_{k'}(\tilde{\mathbf{Z}}_{k'} \in R^{M \times k'})$.

Step 3: Based on the initial clustering centers $\tilde{\mathbf{Z}}_{k'}$, the FCM algorithm is used to iteratively compute the observed data, and when the algorithm converges, a new set of clustering centers $\mathbf{Z}_{k'}$ is obtained, as well as the membership matrix $\boldsymbol{U}_{k'}$ corresponding to the clustering centers.

Step 4: Calculate the ME index $\mathrm{I}(k')$ for the membership matrix $\boldsymbol{U}_{k'}$.

Step 5: Let $k' := k' + 1$, then perform step 2 to get another set of clustering centers and the corresponding ME index $\mathrm{I}(k')$ until $k' = C_{\max}$, at which time $N_{k'}$ ME index values are obtained.

Step 6: Search for the minimum of $N_{k'}$ of $\mathrm{I}(k)$ and use the clustering center $\mathbf{Z}_{k'}$ corresponding to this minimum as an estimate of the mixing matrix.

As can be seen from the above steps, the performance of the FCM algorithm depends on the selection of the initial clustering center. Therefore, before performing FCM clustering on the observed data, if certain measures can be taken to select a better initial clustering center and then use the clustering algorithm to perform clustering, it will improve the accuracy of the mixing matrix estimation.

9.1.3 Sufficiently Sparse Underdetermined Mixing Matrix Estimation Technique Based on Improved K-Means Clustering

Aiming at the problem that the original K-means clustering algorithm needs to set the number of clusters k and the clustering effect depends on the selection of initial cluster centers, some people proposed to select initial cluster centers based on the idea of density and distance optimization to reduce the sensitivity of the algorithm to initial cluster centers. In addition, in order to solve the problem of mixing matrix estimation when the number of source signals is unknown in underdetermined blind source separation, an improved k-means clustering algorithm is proposed, which redefines the radius of the neighborhood of the data object, and then selects the initial clustering centers based on the density and the distance, and determines the optimal number of clusters using the Davies-Bouldin Index (DBI).

1. **Select the Initial Clustering Center Based on Density and Distance**

Let the set of data to be clustered be $\tilde{\mathrm{X}}$, $\tilde{\mathrm{X}} = \{\tilde{x}_i | \tilde{x}_i \in R^M, i = 1, 2, \ldots T\}$, given the following definition.

Definition 9.1 Neighborhood radius of data object.

The neighborhood radius r of data object is the average of the minimum distance between all objects, calculated as follows.

$$r = \frac{1}{T} \times \sum_{i=1}^{T} \min(d(\tilde{\boldsymbol{x}}_i, \tilde{\boldsymbol{x}}_j)), j = 1, 2 \ldots, T, j \neq i \tag{9.13}$$

Definition 9.2 Density parameter of data objects.

The number of data objects in a neighborhood with any data object $\tilde{\boldsymbol{x}}_i$ as the center and r as the radius is called density parameter $B_r(\tilde{\boldsymbol{x}}_i)$ of $\tilde{\boldsymbol{x}}_i$ based on distance r, which is calculated as follows:

$$\begin{cases} B_r(\tilde{\boldsymbol{x}}_i) = \sum_{j=1}^{n} u(r - d(\tilde{\boldsymbol{x}}_i, \tilde{\boldsymbol{x}}_j)) \ , \ i = 1, \ldots, n \\ u(x) = \begin{cases} 1 \ x \geq 0 \\ 0 \ x < 0 \end{cases} \end{cases} \tag{9.14}$$

where $d(\tilde{\boldsymbol{x}}_i, \tilde{\boldsymbol{x}}_j) = \left\| \tilde{\boldsymbol{x}}_i - \tilde{\boldsymbol{x}}_j \right\|$ is the Euclidean distance between any two data objects in set $\tilde{X}$ and $\|\cdot\|$ denotes the 2-norm.

Let the number of source signals $N = 5$ and the number of observed signals $M = 3$. Under the condition that the source signals are sufficiently sparse, the observed signals are shown in Fig. 9.3, which shows that the observed signals have linear clustering characteristics, the number of the straight line is the number of source signals, and the direction of the straight line is determined by the column vectors of the mixing matrix.

The observed signals are projected onto the unit semi-hypersphere, as shown in Fig. 9.4. It can be seen that the projected points of the sampled data also have

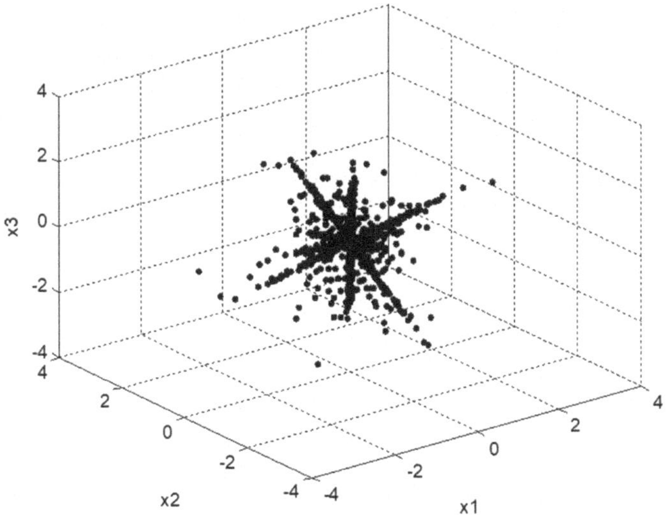

Fig. 9.3 Scatter plot of observed signals for $N = 5$, $M = 3$

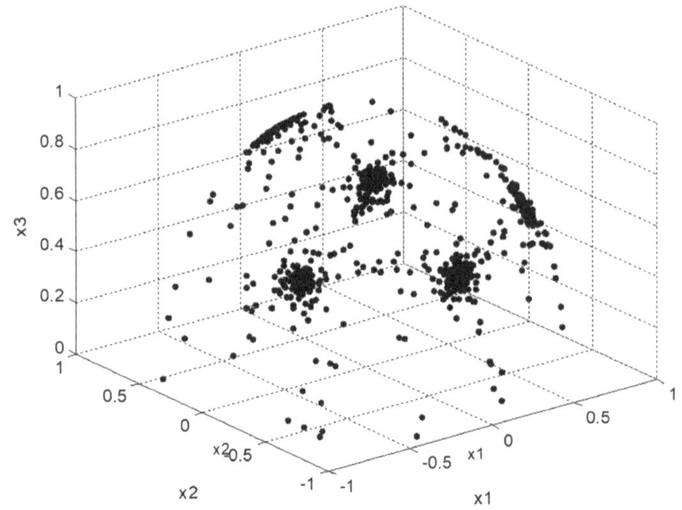

Fig. 9.4 Distribution of projected points of observed signals on the unit hyperplane

clustering characteristics. The projected points corresponding to the column vectors of the mixing matrix are the clustering centers, while the projected points of the sampling data are distributed around the clustering centers, and the number of clusters is the number of source signals.

Because in practice, the source signals cannot have only one value that is not zero at any time, coupled with the influence of noise, there will be some outliers, and the projection points corresponding to outliers and noise are far away from the clustering centers. Under the condition of sufficiently sparse source signals, the number of useful signal sampling points is much greater than the number of outliers. According to Definition 9.2, within the same radius, the density parameter of the projected points corresponding to outliers is smaller than that of sampling points of useful signals. Based on this characteristic, Eq. (9.13) can be employed to calculate the density parameter of each projected point, and subsequently remove the projected points with smaller density parameters to mitigate the impact of noise and outliers on the clustering results. The initial cluster center is selected from the projection point with the larger density parameter. Meanwhile, to prevent the selected initial clustering center from gathering in the region with the highest density of the sampled data, the first initial clustering center is selected from the point with the largest density parameter, and the remaining clustering centers are selected according to the principle of the maximuming the minimum distance, which reduce the sensitivity of the clustering results to the initial cluster center selection.

After removing the low-density projected points, the distribution of the sample data on the unit hypersphere is depicted in Fig. 9.5, which clearly shows five clusters and removes the majority of outliers.

2. **Determining the Optimal Number of Clusters by DBI**

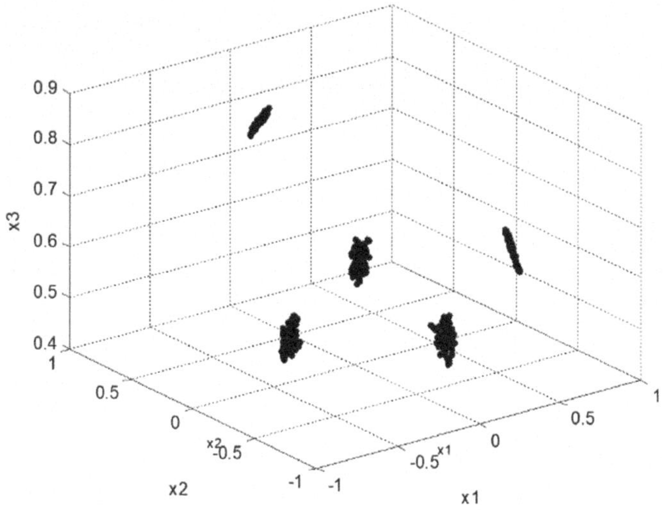

Fig. 9.5 Distribution of observed data after removing low-density points

The Davies-Bouldin Index (DBI) is a non-fuzzy cluster evaluation index, primarily based on the degree of dispersion between clusters and the compact degree of data objects located in the same cluster. The higher the similarity among data objects within the same cluster and the lower the similarity between different clusters, the smaller the value of the DBI and the better classification performance. The similarity between data is measured by the Euclidean distance.

Definition 9.3 Between-class distance

$$M_{ij} = \left\| \tilde{z}_i - \tilde{z}_j \right\| \tag{9.15}$$

where $\tilde{z}_i$ and $\tilde{z}_j$ respectively represent the i-th and j-th clustering center.

Definition 9.4 Within-class distance

$$S_i = \frac{1}{T_i} \sum_{p=1}^{T_i} \left\| \tilde{x}_q - \tilde{z}_i \right\| \tag{9.16}$$

where T_i represents the number of data objects in the i-th class, and $\tilde{x}_q, q = 1, \ldots, T_i$ represents the data objects in the i-th class.

Definition 9.5 DBI

$$I = \frac{1}{K} \sum_{i=1}^{K} \max_j \left(\frac{S_i + S_j}{M_{ij}} \right), j = 1, \ldots K, j \neq i \tag{9.17}$$

When the between-class distance M_{ij} increases, the average distance S_i and S_j between the data objects within the class and the clustering center decreases, resulting in a smaller DBI value. When the DBI value is minimized, it indicates the optimal clustering result for the given number of clusters.

The DBI indicator has the following two advantages:

(1) The indicator is not sensitive to boundary points compared to other methods.
(2) When reclassifying clusters with more than two or more clusters, the DBI index approach can correctly classify the data objects.

Hence, the optimal number of clusters can be determined based on the DBI index. In the underdetermined condition where the number of observed signals M is less than the number of source signals N, the initial number of clustering centers is chosen to be the same as the number of observed signals M, i.e. $K = M$.

3. **Steps of Underdetermined Mixing Matrix Estimation Algorithm Based on Improved k-Means Clustering**

In summary, the steps of the underdetermined mixture matrix estimation algorithm based on improved k-means clustering are as follows:

Step 1: In order to reduce the influence of noise and computational complexity, remove the observed data of $\|\boldsymbol{x}(t)\| < \varepsilon_1$ and the remaining observed data is projected onto the unit semi-hypersphere according to Eq. (9.18).

$$\tilde{\boldsymbol{x}}(t) = sign(x_i(t)) \times \boldsymbol{x}(t)/\|\boldsymbol{x}(t)\|, t = 1, 2, \ldots, T_0 \tag{9.18}$$

denote all the projected points as set $\tilde{\mathrm{X}}$. In general, a threshold $\varepsilon_1 = 0.2\times\overline{E}$ is chosen, where $\overline{E}$ represents the mean energy of the observed signals.

Step 2: Calculate the radius r according to Eq. (9.13), then calculate the density parameter $B_r(\tilde{\boldsymbol{x}}_i)$ for any data object $\tilde{\boldsymbol{x}}_i$ in set $\tilde{\mathrm{X}}$ based on Eq. (9.14). Set thresholds η_1 and η_2, and remove the data objects with density parameter $B_r(\tilde{\boldsymbol{x}}_i) < \eta_1$, obtaining the remaining data object set $\overline{\mathrm{X}}$. Save the data objects with density parameter $B_r(\tilde{\boldsymbol{x}}_i) > \eta_2$, resulting in the high-density data object set D. The values of thresholds η_1 and η_2 are related to the sparsity and signal-to-noise ratio of the source signal, and can be selected through simulation.

Step 3: Select the initial number of cluster centers, denoted as $K = M$.

Step 4: Select initial cluster centers based on density and distance.

Step 4.1: From the high-density data set D, select the data object with the maximum density parameter as the first initial cluster center, denoted as z_1, and choose the data object farthest from z_1 as the second initial cluster center, denoted as z_2.

Step 4.2: For $l = 3, 4, \ldots, K$, calculate the Euclidean distance $d(\tilde{\boldsymbol{x}}_i, z_j)$ from the remaining data objects $\tilde{\boldsymbol{x}}_i$ in the high-density data set D to the already selected $l - 1$ initial clustering centers z_j, where $j = 1, 2, \ldots, l - 1$,

$i \in [1, T_p]$, T_p represent the number of data objects in the high-density data set D. Select the data object with the maximum minimum Euclidean distance to the $l-1$ initial clustering centers as the l-th initial clustering center z_l.

$$z_l = \underset{\substack{\tilde{x}_i \in D \\ \tilde{x}_i \neq z_j}}{\arg\max} \left\{ \min_{\substack{j=1,2,\ldots,l-1 \\ i=1,2,\ldots,T_p}} d(\tilde{x}_i, z_j) \right\} \tag{9.19}$$

Step 5: Calculate the Euclidean distance from all data objects in set $\overline{X}$ to all clustering centers and assign the data to the class with the smallest Euclidean distance. Calculate the average of all data objects within each class as the reclassified clustering centers $\tilde{z}_1 \ldots \tilde{z}_i \ldots \tilde{z}_K$. Repeat this process until the clustering centers no longer change.

Step 6: Based on the clustering results, calculate the Davies-Bouldin Index (DBI) value I_K according to Eqs. (9.15)–(9.17) for the current number of clusters.

Step 7: Let the number of clusters be $K = K + 1$. Repeat steps 4 to 6 until the calculated index value satisfies the iteration termination condition: $I_K > I_{K-1}$ and $I_{K-1} < I_{K-2}$. The optimal number of clusters is $K_{opt} = K - 1$, with the corresponding optimal clustering centers $\tilde{z}_1 \ldots \tilde{z}_i \ldots \tilde{z}_{K-1}$, then the iteration end.

Step 8: Calculate the number of data objects contained in each class C_k, where $k = 1, \ldots, K_{opt}$; to ensure that some of the clusters obtained are not composed of a few outliers and noise, calculate the ratio of the number of data objects contained in each class to the maximum value of the number C, $p_k = C_k / C$, where $C = \max(C_1, C_2, \ldots, C_{K_{opt}})$; set the threshold ε_2, If $p_k < \varepsilon_2$, the class is removed, and the number of clusters retained at the end is the estimated value of the number of source signals, and the corresponding clustering center vectors are the estimated values of the column vectors of the mixing matrix, from which the estimation $\hat{\boldsymbol{A}}$ of the mixing matrix $\boldsymbol{A}$ is obtained; since some noise and outliers are removed by removing low-density projection points, the value of ε_2 should not be too large, generally around 0.1–0.3.

9.1.4 Simulation Results and Analysis

Below are the computer simulation experiment results of the algorithms based on the k-means clustering, fuzzy C-means clustering, and an improved k-means clustering. In the simulation experiment, it is assumed that there are 3 receiving channels and five different radar communication source signals. The first and second source signals are conventional radar signals with a carrier frequency of 5 MHz and 3 MHz respectively. The third and fourth signals are linear frequency modulated radar signals with a carrier frequency of 2 MHz and 1 MHz, and pulse widths of 4 MHz and 5 MHz,

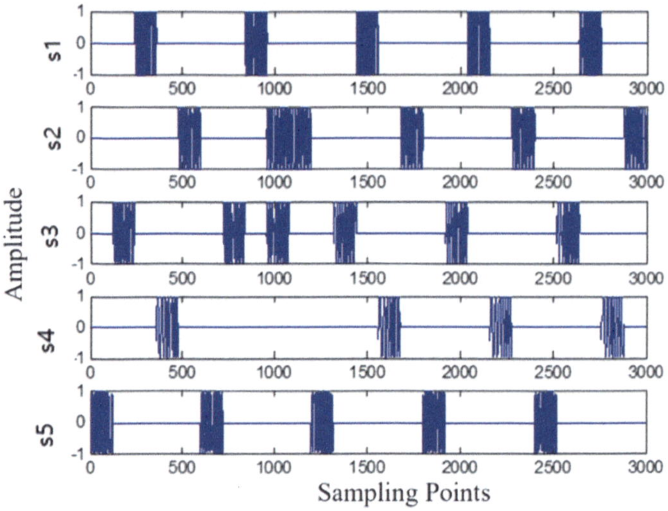

Fig. 9.6 Five different radar communication signals

respectively. The fifth signal is a sinusoidal phase modulation radar signal with a carrier frequency of 6 MHz. The sampling frequency is 30 MHz, and the number of sampling points is 5000. The time-domain waveforms of the five source signals are shown in Fig. 9.6.

Five source signals are sufficiently sparse in most sampling time, while in some sampling time, multiple source signals dominant at the same time. The mixing matrix is randomly generated as follows:

$$A = \begin{bmatrix} 0.4428 & -0.0886 & 0.6550 & 0.1062 & 0.1586 \\ -0.8423 & -0.7358 & -0.4139 & 0.9943 & -0.4827 \\ 0.3075 & -0.6713 & -0.6322 & 0.0126 & 0.8613 \end{bmatrix}$$

Without adding noise, when the given number of clusters $N = 5$, the estimated mixing matrix using the k-means clustering method is:

$$\boldsymbol{A}_{Kmeans} = \begin{bmatrix} 0.4079 & -0.1227 & 0.6441 & 0.1301 & 0.1922 \\ -0.8702 & -0.7337 & -0.4306 & 0.9832 & -0.4612 \\ 0.2765 & -0.6631 & -0.6323 & 0.0565 & 0.8662 \end{bmatrix}$$

The average estimation error of the mixing matrix is $e_{kmeans} \approx 0.99\%$. The estimated mixing matrix using the fuzzy C-means clustering method is:

$$\boldsymbol{A}_{fcm} = \begin{bmatrix} 0.4256 & -0.1012 & 0.6532 & 0.1183 & 0.1922 \\ -0.8568 & -0.7293 & -0.4206 & 0.9930 & -0.4612 \\ 0.2910 & -0.6767 & -0.6296 & 0.0005 & 0.8662 \end{bmatrix}$$

The average estimation error of the mixing matrix is $e_{fcm} \approx 0.55\%$. The estimated mixing matrix using the improved k-means clustering method is:

$$\boldsymbol{A}_{im\_kmeans} = \begin{bmatrix} 0.4428 & -0.0886 & 0.6550 & 0.1062 & 0.1586 \\ -0.8432 & -0.7358 & -0.4139 & 0.9943 & -0.4827 \\ 0.3075 & -0.6713 & -0.6322 & 0.0126 & 0.8613 \end{bmatrix}$$

Average estimation error of the mixing matrix is $e_{im\_kmeans} \approx 0.004\%$.

From the simulation results, it can be seen that without adding noise, both the k-means clustering and fuzzy C-means clustering can obtain estimates of the mixing matrix with small estimation errors. The mixing matrix estimated by the improved k-means algorithm is almost identical to the original mixing matrix, with an estimation error close to 0. This is because in the observed signals, when multiple source signals are dominate simultaneously, the observed signal sampling points no longer have linear clustering characteristics. Directly applying k-means or fuzzy C-means clustering would increase the estimation error of the mixing matrix. In the improved k-means algorithm, due to the addition of pre-processing operations, the sampling points of observed signals when multiple source signals are simultaneously dominant can be removed, and the clustering characteristics of observed signal sampling points are enhanced, so the estimation error of the mixing matrix is small.

With the consideration of noise, Fig. 9.7 shows the simulation results of the estimation error of mixing matrix at different signal-to-noise ratios. For each simulation, a new mixing matrix is randomly generated corresponding to different channel states. The simulations are conducted 300 times, and the average estimation error is calculated. As can be seen from the Fig. 9.7, with the increase of SNR, the estimation errors of all clustering algorithms gradually decrease. At different SNR, the estimation error of the improved k-means algorithm is lower than that of the k-means clustering and fuzzy C-means clustering. Furthermore, the underdetermined mixing matrix estimation algorithm based on improved k-means clustering is applicable when the number of source signals is unknown, while both the k-means clustering and fuzzy C-means clustering algorithms require the number of source signals to be known.

9.2 Estimation of Underdetermined Mixture Matrix Based on Similarity Detection

The underdetermined mixing matrix estimation method based on similarity detection mainly includes three steps: observed signal preprocessing, extraction of high-density points, and estimation of the number of source signals and the mixing matrix. The principles and functions of each step are described below.

As shown in Fig. 9.8 [48], in practical noisy environments, there are often numerous outliers and noise interference in the observed signals, which makes the clustering characteristics of the observed signal not obvious. Therefore, prior to

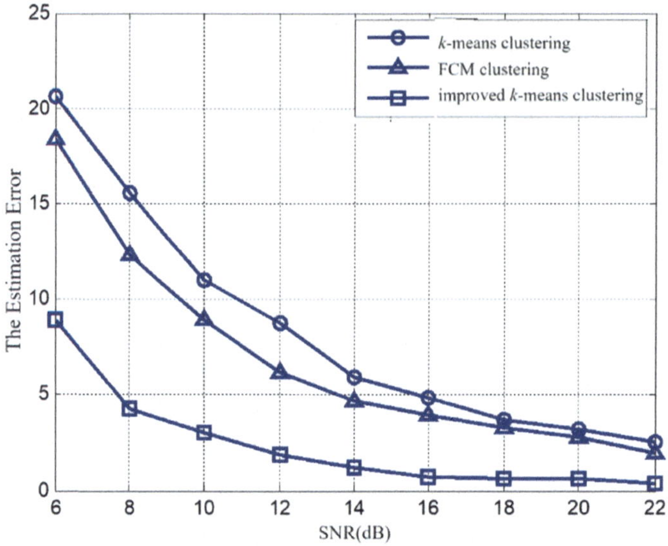

Fig. 9.7 Estimation performance of different clustering algorithms under various SNR

estimating the number of source signals and the mixing matrix, it is necessary to preprocess the observed signals to reduce the impact of noise and outliers, highlight the clustering characteristics of the observed signals, and thereby improve the accuracy of the mixing matrix estimation.

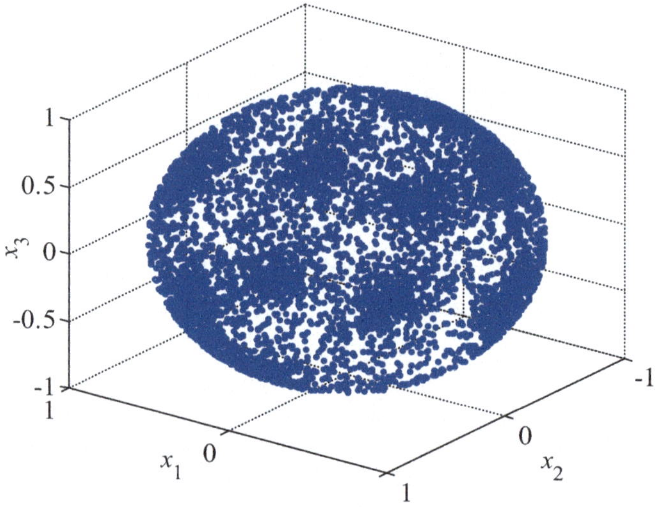

Fig. 9.8 Scatter plot of observed signal distribution in a noisy environment

First, the energy value $E_t = \|x(t)\|_2$ is defined for each sampling point of the observed signal. Low energy sampling points are often more seriously affected by noise than high energy sampling points, which affects the clustering characteristics of observed signal sampling points and is not conducive to the accurate estimate of the number of source signals. Therefore, it is necessary to use a preprocessing method to extract high-energy sampled points with good clustering characteristics from the observed signal before conducting subsequent clustering analysis.

9.2.1 Preprocessing of Observed Signals

For source signals that are temporally sparse or exhibit sparsity in the corresponding transform domain after a linear transformation, the observed signals obtained after passing through a channel often contain numerous outliers and noise due to the influence of the transmission environment and noise. Therefore, prior to clustering the sampling points of the observed signal, it is necessary to perform appropriate preprocessing to reduce the impact of outliers and noise on the estimation of the mixing matrix.

First, the energy value of each sampling point of the observed signal is defined using Eq. (9.20):

$$E_t = \|x(t)\|_2 \ , \quad t = 1, 2, \ldots T \tag{9.20}$$

where E_t represents the energy value of the t-th sampling point of the observed signal, T denotes the number of sampling points, and $\|\bullet\|_2$ represents the 2-norm. In order to estimate the mixing matrix more accurately, Eq. (9.21) is used to remove the low-energy sampling points from the observed signal, retaining the high-energy sampling points to form a matrix of high-energy sampling points.

$$X = \{x(t) | \|x(t)\|_2 \geq \alpha * \max\{\|x(t)\|_2\}\}, t = 1, 2, \ldots, T \tag{9.21}$$

where $\|x(t)\|_2$ denotes 2-norm of the vector $x(t)$, $\alpha \in [0.1, 0.4]$, and each column in X represents a sampling point. The remaining sample points $x(t)$ in X are normalized as follows.

$$x(t) = \frac{x(t)}{\|x(t)\|_2}, t = 1, 2, \ldots, T_2 \tag{9.22}$$

The symbol T_2 represents the total number of columns in X, which corresponds to the number of high-energy sampling points.

If the source signal is sufficiently sparse in a certain domain, in the corresponding sparse domain, normalizing each sampling point of the observed signal will result in a distribution characteristic as shown in Fig. 9.9 due to the linear clustering property. From Fig. 9.9, it can be seen that after normalization, sampling points of observed

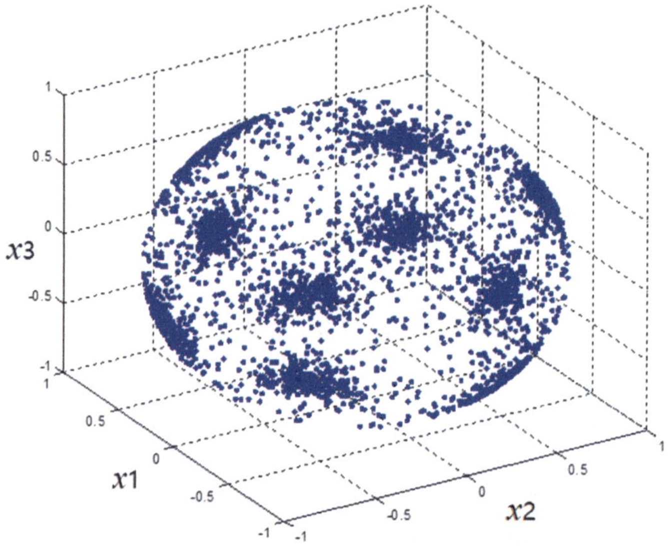

Fig. 9.9 Distribution of sampled points for observed signals when $m = 3$ and $n = 5$

signals in the same direction will converge on a certain region on the unit sphere, which also has clustering characteristics, and the cluster center of each cluster is the column vector $\boldsymbol{a}_i$ of the required mixing matrix.

Because the source signal s_i can take positive or negative values, the sampling points $\boldsymbol{x}(t)$ and $-\boldsymbol{x}(t)$ cluster in the same direction $\boldsymbol{a}_i$. After normalization, the sampling points corresponding to the same source signal on the unit hypersphere are symmetric about the origin, forming two clusters. When estimating the mixing matrix, these two clusters need to be grouped into one class, indicating that they correspond to the same source signal. Therefore, in Fig. 9.9, the number of clusters of the observed signal sampling points will be twice the number of source signals.

9.2.2 *Extracting High-Density Points*

The pre-processing of the observed signal can reduce the influence of noise and outliers on the mixing matrix estimation to some extent, but it can not eliminate the noise and outliers completely. As can be seen from Fig. 9.9, among the extracted high-energy sampling points, there are still many low-density sampling points, which often contain a lot of noise and outliers, which is not conducive to the subsequent clustering process and the accurate estimation of the mixing matrix. Furthermore, on the unit hypersphere, more useful information about the mixing matrix is often contained in the clusters with a larger distribution density of sampling points. Therefore, it is a necessary process to extract high density sampling points from high energy sampling points. The following is a detailed description of the process of extracting high-density sampling points.

First, for any sampling point p of an observed signal, on the unit sphere, the d sampleing points closest to it are formed into a sample set, which is called the neighborhood $\boldsymbol{D}$ of p, where d is the neighborhood capacity of $\boldsymbol{D}$. The neighborhood radius r_d^p is the maximum distance between point p and these d sampling points.

$$r_d^p = \max\{dist(p, q)|q \in \boldsymbol{D}\} \tag{9.23}$$

In this section, we utilize the angle between objects p and q as a measure of their distance, where dist(p, q) represents the angle between p and q. If a sampling point p is a high-density sampling point, then the distribution of sampling points within its neighborhood is relatively dense. This implies that, under certain neighborhood capacity, the neighborhood radius is smaller than that of the low-density sampling points. According to this principle, high density sampling points are extracted.

The angle between each pair of sampling points in the high-energy sampling point matrix $\boldsymbol{X}$.

$$\begin{cases} \theta_{i,j} = (180/\pi)\text{arc cos}\left(\dfrac{\sum\limits_{m=1}^{M} \boldsymbol{x}^m(i)\boldsymbol{x}^m(j)}{\|\boldsymbol{x}(i)\|_2 \|\boldsymbol{x}(j)\|_2}\right) \\ i = 1, 2, \ldots, T_2 \;\; j = 1, 2, \ldots, T_2 \end{cases} \tag{9.24}$$

where $\theta_{i,j}$ represents the angle between the i-th and j-th sampling points in $\boldsymbol{X}$, and m represents the m-th element in each sampling point vector. Since the sampling points have been normalized, we have $\|\boldsymbol{x}(i)\|_2 = 1$, $\|\boldsymbol{x}(j)\|_2 = 1$, so Eq. (9.24) can be rewritten as

$$\theta_{i,j} = (180/\pi)\text{arc cos}\left(\sum_{m=1}^{M} \boldsymbol{x}^m(i)\boldsymbol{x}^m(j)\right) \tag{9.25}$$

Before extracting high-density sampling points, first set a neighborhood capacity d. The value of d should not be too large and is typically $\beta T_2(\beta \in [0.01, 0.06])$. Once the neighborhood capacity d is determined, further find the neighborhood radius of each sampling point and determine if it is a high-density sampling point.

Firstly, the initial angle matrix is calculated by the formula in Eq. (9.25)

$$\boldsymbol{G} = \left\{\theta_{i,j}|1 \le i \le T_2, 1 \le j \le T_2\right\} \tag{9.26}$$

$\boldsymbol{G}$ is a symmetric matrix of size $T_2 \times T_2$, each row represents the angles between each sampling point of observed signal and other sampling points. The rows of $\boldsymbol{G}$ are sorted in ascending order to obtain the sorted angle matrix $\boldsymbol{D}$. Thus, the element values in column $d(d = 1, 2, \ldots T_2)$ of $\boldsymbol{G}$ represent the neighborhood radius of each sampling point when the neighborhood capacity is d. The mean of the d-th column elements in $\boldsymbol{D}$ is calculated to obtain the discriminant threshold $\overline{r}_d$. On the unit sphere, since the observed signal sampling points belonging to the same cluster are densely distributed and have smaller angles between each other, the neighborhood radius of sampling points is smaller when the neighborhood capacity is d. On the other hand, noise points and outliers usually do not belong to any cluster, and their clustering density is generally smaller, i.e., their neighborhood radius values are larger. Based on this principle, observed signal sampling points with a neighborhood radius smaller than the discriminant threshold $\overline{r}_d$ are retained to form the high-density sampling point matrix $\boldsymbol{X}_2$, so as to complete the extraction of high-density sampling points.

9.2.3 Estimate the Number of Source Signals and Mixing Matrix

After preprocessing, the clustering characteristics of the high-density sample points become more obvious. Moreover, in the high-density sampling point matrix $\boldsymbol{X}_2$, when the neighborhood capacity is d, the neighborhood radius $r_d^k(k = 1, 2, \ldots T)$ of each high-density sampling point is always smaller than the discriminant threshold $\overline{r}_d$. Here, T represents the number of observed signal sampling points in the high-density matrix $\boldsymbol{X}_2$. Additionally, due to the small value of d, each high-density sampling point usually exists in the same cluster with other sampling points in its neighborhood. Equation (9.27) is utilized to construct an angle matrix to measure the similarity between each sample point in the high-density matrix $\boldsymbol{X}_2$.

$$\hat{\theta}_{i,j} = (180/\pi)\text{arc cos}\left|\left(\sum_{m=1}^{M} \boldsymbol{x}_2^m(i)\boldsymbol{x}_2^m(j)\right)\right| \tag{9.27}$$

where $\hat{\theta}_{i,j}$ represents the angle between the i-th and j-th sampling points in the high-density matrix $\boldsymbol{X}_2$, $\boldsymbol{x}_2^m(i)$ represents the i-th sample point of m-th component in $\boldsymbol{X}_2$. If $\hat{\theta}_{i,j}$ is less than the discriminant threshold $\overline{r}_d$, it is determined that the i-th sampling point and the j-th sample point in the high-density matrix $\boldsymbol{X}_2$ belong to the same cluster. Based on this principle, the sampling points in $\boldsymbol{X}_2$ are clustered to get a preliminary categorization $\boldsymbol{Y} = \left\{\tilde{\boldsymbol{X}}_1, \tilde{\boldsymbol{X}}_2, \ldots, \tilde{\boldsymbol{X}}_K\right\}$, where $\tilde{\boldsymbol{X}}_k$, $(k = 1, 2, \ldots K)$ represents the k-th cluster obtained after classification, K represents the number of clusters, also the estimated number of source signals.

Taking into account the influence of channel parameters and noise in practical transmission environments, in extreme cases, it is possible for two columns in the mixing matrix to have high similarity. This can lead to the two corresponding clusters being close to each other, or even overlapping, thus forming a large cluster by mistake. Therefore, in the preliminary cluster $\boldsymbol{Y} = \left\{\tilde{\boldsymbol{X}}_1, \tilde{\boldsymbol{X}}_2, \ldots, \tilde{\boldsymbol{X}}_K\right\}$, it is still possible that the number of clusters is too small. Based on this, further evaluation and correction of each cluster in $\boldsymbol{Y} = \{\tilde{\boldsymbol{X}}_1, \tilde{\boldsymbol{X}}_2, \ldots, \tilde{\boldsymbol{X}}_K\}$ is necessary. The evaluation and correction method is as follows.

For the obtained cluster $\boldsymbol{Y} = \{\tilde{\boldsymbol{X}}_1, \tilde{\boldsymbol{X}}_2, \ldots, \tilde{\boldsymbol{X}}_K\}$, calculate the maximum angle $\theta_{\max}$ between sampling points within each cluster $\tilde{\boldsymbol{X}}_i$ $(i = 1, 2, \ldots K)$. The cluster with $\theta_{\max} \leq (\overline{r}_d + \psi)$ remains unchanged, while the cluster with $\theta_{\max} > (\overline{r}_d + \psi)$ is re-divided into two class by using k-means clustering method. $\overline{r}_d$ represents the discriminant threshold obtained during the extraction of high-density points. $\overline{r}_d$ is obtained according to the distribution characteristics of the sampling points of the observed signal, it can adapt to change with the change of the SNR, and can well characterize the distribution dispersion degree of the sampling points of the observed signal. The value of ψ is an empirical value set according to a large number of experimental simulation results, generally $\psi \in [15°, 20°]$.

After correction, a set of updated clusters $\breve{\boldsymbol{Y}} = \left\{\breve{\boldsymbol{X}}_1, \breve{\boldsymbol{X}}_2, \ldots, \breve{\boldsymbol{X}}_{K'}\right\}$ is obtained, where K' represents the number of clusters obtained after reclassification, and $\breve{\boldsymbol{X}}_j$, $j = 1, 2, \ldots K'$ represents the data of the j-th cluster after updating the classification. The clustering center of each cluster can be calculated using Eq. (9.28):

$$\boldsymbol{a}_j = \frac{1}{num_j} \sum \breve{\boldsymbol{X}}_j \; j = 1, 2, \ldots, K' \tag{9.28}$$

where num_j represents the number of sample points in category $\breve{\boldsymbol{X}}_j$. Finally, the estimation of the mixing matrix is obtained as $\boldsymbol{A} = [\boldsymbol{a}_1, \boldsymbol{a}_2, \ldots, \boldsymbol{a}_{K'}]$.

9.2.4 Steps of Underdetermined Mixing Matrix Estimation Algorithm Based on Similarity Detection

In summary, the steps of the underdetermined mixing matrix estimation algorithm based on similarity detection are as follows [49]:

Step 1: Extract high-energy sample points and normalize them based on Eqs. (9.20) and (9.21).

Step 2: Extract high-density sampling points.

(1) Calculate the initial angle matrix **G** based on Eqs. (9.25) and (9.26).
(2) Calculate the sorted angle matrix $\boldsymbol{W} = \text{sort}(\boldsymbol{G})$, sort$(\cdot)$ indicates that each row of the matrix is sorted in ascending order.
(3) Set the neighborhood capacity $d = \beta N$, where $\beta \in [0.01, 0.05]$ and N is the dimension of matrix **G**.
(4) Calculate the discriminant threshold $\overline{r}_d = \mathrm{E}\{\boldsymbol{W}(:, d)\}$, where $\boldsymbol{W}(:, d)$ represents the elements of the **d**-th column of matrix $\boldsymbol{W}$, and E$\{\cdot\}$ represents taking the mean.

(5) Retain the sampling points with a neighborhood radius smaller than the discriminant threshold $\overline{r}_d$, forming the high-density sampling point matrix $\boldsymbol{X}_2$.

Step 3: Perform preliminary clustering:

(1) Construct the angle matrix $\hat{\boldsymbol{\theta}} = \arccos\left(\left|\boldsymbol{X}_2^{\mathrm{T}}\boldsymbol{X}_2\right|\right)$.
(2) Obtain the λ-cut matrix based on $\hat{\boldsymbol{\theta}}$, where $\overline{\lambda} = \overline{r}_d$ represents the threshold for determining the cutoff.
(3) Obtain the preliminary clustering $\boldsymbol{Y} = \left\{\tilde{\boldsymbol{X}}_1, \tilde{\boldsymbol{X}}_2, \ldots, \tilde{\boldsymbol{X}}_{\hat{K}}\right\}$ based on the $\overline{\lambda}$-cut matrix and the high-density sampling point matrix $\boldsymbol{X}_2$.

Step 4: Evaluate and correct each cluster:

(1) Find the maximum angle $\theta_{\max}$ between sampling points within each clustering $\tilde{\boldsymbol{X}}_i$ $\left(i = 1, 2, \ldots, \hat{K}\right)$ according to the angle matrix.
(2) If $\theta_{\max} \le \overline{r}_d + \hat{c}$, keep the clustering unchanged. Otherwise, use the FCM (Fuzzy C-Means) method to reassign the $\tilde{\boldsymbol{X}}_i$ into 2 classes, $\hat{c} \in [12, 18]$

Step 5: Based on the corrected clustering $\breve{\boldsymbol{Y}} = \left\{\breve{\boldsymbol{X}}_1, \breve{\boldsymbol{X}}_2, \ldots, \breve{\boldsymbol{X}}_{K'}\right\}$, calculate the clustering centers using Eq. (9.22) to obtain the estimation of the mixing matrix $\boldsymbol{A} = [\boldsymbol{a}_1, \boldsymbol{a}_2, \ldots, \boldsymbol{a}_{K'}]$.

9.2.5 Simulation Results and Analysis

In this section, two different simulation experiments will be conducted to test the estimation performance of the similarity detection algorithm under different experimental conditions. Simulation Experiment 1 mainly evaluates the estimation performance of the similarity detection algorithm in different application scenarios. Simulation Experiment 2 focuses on testing the estimation performance of the similarity detection algorithm when there are insufficiently sparse sampling points in the source signal. The estimation accuracy of the number of source signals and the estimation error of the mixing matrix will be used to measure the performance of the algorithms in the simulation.

Simulation 1: Estimation Performance in Different Application Scenarios

In the experiments, different scenarios such as $[m, n] = [3, 4]$, $[m, n] = [3, 5]$, $[m, n] = [3, 6]$ and $[m, n] = [6, 7]$, $[m, n] = [6, 8]$, $[m, n] = [6, 9]$ are simulated. Here, m represents the number of received signals, and n represents the number of source signals. The source signals used in the experiments are radar signals with a certain degree of sparsity.

In scenario $[m, n] = [3, 4]$, the first signal and the second signal are conventional radar signals with carrier frequencies of 5 MHz and 3 MHz, respectively. The third signal is a linear frequency modulated (LFM) radar signal with a carrier frequency of 2 MHz and an intra-pulse bandwidth of 4 MHz. The fourth signal is a sinusoidal phase modulation radar signal with a carrier frequency of 6 MHz. The sampling frequency is 30 MHz, and the number of samples is 4800.

In scenario $[m, n] = [3, 5]$, the first signal and the second signal are conventional radar signals with carrier frequencies of 5 MHz and 3 MHz, respectively. The third and fourth signals are linear frequency modulated (LFM) radar signals with carrier frequencies of 2 MHz and 1 MHz, and intra-pulse bandwidths of 4 MHz and 5 MHz, respectively. The fifth signal is a sinusoidal phase modulation radar signal with a carrier frequency of 6 MHz. The sampling frequency is 30 MHz, and the number of samples is 4800.

In scenario $[m, n] = [3, 6]$, the first signal and the second signal are conventional radar signals with carrier frequencies of 5 MHz and 3 MHz, respectively. The third and fourth signals are linear frequency modulated (LFM) radar signals with carrier frequencies of 2 MHz and 1 MHz, and intra-pulse bandwidths of 4 MHz and 5 MHz, respectively. The fifth and sixth signals are sinusoidal phase modulation radar signals with carrier frequencies of 6 MHz and 4 MHz, respectively. The sampling frequency is 30 MHz, and the number of samples is 5760.

In scenario $[m, n] = [6, 7]$, the first signal and the second signal are conventional radar signals with carrier frequencies of 5 MHz and 3 MHz, respectively. The third and fourth signals are linear frequency modulated (LFM) radar signals with carrier frequencies of 2 MHz and 1 MHz, and intra-pulse bandwidths of 4 MHz and 5 MHz, respectively. The fifth and sixth signals are sinusoidal phase modulation radar signals with carrier frequencies of 6 MHz and 4 MHz, respectively. The seventh signal is a non-linear frequency modulated signal with a starting frequency of 2 MHz and an ending frequency of 4 MHz. The sampling frequency is 30 MHz, and the number of samples is 6720.

In scenario $[m, n] = [6, 8]$, the first signal and the second signal are conventional radar signals with carrier frequencies of 5 MHz and 3 MHz, respectively. The third and fourth signals are linear frequency modulated (LFM) radar signals with carrier frequencies of 2 MHz and 1 MHz, and intra-pulse bandwidths of 4 MHz and 5 MHz, respectively. The fifth and sixth signals are sinusoidal phase modulation radar signals with carrier frequencies of 6 MHz and 4 MHz, respectively. The seventh and eighth signals are non-linear frequency modulated signals with starting frequencies of 2 MHz and 3 MHz, and ending frequencies of 6 MHz and 4 MHz, respectively. The sampling frequency is 30 MHz, and the number of samples is 7680.

In scenario $[m, n] = [6, 9]$, the first signal and the second signal are conventional radar signals with carrier frequencies of 5 MHz and 3 MHz, respectively. The third and fourth signals are linear frequency modulated (LFM) radar signals with carrier frequencies of 2 MHz and 1 MHz, and intra-pulse bandwidths of 4 MHz and 5 MHz, respectively. The fifth and sixth signals are sinusoidal phase modulation radar signals with carrier frequencies of 6 MHz and 4 MHz, respectively. The seventh and eighth signals are non-linear frequency modulated signals with starting frequencies of 2 MHz and 3 MHz, and ending frequencies of 6 MHz and 4 MHz, respectively. The ninth signal is a conventional radar signal with a carrier frequency of 2 MHz. The sampling frequency is 30 MHz, and the number of samples is 5400.

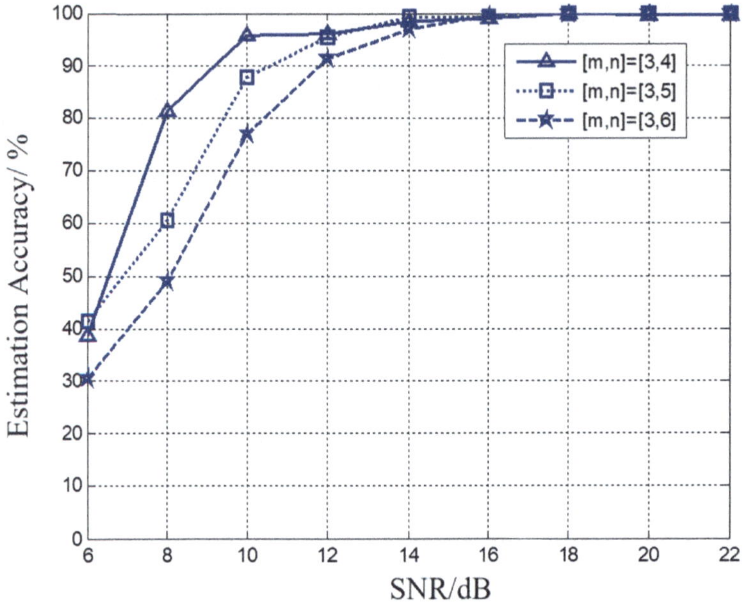

Fig. 9.10 Esitmation accuracy of the source number

Figures 9.10 and 9.11 show the simulation results of the similarity detection algorithm when $m = 3$. These results include the estimation accuracy of the number of source signals and the estimation error of the mixing matrix when the source signals are accurately estimated. The simulations were conducted 300 times, and in each simulation, the mixing matrix was randomly generated to represent different channel conditions.

It can be seen that in different simulation scenarios, the proposed method can accurately estimate the number of source signals and the mixing matrix. When the SNR is 12 dB, the estimation accuracy of the number of source signals is higher than 90% in all scenarios, and the estimation error of the mixing matrix is less than 4%. Additionally, as the number of source signals decreases, the estimation accuracy of the number of source signals increases. In the scenario [m, n] = [3, 4], when the SNR is 8 dB, the estimation accuracy of the number of source signals reaches around 80%, and the estimation error of the mixing matrix when the number of source signals is accurately estimated is less than 8%.

To further validate the universality of the similarity detection algorithm, the simulation results when $m = 6$ are shown in Figs. 9.12 and 9.13.

It can be seen from Figs. 9.12 and 9.13 that the similarity detection algorithm is still applicable when $m = 6$, and its estimation performance is even better than that in the $m = 3$ scenario, especially at low signal-to-noise ratios (SNR). When the SNR is greater than 12 dB, the estimation accuracy of the number of source signals is close

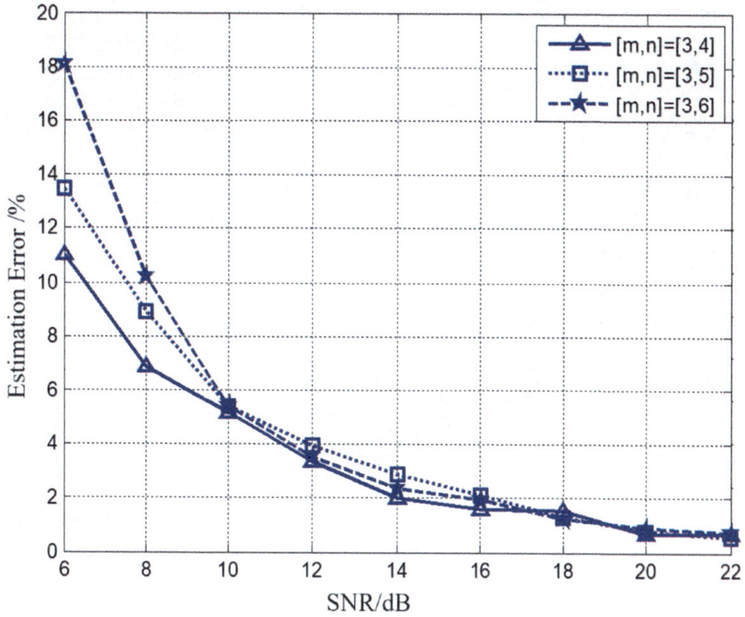

Fig. 9.11 Estimation error of the mixing matrix when the source number are accurately estimated

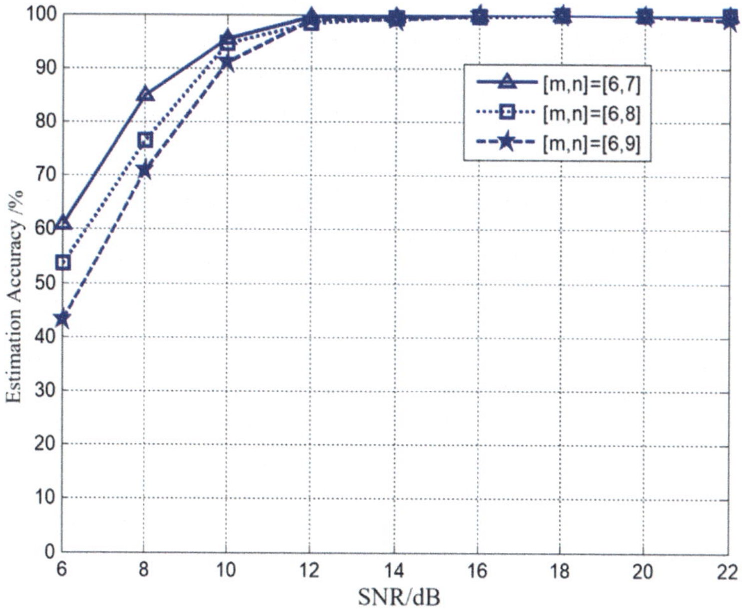

Fig. 9.12 Accuracy of source estimation in different scenarios for m = 6

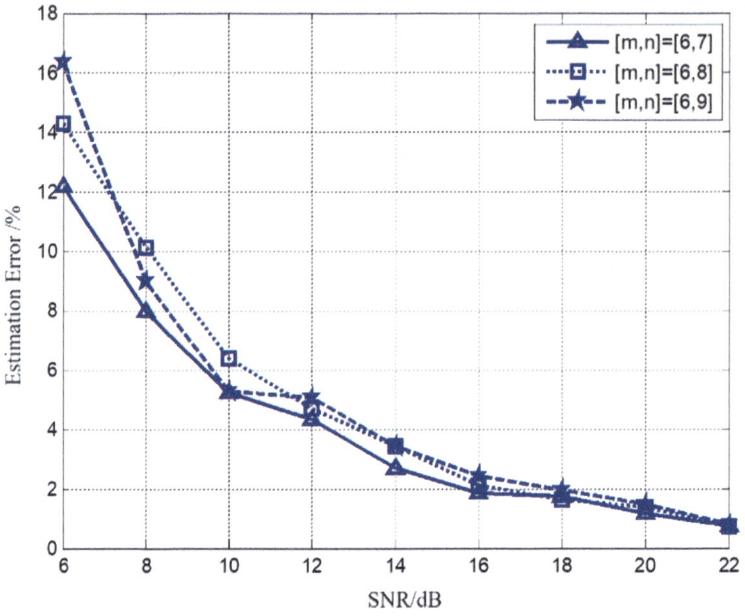

Fig. 9.13 estimation error of the mixing matrix when the number of sources is accurately estimated

to 100% in all scenarios, and the estimation error of the mixing matrix is less than 5%.

Simulation 2: Estimation Performance When There Are Insufficient Sparse Sampling Points

In simulation 2, the performance of the similarity detection algorithm was tested when there are partially insufficient sparse samples in the source signals. The types of source signals used in this simulation are the same as those used in simulation 1.

Figures 9.14 and 9.15 compare the estimation performance when the source signals are completely sufficiently sparse and when there are partially insufficient sparse sampling points in the source signals. From the figures, it can be seen that the estimation accuracy of the number of source signals and estimation error of the mixing matrix are affected when there are partially insufficient sparse sampling points in the source signals, especially at low SNR. However, as the SNR increases, the impact decreases. When the SNR is 16 dB, the estimation accuracy of the number of source signals is above 90% in all cases, and the estimation error of the mixing matrix is less than 5%.

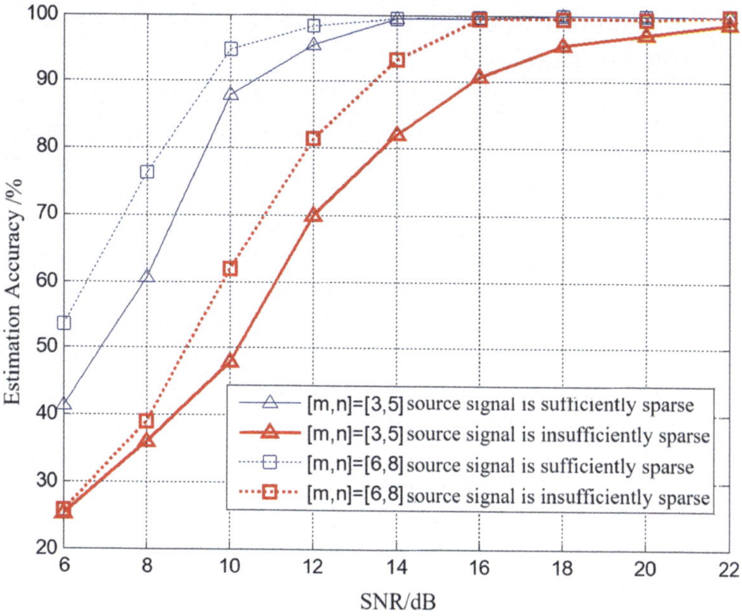

Fig. 9.14 Estimation accuracy of source number in various scenarios

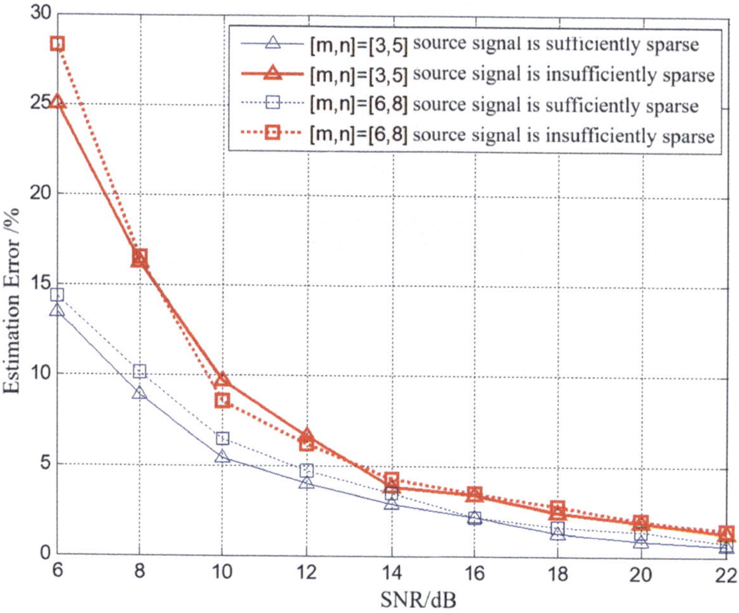

Fig. 9.15 Estimation error of the mixing matrix when the number of sources is accurately estimated in various scenarios

9.3 Estimation of Underdetermined Mixing Matrix Based on Bullseye Retrieval

Under certain signal-to-noise ratio (SNR), after the preliminary preprocessing steps, the high-density sampling points retained on the unit sphere can be clearly divided into several distinct clusters. However, when the directions of two column vectors in the mixing matrix are close to each other, it can cause the corresponding clusters to be close to each other, resulting in blurred cluster boundaries and even overlapping, as shown in Fig. 9.16. In such cases, conventional clustering algorithms like fuzzy C-means clustering may struggle to distinguish between these clusters and mistakenly estimate them as a single cluster. In the mixing matrix estimation method based on similarity detection, the steps of clustering center evaluation and correction are added, which can distinguish these different clusters to a certain extent, and can better adapt to the change of sampling point distribution of observed signals under different SNR. In this section, a mixing matrix estimation method based on bullseye retrieval is presented, which has better estimation performance for the situation of cluster boundary ambiguity in the observed signal distribution.

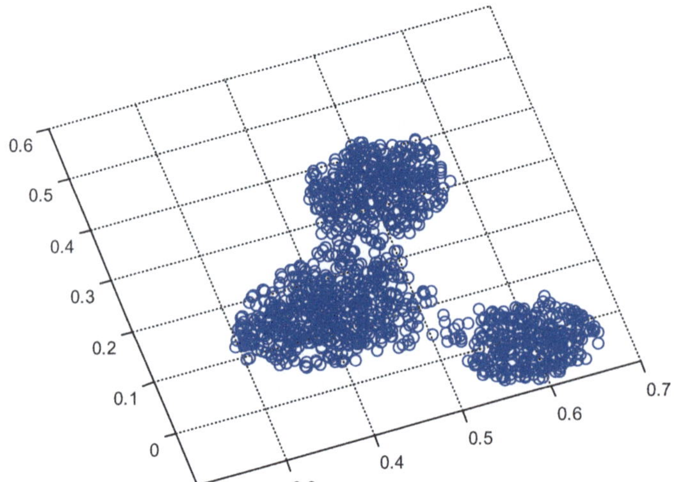

Fig. 9.16 Scatter plot when two clusters are close together

9.3.1 Preprocessing of Observed Signals and Extraction of High-Density Sampling Points

Before clustering the sampling points of the observed signals, it is necessary to preprocess the observed data matrix and extract high-density sampling points to reduce the impact of noise and outliers. First, according to Eqs. (9.29) and (9.30), the high-energy sampling points in the observed signal sampling points are extracted to form a high-energy point matrix $\boldsymbol{X}_1$.

$$\begin{cases} \mathbf{Z} = \{\boldsymbol{x}(t) | \|\boldsymbol{x}(t)\|_2 \geq a\} \\ s.t. \quad a = E\{\|\boldsymbol{x}(t)\|_2\}, \quad t = 1, 2, \ldots, T \end{cases} \tag{9.29}$$

$$\begin{cases} X_1 = \{z(k) | \|z(k)\|_2 \geq a + \alpha * (E\{\|z(k)\|_2\} - a)\} \\ k = 1, 2, \ldots, T_1 \end{cases} \tag{9.30}$$

where $E(\cdot)$ represents the expectation operation, $z(k)$ represents the k-th column in the matrix $\mathbf{Z}$, T_1 represents the total number of columns in $\mathbf{Z}$, and $\alpha \in (0, 1)$. Normalize each column of the high-energy matrix $\boldsymbol{X}_1$, which corresponds to each observed signal sampling point:

$$\boldsymbol{x}_1(l) := \frac{\boldsymbol{x}_1(l)}{\|\boldsymbol{x}_1(l)\|_2,} \quad l = 1, 2, \ldots, T_3 \tag{9.31}$$

where $\boldsymbol{x}_1(l)$ represents the l-th column in $\boldsymbol{X}_1$, and T_3 represents the total number of columns in $\boldsymbol{X}_1$, also the number of extracted high-energy observed signal sampling points.

The extraction of high-density sampling points is carried out using the same method as described in Sect. 9.2.2, so it will not be repeated. After extracting the high-density sampling points, we obtain the high-density sampling point matrix $\boldsymbol{X}_2$ and the discriminant threshold $\overline{r}_d$.

9.3.2 Estimation of the Number of Source Signals and Mixing Matrix

As shown in Fig. 9.17, on the unit hypersphere, the observed signal sampling points gather around the points determined by the column vectors of the mixing matrix, forming a dense cluster. The clustering center represents the desired column vectors of the mixing matrix. For the observed signal sampling points belonging to the same cluster, within a certain neighborhood radius δ in the L_1 norm, each sampling point has a large neighborhood capacity, and the sampling point with the maximum neighborhood capacity is often located inside the cluster and closer to the cluster center.

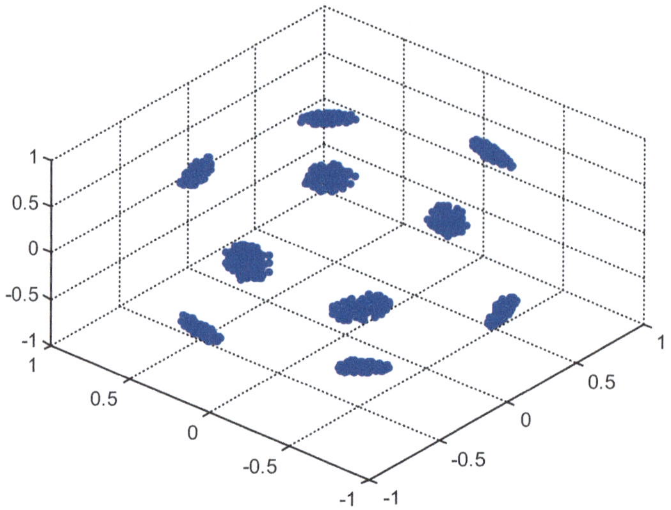

Fig. 9.17 Distribution of high-density sampling points

Therefore, by searching for the position of the sampling point with the maximum neighborhood capacity in each cluster when the neighborhood radius is set to δ, and extracting all the observed signal sampling points within the neighborhood of this sampling point, we can calculate the clustering center by taking the mean, which serves as an estimation of the column vectors of the mixing matrix. When all clusters on the unit hypersphere have been searched, the estimated number of source signals and the estimation of the mixinng matrix can be obtained.

The selection of a neighborhood radius of δ can be set as $\delta = \overline{r}_d + \sigma$, where $\overline{r}_d$ represents the discriminant threshold obtained during the extraction of high-density sampling points, and σ represents a small constant value that can be chosen between 1° and 5°. The reasons for this selection are as follows:

(1) With $\overline{r}_d$ as the neighborhood radius, all sampling points in the neighborhood can be guaranteed to be in the same cluster.
(2) The discrimination threshold $\overline{r}_d$ is obtained according to the distribution characteristics of the sampling points of the observed signal, which can well characterize the dispersion degree of the distribution of the observed signal sampling points, and adapt to the change of the signal-to-noise ratio. When the signal-to-noise ratio is lower, the observed signal sampling points tend to be more scattered, leading to a larger value of $\overline{r}_d$. Conversely, when the signal-to-noise ratio is higher, the observed signal sampling points tend to be more densely clustered, resulting in a smaller value of $\overline{r}_d$.
(3) When the signal-to-noise ratio tends to infinity, the value of $\overline{r}_d$ approaches 0. However, to avoid the neighborhood radius becoming too small and approaching 0, a small σ is added to $\overline{r}_d$, resulting in $\delta = \overline{r}_d + \sigma$. This adjustment prevents the neighborhood radius from approaching 0.

Therefore, when a neighborhood radius of $\delta = \overline{r}_d + \sigma$ is given, the method for searching the sampling point with the maximum neighborhood capacity and its neighborhood is as follows.

First, construct an angle matrix (matrix $\overline{\boldsymbol{\theta}}$) that measures the angular relationship between the sampling points in the high-density sampling matrix $\boldsymbol{X}_2$.

$$\overline{\boldsymbol{\theta}} = \begin{pmatrix} \overline{\theta}_{11} & \overline{\theta}_{12} & \cdots & \overline{\theta}_{1M'} \\ \overline{\theta}_{21} & \overline{\theta}_{22} & \cdots & \overline{\theta}_{2M'} \\ \vdots & \vdots & & \vdots \\ \overline{\theta}_{M'1} & \overline{\theta}_{M'2} & \cdots & \overline{\theta}_{M'M'} \end{pmatrix} \tag{9.32}$$

where

$$\overline{\theta}_{i,j} = (180/\pi)\text{arc cos}\left(\sum_{k=1}^{m} x_k(i)x_k(j)\right), \quad i,j = 1, 2, \ldots, M' \tag{9.33}$$

Matrix $\overline{\boldsymbol{\theta}}$ is a symmetric matrix of size $M' \times M'$. $\overline{\theta}_{i,j}$ represents the angle between the i-th and j-th sampling points in the high-density sampling matrix $\boldsymbol{X}_2$. M' represents the number of observed signal sampling points in the high-density sampling matrix $\boldsymbol{X}_2$.

Definition 9.6 λ-cut matrix: For a given matrix $\boldsymbol{R}$, the elements of the matrix less than λ are set to 1, and the elements greater than or equal to λ are set to 0, and a new matrix $\tilde{\boldsymbol{R}}$ is obtained, which is called the λ-cut matrix of the matrix $\boldsymbol{R}$.

According to Definition 9.6, we can obtain the λ-cut matrix $\tilde{\boldsymbol{\theta}}$ of the angle matrix $\overline{\boldsymbol{\theta}}$ by setting the value of λ equal to the neighborhood radius δ.

$$\tilde{\boldsymbol{\theta}} = \begin{pmatrix} 1 & 0 & \cdots & 0 \\ 0 & 1 & \cdots & 1 \\ \vdots & \vdots & & \vdots \\ 0 & 1 & \cdots & 1 \end{pmatrix} \tag{9.34}$$

In the cut matrix $\tilde{\boldsymbol{\theta}}$, each column (or row) represents the angular relationship between each observed signal sampling point and other sampling points in the high-density sampling matrix $\boldsymbol{X}_2$. If $\tilde{\theta}_{i,j} = 1$ in the j-th column, it indicates that the angle between the i-th sampling point and the j-th sampling point in the high-density sampling matrix $\boldsymbol{X}_2$ is less than the neighborhood radius δ. This means that the j-th sampling point is within the neighborhood of the i-th sampling point.

Therefore, by summing the elements in each column of the cut matrix $\tilde{\boldsymbol{\theta}}$, we can obtain the neighborhood capacity or density of each sampling point in the high-density sampling matrix $\boldsymbol{X}_2$. The neighborhood capacity of each sampling point in the high-density matrix $\boldsymbol{X}_2$ is calculated by Eq. (9.35), and the sampling point with the largest neighborhood capacity and its serial number in $\boldsymbol{X}_2$ are obtained.

$$\gamma = \arg\max \sum_{in\ col} \tilde{\boldsymbol{\theta}} \tag{9.35}$$

where, $\sum_{in\ col}(\bullet)$ represents the summation of each column in the matrix. According to the cut matrix $\tilde{\boldsymbol{\theta}}$ and Eq. (9.35), we extract the γ-th sampling point from the high-density sampling matrix $\boldsymbol{X}_2$, along with all the sampling points in its neighborhood, to form a clustering set $\boldsymbol{Y}$. Then, the clustering center of $\boldsymbol{Y}$ is calculated using Eq. (9.36)

$$\boldsymbol{a} = \frac{1}{num'} \sum_{in\ row} \boldsymbol{Y} \tag{9.36}$$

where $\sum_{in\ row}(\bullet)$ represents the summation of each row in the matrix, $\boldsymbol{a}$ represents the clustering center, and num' represents the number of sampling points in $\boldsymbol{Y}$.

After obtaining the first clustering center, the same method is used to search for the next sampling point with the maximum neighborhood capacity and calculate the new clustering center. At this time, in the high-density sampling matrix $\boldsymbol{X}_2$, all the sampling points that have been extracted are marked as extracted points. For the cluster whose clustering center has been calculated, the extracted points do not need to be searched again to avoid getting the clustering center of the same cluster repeatedly. Therefore, when calculating the clustering center of the next cluster, in order to avoid that the sampling point with the maximum neighborhood capacity found is the extracted point, the rows and columns corresponding to the extracted point are also marked accordingly in the cut matrix $\tilde{\boldsymbol{\theta}}$. When searching the sampling point with the maximum neighborhood capacity, only the column vectors that are not marked in the cut matrix $\tilde{\boldsymbol{\theta}}$ are calculated and searched. The marked rows and columns are no longer involved in the computation or search. This process is repeated, using the same method, to calculate the next clustering center, until a previously calculated clustering center is encountered within the neighborhood range of the sampling point with the maximum neighborhood capacity in the high-density sampling matrix $\boldsymbol{X}_2$, and the proportion of non-extracted points is less than η, typically taken as $\eta \in (0.3, 0.8)$. This indicates that the clustering center of the cluster to which the sampling point with the maximum neighborhood capacity belongs has already been calculated. At this time, the search is stopped, and all the clusters in the observed signal sampling points have been retrieved.

All the obtained clustering centers are assembled into matrix $\boldsymbol{B} = [\boldsymbol{a}_1, \boldsymbol{a}_2 \ldots \boldsymbol{a}_K]$, where $\boldsymbol{a}_i$ represents the clustering centers and K represents the number of clustering centers. From the previous analysis, it is known that the number of clustering centers is twice the number of source signals. Therefore, the clustering centers in matrix $\boldsymbol{B}$ that correspond to the same source signal need to be merged into one class. The estimated number of source signals is given by [K/2], where $[\cdot]$ denotes the ceiling operation.

When estimating the mixing matrix, the column vectors in matrix $\boldsymbol{B}$ with negative elements in the first row are first flipped using Eq. (9.37). This is equivalent to projecting the clustering center points located on the lower hemisphere of the unit hypersphere onto the upper hemisphere. Ideally, the two clustering centers corresponding to the same source signal will be located at the same position on the hypersphere.

$$\boldsymbol{a}_j = \begin{cases} \boldsymbol{a}_j, & \boldsymbol{a}_j(1) \geq 0 \\ -\boldsymbol{a}_j, & \boldsymbol{a}_j(1) < 0 \end{cases} \quad j = 1, 2, \ldots K \tag{9.37}$$

Finally, the clustering centers in matrix $\boldsymbol{B}$ are divided into [K/2] classes by k-means clustering method, and the classification results are used as the estimation of the mixing matrix.

In summary, the steps of the underdetermined blind source separation method based on bullseye retrieval are as follows.

Step 1: Extract high-energy sampling points and normalize them to obtain matrix $\boldsymbol{X}_1$.

Step 2: Extract high-density sampling points.

(1) Calculate the initial angle matrix $\boldsymbol{G} = \arccos\left(\boldsymbol{X}_1^{\mathrm{T}}\boldsymbol{X}_1\right)$.
(2) Calculate the sorted angle matrix $\boldsymbol{W} = \mathrm{sort}(\boldsymbol{G})$, where $\mathrm{sort}(\cdot)$ represents sorting each row of the matrix in ascending order.
(3) Set the neighborhood capacity $d = \beta N$, where $\beta \in [0.01, 0.05]$ and N are the dimensions of **G**.
(4) Calculate the discrimination threshold $\overline{r}_d = \mathrm{E}\{\boldsymbol{W}(:, d)\}$, where $\boldsymbol{W}(:, d)$ represents the d-th column element of matrix $\boldsymbol{W}$.
(5) Retain the sampling points with a neighborhood radius smaller than the discrimination threshold $\overline{r}_d$ and form the high-density sampling matrix $\boldsymbol{X}_2$.

Step 3: Construct the angle matrix $\hat{\boldsymbol{\theta}} = \arccos\left(\boldsymbol{X}_2^{\mathrm{T}}\boldsymbol{X}_2\right)$ for the sampling points in matrix $\boldsymbol{X}_2$. According to the matrix $\hat{\boldsymbol{\theta}}$, obtain the $\overline{\lambda}$-cut matrix, $\overline{\lambda} = \overline{\delta} = \overline{r}_d + \hat{\sigma}_d$ and $\hat{\sigma} \in [1°, 5°]$.

Step 4: Apply Eqs. (9.34) and (9.35) to search for the bullseye position and extract the set of sampling points $\boldsymbol{Y}$ within the neighborhood range of the bullseye.

Step 5: Determine if the proportion of non-extracted points in the set $\boldsymbol{Y}$ is greater than η, where $\eta \in [0.6, 1]$.

(1) If the proportion of non-extracted points is greater than η, calculate the clustering centers of the set $\boldsymbol{Y}$ using Eq. (9.36), and store them in matrix $\hat{\boldsymbol{B}}$.
(2) If the proportion of non-extracted points is less than η, the algorithm stops.

Step 6: Apply Eq. (9.37) and k-means clustering to further partition matrix $\hat{\boldsymbol{B}}$, resulting in the final estimation of the mixing matrix.

9.3.3 Simulation Results and Analysis

In this simulation, two distinct experiments were conducted to assess the estimation performance of the proposed method under different experimental conditions and to examine the impact of the presence of insufficiently sparse components in the source signals on the estimation performance. Simulation 1 primarily tested the estimation performance of the proposed method in various application scenarios (different numbers of source signals). Simulation 2 mainly tests the effect of insufficiently sparse source signal on the performance of the estimation method.

In both Simulation 1 and Simulation 2, the source signals are consistent with the source signals described in Sect. 9.2.4.

Simulation 1: Estimation Performance in Different Application Scenarios

Figures 9.18 and 9.19 respectively depict the estimation accuracy of the number of source signals and the estimation error of the mixing matrix when the source signals are accurately estimated using the estimation method introduced in this section, with $m = 3$. The simulations were conducted 300 times, with each simulation utilizing a randomly generated Gaussian matrix as the mixing matrix.

It can be seen that the proposed method have better estimation performance in different simulation scenarios. Specifically, when SNR = 10 dB, the estimation accuracy of the number of sources exceeds 80% in all scenarios, while the estimation error of the mixing matrix when the number of sources is correctly estimated remains

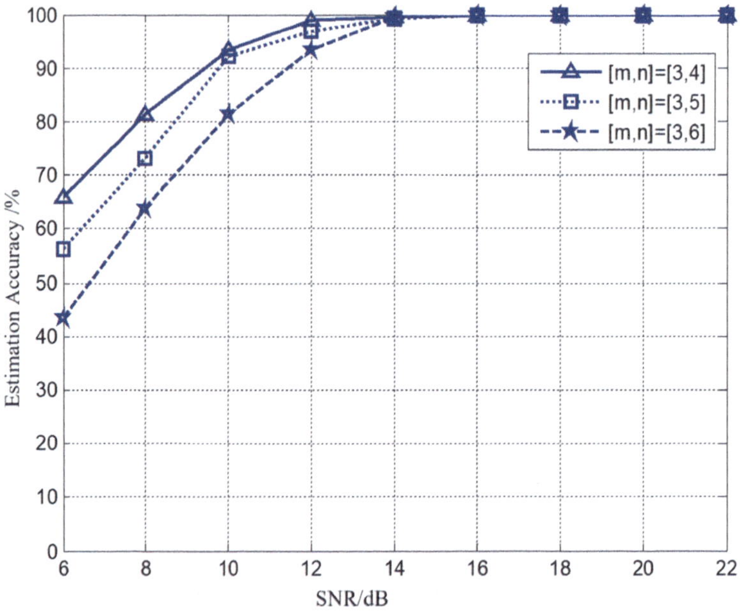

Fig. 9.18 Estimation accuracy of source number in different scenarios for $m = 3$

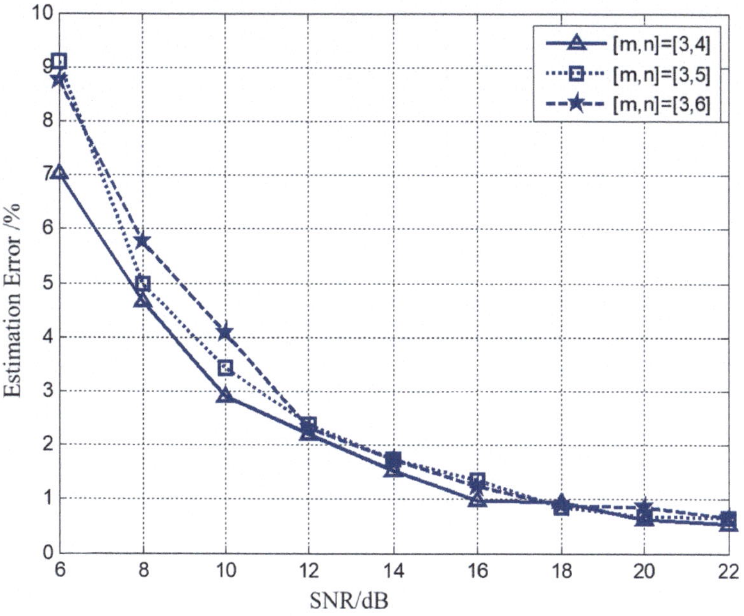

Fig. 9.19 Estimation error of the mixing matrix when the number of sources is accurately estimated

below 2%. Furthermore, a lower number of source signals corresponds to higher accuracy in the estimation of the number of sources. In the [3, 4] scenario, when SNR = 8 dB, the accuracy of the source number estimation has already reached approximately 80%, with the estimation error of the mixing matrix when the number of sources is accurately estimated being less than 4%.

Simulation 2: Estimation Performance When There Are Insufficiently Sparse Sampling Points in the Source Signal

Figures 9.20 and 9.21 compare the estimation performance when the source signals are completely sufficiently sparse and insufficiently sparse. Here, "insufficiently sparse" means that in some time periods, the number of dominant (or non-zero) source signals is greater than 1.

The results indicate that when there are insufficiently sparse source signals, both estimation of the number of source signals and mixing matrix are affected, regardless of whether $[m, n] = [3, 5]$ or $[m, n] = [6, 8]$. The effect is less when $m = 6$ than when $m = 3$. As the signal-to-noise ratio (SNR) increases, these impacts gradually diminish. When SNR = 14 dB, the estimation accuracy of the number of source signals is above 90% in different scenarios, and the mixing matrix estimation error is less than 3%. In general, compared with other algorithms, the insufficiently sparse source signal has less impact on the performance of the algorithm based on maximum density detection.

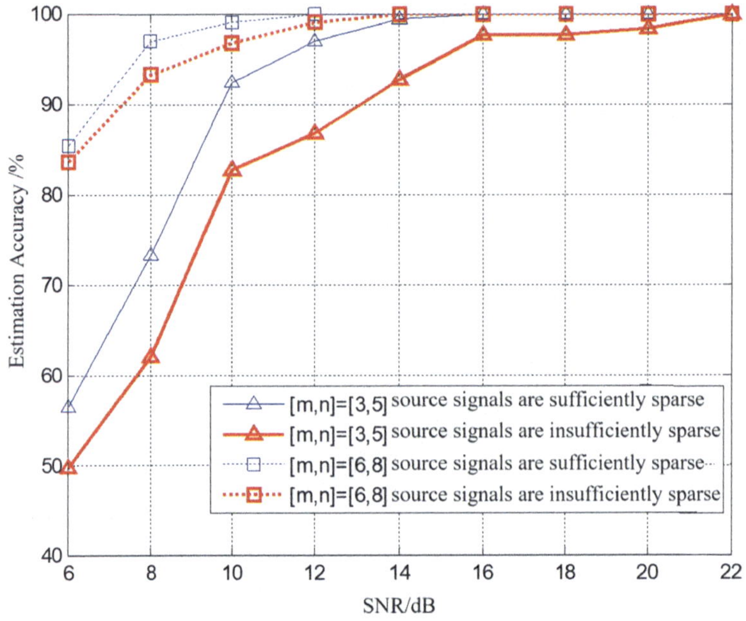

Fig. 9.20 Estimation accuracy of the number of sources in different scenarios

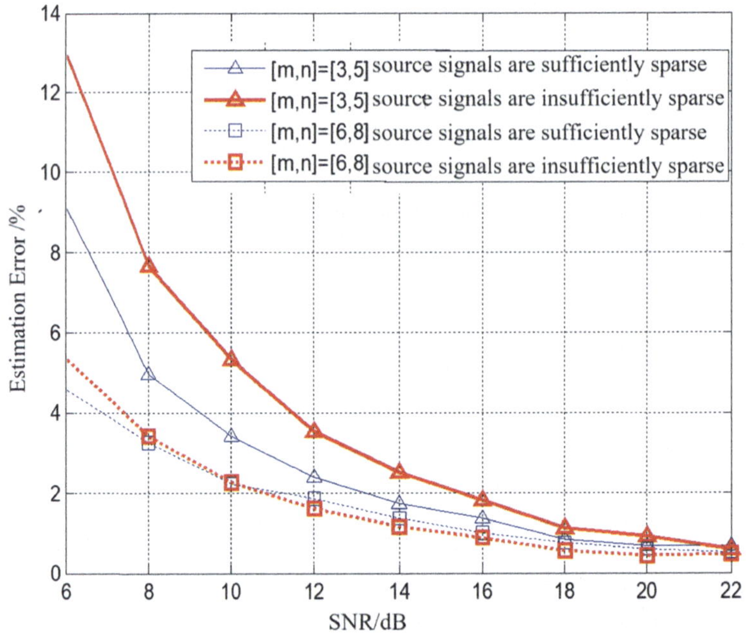

Fig. 9.21 Estimation error of the mixing matrix when the number of sources is accurately estimated

Summary

This chapter focuses on the underdetermined mixing matrix estimation problem for sufficiently sparse source signals. It first investigates clustering-based underdetermined mixing matrix estimation, including the k-means clustering algorithm, the fuzzy C-means clustering algorithm, and an improved k-means clustering algorithm. The chapter provides a detailed introduction to the principles and implementation steps of these algorithms, and evaluates their performance through simulation and analysis. The results indicate that the improved k-means clustering algorithm yields the smallest estimation error of the mixing matrix.

Then, two mixing matrix estimation algorithms with good anti-noise ability are studied, namely underdetermined mixing matrix estimation algorithm based on similarity detection and that based on bullseye retrieval. The specific principles and implementation processes of the algorithms are introduced in detail, and the performance of the algorithms under different conditions is simulated and analyzed. The results show that the two algorithms presented in this chapter have the advantages of high estimation accuracy of source signal number and small error estimation of the mixing matrix.

Chapter 10
The Estimation of Insufficiently Sparse Underdetermined Mixing Matrix

For the linear instantaneous mixing model, when the source signals are not sufficiently sparse, the spatial distribution of the sampling points of the observed signal no longer meets the linear clustering characteristics, but is distributed on some hyperplanes in the space, showing the characteristics of plane clustering. To estimate the mixing matrix, we first estimate each hyperplane composed of observed signals in space, and then estimate the intersection direction of the plane as the estimation of the column vector of the mixed matrix.

This chapter introduces the mixing matrix estimation techniques under insufficiently sparse conditions, including the mixing matrix estimation algorithms based on k-dimension subspace, parameter estimation, plane clustering potential function, homogeneous polynomial representation and single source point detection. The principles and implementation steps of these algorithms are given, and the performance of some algorithms is simulated and analysed [50].

10.1 Blind Estimation of Mixing Matrix Based on k-Dimensional Subspace

10.1.1 Algorithm Principles

A k-dimensional subspace is a space defined by k vectors. In the underdetermined blind source separation technique, it is assumed that the number of source signals is N and the number of observed signals is M. When the source signal is not sufficiently sparse, blind separation can be achieved when the number of sources that play a leading role simultaneously is not greater than $M - 1$. Let's set the number of active source signals at the same time to k. Then the observed signals are gathered in the space composed of k column vectors, which is called k-dimensional subspace, and

W. Fu, *Principle and Application of Blind Source Separation Technology*,
https://doi.org/10.1007/978-981-96-0022-9_10

also called cluster subspace or cluster sub-plane. These k column vectors belong to a mixing matrix $\boldsymbol{A}$. The k-dimensional subspace method involves first estimating the k-dimensional subspace from the observed signal, and then based on the estimated subspace, finding the intersection lines and estimate the column vectors of the mixing matrix.

To facilitate the description of the algorithm principle in this section, Eq. (2.3) is rewritten as follows

$$\boldsymbol{x}(t) = s_1(t)\boldsymbol{a}_1 + s_2(t)\boldsymbol{a}_2 + \cdots + s_N(t)\boldsymbol{a}_N \tag{10.1}$$

where, $\boldsymbol{x}(t)$ is the observed signal vector at the current sampling time t, $s_i(t)(i = 1, \ldots, N)$ is the value of the i-th source signal at the sampling time t, and $\boldsymbol{a}_i(i = 1, \ldots, N)$ is the i-th column vector of the mixing matrix $\boldsymbol{A}$. In the case of insufficiently sparse source signals, the number of active source signals $k > 1$, and the values of other $N - k$ source signals are close to 0. Therefore, Eq. (10.1) becomes

$$\boldsymbol{x}(t) \approx s_{i_1}(t)\boldsymbol{a}_{i_1} + s_{i_2}(t)\boldsymbol{a}_{i_2} + \cdots + s_{i_k}(t)\boldsymbol{a}_{i_k} \tag{10.2}$$

where $\{i_1, i_2, \ldots, i_k\} \subset \{1, 2, \ldots, N\}$. Obviously, the observed signal vector $\boldsymbol{x}(t)$ is in the k-dimensional subspace (hyperplane) spanned by $\boldsymbol{a}_{i_1}, \boldsymbol{a}_{i_2}, \ldots, \boldsymbol{a}_{i_k}$. It can be seen that the observed vectors are distributed in N_k ($N_k = C_N^k$) k-dimensional subspaces. That is, there are C_N^k k-dimensional subspaces in the observed signal space, and the k-dimensional subspace clustering is also named for this. Mixing matrix estimation method based on k-dimensional subspaces first estimates these k-dimension subspaces, and then uses them to estimate the individual column vectors of the mixing matrix.

- Estimation of k-dimensional subspaces

Let the k-dimensional subspace be represented by a matrix $\boldsymbol{B} \in \mathrm{R}^{M\times k}$ of M rows and k columns, $\boldsymbol{B} = [\boldsymbol{b}_1, \boldsymbol{b}_2, \ldots, \boldsymbol{b}_k]$ consists of a set of orthonormal bases of the k-dimensional subspace. Define the potential function as

$$f(\boldsymbol{B}) = \sum_{t=1}^{T} \exp\left(-\frac{d^2(\boldsymbol{x}(t), \boldsymbol{B})}{2\sigma_1^2}\right) \tag{10.3}$$

where $d(\boldsymbol{x}(t), \boldsymbol{B})$ represents the distance from the observed signal vector $\boldsymbol{x}(t)$ to the sub-plane/subspace $\boldsymbol{B}$.

Let an M-dimensional column vector $\boldsymbol{v}$ represent an observed signal point. $\boldsymbol{B} = [\boldsymbol{b}_1, \boldsymbol{b}_2, \ldots, \boldsymbol{b}_k]$ represents a k-dimensional sub-plane of an M-dimensional space. Then the distance from the column vector $\boldsymbol{v}$ to the sub-plane $\boldsymbol{B}$ is defined as follows.

$$d(\boldsymbol{v}, \boldsymbol{B}) = \sqrt{1 - [(\boldsymbol{v}, \boldsymbol{b}_1)^2 + \cdots + (\boldsymbol{v}, \boldsymbol{b}_k)^2]} \tag{10.4}$$

where $(\boldsymbol{v}, \boldsymbol{b}_i)$ represents the inner product of $\boldsymbol{v}$ and $\boldsymbol{b}_i$ let the other plane is $\hat{\boldsymbol{B}} = [\hat{\boldsymbol{b}}_1, \hat{\boldsymbol{b}}_2, \ldots, \hat{\boldsymbol{b}}_k]$. In the M-dimensional observed signal space, the distance between the two planes is calculated by formula (10.5).

$$d\left(\boldsymbol{B}, \hat{\boldsymbol{B}}\right) = \sqrt{d^2\left(\boldsymbol{b}_1, \hat{\boldsymbol{B}}\right) + \cdots + d^2\left(\boldsymbol{b}_k, \hat{\boldsymbol{B}}\right)} \tag{10.5}$$

where $d\left(\boldsymbol{b}_i, \hat{\boldsymbol{B}}\right)$ indicates the distance from $\boldsymbol{b}_i$ to plane $\hat{\boldsymbol{B}}$.

If the value of $d(\boldsymbol{x}(t), \boldsymbol{B})$ is small relative to the value of σ_1, then the value of $\exp(-d^2(\boldsymbol{x}(t), \boldsymbol{B})/2\sigma_1^2)$ is close to 1; If the value of $d(\boldsymbol{x}(t), \boldsymbol{B})$ is large relative to the value of σ_1, then the value of $\exp(-d^2(\boldsymbol{x}(t), \boldsymbol{B})/2\sigma_1^2)$ is close to 0. Thus, for the sufficiently small value of σ_1, the value of the function (10.3) is roughly equal to the number of observed signal points distributed on the plane $\boldsymbol{B}$. Further, if a set of data points is distributed over several different k-dimensional subspaces/sub-planes, then the function f takes a local maximum at each k-dimensional subspace/sub-plane.

For σ_1, if the value is too small, many local maxima will appear in f function under the influence of noise. Most of the local maxima are caused by noise, not what we are looking for, and these local maxima affect the selection of the real local maximum. If the value is too large, two or more similar local maxima may be smoothed into a single local maximum, thus missing some of the local maximum we need.

In order to solve this problem, you can first use a large value of σ_1, and then use a relatively small value of σ_1, that is, take the form of a descending sequence of σ_1. For example, let $\sigma_1 = [r_1, \ldots, r_c]$, according to experimental experience, r_1 is generally between 0.05 and 0.5, $r_{i+1} = r_i/2$.

- Estimate of mixing matrix

Now assume that all k-dimensional subspaces/sub-planes have been estimated, denoted by $\boldsymbol{B}_i, i = 1, \ldots, N_p$. We can use these k-dimensional subspaces/sub-planes to estimate the column vectors of mixing matrix.

In the k-dimensional subspace estimating, we know that each k-dimensional subspace is spanned by k M-dimensional column vectors. For a column vector in a mixing matrix $\boldsymbol{A}$, it is located at the intersection of C_{N-1}^{k-1} subspaces. Let's say $\boldsymbol{v}$ represents a vector in M-dimensional space, and we define the function:

$$g(\boldsymbol{v}) = \sum_{i=1}^{N_p} \exp\left(-\frac{d^2(\boldsymbol{v}, \boldsymbol{B}_i)}{2\sigma_2^2}\right) \tag{10.6}$$

where $d(\boldsymbol{v}, \boldsymbol{B}_i)$ represents the distance from vector $\boldsymbol{v}$ to the i-th k-dimensional subspace.

If the value of $d(\boldsymbol{v}, \boldsymbol{B}_i)$ is small relative to the value of σ_2, then the value of $\exp(-d^2(\boldsymbol{v}, \boldsymbol{B}_i)/2\sigma_2^2)$ is close to 1; If the value of $d(\boldsymbol{v}, \boldsymbol{B}_i)$ is large relative to the value of σ_2, then the value of $\exp(-d^2(\boldsymbol{v}, \boldsymbol{B}_i)/2\sigma_2^2)$ is close to 0.

Thus, for sufficiently small values of σ_2, the value of the function (10.6) is roughly equal to the number of subspaces that intersect the vector $\boldsymbol{v}$. Further, the function g takes a local maximum at each column vector of the mixing matrix $\boldsymbol{A}$.

For the value of σ_2, it is the same as σ_1, also adopts the form of a descending sequence, that is, $\sigma_2 = [R_1, \ldots, R_c]$. According to experimental experience, R_1 is generally between 0.05 and 0.5, $R_{i+1} = R_i/2$.

In general, for each column vector of the mixing matrix $\boldsymbol{A}$, it is located on the intersection of C_{N-1}^{k-1} subspaces. There may be other vectors that lie at the intersections of the $q_0 < \mathrm{C}_{n-1}^{k-1}$ subspaces. If we set $q = \mathrm{C}_{n-1}^{k-1}$, we need to detect the vectors at the intersections of the q subspaces. This is the ideal case, but in fact, under the influence of noise, the column vectors of the mixing matrix $\boldsymbol{A}$ may lie at the intersections of P $<< \mathrm{C}_{n-1}^{k-1}$ subspaces. Therefore, in practice, $q < \mathrm{C}_{n-1}^{k-1}$ should be set.

In order to find the column vectors of the mixing matrix more accurately, we use the function of Eq. (10.7) to replace Eq. (10.6).

$$h(\boldsymbol{v}) = \sum_{1 \le i_1 < \cdots < i_q \le N_p} u\left(\boldsymbol{v}, \boldsymbol{B}_{i_1}\right) \ldots u\left(\boldsymbol{v}, \boldsymbol{B}_{i_q}\right) \tag{10.7}$$

where $u(\boldsymbol{v}, \boldsymbol{B}) = \exp(-d^2(\boldsymbol{v}, \boldsymbol{B})/2\sigma_2^2)$, $d(\boldsymbol{v}, \boldsymbol{B})$ represents the distance from vector $\boldsymbol{v}$ to subspace $\boldsymbol{B}$. When σ_2 is sufficiently small, the value $u(\boldsymbol{v}, \boldsymbol{B})$ is close to 1 if the vector $\boldsymbol{v}$ is on the plane $\boldsymbol{B}$, otherwise, it is close to zero. So we get

$$u\left(\boldsymbol{v}, \boldsymbol{B}_{i_1}\right) \ldots u\left(\boldsymbol{v}, \boldsymbol{B}_{i_q}\right) \approx \begin{cases} 1 & \text{if } \boldsymbol{v} \text{ lines } \boldsymbol{B}_{i_1} \ldots \boldsymbol{B}_{i_q} \\ 0 & \text{else} \end{cases} \tag{10.8}$$

Therefore, we can use the formula (10.7) to estimate the column vectors of the mixing matrix $\boldsymbol{A}$.

10.1.2 Algorithm Implementation Steps

In summary, the steps of mixing matrix estimation method based on k-dimensional subspace are as follows:

Step 1 Remove the low energy points and perform 2-norm normalization on all observed vectors.

Step 2 Set the descending sequence $[\sigma_1, \sigma_2, \ldots, \sigma_\tau]$ for the potential function parameter σ, take $\sigma_1 \le 0.3$, and the recursion formula of the parameter is $\sigma_{i+1} = \sigma_i / 2(i = 1, 2, \ldots, \tau - 1)$.

Step 3 For $l = 1, 2, \ldots, L_1$ (L_1 is the number of k-dimensional subspaces. In order to estimate as many local extreme points as possible, L_1 is generally larger, $L_1 \ge 400$).

(1) Initialize a $M \times k$-dimensional random matrix $\boldsymbol{B}_l$, normalize and orthogonalize the column vectors of the matrix;
(2) Set $j = 1$;
(3) Take $\boldsymbol{B}_l$ as the initial iteration point, optimize the formula (10.3) by using the steepest ascent method, and obtain a new $\boldsymbol{B}_l$ after iteration convergence, normalize and orthogonalize the column vector of $\boldsymbol{B}_l$;
(4) If $j < \tau$, update j: $j \to j + 1$, return to (3) in Step 3.

End

Step 4 Remove the subspace of repeated estimates. (Since L_1 is greater than the number of k-dimensional subspaces that actually exist, the estimate of $\boldsymbol{B}_l$ in step 3 will be repeated).

Step 5 From the subspace that has been estimated, select the N_σ matrices $\boldsymbol{B}_l$ that make Eq. (10.3) the largest and N_σ less than L_1. (k-dimensional subspace clustering is completed and N_σ subspaces are obtained).

Step 6 Set the descending sequence $[\overline{\sigma}_1, \overline{\sigma}_2, \ldots, \overline{\sigma}_{\overline{\tau}}]$ of parameter $\overline{\sigma}$.

Step 7 for $l' = 1, 2, \ldots, L_2$ (In order to estimate all column vectors of the mixing matrix as completely as possible, take $L_2 \geq N$, N is the number of source signals).

(1) Initialize a random vector $v_{l'}$;
(2) Set $j' = 1$;
(3) $v_{l'}$ is taken as the initial point, and the formula (10.6) is optimized by the steepest ascent method, and a new $v_{l'}$ is obtained after iteration convergence;
(4) If $j' < \overline{\tau}$, add j' by 1 and return to (3) in Step 7.

End

Step 8 Remove $v_{l'}$ that makes the value of Eq. (10.7) less than ψ, where ψ is the threshold value, which is generally 0.8 times the maximum value of Eq. (10.7).

Step 9 Similar to Step 4, remove multiple estimating mixing matrix column vectors.

From the steps of the mixing matrix estimation method based on k-dimensional subspace clustering, we can see that the method uses double steepest ascent methods, the calculation is complex, and the selection of parameters is not instructive. In addition, in practical application, due to the interference of noise, there will be many local extreme points of interference, which is easy to cause the error of mixing matrix estimation.

10.2 Mixing Matrix Estimation Algorithm Based on Parameter Estimation

10.2.1 Algorithm Principles

The mixing matrix estimation precision of the k-dimensional subspace method described in the previous section is closely related to the selection of the parameters σ_1 and σ_2 of the potential function to some extent. To solve this problem, the mixing matrix estimation method based on parameter estimation adopts the adaptive parameter estimation method to further improve the accuracy of the mixing matrix estimation.

Similar to the k-dimensional subspace method, the mixing matrix estimation method based on parameter estimation first performs plane clustering and then estimates the mixing matrix. When plane clustering is carried out, the potential function shown in Eq. (10.3) in the above section is used, but the accuracy of plane clustering is greatly affected by the value of parameter σ_1. If the value of σ_1 is larger, the smoothness of Eq. (10.3) is better, which is conducive to optimization, but important local extremum points may be missing, resulting in incomplete plane clustering. If the value of σ_1 is too small, the local extremum features are more obvious, but the function smoothness is poor, which is not conducive to optimization. Therefore, the mixing matrix estimation method based on parameter estimation first estimates the reference value

$$r_1 = \frac{1}{T}\sum_{t=1}^{T}\frac{1}{M}\sum_{m=1}^{M}\left|x_m(t) - \overline{\upsilon}_{\boldsymbol{x}(t)}\right| \tag{10.9}$$

where $x_m(t)$ is the m-th element of the observed signal vector $\boldsymbol{x}(t)$ at sampling time t, and $\overline{\upsilon}_{\boldsymbol{x}(t)}$ is the average of all elements in $\boldsymbol{x}(t)$. According to the influence of σ_1 value on the clustering plane estimation effect, we adopt the idea of gradually approaching the local maximum. For example, random initialize $\boldsymbol{B}_0 \subset \mathrm{R}^{M\times k}$, r_1 obtained from Eq. (10.9) is first used as the scale parameter, and $\boldsymbol{B}_1$ is obtained for some local maximum value of $f(\boldsymbol{B})$ by using optimization method, and $\boldsymbol{B}_1$ is used as the initial value. Then $r_1/2$ is selected as the scale parameter and $\boldsymbol{B}_2$ for the local maximum of $f(\boldsymbol{B})$ obtained by optimization method. By analogy, an accurate estimate of the clustering plane is finally obtained. Therefore, the scale parameter r_1 selected is the descending sequence, that is $\sigma_1 = [r_1, r_1/2, r_1/4, r_1/8]$. In this way, the influence of scale parameters on the estimation effect can be reduced, and more accurate clustering plane estimation can be obtained.

When there are noise and outliers, wrong local maximum value can still occur despite the use of scale parameters in the form of descending sequence, but the potential of these local maximum value (i.e. the value of $f(\boldsymbol{B})$) is relatively small, so it can be screened out by setting a threshold.

Slightly different from the mixing matrix estimation in k-dimensional subspace method, the method based on parameter estimation uses the normal vector of the estimated subspace (hyperplane) to represent the subspace when estimating the column vector of the mixing matrix, so the potential function is

$$\overline{g}_{\sigma_2}(\boldsymbol{v}) = \sum_{j=1}^{L_c} \exp\left(-\frac{<\boldsymbol{v}, \boldsymbol{n}_j>^2}{2\sigma_2^2}\right) \tag{10.10}$$

where $\boldsymbol{n}_j$ represents the normal vector of the j-th subspace (hyperplane), L_c represents the estimated number of subspaces, and $\langle \boldsymbol{v}, \boldsymbol{n}_j \rangle$ represents the dot product of vectors, i.e. the sum of the multiplication of corresponding elements. If the vector $\boldsymbol{v}$ is in the j-th subspace, then $\boldsymbol{v}$ is orthogonal to $\boldsymbol{n}_j$, $\langle \boldsymbol{v}, \boldsymbol{n}_j \rangle$ has the value 0, and $\exp\left(-\langle \boldsymbol{v}, \boldsymbol{n}_j \rangle^2 / 2\sigma_2^2\right)$ in the Eq. (10.10) has the value 1. If $\boldsymbol{v}$ is not in the j-th subspace, then the value of $\langle \boldsymbol{v}, \boldsymbol{n}_j \rangle$ is not 0, and if $\sigma_2 \ll \langle \boldsymbol{v}, \boldsymbol{n}_j \rangle$, then the value of $\exp\left(-\langle \boldsymbol{v}, \boldsymbol{n}_j \rangle^2 / 2\sigma_2^2\right)$ is close to 1. Thus, the Eq. (10.9) is approximately equal to the number of subspaces in which the vector $\boldsymbol{v}$ is located. Function $\overline{g}_{\sigma_2}(\boldsymbol{v})$ has multiple local maximum points. By optimizing function $\overline{g}_{\sigma_2}(\boldsymbol{v})$ and searching for local maximum points, each column vector of the mixing matrix can be found.

When the normal vector of the clustering plane is known, the estimation accuracy of the column vector of the mixing matrix mainly depends on the scale parameter σ_2. When σ_2 is too large, the curve $\overline{g}_{\sigma_2}(\boldsymbol{v})$ is relatively smooth and some maximum points will be lost. When σ_2 is too small, some false local maximum values of function $\overline{g}_{\sigma_2}(\boldsymbol{v})$ will be generated under the influence of noise and outliers. For the selection of σ_2 values, we first estimate the reference value r_2 according to the normal vector estimated by the plane clustering, and the estimation method is shown in Eq. (10.11).

$$r_2 = \frac{1}{c} \sum_{j=1}^{c} \left(\frac{1}{m} \sum_{i=1}^{m} \left| z_j(i) - u_{z_j} \right| \right) \tag{10.11}$$

where c is the estimated number of normal vectors, $z_j(i)$ is the i-th element of the j-th normal vector, and u_{z_j} is the mean value of each element of the vector z_j.

The influence of σ_2 on the estimation effect of the column vector of the mixing matrix is the same as that of σ_1 on the estimation effect of the clustering plane. Therefore, the scale parameter is selected as a descending sequence based on r_2, that is, $\sigma_2 = [r_2, r_2/2, r_2/4, r_2/8]$, which can reduce the influence of the scale parameter on the estimation effect and improve the accuracy of the estimation of the mixing matrix.

10.2.2 Algorithm Implementation Steps

The mixing matrix estimation algorithm based on parameter estimation is introduced in this section, including the estimation of clustering plane based on parameter estimation and the estimation of normal vector based on parameter estimation. The implementation steps of the algorithm are given below.

- Steps for clustering planes estimation

Step 1 Remove the observed signal that are closer to the origin and normalize the observed signal.

Step 2 According to the observed signal, r_1 is estimated by Eq. (10.9). Determine the scale parameter σ_1, $\sigma_1 = [r_1, r_1/2, r_1/4, r_1/8]$, the length of σ_1 is L.

Step 3 For $j = 1 : K$ (K is the initial number of clustering planes, $K >> C_n^k$).

(a) Let $i = 1$, generate a random orthogonalized matrix $\boldsymbol{R}_0$ as the initial matrix.

(b) Let $\boldsymbol{B}_{j,i}^0 = \boldsymbol{R}_0$, $\boldsymbol{B}_{j,i}^0$ represents the initial value of $\boldsymbol{B}_{j,i}$.

(c) The method of steepest ascent is used to find $\boldsymbol{B}_{j,i}$ that maximizes $f\left(\boldsymbol{B}_{j,i}\right)$. That is, the iterative method is adopted to solve the problem, and the formula is shown as Eq. (10.12):

$$\boldsymbol{B}_{j,i}^{d+1} = \boldsymbol{B}_{j,i}^{d} + u\frac{\partial f(\boldsymbol{B}_{j,i}^{d})}{\partial \boldsymbol{B}_{j,i}^{d}} \tag{10.12}$$

The resulting $\boldsymbol{B}_{j,i}^{d+1}$ is orthogonalized after each iteration. Where d is the number of iterations, u is the step-size, and $\boldsymbol{B}_{j,i}^{d}$ represents $\boldsymbol{B}_{j,i}$ obtained by the d-th iteration. When $d(\boldsymbol{B}_{j,i}^{d+1}, \boldsymbol{B}_{j,i}^{d}) < e_1$, the iteration is terminated and $\boldsymbol{B}_{j,i}^{opt} = \boldsymbol{B}_{j,i}^{d}$. $\boldsymbol{B}_{j,i}^{opt}$ represents the optimal solution obtained at j-th optimization of $f\left(\boldsymbol{B}_{j,i}\right)$, and the scale parameter used in the potential function is $\sigma_1(i)$. Where $d(\boldsymbol{B}_{j,i}^{d+1}, \boldsymbol{B}_{j,i}^{d})$ represents the angle between the planes obtained by two iterations, and e_1 is threshold.

(d) if $i < L$, take the $\boldsymbol{B}_{j,i}^{opt}$ obtained in step (c) as the initial value, i.e. $\boldsymbol{R}_0 = \boldsymbol{B}_{j,i}^{opt}$, and add 1 to i, returning to step (b); Otherwise, $\boldsymbol{B}_{j}^{opt} = \boldsymbol{B}_{j,i}^{opt}$, and let $j = j + 1$, go back to step (a), where $\boldsymbol{B}_{j}^{opt}$ represents the optimal solution of the j-th cycle, i.e. the clustering plane estimated at the j-th cycle.

Step 4 If duplicate planes exist in the estimated plane, remove the duplicate planes.

Step 5 Calculate the potential $f(\boldsymbol{B}_{j}^{opt})$ of each plane and remove the plane with relatively small potential. The remaining plane is the estimate of the clustering plane.

- Steps for estimating mixing matrix

After the normal vector of the clustering plane is estimated, the column vectors of the mixing matrix can be estimated. The detailed steps are as follows.

Step 1 Estimate r_2 by Eq. (10.11) according to the normal vector of the clustering plane. Determine the descending sequence $\sigma_2 = [r_2, r_2/2, r_2/4, r_2/8]$, the length of σ_2 is L.

Step 2 For $j = 1 : P(P$ is the initial number of column vectors of mixing matrix, $P >> N)$.

(a) Let $i = 1$, generate a random unit vector $\boldsymbol{q}_0$.

(b) Let $\boldsymbol{v}_{j,i}^0 = q_0$, $\boldsymbol{v}_{j,i}^0$ represents the initial value of $\boldsymbol{v}_{j,i}$.

(c) Use the steepest ascent method to finds $\boldsymbol{v}_{j,i}$ where $\overline{g}_{\sigma_2}(\boldsymbol{v})$ reaches its maximum value. That is, the iterative method is adopted to solve the problem, and the formula is shown as Eq. (10.13):

$$\boldsymbol{v}_{j,i}^{d+1} = \boldsymbol{v}_{j,i}^{d} + u\frac{\partial g_{\sigma_2}(\boldsymbol{v}_{j,i}^{d})}{\partial \boldsymbol{v}_{j,i}^{d}} \tag{10.13}$$

The resulting $\boldsymbol{v}_{j,i}^{d+1}$ is unitized after each iteration. Where d is the number of iterations and u is the step-size. When $d(\boldsymbol{v}_{j,i}^{d+1}, \boldsymbol{v}_{j,i}^{d}) < e_2$, the iteration is terminated and $\boldsymbol{v}_{j,i}^{opt} = \boldsymbol{v}_{j,i}^{d}$, then $\boldsymbol{v}_{j,i}^{opt}$ represents the optimal solution obtained at the j-th optimization $g(\boldsymbol{v})$, and the scale parameter used in the potential function $\overline{g}_{\sigma_2}(\boldsymbol{v})$ is $\sigma_2(i)$. Where $d(\boldsymbol{v}_{j,i}^{d+1}, \boldsymbol{v}_{j,i}^{d})$ represents the difference between the vectors obtained through two iterations. e_2 is the threshold.

(d) If $i < L$, take the $\boldsymbol{v}_{j,i}^{opt}$ obtained in step (b) as the initial value, that is, $\boldsymbol{v}_{j,i}^{opt} = q_0$, go back to step (b), and add 1 to i; Otherwise, let $\boldsymbol{v}_{j}^{opt} = \boldsymbol{v}_{j,i}^{opt}$, $j = j + 1$ go back to step (a), where $\boldsymbol{v}_{j}^{opt}$ represents the optimal solution of the j-th cycle, i.e. the column vector estimated at the j-th cycle.

Step 3 Remove the estimated column vectors if they are duplicated.

Step 4 Calculate the potential $g_{\sigma_2}(\boldsymbol{v}_{j}^{opt})$ of each vector and remove the vector with relatively small potential. The remaining vectors are the estimates of the column vectors of the mixing matrix.

10.3 Mixing Matrix Estimation Algorithm Based on Plane Clustering Potential Function

10.3.1 Principle of Plane Clustering Potential Function Method

The idea of plane clustering potential function method is to estimate the clustering plane first, and then estimate the mixing matrix according to the estimated clustering plane. The following two parts introduce the principle of plane clustering potential function method.

- Estimation of the clustering plane

Suppose the number of source signals is N, the number of observed signals is M, and the number of non-zero sources at any given time is k, then the observed signals are distributed on C_N^k planes. To simplify the calculation, we use the normal of the plane to represent the plane, then the estimate of the plane is converted to the estimate of the normal. Define the potential function of (10.14).

$$f(z) = \sum_{t=1}^{T} \exp\left(-\frac{(\boldsymbol{x}(t), z)}{b}\right) \tag{10.14}$$

where z is the normal vector of the clustering plane. $(\boldsymbol{x}(t), z)$ is the inner product of $\boldsymbol{x}(t)$ and z, reflecting the distance from $\boldsymbol{x}(t)$ to the plane determined by z. If $\boldsymbol{x}(t)$ is in the plane determined by z, then the value of $(\boldsymbol{x}(t), z)$ is 0. b is a scale parameter whose value is obtained by the following formula:

$$b = \frac{1}{T}\sum_{t=1}^{T}\left(\frac{1}{M}\sum_{i=1}^{M}\left|x_i(t) - u_{x_i}\right|\right) \tag{10.15}$$

where u_{x_i} is the mean value of observed signal $x_i(t)$. From the formula (10.14), it can be seen that when the value of b is small enough, then the value of $\exp(-(\boldsymbol{x}(t), z)/b)$ is 1 as long as $\boldsymbol{x}(t)$ is in the plane determined by z; otherwise, the value of $\exp(-(\boldsymbol{x}(t), z)/b)$ is 0. Therefore, the value of the function $f(z)$ reflects the number of observed signal points gathered on the plane z. Further, the function $f(z)$ obtains a local maximum value at each plane where the observed signal is gathered.

The scale parameter b has a great influence on the estimation effect. If the value b is too small, many false local maxima will be generated under the influence of noise. If the value of b gets too large, then the curve of the function $f(z)$ is relatively smooth, and many local maxima disappear. In order to reduce the influence of b on the estimation, we extend the formula (10.14) into the form (10.16).

$$J(z) = \sum_{j=1}^{K} \sum_{t=1}^{T} (\exp(-(\boldsymbol{x}(t), z_j)^2)/b)^{\gamma} \tag{10.16}$$

where K is the initial number of normal vectors, and the value of K is generally much larger than C_N^k. $J(z)$ reflects the distribution of the observed data and obtains the local maximum value at the normal vector of each clustering plane of the observed signal. In order to further reduce the influence of b on the estimation effect, parameter γ is introduced. The value of γ is estimated by correlation comparison algorithm, which is as follows.

To estimate γ, define

$$\tilde{J}(x(t_1))_l = \sum_{t=1}^{T} e^{\left(\frac{-D(\boldsymbol{x}(t),\boldsymbol{x}(t_1))}{b}\right)^{\gamma_l}}, t_1 = 1, 2, \ldots, T \tag{10.17}$$

where $\gamma_l = 2l, l = 1, 2, \ldots$. Set $l = 1$, and the correlation coefficients of $\tilde{J}(x(t_1))_l$ and $\tilde{J}(x(t_1))_{l+1}$ are calculated. If their correlation coefficients are greater than a suitable initial value ξ_0, $\gamma = \gamma_l$; otherwise,$l = l + 1$, the correlation coefficients of $\tilde{J}(x(t_1))_l$ and $\tilde{J}(x(t_1))_{l+1}$ are re-calculated until their correlation coefficients are greater than ξ_0, and ξ_0 is generally taken as 0.97–0.999.

After γ is estimated, the gradient method can be used to estimate the local maximum value of $J(z)$. The iterative formula for z is as follows:

$$z_j^{d+1} \leftarrow z_j^d + \eta_j \frac{\partial J(z)}{\partial z_j} / || \frac{\partial J(z)}{\partial z_j} || \tag{10.18}$$

$$\frac{\partial J(z)}{\partial z_j} = \frac{1}{T} \sum_{t=1}^{T} \left\{ -\frac{2\gamma}{b} e^{-\gamma \langle \boldsymbol{x}(t), z_j \rangle^2 / b} \langle \boldsymbol{x}(t), z_j \rangle \boldsymbol{x}(t) \right\} \tag{10.19}$$

After the algorithm converges, the normal vector z of the original plane condenses at each local maximum of $J(z)$, forming several new clustering planes. Without the influence of outliers, these new clustering planes are the clustering planes to be estimated.

For the obtained K normal vectors, if there are any duplicates, the duplicates are removed. The number of remaining normal vectors is still much larger than C_n^k. There are many local maximum points formed due to noise. For these false maximum points, the values of $J(z_j)$ are generally small. In order to remove these points, we first find the maximum value of $J(z_j)$, and then for each $J(z_j)$, find $p_j = J(z_j)/\max(J(z_j))$. Removed the normal vector z_j whose p_j is less than a certain threshold value e_2 and the rest is the estimate of the normal vector that we need. The value of e_2 is generally between 0.4 and 0.6.

- Estimation of mixing matrix

After estimating the normal vector corresponding to the observed signal cluster plane, the mixing matrix can be estimated on the basis of the normal vector. Because for a column vector of the mixing matrix, it is located on C_{N-1}^{k-1} cluster planes, then the normal vectors corresponding to C_{N-1}^{k-1} planes are perpendicular to this column vector of the mixing matrix. Using a method similar to that used to estimate normal vectors, define functions as follow.

$$g(v) = \sum_{j=1}^{C} \exp\left(-\frac{(\boldsymbol{v}, z_j)^2}{\sigma^2}\right) \tag{10.20}$$

where z_j is the estimated normal vector, C is the number of normal vectors, $\boldsymbol{v}$ is the M-dimensional column vector, $(\boldsymbol{v}, z_j)$ represents the inner product of $\boldsymbol{v}$ and z_j. σ is the scale parameter. When σ is small enough, if $\boldsymbol{v}$ is perpendicular to z_j, the value of $(\boldsymbol{v}, z_j)$ is 0 and $\exp(-(\boldsymbol{v}, z_j)^2/\sigma^2)$ is close to 1; If $\boldsymbol{v}$ is not perpendicular to z_j, the value of $(\boldsymbol{v}, z_j)$ is not 0, and the value of $\exp(-(\boldsymbol{v}, z_j)^2/\sigma^2)$ is close to 0. Therefore, the value of the function g reflects the number of clustering planes containing the vector $\boldsymbol{v}$, and the function g takes local maximum at each column vector of the mixing matrix $\boldsymbol{A}$. The value of σ is estimated according to Eq. (10.21).

$$\sigma = \frac{1}{C}\sum_{j=1}^{C}\left(\frac{1}{M}\sum_{i=1}^{M}\left|z_j(i) - u_{z_i}\right|\right) \tag{10.21}$$

where u_{z_i} is the mean of $z_j(i)$. The value of σ has a great influence on the estimation of the mixing matrix. In order to reduce the influence of σ on the estimation effect and improve the estimation accuracy of the mixing matrix, we extend the Eqs. (10.20) to (10.22).

$$h(\boldsymbol{v}) = \sum_{j=1}^{C}\sum_{i=1}^{N_0}\left(\exp(-(\boldsymbol{v}_i, z_j)^2/\sigma)\right)^{\lambda} \tag{10.22}$$

where N_0 is the initial value of the number of column vectors, generally $N_0 \geq N$. Since a layer of summation is added and the parameter λ is introduced, the formula (10.20) is more robust than the formula (10.22). The value of parameter λ can also be estimated according to the correlation comparison algorithm. When $h(\boldsymbol{v})$ converges, for the obtained N column vectors, if there are duplicates, the duplicates are removed. The remaining number of column vectors is still much larger than N. Because there are many maximum points formed due to noise, for these wrong maximum points, $h(\boldsymbol{v}_i)$ is generally small. In order to remove these points, we first find the maximum of $h(\boldsymbol{v}_i)$, and then for each $h(\boldsymbol{v}_i)$, find $q_i = h(\boldsymbol{v}_i)/\max(h(\boldsymbol{v}_i))$. Remove the $\boldsymbol{v}_i$ whose q_i less than a certain threshold value. And remains is the estimate of column vectors of the mixing matrix that we need.

10.3.2 Steps of the Plane Clustering Potential Function Method

According to the principle of plane clustering potential function method, we get the estimation steps of clustering plane and column vectors of mixing matrix.

- Steps for estimating clustering planes

(1) The observed signal points that are relatively close to the origin of coordinates are removed and the observed signal is normalized.
(2) Randomly initialize K normal vectors $z_j \in \boldsymbol{R}^{M \times 1}, j = 1, \ldots, K$, and normalize z_j.
(3) Parameter γ is estimated according to the correlation comparison algorithm.
(4) For each $z_j \in \boldsymbol{R}^{M \times 1}, j = 1, \ldots, K$, the steepest ascent method shown in (10.18) and (10.19) is used to update the normal vector.
(5) If condition $\left\| z_j^{d+1} - z_j^d \right\| < e_1$ is satisfied, the iteration ends and z_j is the estimated normal vector; Otherwise, go back to step (4) and continue iterating until condition $\left\| z_j^{d+1} - z_j^d \right\| < e_1$ is satisfied.
(6) For the estimated normal vector, the repeated ones are removed, and the number of remaining normal vectors is D.
(7) In D normal vectors, find the maximum value of $J(z_j)$.
(8) For $J(z_j), j = 1, \ldots, D$, calculate $p_j = J(z_j)/\max(J(z_j))$.
(9) For $p_j, j = 1, \ldots, D$, if $p_j < e_2$, then the corresponding z_j is removed. The rest is the estimate of the normal vectors that we need.

- Steps for estimating mixing matrix

After estimating the normal vectors of the clustering plane, we can estimate the column vectors of the mixing matrix. The detailed steps are as follows.

(1) Randomly initialize N_0 normal vectors $\boldsymbol{v}_i \in \mathrm{R}^{M \times 1}, i = 1, \ldots, N_0$, and normalize $\boldsymbol{v}_i$.
(2) Parameter λ is estimated according to the correlation comparison algorithm.
(3) For each $\boldsymbol{v}_i \in \mathrm{R}^{M \times 1}, i = 1, \ldots, N$, the steepest ascent method is used to estimate the local maximum value of $h(\boldsymbol{v})$ and the formula used is as follows.

$$\boldsymbol{v}_i^{d+1} = \boldsymbol{v}_i^d + \eta_2 \frac{\partial h(\boldsymbol{v})}{\partial \boldsymbol{v}_i} \tag{10.23}$$

where η_2 is the step-size, d is the number of iterations, and the new $\boldsymbol{v}_i$ obtained by each iteration is normalized.
(4) If condition $\left\| \boldsymbol{v}_i^{d+1} - \boldsymbol{v}_i^d \right\| < e_3$ is satisfied, the iteration ends, and the obtained $\boldsymbol{v}_i$ is an estimated mixing matrix column vector; Otherwise, go back to step (3) and continue iterating with the formula (10.23) until condition $\left\| \boldsymbol{v}_i^{d+1} - \boldsymbol{v}_i^d \right\| < e_3$ is satisfied.

(5) For the obtained column vectors, the repeated ones are removed, and the number of remaining column vectors is F.
(6) In F vectors, find the maximum value of $h(\boldsymbol{v}_i)$, denoted by $\max(h(\boldsymbol{v}_i))$.
(7) For $h(\boldsymbol{v}_i), j = 1, \ldots, F$, calculate $q_i = h(\boldsymbol{v}_i)/\max(h(\boldsymbol{v}_i))$.
(8) For $q_i, j = 1, \ldots, F$, if $q_i < e_4$, the corresponding $\boldsymbol{v}_i$ is removed. The rest is the estimate of the column vectors of the mixing matrix that we need.

10.4 Underdetermined Mixing Matrix Blind Estimation Method Based on Homogeneous Polynomial Representation

The above estimating methods of mixing matrix use the steepest ascent method to find the maximum points. Due to the interference of many local maximum points, the estimation accuracy is easy to be affected, and a large number of initial points are needed to estimate the maximum points with large values completely. To solve the above problems, an underdetermined mixing matrix blind estimation method based on homogeneous polynomial representation is introduced in this section. This method uses generalized principal components analysis (GPCA) method on the basis of algebraic geometry theory to realize the clustering of hyperplanes, and then the hyperplane intersection lines are found by singular value decomposition to estimate each column vector of the mixing matrix. The mixing matrix estimation method introduced in this section uses another clustering method—spectral clustering. So before introducing the principle of algorithm, we first introduce the related concepts of spectral clustering.

10.4.1 Spectral Clustering

Spectral clustering is an algorithm evolved from graph theory and has been widely used in clustering. The main idea is to treat all data as points in space that can be connected by edges. The edge weight value between two points that are farther away is lower, while the edge weight value between two points that are closer is higher. By cutting the graph composed of all data points, sum of the edge weights between different subgraphs after cutting the graph is as low as possible, while sum of the edge weight in the subgraph is as high as possible, so as to achieve the purpose of clustering.

1. **Undirected Weight Graph**

Since spectral clustering is based on graph theory, let's first review the concept of graph. A graph G is usually described by the set of points V and the set of edges E, i.e. $G(V, E)$. Where V is all data set $(v_1, v_2, \ldots, v_n)$. For any two points in V, you can have an edge connection or no edge connection. We define w_{ij} as the weight between v_i and v_j. Since we are undirected graphs, so $w_{ij} = w_{ji}$.

For the two points v_i and v_j connected by edges, $w_{ij} > 0$. For two points v_i and v_j without an edge connection, $w_{ij} = 0$. For any point v_i in the graph, its degree d_i means the sum of the weights of all the edges connected to it, i.e.

$$d_i = \sum_{j=1}^{n} w_{ij} \tag{10.24}$$

Using the definition of degree of each point, we can get an $n \times n$ degree matrix $\boldsymbol{D}$, which is a diagonal matrix, d_i corresponds to the degree of the i-th point. $\boldsymbol{D}$ is defined as follows:

$$\boldsymbol{D} = \begin{bmatrix} d_1 & & & \\ & d_2 & & \\ & & \ddots & \\ & & & d_n \end{bmatrix} \tag{10.25}$$

Using the weights between all the points, we can get the adjacency matrix $\boldsymbol{W}$ of the graph, which is also an $n \times n$ matrix, and the j-th value of the i-th row corresponds to the weight w_{ij}.

In addition, for a subset A of point set V,$A \subset V$, we use $|A|$ to represent the number of points in subset A, and define $\mathrm{vol}(A)$ as follows.

$$\mathrm{vol}(A) := \sum_{i \in A} d_i \tag{10.26}$$

2. **Similarity Matrix**

Above we talked about the adjacency matrix $\boldsymbol{W}$, which is a matrix composed of the weight values w_{ij} between any two points. In spectral clustering, we only have the definition of the data points and do not give the adjacency matrix directly, so how do we get this adjacency matrix?

The basic idea is that the edge weights between the two points that are farther apart are lower, and the edge weights between the two points that are closer together are higher. In general, we can obtain the adjacency matrix $\boldsymbol{W}$ from the similarity matrix $\boldsymbol{S}$ of the distance measurement between sample points.

There are three types of methods for constructing adjacency matrix $\boldsymbol{W}$. ε-nearest neighbor method, K-nearest neighbor method and full connection method.

For the ε-nearest neighbor method, it sets a distance threshold ε, and then measures the distance between any two points x_i and x_j in Euclidean distance s_{ij}. That is, $s_{ij} = \left\|x_i - x_j\right\|_2^2$, and then according to magnitude of s_{ij} and ε, the adjacency matrix $\boldsymbol{W}$ is defined as follows:

$$w_{ij} = \begin{cases} 0 & s_{ij} > \varepsilon \\ \varepsilon & s_{ij} \le \varepsilon \end{cases} \tag{10.27}$$

As can be seen from the above equation, the weight between two points is ε or 0, there is no other information. Distance measurement is very imprecise, so in practical applications, we rarely use the ε-nearest neighbor method.

The second method to define the adjacency matrix $\boldsymbol{W}$ is the K- nearest neighbor method, which uses the KNN algorithm to traverse all sample points, and takes the nearest K points of each sample as the nearest neighbors, only K points that are nearest the sample point have $w_{ij} > 0$. However, this method will cause asymmetry of the adjacency matrix $\boldsymbol{W}$, but our later algorithm needs symmetric adjacency matrix. In order to solve this problem, one of the following two methods is generally adopted:

The first K-nearest neighbor method preserves s_{ij} as long as a point is in the K-nearest neighbor of another point

$$w_{ij} = w_{ji} = \begin{cases} 0 & x_i \notin KNN(x_j) \text{ and } x_j \notin KNN(x_i) \\ \exp\left(-\frac{\|x_i - x_j\|_2^2}{2\sigma^2}\right) & x_i \in KNN(x_j) \text{ or } x_j \in KNN(x_i) \end{cases} \tag{10.28}$$

The second K-nearest neighbor method is that two points must be K nearest neighbors to each other in order to preserve $s_{ij}s_{ij}$

$$w_{ij} = w_{ji} = \begin{cases} 0 & x_i \notin KNN(x_j) \text{ or } x_j \notin KNN(x_i) \\ \exp\left(-\frac{\|x_i - x_j\|_2^2}{2\sigma^2}\right) & x_i \in KNN(x_j) \text{ and } x_j \in KNN(x_i) \end{cases} \tag{10.29}$$

The third method to define the adjacency matrix $\boldsymbol{W}$ is the full connection method. Compared with the first two methods, the weight values between all points in the third method are greater than 0, so it is called the full connection method. Different kernel functions can be selected to define edge weights, commonly used are polynomial kernel functions, Gaussian kernel functions, and Sigmoid kernel functions. The most common is the Gaussian kernel function, where the similar matrix and the adjacency matrix are the same:

$$w_{ij} = s_{ij} = \exp\left(-\frac{\|x_i - x_j\|_2^2}{2\sigma^2}\right) \tag{10.30}$$

In practical applications, it is most common to use the third full connection method to establish the adjacency matrix, and the Gaussian radial kernel function is the most common in the full connection method.

3. **Laplace Matrix**

The Laplace matrix is introduced separately because the following algorithms are closely related to the properties of this matrix. Its definition is very simple, the

Laplace matrix $\boldsymbol{L} = \boldsymbol{D} - \boldsymbol{W}$. $\boldsymbol{D}$ is the degree matrix, which is a diagonal matrix. And $\boldsymbol{W}$ is the adjacency matrix, which can be constructed by our method in Sect. 10.2.

Laplacian matrix has some nice properties as follows:

(1) Laplace matrix is a symmetric matrix, which can be derived from the fact that $\boldsymbol{D}$ and $\boldsymbol{W}$ are symmetric matrices.
(2) Since the Laplace matrix is symmetric, all its eigenvalues are real.
(3) For any vector $\boldsymbol{f}$, we have

$$\boldsymbol{f}^{\mathrm{T}}\boldsymbol{L}\boldsymbol{f} = \frac{1}{2}\sum_{i,j=1}^{n} w_{ij}\left(f_i - f_j\right)^2 \tag{10.31}$$

It is easy to prove using the definition of the Laplace matrix as follows.

$$\begin{aligned} \boldsymbol{f}^{\mathrm{T}}\boldsymbol{L}\boldsymbol{f} &= \boldsymbol{f}^{\mathrm{T}}\boldsymbol{D}\boldsymbol{f} - \boldsymbol{f}^{\mathrm{T}}\boldsymbol{W}\boldsymbol{f} = \sum_{i=1}^{n} d_i f_i^2 - \sum_{i,j=1}^{n} w_{ij} f_i f_j \\ &= \frac{1}{2}\left(\sum_{i=1}^{n} d_i f_i^2 - 2\sum_{i,j=1}^{n} w_{ij} f_i f_j + \sum_{j=1}^{n} d_j f_j^2\right) = \frac{1}{2}\sum_{i,j=1}^{n} w_{ij}\left(f_i - f_j\right)^2 \end{aligned} \tag{10.32}$$

(4) The Laplace matrix is semidefinite, and the corresponding n real eigenvalues are all greater than or equal to 0, that is $0 = \lambda_1 \leq \lambda_2 \leq \cdots \leq \lambda_n$, and the smallest eigenvalue is 0, which is easily derived from property (3).

4. **Undirected Cut Graph**

For a cut graph of an undirected graph G, our goal is to cut the graph $G(V, E)$ into k subgraphs that are not connected to each other, and the set of points in each subgraph is: $A_1, A_2, \ldots, A_k$, they satisfy $A_i \cap A_j = \emptyset$, and $A_1 \cup A_2 \cup \cdots \cup A_k = V$.

For any two sets $A, B \subseteq V, A \cap B = \emptyset$, we define the weight of cut graph between A and B as:

$$W(A, B) = \sum_{i \in A, j \in B} w_{ij} \tag{10.33}$$

Then for k sets $A_1, A_2, \ldots, A_k$ of subgraph points, we define the cutting graph as:

$$\operatorname{cut}(A_1, A_2, \ldots A_k) = \frac{1}{2}\sum_{i=1}^{k} W\left(A_i, \overline{A}_i\right) \tag{10.34}$$

where $\overline{A}_i$ is the complementary set of A_i, meaning the union of subsets of V except subsets of A_i.

So how to make the sum of weights of points in the subgraphs high and the sum of weights of point between the subgraphs low? A natural idea is to minimize $\mathrm{cut}(A_1, A_2, \ldots A_k)$, but this minimized cutting graph is not the optimal cutting graph. How to avoid using this minimized cutting graph and find the optimal cutting graph? The cut graph methods used in spectral clustering are described below.

5. **Cutting Graph Clustering**

In order to avoid the poor cutting effect caused by minimized cutting graph, it is necessary to limit the scale of each subgraph. Generally speaking, there are two ways of cutting graph, the first is RatioCut and the second is Ncut.

(1) **RatioCut**

In order to avoid the minimized cutting graph, for each cutting graph, RatioCut considers not only minimizing $\mathrm{cut}(A_1, A_2, \ldots A_k)$, but also maximizing of the number of each subgraph points, namely:

$$\mathrm{RatioCut}(A_1, A_2, \ldots A_k) = \frac{1}{2}\sum_{i=1}^{k}\frac{W\left(A_i, \overline{A}_i\right)}{|A_i|} \tag{10.35}$$

So how to minimize the RatioCut function? It is found that the RatioCut function can be represented in the following way.

Introduce the indicator vector $\boldsymbol{h}_j \in \{\boldsymbol{h}_1, \boldsymbol{h}_2, \ldots \boldsymbol{h}_k\}$, any vector $\boldsymbol{h}_j$ is an n-dimensional vector (n is the number of samples), and define h_{ij} as:

$$h_{ij} = \begin{cases} 0 & v_j \notin A_i \\ \frac{1}{\sqrt{|A_i|}} & v_j \in A_i \end{cases} \quad (i = 1, 2, \ldots, k; j = 1, 2, \ldots, n) \tag{10.36}$$

Then for $\boldsymbol{h}_i^{\mathrm{T}}\boldsymbol{L}\boldsymbol{h}_i$, there is:

$$\begin{aligned}
\boldsymbol{h}_i^{\mathrm{T}}\boldsymbol{L}\boldsymbol{h}_i &= \frac{1}{2}\sum_{p=1}^{n}\sum_{q=1}^{n} w_{pq}\left(h_{ip} - h_{iq}\right)^2 \\
&= \frac{1}{2}\left(\sum_{p\in A_i, q\notin A_i} w_{pq}\left(\frac{1}{\sqrt{|A_i|}} - 0\right)^2 + \sum_{p\notin A_i, q\in A_i} w_{pq}\left(0 - \frac{1}{\sqrt{|A_i|}}\right)^2\right) \\
&= \frac{1}{2}\left(\sum_{p\in A_i, q\notin A_i} w_{pq}\frac{1}{|A_i|} + \sum_{p\notin A_i, q\in A_i} w_{pq}\frac{1}{|A_i|}\right) \\
&= \frac{1}{2}\left(cut\left(A_i, \overline{A}_i\right)\frac{1}{|A_i|} + \mathrm{cut}\left(\overline{A}_i, A_i\right)\frac{1}{|A_i|}\right) \\
&= \frac{\mathrm{cut}\left(A_i, \overline{A}_i\right)}{|A_i|}
\end{aligned} \tag{10.37}$$

for a certain subgraph i, its RatioCut corresponds to $\boldsymbol{h}_i^{\mathrm{T}}\boldsymbol{L}\boldsymbol{h}_i$, then the RatioCut function corresponding to k subgraphs is

$$\mathrm{RatioCut}(A_1, A_2, \ldots A_k) = \sum_{i=1}^{k} \boldsymbol{h}_i^{\mathrm{T}}\boldsymbol{L}\boldsymbol{h}_i = \sum_{i=1}^{k} \left(\boldsymbol{H}^{\mathrm{T}}\boldsymbol{L}\boldsymbol{H}\right)_{ii} = \mathrm{tr}\left(\mathbf{H}^{\mathrm{T}}\boldsymbol{L}\boldsymbol{H}\right) \quad (10.38)$$

where $\mathrm{tr}\left(\mathbf{H}^{T}\boldsymbol{L}\boldsymbol{H}\right)$ is the trace of the matrix. In other words, RatioCut is actually minimized $\mathrm{tr}\left(\mathbf{H}^{\mathrm{T}}\boldsymbol{L}\boldsymbol{H}\right)$. If $\boldsymbol{H}^{\mathrm{T}}\boldsymbol{H} = \boldsymbol{I}$, then the optimization objective of cutting graph is

$$\underbrace{\arg\min}_{\boldsymbol{H}} \mathrm{tr}\left(\mathbf{H}^{\mathrm{T}}\boldsymbol{L}\boldsymbol{H}\right) \text{ s.t. } \boldsymbol{H}^{\mathrm{T}}\boldsymbol{H} = \boldsymbol{I} \quad (10.39)$$

Because every indicator vector in matrix $\boldsymbol{H}$ is n-dimensional, and every element in the vector has a value of 0 or $\frac{1}{\sqrt{|A_j|}}$, there are 2^n values. If there are k subgraphs and k indicator vectors, then there are $k \cdot 2^n$ kinds of $\boldsymbol{H}$. Therefore, finding $\boldsymbol{H}$ satisfying the above optimization objective is a nondeterministic polynomial hard (NP-hard) problem.

For each optimization sub-target $\boldsymbol{h}_i^{\mathrm{T}}\boldsymbol{L}\boldsymbol{h}_i$ in $\mathrm{tr}\left(\mathbf{H}^{\mathrm{T}}\boldsymbol{L}\boldsymbol{H}\right)$, where $\boldsymbol{h}_i$ is the unit orthogonal basis and $\boldsymbol{L}$ is the symmetric matrix, and the maximum value of $\boldsymbol{h}_i^{\mathrm{T}}\boldsymbol{L}\boldsymbol{h}_i$ is the maximum eigenvalue of $\boldsymbol{L}$, and the minimum value is the minimum eigenvalue of $\boldsymbol{L}$. For $\boldsymbol{h}_i^{\mathrm{T}}\boldsymbol{L}\boldsymbol{h}_i$, the optimization goal is to find the smallest eigenvalue of $\boldsymbol{L}$, and for $\mathrm{tr}\left(\mathbf{H}^{\mathrm{T}}\boldsymbol{L}\boldsymbol{H}\right) = \sum_{i=1}^{k} \boldsymbol{h}_i^{\mathrm{T}}\boldsymbol{L}\boldsymbol{h}_i$, the goal is to find k smallest eigenvalues. In general, k is much less than n, that is, by reducing the dimension from n to k, this NP-hard problem can be solved approximately.

By finding the smallest k eigenvalues of $\boldsymbol{L}$, the corresponding k eigenvectors can be obtained, and these k eigenvectors form an $n \times k$ matrix, which is $\boldsymbol{H}$. It is usually necessary to normalize the matrix $\boldsymbol{H}$ by row, i.e.

$$h_{ij}^{*} = \frac{h_{ij}}{\left(\sum_{t=1}^{k} h_{it}^{2}\right)^{1/2}} \quad (10.40)$$

After obtaining the $n \times k$ matrix $\boldsymbol{H}$, it is also necessary to perform a traditional clustering for each row, such as K-Means clustering.

(2) **Ncut**

Ncut graphs are similar to RatioCut graphs. Ncut graph can be obtained by replacing the denominator $|A_i|$ of RatioCut with $vol(A_i)$. The number of subgraph samples is large, the weight is not necessarily large. Therefore, cutting the graph based on weight is more in line with the expected goal, generally speaking, Ncut graphs are better than RatioCut graphs, and the calculation formula of Ncut graphs is as follows.

$$\mathrm{NCut}(A_1, A_2, \ldots A_k) = \frac{1}{2}\sum_{i=1}^{k} \frac{W\left(A_i, \overline{A}_i\right)}{\mathrm{vol}(A_i)} \tag{10.41}$$

Correspondingly, the Ncut graph has improved the indicator vector $\boldsymbol{h}$, and the Ncut graph uses the subgraph weight $\frac{1}{\sqrt{vol(A_j)}}$ to mark the indicator vector $\boldsymbol{h}$, which is defined as follows.

$$h_{ij} = \begin{cases} 0 & v_i \notin A_j \\ \frac{1}{\sqrt{\mathrm{vol}(A_j)}} & v_i \in A_j \end{cases} \tag{10.42}$$

Then

$$\boldsymbol{h}_i^{\mathrm{T}} \boldsymbol{L} \boldsymbol{h}_i = \frac{1}{2}\sum_{p=1}^{n}\sum_{q=1}^{n} w_{pq}\left(h_{ip} - h_{iq}\right)^2 = \frac{\mathrm{cut}\left(A_i, \overline{A}_i\right)}{\mathrm{vol}(A_i)} \tag{10.43}$$

The derivation is exactly the same as RatioCut. In other words, the optimization goal is still

$$\mathrm{NCut}(A_1, A_2, \ldots A_k) = \sum_{i=1}^{k} \boldsymbol{h}_i^{\mathrm{T}} \boldsymbol{L} \boldsymbol{h}_i = \sum_{i=1}^{k} \left(\boldsymbol{H}^{\mathrm{T}} \boldsymbol{L} \boldsymbol{H}\right)_{ii} = \mathrm{tr}\left(\boldsymbol{H}^{\mathrm{T}} \boldsymbol{L} \boldsymbol{H}\right) \tag{10.44}$$

But now $\boldsymbol{H}^{\mathrm{T}}\boldsymbol{H} \neq \boldsymbol{I}$, and $\boldsymbol{H}^{\mathrm{T}}\boldsymbol{D}\boldsymbol{H} = \boldsymbol{I}$. The derivation is as follows.

$$\boldsymbol{h}_i^{\mathrm{T}} \boldsymbol{D} \boldsymbol{h}_i = \sum_{j=1}^{n} h_{ij}^2 d_j = \frac{1}{\mathrm{vol}(A_i)} \sum_{j \in A_i} d_j = \frac{1}{\mathrm{vol}(A_i)} \mathrm{vol}(A_i) = 1 \tag{10.45}$$

Therefore, the optimization goal is finally

$$\underbrace{\arg\min}_{\boldsymbol{H}} H\,\mathrm{tr}\left(\boldsymbol{H}^{\mathrm{T}} \boldsymbol{L} \boldsymbol{H}\right) \ \mathrm{s.t.}\, \boldsymbol{H}^{\mathrm{T}} \boldsymbol{D} \boldsymbol{H} = \boldsymbol{I} \tag{10.46}$$

In this case, the indicator vector $\boldsymbol{h}$ in $\boldsymbol{H}$ is not orthonormal basis, so the dimensionality reduction idea in RatioCut cannot be used directly. Set $\boldsymbol{H} = \boldsymbol{D}^{-1/2}\boldsymbol{F}$, then

$$\boldsymbol{H}^{\mathrm{T}} \boldsymbol{L} \boldsymbol{H} = \boldsymbol{F}^{\mathrm{T}} \boldsymbol{D}^{-1/2} \boldsymbol{L} \boldsymbol{D}^{-1/2} \boldsymbol{F} \tag{10.47}$$

$$\boldsymbol{H}^{\mathrm{T}} \boldsymbol{D} \boldsymbol{H} = \boldsymbol{F}^{\mathrm{T}} \boldsymbol{F} = \boldsymbol{I} \tag{10.48}$$

In other words, the optimization goal becomes

$$\underbrace{\arg\min}_{F} \operatorname{tr}(\boldsymbol{F}^{\mathrm{T}}\boldsymbol{D}^{-1/2}\boldsymbol{L}\boldsymbol{D}^{-1/2}\boldsymbol{F}) \text{ s.t. } \boldsymbol{F}^{\mathrm{T}}\boldsymbol{F} = \boldsymbol{I} \tag{10.49}$$

You can see that this formula is basically the same as RatioCut, except that the $\boldsymbol{L}$ in the middle changes to $\boldsymbol{D}^{-1/2}\boldsymbol{L}\boldsymbol{D}^{-1/2}$. In this way, we can continue to calculate the first k minimum eigenvalues of $\boldsymbol{D}^{-1/2}\boldsymbol{L}\boldsymbol{D}^{-1/2}$ according to the RatioCut idea, and then calculate the corresponding eigenvectors, and standardize them to obtain the final eigenmatrix $\boldsymbol{F}$, and finally perform a traditional clustering of $\boldsymbol{F}$, such as K-Means algorithm.

6. **Steps of Spectral Clustering Algorithm**

Generally speaking, the main attention points of spectral clustering are the generation of similar matrix, the way of cutting graph and the final clustering method. The most common method of generating similar matrix is the full connection method based on Gaussian kernel function, and the most common method of cutting graph is Ncut. Finally, the common clustering method is K-Means. The steps of spectral clustering algorithm are summarized as follows.

Input: Sample set $X(x_1, x_2, \ldots, x_n)$, generation method of similar matrix, dimension k_1 after dimensionality reduction, clustering method, dimension k_2 after clustering;

Output: Cluster partition $C(c_1, c_2, \ldots c_{k_2})$;

Step 1: The similarity matrix $\boldsymbol{S}$ of the sample is constructed according to generation method of the similarity matrix.

Step 2: Construct adjacency matrix $\boldsymbol{W}$ and degree matrix $\boldsymbol{D}$ according to similarity matrix $\boldsymbol{S}$.

Step 3: Calculate the Laplace matrix $\boldsymbol{L}$.

Step 4: Construct the standardized Laplace matrix $\boldsymbol{D}^{-1/2}\boldsymbol{L}\boldsymbol{D}^{-1/2}$.

Step 5: Calculate the corresponding eigenvectors $\boldsymbol{f}$ of the k_1 smallest eigenvalues of $\boldsymbol{D}^{-1/2}\boldsymbol{L}\boldsymbol{D}^{-1/2}$.

Step 6: The matrix composed of the corresponding eigenvectors $\boldsymbol{f}$ is normalized by row, and the $n \times k_1$ eigenmatrix $\boldsymbol{F}$ is finally formed.

Step 7: Each row in $\boldsymbol{F}$ is taken as a k_1-dimensional sample, a total of n samples are clustered using the input clustering method, and the clustering dimension is k_2.

Step 8: Get cluster partition $C(c_1, c_2, \ldots c_{k_2})$.

10.4.2 Principles of the Algorithm

1. **Polynomial Representation of Multiple Subspaces**

The GPCA uses an algebraic-geometric method to cluster subspaces in the high-dimensional space. The principle of the GPCA is to acquire the basis vectors of the orthogonal complementary space of each subspace through polynomial fitting and differentiation, thus completing the subspace clustering. The union of n subspaces can be expressed by homogeneous polynomials. The observed signal vector is expressed as $\boldsymbol{x} = [x_1, x_2, \ldots, x_M]^{\mathrm{T}}$, and then the basis of the n-order homogeneous polynomial space is the set $\{x_1^n, \ldots, x_1^{n_1} x_2^{n_2} \ldots x_M^{n_M}, \ldots, x_M^n\}$, which has $D_n(M) = C_{n+M-1}^{M-1}$ elements, and each element of the set is $x_1^{n_1} x_2^{n_2} \ldots x_M^{n_M}$, where $1 \leq n_z \leq M (z = 1, 2, \ldots M)$ and $n_1 + n_2 + \ldots n_M = n$. Therefore, by using the linear combination of bases, any homogeneous polynomial in the space can be expressed as

$$p_n(\boldsymbol{x}) = \boldsymbol{c}_n^{\mathrm{T}} \mathrm{v}_n(\boldsymbol{x}) = \sum_i^{D_n(M)} c_i \mathrm{v}_n^i(\boldsymbol{x}) \tag{10.50}$$

where $\boldsymbol{c}_n = [c_1^{(n)}, c_2^{(n)}, \ldots, c_{D_n(M)}^{(n)}]^{\mathrm{T}}$ is the coefficient vector of the basis, $v_n^i(\boldsymbol{x})$ is the i-th element of $\boldsymbol{v}_n(\boldsymbol{x})$, which is the n-order Veronese mapping, and $\boldsymbol{v}_n(\boldsymbol{x})$ is defined as $[x_1, x_2, \ldots, x_M]^{\mathrm{T}} \rightarrow [x_1^n, \ldots, x_1^{n_1} x_2^{n_2} \ldots x_M^{n_M}, \ldots, x_M^n]$. For example, if three sensors are to receive the source signals and there are two subspaces in the observed signal space, we have $D_n(M) = D_2(3) = C_{2+3-1}^{3-1} = 6$, and the 2-order Veronese mapping is $\boldsymbol{v}_2(\boldsymbol{x}) = \left[x_1^2, x_1x_2, x_1x_3, x_2^2, x_2x_3, x_3^2\right]^{\mathrm{T}} \in \boldsymbol{R}^6$.

The n subspaces in the observed signal space are denoted as $S_i \subset \boldsymbol{R}^M$, $i = 1, 2, \ldots, n$, the dimension of the i-th subspace is $d_i (0 < d_i < M)$, and the corresponding orthogonal complementary space $S_i^{\perp}$ is denoted by a matrix $\boldsymbol{B}_i = [b_{i1}, b_{i2}, \ldots, b_{i(M-d_i)}] \in \boldsymbol{R}^{M \times (M-d_i)}$. In order to represent the union set of multiple subspaces in terms of homogeneous polynomials, theorem 10.1 is introduced.

Theorem 10.1 *The union set of multiple subspaces can be represented by a set of homogeneous polynomials, and the vectors in each subspace satisfy this set of homogeneous polynomials, and each homogeneous polynomial can be represented as*

$$p_{n\ell}(\boldsymbol{x}) = \prod_{i=1}^{n} \left(\boldsymbol{b}_{i\ell}^{\mathrm{T}} \boldsymbol{x}\right) = \boldsymbol{c}_{n\ell}^{\mathrm{T}} \boldsymbol{v}_n(\boldsymbol{x}) = 0 \tag{10.51}$$

where, the coefficient vector space spanned by the coefficient vector $\boldsymbol{c}_{n\ell}$ is denoted by W_n, $\ell \in \{1, 2, \ldots, dim(W_n)\}$, $\dim(\cdot)$ represents the dimension of the linear space, $\boldsymbol{b}_{i\ell} \in \boldsymbol{R}^M$ is a column vector in $\boldsymbol{B}_i$, that is, $\boldsymbol{b}_{i\ell}$ is a normal vector of the i-th subspace. However, for planar clustering, it is only necessary to find a vector $\boldsymbol{c}_{n\ell}$ in the coefficient vector space $\boldsymbol{W}_n$, which is expressed by $\boldsymbol{c}_n$ in the following paragraphs.

According to Theorem 10.1, each subspace can be identified by solving the normal vector or coefficient vector $\boldsymbol{c}_n$ of each subspace, that is, a set of polynomials can be obtained to represent the union set of multiple subspaces in the observed signal space. Next, we discuss how to solve polynomials using only the observed signal, i.e. homogeneous polynomial fitting.

Suppose that there are T observed vectors $\boldsymbol{x}_t (t = 1, 2, \ldots, T)$, and all of them satisfy the formula (10.51), so

$$p_{n\ell}(\boldsymbol{x}) = \prod_{i=1}^{n} \left(\boldsymbol{b}_{i\ell}^{\mathrm{T}} \boldsymbol{x}\right) = \boldsymbol{c}_{n\ell}^{\mathrm{T}} \, \mathrm{v}_n(\boldsymbol{x}) = 0 \tag{10.52}$$

Obviously, $\boldsymbol{c}_n \in \mathrm{null}(\boldsymbol{V}_n(T))$, $\boldsymbol{W}_n \subseteq \mathrm{null}(\boldsymbol{V}_n(T))$, $\mathrm{null}(\cdot)$ represent the null space of the matrix. When the number of sample T is large enough, $\dim(\boldsymbol{W}_n) = \dim(\mathrm{null}(\boldsymbol{V}_n(T)))$. According to the theory of linear algebra, $\boldsymbol{W}_n$ and $\mathrm{null}(\boldsymbol{V}_n(T))$ are two identical linear spaces, so the coefficient vector $\boldsymbol{c}_n$ of the polynomial can be obtained by solving the null space of $\boldsymbol{V}_n(T)$, and homogeneous polynomial fit the union set of multiple subspaces can be realized.

2. **Acquiring the Bases of the Complementary Space with Polynomial Differentiation**

The previous section completes the homogeneous polynomial fitting of the union of n subspaces, resulting in a set of homogeneous polynomials. With these polynomials, we can get a set of bases of an orthogonal complementary space of each subspace and the dimension d_i of each subspace. Theorem 10.2 is introduced for this purpose.

Theorem 10.2 *Takes the vector $\boldsymbol{y}_i \in S_i$, and for $i \neq j$, $\boldsymbol{y}_i \notin S_j$, if $\dim(W_n) = \dim(\mathrm{null}(\boldsymbol{V}_n(T))) = w_n$, then the orthogonal complement of each subspace S_i is*

$$S_i^{\perp} = \mathrm{span}\left\{ \frac{\partial}{\partial \boldsymbol{x}} \boldsymbol{c}_n^{\mathrm{T}} \boldsymbol{v}_n(\boldsymbol{x}) \big|_{x=y_i}, \forall \boldsymbol{c}_n \in \mathrm{null}(\boldsymbol{V}_n(T)) \right\} \tag{10.53}$$

and the dimension of each subspace S_i is

$$d_i = M - \mathrm{rank}\left(D\overline{P}_n(\boldsymbol{y}_i)\right), \quad i = 1, 2, \ldots, n \tag{10.54}$$

where D is a differential operator and $\mathrm{rank}(\bullet)$ *denotes the rank of the matrix,* $\overline{P}_n(\boldsymbol{y}_i) = \left[p_{n1}(\boldsymbol{y}_i), p_{n2}(\boldsymbol{y}_i), \ldots, p_{nw_n}(\boldsymbol{y}_i)\right]$, $D\overline{P}_n(\boldsymbol{y}_i) = \left[Dp_{n1}(\boldsymbol{y}_i), Dp_{n2}(\boldsymbol{y}_i), \ldots, Dp_{nw_n}(\boldsymbol{y}_i)\right] \in \boldsymbol{R}^{M \times w_n}$.

According to Theorem 10.2, if the homogeneous polynomial fitting is accomplished, with the method of polynomial differentiation, only one vector is needed in any subspace to obtain a set of bases of the orthogonal complementary space of the subspace. To explain it intuitively, take Fig. 3.5 as an example. In Fig. 3.5, there are six subspaces (hyperplanes), and the dimension of the orthogonal complementary space of each subspace is 1. Six subspaces can be represented by 6 normal vectors, i.e. $\boldsymbol{b}_i (i = 1, 2, \ldots, 6)$. Therefore, it is easy to find that the polynomial representation of 6 subspaces is $p_6(\boldsymbol{x}) = \left(\boldsymbol{b}_1^{\mathrm{T}} \boldsymbol{x}\right)\left(\boldsymbol{b}_2^{\mathrm{T}} \boldsymbol{x}\right) \ldots \left(\boldsymbol{b}_6^{\mathrm{T}} \boldsymbol{x}\right) = \boldsymbol{c}_6^{\mathrm{T}} \, \mathrm{v}_6(\boldsymbol{x})$, where $\boldsymbol{c}_6 \in \mathrm{null}(\boldsymbol{V}_6(T))$ when T is large enough, the dimension of null space of $\boldsymbol{V}_6(T)$ is 1, that is, $\boldsymbol{c}_6$ is uniquely determined by solving the kernel space of $\boldsymbol{V}_6(T)$. The bases of the orthogonal complementary space of each subspace are solved by polynomial differentiation.

The derivative of $p_6(\boldsymbol{x})$ for the observed signal vector $\boldsymbol{x}_t \in S_i$ is

$$Dp_6(\boldsymbol{x})\big|_{x=x_t} = \frac{\partial p_6(\boldsymbol{x})}{\partial \boldsymbol{x}}\big|_{x=x_t} = \frac{\partial}{\partial \boldsymbol{x}} \prod_{k=1}^{6} (\boldsymbol{b}_k^{\mathrm{T}} \boldsymbol{x})\big|_{x=x_t} = \sum_{k=1}^{6} (\boldsymbol{b}_k) \prod_{j \neq k} \left(\boldsymbol{b}_j^{\mathrm{T}} \boldsymbol{x}\right)\big|_{x=x_t} \tag{10.55}$$

Since $\boldsymbol{b}_i^{\mathrm{T}} \boldsymbol{x}_t = 0$, the Eq. (10.55) becomes

$$Dp_6(\boldsymbol{x}_t) = \boldsymbol{b}_i \prod_{j \neq i} \left(\boldsymbol{b}_j^{\mathrm{T}} \boldsymbol{x}_t\right) \tag{10.56}$$

By refreshing the formula (10.56), the normalized normal vectors of each subspace can be obtained as

$$\frac{\boldsymbol{b}_i}{\|\boldsymbol{b}_i\|} = \frac{Dp_6(\boldsymbol{x}_t)}{\|Dp_6(\boldsymbol{x}_t)\|} \tag{10.57}$$

Therefore, we can obtain the normal vector using $Dp_n(\boldsymbol{x}_t)$. The following is a method for solving $Dp_n(\boldsymbol{x}_t)$. Considering a simple example first, the second-order Veronese mapping is $\boldsymbol{v}_2(\boldsymbol{x}) = \left[x_1^2, x_1x_2, x_1x_3, x_2^2, x_2x_3, x_3^2\right]^{\mathrm{T}} \in \boldsymbol{R}^6$, and we define each element of the power matrix $\hat{\boldsymbol{P}}_n \in \boldsymbol{R}^{D_n(M) \times M}$ as the power of each component $x_1, \ldots, x_M$ of $\boldsymbol{x}$ in the Veronese mapping. For the Veronese mapping $\boldsymbol{v}_2(\boldsymbol{x})$, we have

$$\hat{\boldsymbol{P}}_2 = \begin{bmatrix} 2\ 1\ 1\ 0\ 0\ 0 \\ 0\ 1\ 0\ 2\ 1\ 0 \\ 0\ 0\ 1\ 0\ 1\ 2 \end{bmatrix}^{\mathrm{T}} \tag{10.58}$$

The corresponding homogeneous polynomial is

$$p_2(\boldsymbol{x}) = \boldsymbol{c}_2^{\mathrm{T}} \boldsymbol{v}_2(\boldsymbol{x}) = c_1x_1^2 + c_2x_1x_2 + c_3x_1x_3 + c_4x_2^2 + c_5x_2x_3 + c_6x_3^2 \tag{10.59}$$

Further, the first-order derivative of $p_2(\boldsymbol{x})$ is

$$Dp_2(\boldsymbol{x}) = \begin{bmatrix} 2c_1x_1 + c_2x_2 + c_3x_3 \\ c_2x_1 + 2c_4x_2 + c_5x_3 \\ c_3x_1 + c_5x_2 + 2c_6x_3 \end{bmatrix} = \begin{bmatrix} 2c_1 & c_2 & c_3 \\ c_2 & 2c_4 & c_5 \\ c_3 & c_5 & 2c_6 \end{bmatrix} \begin{bmatrix} x_1 \\ x_2 \\ x_3 \end{bmatrix} \tag{10.60}$$

Formula (10.60) is further arranged

$$\begin{aligned} Dp_2(\boldsymbol{x}) &= \text{nonzero}\big((\boldsymbol{c}_2 \boldsymbol{1}_3) \odot \hat{\boldsymbol{P}}_2\big)\boldsymbol{v}_1(\boldsymbol{x}) \\ &= \text{nonzero}\big((\boldsymbol{c}_2 \boldsymbol{1}_3) \odot \hat{\boldsymbol{P}}_2\big)\boldsymbol{v}_1(\boldsymbol{x}) \\ &= \boldsymbol{D}_2 \boldsymbol{v}_1(\boldsymbol{x}) \end{aligned} \tag{10.61}$$

where, $\boldsymbol{D}_2 = \text{nonzero}\left(\left(\boldsymbol{c}_2 \mathbf{1}_3\right) \odot \hat{\boldsymbol{P}}_2\right)$, nonzero$(\cdot)$ denotes removing zero elements arranged in the columns of the matrix, $\mathbf{1}_3 = [1, 1, 1]$, $\odot$ denotes the multiplication of the corresponding elements in the matrix, and $\boldsymbol{v}_1(\boldsymbol{x}) = \left[x_1\ x_2\ x_3\right]$ represents the 1-order Veronese mapping. The expression (10.61) is extended into M-dimensional space, for the n-order Veronese mapping, we have

$$Dp_n(\boldsymbol{x}) = \boldsymbol{D}_n \boldsymbol{v}_{n-1}(\boldsymbol{x}) \tag{10.62}$$

where, $\boldsymbol{D}_n = \text{nonzero}\left(\left(\boldsymbol{c}_n \mathbf{1}_M\right) \odot \hat{\boldsymbol{P}}_n\right)$, $\mathbf{1}_M = [1, \ldots, 1] \in \boldsymbol{R}^{1\times M}$, and M is the dimension of the observed signal space (the number of sensors). Therefore, the normal vector of each subspace can be obtained by calculating Eq. (10.62).

3. **Spectral Clustering**

Using Eq. (10.62) to compute the derivative $Dp_n(x_t)$ of $p_n(\boldsymbol{x})$ at all observed vectors $x_t (t = 1, \ldots, T)$, $\overline{\boldsymbol{D}}p_n(\boldsymbol{x}) = \left[Dp_n(x_1), Dp_n(x_2), \ldots, Dp_n(x_T)\right] \in \boldsymbol{R}^{M\times T}$ is obtained. Since $Dp_n(x_t)$ is the normal vector of the subspace in which x_t is located, the two column vectors in the matrix $\overline{\boldsymbol{D}}p_n(\boldsymbol{x})$ whose correlation is close to 1 belong to the same normal vector of the same subspace. We can use $\overline{\boldsymbol{D}}p_n(x)$ to establish an undirected graph $\mathrm{G} = (\Omega, \mathrm{E})$, where Ω is the vertex set of graph G, consisting of each vector of the $\overline{\boldsymbol{D}}p_n(x)$, and the vertex edge weight $\varphi\left(\Omega_{t_1}, \Omega_{t_2}\right)$, which belongs to $\boldsymbol{E}$, is defined as

$$\varphi\left(\Omega_{t_1}, \Omega_{t_2}\right) = \frac{Dp_n\left(\boldsymbol{x}_{t_1}\right)^{\mathrm{T}} Dp_n\left(\boldsymbol{x}_{t_2}\right)}{\left\|Dp_n\left(\boldsymbol{x}_{t_1}\right)\right\| \left\|Dp_n\left(\boldsymbol{x}_{t_2}\right)\right\|} \tag{10.63}$$

where $t_1, t_2 \in \{1, 2, \ldots, T\}$. Obviously, $\varphi\left(\Omega_{t_1}, \Omega_{t_2}\right)$ measures the degree of correlation between $Dp_n\left(\boldsymbol{x}_{t_1}\right)$ and $Dp_n\left(\boldsymbol{x}_{t_2}\right)$. For $\boldsymbol{G}$, the subspace clustering can be realized by spectral clustering.

4. **Estimation of the Number of Subspaces (Hyperplanes)**

In the previous analysis, it is assumed that the number of subspaces (planes) is known. But in practice there is no prior information, and thus it is necessary to estimate the number of hyperplanes based on the observed signals.

For analysis convenience, suppose there are n planes whose n-order homogeneous polynomial is $p_n(\boldsymbol{x}) = \boldsymbol{c}_n^{\mathrm{T}} \boldsymbol{v}_n(\boldsymbol{x}) = 0$. It is clear that $\boldsymbol{c}_n^{\mathrm{T}}$ is unique, and hence $\text{rank}(\boldsymbol{V}_n(T)) = D_n(M) - 1$. If the order of the polynomial $i < n, p_i(\boldsymbol{x}) = \boldsymbol{c}_i^{\mathrm{T}} \boldsymbol{v}_i(\boldsymbol{x}) = 0$ is no longer satisfied, so the $\boldsymbol{V}_i(T)$ column is full rank, i.e. $\text{rank}(\boldsymbol{V}_n(T)) = D_n(M)$. If the order of polynomial $i > n$, then there are multiple i-order homogeneous polynomial,$\text{rank}(\boldsymbol{V}_i(T)) < D_n(M) - 1$. Therefore, we have

$$\text{rank}(\boldsymbol{V}_i(T)) \begin{cases} > D_i(M) - 1, & i < n \\ = D_i(M) - 1, & i = n \\ < D_i(M) - 1, & i > n \end{cases} \tag{10.64}$$

According to the formula (10.64), the number of subspaces (hyperplanes) can be estimated by the following equation

$$n = \arg\min_{i}\{\mathrm{rank}(\boldsymbol{V}_i(T)) = D_i(M) - 1\} \tag{10.65}$$

According to the Eq. (10.65), the iterative judgment starts from $i = 1$ until $\mathrm{rank}(\boldsymbol{V}_i(T)) = D_i(M) - 1$ is satisfied, and we get $n = i$. However, in the presence of noise, $\boldsymbol{V}_i(T)$ may be numerically full rank. Hence, the rank of $\boldsymbol{V}_i(T)$ should be determined by decomposing singular values and counting the number of remarkably large singular values. The rank of $\boldsymbol{V}_i(T)$ is generally determined according to the formula (10.65)

$$\mathrm{rank}(\boldsymbol{V}_i(T)) = \arg\min_{m}\left\{\frac{\sigma_{m+1}^2(\boldsymbol{V}_i(T))}{\sum_{j=1}^{m}\sigma_j^2(\boldsymbol{V}_i(T))} + \omega * m\right\} \tag{10.66}$$

where $\sigma_j(\boldsymbol{V}_i(T))$ represents the j-th singular value of $\boldsymbol{V}_i(T)$, $\sigma_j(\boldsymbol{V}_i(T)) \geq \sigma_{j+1}(\boldsymbol{V}_i(T))$. ω is model selection parameter, its selection is related to noise intensity, and it is selected by experience in practical application. In the simulation experiments of this chapter, if the SNR is greater than 10 dB, we can select $\omega \leq 0.001$, and if the SNR is less than 10 dB, we can select $\omega \approx 0.01$.

5. **Estimation of the Column Vectors of the Mixing Matrix**

After the number of subspaces (hyperplanes) in the observed signal space is estimated by using the method described in the previous section, the homogeneous polynomial representing the subspaces can be obtained by polynomial fitting, and the bases of the complementary space can be obtained by polynomial differentiation. The subspaces are finally obtained by clustering all the base vectors using the spectral clustering method. Each column vectors of the mixing lies in the intersections of C_{N-1}^{k-1}, where k is the number of active source signals. Therefore, the mixing matrix estimation can be accomplished by identifying the intersections of C_{n-1}^{k-1} planes.

Theorem 10.3 *The normal vectors of n planes with the same intersecting line are in the same plane, and the normal vector of the plane is the intersection line of these n planes.*

Proof Let n planes $\alpha_1, \alpha_2, \ldots, \alpha_n$ intersect the same line ℓ, and $\boldsymbol{b}_1, \boldsymbol{b}_2, \ldots, \boldsymbol{b}_n$ are the normal vectors of these planes. Assuming the plane determined by $\boldsymbol{b}_1, \boldsymbol{b}_2$ is β. for $\ell \in \alpha_1$ and $\ell \in \alpha_2$, so $\ell \perp \boldsymbol{b}_1$ and $\ell \perp \boldsymbol{b}_2$, that is, ℓ is the normal vector of the plane β. Now suppose that $\boldsymbol{b}_3$ is not in plane β, i.e. $\boldsymbol{b}_3 \notin \beta$, then $\boldsymbol{b}_3$ will not be perpendicular to ℓ. However, due to $\ell \in \alpha_3$, $\ell \perp \boldsymbol{b}_3$, there is a contradiction. Therefore,$\boldsymbol{b}_1, \boldsymbol{b}_2, \boldsymbol{b}_3$ are coplanar. In general cases, $\boldsymbol{b}_1, \boldsymbol{b}_2, \ldots, \boldsymbol{b}_n$ are coplanar, and the normal vector of this plane lies in the intersection of $\alpha_1, \alpha_2, \ldots, \alpha_n$.

According to Theorem 10.3, the normal vector $\boldsymbol{b}_i$ of each subspace (plane) can first be obtained by solving the null space of the data subspace. The right singular vector corresponding to the minimum singular value in the singular value decomposition can be used as the normal vector of the subspace, and then seek the null space of the matrix $\boldsymbol{B} = \left[\boldsymbol{b}_i, \boldsymbol{b}_j\right](i \neq j)$ composed of any two different normal vectors to obtain the normal vector $\boldsymbol{a}$ of the plane determined by $\boldsymbol{b}_i, \boldsymbol{b}_j$. And finally, all normal vectors $\boldsymbol{a}$ are clustered to get the column vectors of the mixing matrix.

10.4.3 Algorithm Steps

To summarize, the algorithm steps of mixing matrix estimation in UBSS based on polynomial representation are as follows [51]:

Step 1: Initialize $i = 1$, construct $\boldsymbol{V}_1(T)$, and solve Eq. (10.66) to obtain $\text{rank}(\boldsymbol{V}_1(T))$.

Step 2: While $\text{rank}(\boldsymbol{V}_i(T)) \neq D_i(M) - 1$, perform Steps 2.1 and 2.2 repeatedly.

2.1 $i = i + 1$;
2.2 $\text{rank}(\boldsymbol{V}_i(T))$ is obtained by solving Eq. (10.66).

Step 3: Let $n = i$, perform singular value decomposition $\boldsymbol{V}_n(T) = [\boldsymbol{u}_1, \boldsymbol{u}_2, \ldots, \boldsymbol{u}_T]\sum\left[\boldsymbol{v}_1, \boldsymbol{v}_2, \ldots, \boldsymbol{v}_{D_n(M)}\right]^{\mathrm{T}}$ on $\boldsymbol{V}_n(T)$, and let $\boldsymbol{c}_n = \boldsymbol{v}_{D_n(M)}$, which corresponds to the minimum singular value. (Note: this step is to obtain a vector in the null space of $\boldsymbol{V}_n(T)$).

Step 4: Calculate $\overline{\boldsymbol{D}}p_n(\boldsymbol{x}) = \left[Dp_n(\boldsymbol{x}_1), Dp_n(\boldsymbol{x}_2), \ldots, Dp_n(\boldsymbol{x}_T)\right]$ using Eq. (10.61).

Step 5: Establish the undirected graph $\boldsymbol{G}$ according to Eq. (10.63). Use the spectral clustering method to obtain the matrix $\boldsymbol{S}_i = \left[\boldsymbol{x}_1, \ldots, \boldsymbol{x}_{T_i}\right]$ of n subspace, where $(i = 1, 2, \ldots, n)$,T_i is the number of observed signal vectors in the i-th subspace.

Step 6: Perform singular value decomposition on the signal matrix $\boldsymbol{S}_i$ for each subspace, obtain $\boldsymbol{S}_i = \boldsymbol{U}\sum\boldsymbol{V}^{\mathrm{H}}$, and take the right singular vector corresponding to the minimum singular value as the normal vector $\boldsymbol{b}_i$.

Step 7: Construct the matrix $\boldsymbol{B} = (\boldsymbol{b}_i, \boldsymbol{b}_j)(i \neq j)$, where $\boldsymbol{b}_i, \boldsymbol{b}_j$ are any two normal vectors. Similar to step 6, singular value decomposition is done on matrix $\boldsymbol{B}$. Denote the right singular vector corresponding to the minimum singular value as $\boldsymbol{a}$, and cluster all $\boldsymbol{a}$ to obtain the column vectors of the mixing matrix.

10.4.4 Computer Simulation

1. **Blind Estimation of Underdetermined Mixing Matrix Without Noise**

In order to verify the performance of the underdetermined mixing matrix blind estimation algorithm based on polynomial representation, we compared it with the method based on parameter estimation and the k-dimension subspace method. The source signal is a Gaussian random signal, the number of sensors is 3, the number of source signals is 4, the number of source signals active at each sampling time is 2, the time length T is 2000, and the resulting mixing matrix is $\boldsymbol{A} \in \boldsymbol{R}^{3\times4}$ Gaussian matrix. The scatter plot of the observed signals is shown in Fig. 3.5, and six clear subspaces (planes) are obtained by using the subspace clustering method based on homogeneous polynomial representation, as shown in Fig. 10.1.

In order to measure the estimation accuracy of the mixing matrix, the estimation error ε (defined in Sect. 3.5.4) is used as the measurement index. The simulation results are analyzed below.

The three algorithms based on k-dimension subspace, parameter estimation and homogeneous polynomial representation were respectively run 1000 times (under the same software and hardware environment) to obtain all the results (no noise) in Table 10.1. It can be seen from the table that the average estimation error of the underdetermined mixing matrix estimation algorithm based on homogeneous polynomial representation is significantly lower than that of the other two traditional algorithms. However, the running time required is higher than the other two traditional algorithms. For k-dimension subspace method and parameter estimation method, even in the ideal conditions without noise, it is difficult to converge to the

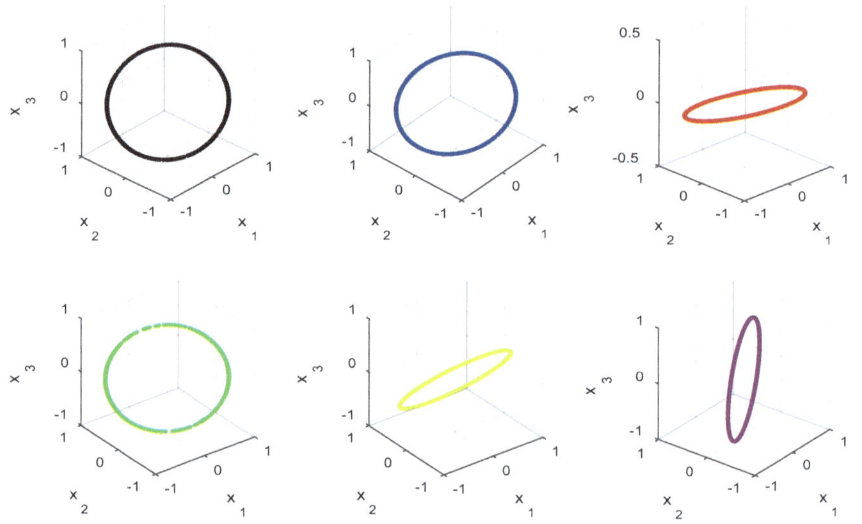

Fig. 10.1 Subspace clustering scatterplot based on polynomial representation

exact solution because there are many local extreme points in the objective function, and accuracy of the algorithm is difficult to improve. For the blind estimation of underdetermined mixing matrix based on homogeneous polynomial representation, by using the algebraic-geometry method, the mixing matrix can be accurately estimated without noise (from the numerical value, it can be seen that the estimation error reaches 10^{-16} orders of magnitude, almost close to 0). However, because of the use of singular value decomposition, it needs to consume a certain amount of operation time. In addition, $P(\varepsilon < 0.005)$ in Table 10.1 represents the proportion of simulation times whose estimation error is less than 0.005 in the total number of times. It can be seen that the reliability of the underdetermined mixing matrix estimation method based on homogeneous polynomials is also far greater than that of traditional methods.

2. **Blind Estimation of Underdetermined Mixing Matrix with Noise**

In this experiment, the performance of each algorithm is verified in the presence of additive noise. In the noisy case, the mathematical model of UBSS becomes

$$\boldsymbol{x}_t = \boldsymbol{A}\boldsymbol{s}(t) + \mu \boldsymbol{v}(t) \tag{10.67}$$

where, $\boldsymbol{A}$ is a Gaussian mixing matrix, and $\boldsymbol{s}(t)$ is a random signal generated by matlab; $\boldsymbol{v}(t) \sim N(0, 1)$ denotes Gaussian noise with mean 0 and variance 1 at time t. μ is the noise amplitude. The experiment is repeated 1000 times. As shown in Fig. 10.2, the comparison results of three different algorithms under different additive noise intensities are demonstrated the horizontal coordinate is μ, the vertical coordinate is interference-to-signal ratio (ISR), which is defined as $ISR = 20\log_{10}(\varepsilon)$, and ε is the estimation error. The smaller the ISR, the stronger the noise immunity.

It can be seen from Fig. 10.2 that mixing matrix estimation algorithm based on polynomial representation is better than the other two algorithms. For example, when $\mu = 0.04$, ISR of the algorithm based on polynomial representation is about 4 dB and 9 dB lower than that of the parameter estimation method and k-dimension subspace method, respectively.

3. **Mixing Matrix Estimation for Single-Source and Multiple-Source Points**

This experiment considers that there are both single-source points and multiple-source points in the observed signal, so the characteristics of line clustering and plane

Table. 10.1 Comparison of estimation error and running times of different algorithms

Algorithm	k-dimension subspace	Parameter estimation method	Method based on polynomial representation
Average estimation error ε	$5.1 * 10^{-3}$	$1.8 * 10^{-3}$	$8.54 * 10^{-16}$
Run time /s	3.71	2.24	5.94
$P(\varepsilon < 0.005)$	51%	74%	89%

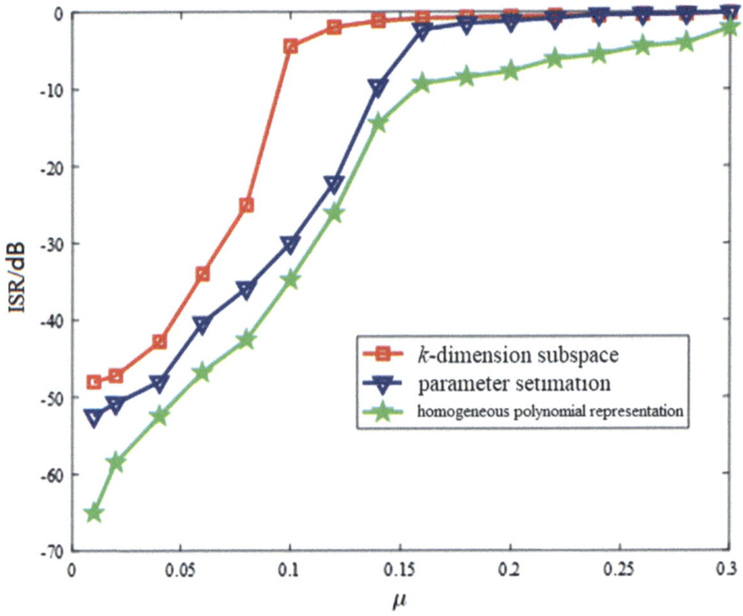

Fig. 10.2 Comparison of three mixing matrix estimation algorithms

clustering are presented in the observed signal space simultaneously. As shown in Fig. 10.3, there are both straight lines and planes in the observed signal space. The results obtained after using subspace clustering based on polynomial representation are shown in Fig. 10.4. It can be seen that multiple source points and single source points can be detected simultaneously. Obviously, the line direction of the single source point corresponds to a certain column vector of the mixing matrix. The partial column vectors of the mixing matrix can be obtained by finding the center of the single source point. The other column vectors of the mixing matrix lie in the plane intersection lines.

The model of Eq. (10.67) is adopted, and the noise amplitude is set to 0.1. The mixing matrix $\boldsymbol{A}$ is

$$\boldsymbol{A} = \begin{bmatrix} 0.7432 & 0.5391 & 0.3996 & 0.9197 \\ 0.6522 & 0.6862 & 0.9167 & 0.3240 \\ 0.1495 & -0.4883 & -0.0032 & -0.2217 \end{bmatrix}$$

The mixing matrix obtained by using the method based on polynomial representation is

$$\hat{\boldsymbol{A}} = \begin{bmatrix} 0.7430 & 0.5387 & 0.3994 & 0.9196 \\ 0.6524 & 0.6863 & 0.9167 & 0.3240 \\ 0.1493 & -0.4887 & -0.0037 & -0.2223 \end{bmatrix}$$

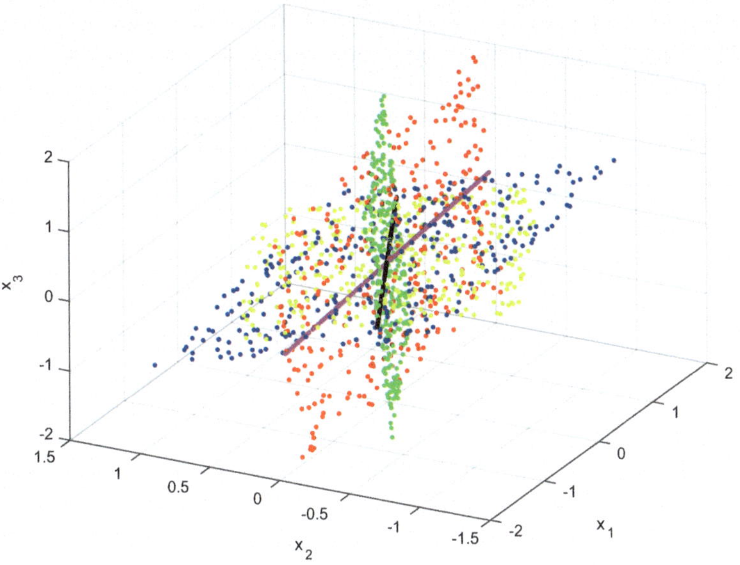

Fig. 10.3 Scatterplot of observed signals when there are both single source points and multiple source points simultaneously

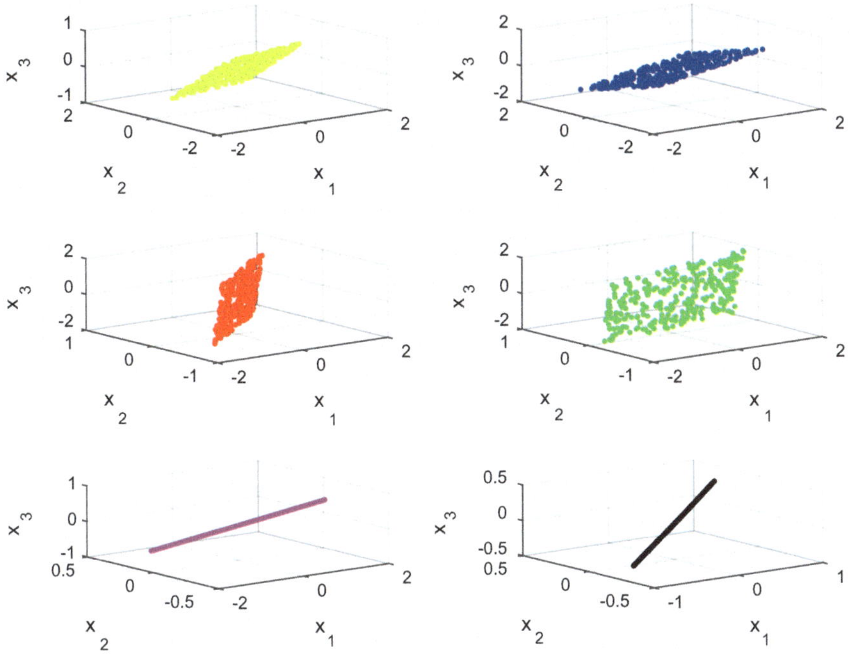

Fig. 10.4 Scatterplots of the observed signals after subspace clustering

Average estimation error $\varepsilon = 8.5167 \times 10^{-4}$. The mixing matrices estimated by the k-dimension subspace method and parameter estimation method are

$$\hat{A}' = \begin{bmatrix} 0.4024 & 0.5426 & 0.4024 & 0.9185 \\ 0.9155 & 0.6830 & 0.9155 & 0.3276 \\ -0.0041 & -0.4889 & -0.0041 & -0.2216 \end{bmatrix}$$

and

$$\hat{A}'' = \begin{bmatrix} 0.6633 & 0.5434 & 0.3996 & 0.9005 \\ 0.7418 & 0.6575 & 0.9167 & 0.3808 \\ -0.0994 & -0.5219 & -0.0038 & -0.2101 \end{bmatrix}$$

The average estimation errors are 0.2286 and 0.1432, respectively. It can be seen that the estimation error of the mixing matrix estimation algorithm based on polynomial representation is much smaller than that of the other two algorithms, because k-dimension subspace method and parameter estimation method can only estimate part of the column vectors of the mixing matrix.

In order to verify the performance of the algorithm more directly, Gaussian matrix and source signal were randomly generated, and the experiment was repeated 1000 times under different noise intensity environments and the experimental results were averaged. The results are shown in Fig. 10.5, where the y-axis is the estimation error ε and x-axis is the noise intensity μ.

As can be seen from the figure, when the noise intensity is less than 0.3, the estimation error of the algorithm based on polynomial representing is less than 0.1, and when the noise intensity is less than 0.01, the estimation error is less than 10^{-4} orders of magnitude. The estimation errors of the other two traditional methods are greater than 0.1 under different noise intensity environments.

Therefore, the k-dimension subspace method and parameter estimation method cannot effectively deal with the case of single source points. However, the underdetermined mixing matrix estimation algorithm based on polynomial representation can effectively deal with situation when there are both single source points and multiple source points simultaneously.

10.5 Underdetermined Mixing Matrix Estimation Algorithm Based on Single Source Point (SSP) Detection

In the actual application scenario, although the source signal is not sufficiently sparse, when the time–frequency analysis of the signal is carried out, we will find that only one source signal plays a dominant role in some time–frequency points, which is called single source point (SSP). If all single source points can be extracted, it will

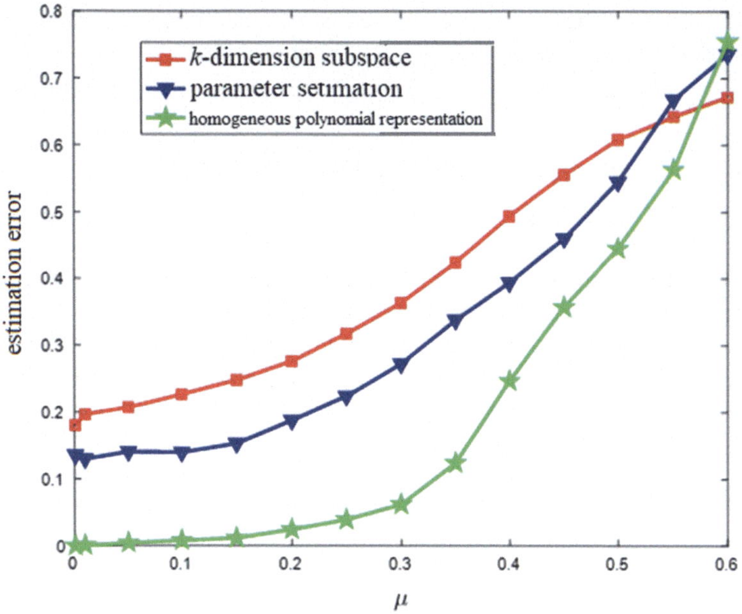

Fig. 10.5 Performance of mixed matrix estimation when there are both single source points and multiple source points simultaneously

be found that these single source points have the characteristics of line clustering. And the direction of these lines is the direction of the column vector of the mixing matrix. The mixing matrix estimation algorithm introduced in this section first extracts all the single source points by a certain method, and then estimate the column vectors of the mixing matrix by using the clustering algorithm introduced in the previous section. The key of this method is whether it can accurately extract as many single source points as possible, so the focus of this section is the detection of single source points.

To solve this problem, the Time–Frequency Ratio of Mixtures (TIFROM) method uses the time–frequency observed signals to create the time–frequency ratio matrix, determines the single-source region by searching for the minimum variance, and then clusters all single-source regions to obtain the estimation of the mixing matrix. This method has certain anti-noise property. However, it requires that only one source signal is active in the adjacent time–frequency window. The method of single-source time–frequency point identification is to use the angle between the real part and the imaginary part of observed signal to detect the single source points in the time–frequency domain, which reduces the requirement of sparsity and improves the estimation performance of the mixing matrix to some extent. In order to further improve the accuracy of mixing matrix estimation, the improved mixing matrix estimation method of identifying time–frequency SSP uses Gabor transform to achieve higher time–frequency resolution, so that SSP can be obtained more effectively and the estimation accuracy of mixing matrix can be improved [50].

10.5.1 Time–Frequency Ratio of Mixtures (TIFROM) Method

Time–Frequency Ratio of Mixtures (TIFROM) is an underdetermined mixing matrix estimation method based on time–frequency domain. By using the sparsity of the source signal in the time–frequency domain, this method can automatically search out the single source point region in the time–frequency domain when only one source signal is active, and then use the obtained single source point to estimate each column vector of the mixing matrix.

First, assuming that the number of source signals is N and the number of observing channels is 2, the received linear instantaneous mixed signals can be expressed as:

$$\begin{cases} x_1(t) = \sum_{i=1}^{N} a_{1i} s_i(t) \\ x_2(t) = \sum_{i=1}^{N} a_{2i} s_i(t) \end{cases} \tag{10.68}$$

where, $x_1(t)$ represents the first received signal, $x_2(t)$ represents the second received signal, a_{1i} and a_{2i} represent the elements of row 1 column i and row 2 column i in the mixing matrix $\boldsymbol{A}$, respectively. The Short-Time Fourier Transform (STFT) is used to transform the time-domain signal into the time–frequency domain, then

$$\begin{cases} X_1(\tau, \omega) = \sum_{i=1}^{N} a_{1i} S_i(\tau, \omega) \\ X_2(\tau, \omega) = \sum_{i=1}^{N} a_{2i} S_i(\tau, \omega) \end{cases} \tag{10.69}$$

where, $X_1(\tau, \omega)$, $X_2(\tau, \omega)$ and $S_i(\tau, \omega)$ represent the STFT coefficients of the first observed signal, the second observed signal and the source signal $\boldsymbol{s}_i$ respectively, τ and ω represent the time point and frequency point in the time–frequency domain respectively.

In the TIFROM method, for each source signal $\boldsymbol{s}_i$, it is required that in a short time interval Δt(from t_{i_1} to t_{i_p}) in the time–frequency domain, there are adjacent time–frequency points $(\tau_{\Delta t}, \omega_k)$ as shown in Fig. 10.6, so that only the source signal $\boldsymbol{s}_i$ plays a leading role at these time–frequency points, where $\tau_{\Delta t} = t_{i_i}, t_{i_2}, \ldots, t_{i_p}$ and ω_k represent a fixed frequency point in the time–frequency domain.

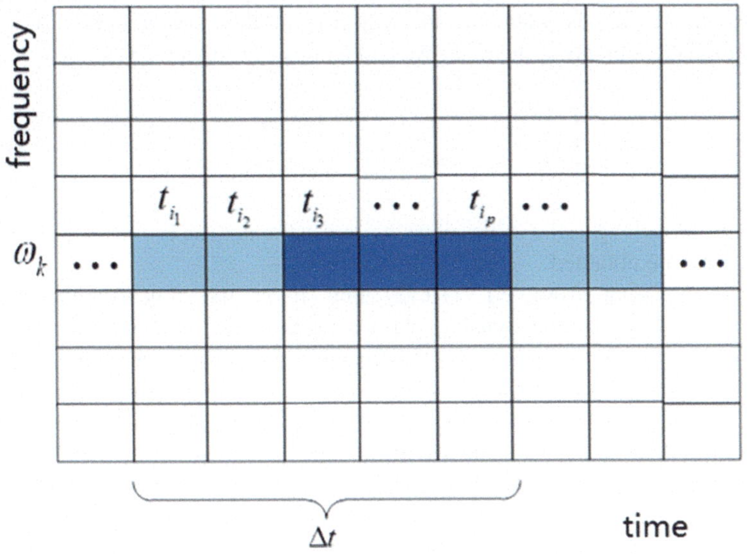

Fig. 10.6 Schematic diagram of single source signal at fixed frequency points and corresponding time intervals

Calculate the time–frequency ratio for $X_1(\tau, \omega)$ and $X_2(\tau, \omega)$, and get

$$\alpha(\tau, \omega) = \frac{X_1(\tau, \omega)}{X_2(\tau, \omega)} \tag{10.70}$$

Substitute the Eq. (10.69) into (10.70) to get

$$\alpha(\tau, \omega) = \frac{\sum_{i=1}^{N} a_{1i} S_i(\tau, \omega)}{\sum_{i=1}^{N} a_{2i} S_i(\tau, \omega)} \tag{10.71}$$

At the time–frequency window $(\tau_{\Delta t}, \omega_k)$, if only one source signal $\boldsymbol{s}_i$ plays a leading role while other source signals take smaller values, that is, $S_i(\tau_{\Delta t}, \omega) \neq 0$, $S_j(\tau_{\Delta t}, \omega) = 0 (j \neq i)$, then the Eq. (10.71) becomes

$$\alpha(\tau_{\Delta t}, \omega_k) = \frac{a_{1i} S_i(\tau_{\Delta t}, \omega)}{a_{2i} S_i(\tau_{\Delta t}, \omega)}, \quad \tau_{\Delta t} = t_{i_1}, t_{i_2}, \ldots, t_{i_p} \tag{10.72}$$

Further simplification, we get

$$\alpha\left(\tau_{i_1}, \omega_k\right) = \alpha\left(\tau_{i_2}, \omega_k\right) = \cdots = \alpha\left(\tau_{i_p}, \omega_k\right) = \frac{a_{1i}}{a_{2i}} \tag{10.73}$$

If we divide all the elements in each row of the mixing matrix $\boldsymbol{A}$ by the corresponding elements in the first row, and let $c_i = \frac{a_{1i}}{a_{2i}}$, then we get

$$\hat{\boldsymbol{A}} = \begin{bmatrix} 1 & 1 & \cdots & 1 \\ \frac{a_{21}}{a_{11}} & \frac{a_{22}}{a_{12}} & \cdots & \frac{a_{2N}}{a_{1N}} \end{bmatrix} = \begin{bmatrix} 1 & 1 & \cdots & 1 \\ 1/c_1 & 1/c_2 & \cdots & 1/c_N \end{bmatrix} \quad (10.74)$$

Therefore, as long as the values of each c_i are estimated, an estimate of the mixing matrix $\mathbf{A}$ can be obtained.

It can be seen from Eq. (10.72) that if at the time–frequency point $(\tau_{\Delta t}, \omega_k)$, only one source signal $\boldsymbol{s}_i$ plays a leading role, each element in vector $\left[\alpha\left(\tau_{i_1}, \omega_k\right), \alpha\left(\tau_{i_2}, \omega_k\right), \ldots, \alpha\left(\tau_{i_p}, \omega_k\right)\right]$ is equal to the mean value, that is, the variance of $\alpha(\tau_{\Delta t}, \omega_k)$ is 0. However, when there are multiple active source signals, the Eq. (10.72) no longer holds, that is, the variance of $\alpha(\tau_{\Delta t}, \omega_k)$ is no longer 0. According to this feature, Δt is taken as a time interval to divide each row vector of the observed time–frequency matrix of each channel, and there is a 50% overlap between each interval, then each row vector is divided into $(L - 0.5\Delta t)/(\Delta t - 0.5\Delta t)$ time windows. Then, for each time window and each frequency point ω_k, the variance of $\alpha(\tau_{\Delta t}, \omega_k)$ is calculated, and find the first N $\alpha(\tau_{\Delta t}, \omega_k)$ with the lowest variance, and its mean is taken as the $c_j (j = 1, 2, \ldots, N)$. Finally, each column of the mixing matrix is calculated according to Eq. (10.74) to complete the estimation of the mixing matrix. The specific steps of the time–frequency ratio of mixtures method are summarized as follows:

Step 1 Perform STFT operations on the received signals $\boldsymbol{x}_1(t)$ and $\boldsymbol{x}_2(t)$ respectively to obtain $X_1(\tau, \omega)$ and $X_2(\tau, \omega)$.

Step 2 Calculate the mixed time–frequency ratio $\alpha(\tau, \omega)$ of each time–frequency point according to Eq. (10.70).

Step 3 Each row vector of the observed time–frequency matrix is divided by taking Δt as the length of the time window and the overlap rate between adjacent windows is 50% to obtain N_{tf} time–frequency windows $\left(\tau_{\Delta t}^i, w_k\right)$, where $N_{tf} = W(L - 0.5\Delta t)/(\Delta t - 0.5\Delta t)$, $i, k \in \{1, 2, \ldots,\}$, and W is the number of the frequency points of STFT.

Step 4 For each time–frequency window $\left(\tau_{\Delta t}^i, \omega_k\right)$, calculate the variance of the mixed time–frequency ratio $\alpha\left(\tau_{\Delta t}^i, \omega_k\right)$, and sort it from smallest to largest, find the first N $\alpha\left(\tau_{\Delta t}^i, \omega_k\right)$ with the smallest variance, and denote it as $\alpha_j\left(\tau_{\Delta t}^i, \omega_k\right), j = 1, 2, \ldots, N$.

Step 5 Set $c_j = \overline{\alpha}_j\left(\tau_{\Delta t}^i, \omega_k\right)$, where $\overline{\alpha}_j\left(\tau_{\Delta t}^i, \omega_k\right)$ represents the mean value of $\alpha_j\left(\tau_{\Delta t}^i, \omega_k\right)$.

Step 6 According to Eq. (10.74), obtain the estimation $\hat{\boldsymbol{A}}$ of the mixing matrix.

Above steps are for the case where there are two observing channels, and the method can be easily generalized to the case of more observing channels, which will not be repeated here.

10.5.2 Underdetermined Mixing Matrix Estimation Method Based on Identification Time–Frequency SSP

The time–frequency ratio of mixtures method requires that only one source signal is active in the adjacent time period after the short-time Fourier transform of the observed signal, which is difficult to meet in practical application. In order to reduce the requirement of observed signal time–frequency distribution, the underdetermined mixing matrix estimation method based on identification time–frequency SSP uses the ratio of real and imaginary parts of time–frequency coefficients to identify single source points, and then performs clustering on single source points to obtain the estimation of mixing matrix. The following describes the principles and steps of the identification time–frequency SSP method.

First, STFT is performed on both sides of Eq. (2.3)

$$\boldsymbol{X}(t,\omega)=\boldsymbol{A}\boldsymbol{S}(t,\omega) \tag{10.75}$$

where, $\boldsymbol{X}(t,\omega) = [X_1(t,\omega), X_2(t,\omega), \ldots, X_M(t,\omega)]^{\mathrm{T}}, \boldsymbol{S}(t,\omega) = [S_1(t,\omega), S_2(t,\omega), \ldots, S_N(t,\omega)], X_i(t,\omega)(i=1,2,\ldots,M)$ and $S_j(t,\omega)(j=1,2,\ldots,N)$ are the value of the i-th observed signal and j-th source signal at the time–frequency point (t,ω) respectively. If the mixing matrix $\boldsymbol{A}$ is written in vector form, then the Eq. (10.75) is rewritten as

$$\boldsymbol{X}(t,\omega)=\boldsymbol{a}_1S_1(t,\omega)+\boldsymbol{a}_2S_2(t,\omega)+,\cdots,+\boldsymbol{a}_NS_N(t,\omega) \tag{10.76}$$

where,$\boldsymbol{a}_i(i=1,2,\ldots,N)$ is the i-th column vector of the mixing matrix $\boldsymbol{A}$, and $S_i(t,\omega)$ represents the i-th component of the source signal vector $\boldsymbol{S}(t,w)$ in the time–frequency domain. Since the time–frequency domain signal after STFT is a complex signal, the Eq. (10.76) can be written

$$\mathrm{Re}\{\boldsymbol{X}(t,\omega)\}=\boldsymbol{a}_1\mathrm{Re}\{S_1(t,\omega)\}+\boldsymbol{a}_2\mathrm{Re}\{S_2(t,\omega)\}+,\cdots,+\boldsymbol{a}_N\mathrm{Re}\{S_N(t,\omega)\} \tag{10.77}$$

$$\mathrm{Im}\{\boldsymbol{X}(t,\omega)\}=\boldsymbol{a}_1\mathrm{Im}\{S_1(t,\omega)\}+\boldsymbol{a}_2\mathrm{Im}\{S_2(t,\omega)\}+,\cdots,+\boldsymbol{a}_N\mathrm{Im}\{S_N(t,\omega)\} \tag{10.78}$$

If the observed signal has some single source points in the time–frequency domain, that is, at some fixed time–frequency points, only one source signal plays a dominant role, and the value of other source signals is 0 or approximately 0. Assuming that at time–frequency point (t,ω), only the i-th source signal is dominant, then Eqs. (10.77) and (10.78) become

$$\mathrm{Re}\{\boldsymbol{X}(t,\omega)\}=\boldsymbol{a}_i\mathrm{Re}\{S_i(t,\omega)\} \tag{10.79}$$

$$\mathrm{Im}\{\boldsymbol{X}(t,\omega)\}=\boldsymbol{a}_i\mathrm{Im}\{S_i(t,\omega)\} \tag{10.80}$$

where Re{•} represents the real part of each element, and Im{•} represents the imaginary part of each element. Now assume the angle between vector $\mathrm{Re}\{\boldsymbol{X}(t,\omega)\}$ and vector $\mathrm{Im}\{\boldsymbol{X}(t,\omega)\}$ is θ, then

$$\cos(\theta) = \frac{\left(\mathrm{Re}\{\boldsymbol{X}(t,\omega)\}^{\mathrm{T}}\,\mathrm{Im}\{\boldsymbol{X}(t,\omega)\}\right)}{\|\mathrm{Re}\{\boldsymbol{X}(t,\omega)\}\|\;\|\mathrm{Im}\{\boldsymbol{X}(t,\omega)\}\|} \tag{10.81}$$

Take absolute values of both sides of Eq. (10.81), we get

$$\begin{aligned}|\cos(\theta)| &= \left|\frac{\left(\mathrm{Re}\{\boldsymbol{X}(t,\omega)\}^{\mathrm{T}}\,\mathrm{Im}\{\boldsymbol{X}(t,\omega)\}\right)}{\|\mathrm{Re}\{\boldsymbol{X}(t,\omega)\}\|\;\|\mathrm{Im}\{\boldsymbol{X}(t,\omega)\}\|}\right| \\ &= \frac{\left|\left(\boldsymbol{a}_i\mathrm{Re}\{S_i(t,\omega)\}^{\mathrm{T}}\boldsymbol{a}_i\mathrm{Im}\{S_i(t,\omega)\}\right)\right|}{\|\boldsymbol{a}_i\mathrm{Re}\{S_i(t,\omega)\}\|\;\|\boldsymbol{a}_i\mathrm{Im}\{S_i(t,\omega)\}\|} \\ &= \frac{|\mathrm{Re}\{S_i(t,\omega)\}||\mathrm{Im}\{S_i(t,\omega)\}|\|\boldsymbol{a}_i\|^2}{|\mathrm{Re}\{S_i(t,\omega)\}||\mathrm{Im}\{S_i(t,\omega)\}|\|\boldsymbol{a}_i\|\,\|\boldsymbol{a}_i\|} \\ &= 1\end{aligned} \tag{10.82}$$

According to the Eq. (10.81), for a single source point in the time–frequency domain, the angle between the real part and the imaginary part of the observed vector is 0° or 180°, that is, the direction of the real part and the imaginary part of the observed vector is opposite or consistent. The case of multiple source points is discussed below. Assuming that at the time–frequency point (t,ω), the first source signal and the second source signal are active simultaneously, and the value of other source signals is approximately 0, then Eqs. (10.77) and (10.78) become

$$\mathrm{Re}\{\boldsymbol{X}(t,\omega)\} = \boldsymbol{a}_1\mathrm{Re}\{S_1(t,\omega)\} + \boldsymbol{a}_2\mathrm{Re}\{S_2(t,\omega)\} \tag{10.83}$$

$$\mathrm{Im}\{\boldsymbol{X}(t,\omega)\} = \boldsymbol{a}_1\mathrm{Im}\{S_1(t,\omega)\} + \boldsymbol{a}_2\mathrm{Im}\{S_2(t,\omega)\} \tag{10.84}$$

According to Eqs. (10.83) and (10.84), if the following formula is true

$$\frac{\mathrm{Re}\{S_1(t,\omega)\}}{\mathrm{Im}\{S_1(t,\omega)\}} = \frac{\mathrm{Re}\{S_2(t,\omega)\}}{\mathrm{Im}\{S_2(t,\omega)\}} \tag{10.85}$$

Then $|\cos(\theta)| = 1$, But in general, the probability of formula (10.85) being true is very low.

Therefore, the single source point identification is carried out by verifying $|\cos(\theta)| = 1$, and then the clustering centers obtained by clustering all single source points are the column vectors of the mixing matrix. The steps of the mixing matrix blind estimation method based on identification time–frequency SSP are as follows.

Step 1 Perform STFT on the observed signal to obtain time–frequency observed signal $\boldsymbol{X}(t,\omega)$.

Step 2 For each time–frequency point (t, ω), calculate the absolute value of the cosine of the angle between the imaginary part and the real part of observed signal vector as follows.

$$|\cos(\theta)| = \left| \frac{\left(\mathrm{Re}\{\boldsymbol{X}(t,\omega)\}^{\mathrm{T}}\,\mathrm{Im}\{\boldsymbol{X}(t,\omega)\}\right)}{\|\mathrm{Re}\{\boldsymbol{X}(t,\omega)\}\|\ \|\mathrm{Im}\{\boldsymbol{X}(t,\omega)\}\|} \right| \tag{10.86}$$

If $|\cos(\theta)| \geq \varepsilon$, where $0.9 \leq \varepsilon < 1$, then $\boldsymbol{X}(t, \omega)$ is seen as the single source point.

Step 3 Eliminate the low energy points in all single source points and perform normalization to obtain the single source point set $\overline{\boldsymbol{X}}(t, \omega)$.

Step 4 Perform point clustering on $\overline{\boldsymbol{X}}(t, \omega)$ to obtain the estimation $\hat{\boldsymbol{A}}$ of the mixing matrix.

10.5.3 Underdetermined Mixing Matrix Estimation Method Based on Improved Identification Time–Frequency SSP

The above method uses the real and imaginary parts of the time–frequency signal in SSP identification, and the detection performance is inevitably affected by the STFT time resolution and frequency resolution. The following is a further analysis of the time–frequency SSP identification method. The essence of this method is to find sparse points and maintain the linear characteristics of the UBSS model when carrying out the time–frequency domain transformation, that is

$$\kappa[\boldsymbol{X}] = \kappa[\boldsymbol{AS}] = \boldsymbol{A}\kappa[\boldsymbol{S}] \tag{10.87}$$

where $\kappa[\cdot]$ represents some kind of linear time–frequency transformation. At the same time, ensure that the signal in transform domain is a complex signal, that is

$$\begin{aligned}
\kappa[\boldsymbol{X}] &= \boldsymbol{A}\kappa[\boldsymbol{S}] \\
&= \boldsymbol{A}[\mathrm{Re}\{\kappa[\boldsymbol{S}]\} + \mathrm{Im}\{\kappa[\boldsymbol{S}]\}] \\
&= \boldsymbol{A}\mathrm{Re}\{\kappa[\boldsymbol{S}]\} + \boldsymbol{A}\mathrm{Im}\{\kappa[\boldsymbol{S}]\} \\
&= \mathrm{Re}\{\kappa[\boldsymbol{AS}]\} + \boldsymbol{A}\mathrm{Im}\{\kappa[\boldsymbol{AS}]\} \\
&= \mathrm{Re}\{\kappa[\boldsymbol{X}]\} + \mathrm{Im}\{\kappa[\boldsymbol{X}]\}
\end{aligned} \tag{10.88}$$

If there is a single source point in the time–frequency domain, for example, only the i-th source signal is active at the time–frequency point (t, ω), then the formula (10.93) becomes

$$\boldsymbol{X}(t, \omega) = \boldsymbol{a}_i \mathrm{Re}\{S_i(t, \omega)\} + \boldsymbol{a}_i \mathrm{Im}\{S_i(t, \omega)\} \tag{10.89}$$

where, $\boldsymbol{a}_i$ is the i-th column of the mixing matrix $\boldsymbol{A}$, and $S_i(t, \omega)$ is the amplitude value of the i-th source signal at the time frequency point (t, ω) after the time–frequency transformation. According to Eq. (10.89), for a single source point, the real part and the imaginary part are strictly proportional, which is consistent with the result of Eq. (10.82). Therefore, a single source point can be identified using both imaginary and real parts. In the time–frequency analysis method, the Wigner-Ville distribution can better describe the local characteristics of the signal and further improve the time–frequency resolution, but it is a nonlinear transformation and does not satisfy the formula (10.86). Wavelet transform is a linear transform, but there is no real part and imaginary part after wavelet transform, single source point detection cannot be carried out. The Gabor transform is a windowed Fourier transform, which satisfies the Eqs. (10.87) and (10.88), and can further improve the time–frequency resolution compared with the general STFT. Therefore, Gabor transform is introduced to improve the accuracy of mixing matrix estimation before identifying single source points. The underdetermined mixing matrix estimation method based on improved identification time–frequency SSP is introduced in the following.

First, the average time and average frequency of the i-th observed signal $x_i(t)$ are defined

$$\bar{t}_x = \int_{-\infty}^{\infty} t|x_i(t)|^2 dt \tag{10.90}$$

$$\overline{\omega}_x = \int_{-\infty}^{\infty} \omega|X_i(\omega)|^2 d\omega \tag{10.91}$$

where $\bar{t}$ is the average time, $\overline{\omega}$ is the average frequency, and $X_i(\omega)$ is the Fourier transform of $x_i(t)$.

Then the time width $\hat{t}_x$ and bandwidth $\hat{\omega}_x$ of the observed signal $x_i(t)$ are defined as follows.

$$\hat{t}_x^2 = \int_{-\infty}^{\infty} \left(t - \bar{t}_x\right)^2 |x_i(t)|^2 dt \tag{10.92}$$

$$\hat{\omega}_x^2 = \int_{-\infty}^{\infty} (\omega - \overline{\omega}_x)^2 |X_i(\omega)|^2 d\omega \tag{10.93}$$

According to the uncertainty principle, for any signal $x(t)$ that is square integrable or absolutely summable, the product of its time width and bandwidth satisfies

$$\hat{t}_x \hat{\omega}_x \geq 1/2 \tag{10.94}$$

The Eq. (10.94) is called Heisenberg's inequality.

Equation (10.94) shows that the time width and bandwidth of the signal cannot be arbitrarily small simultaneously, that is, the time resolution and the frequency resolution are contradictory. STFT is the inner product of the time domain signal

$x(t)$ and the modulated window function, i.e.

$$\begin{aligned}\mathrm{STFT}_x(t,\omega) &= \int_{-\infty}^{\infty} x(u)g^*(u-t)e^{-j\omega u}du \\ &= \int_{-\infty}^{\infty} x(u)g^*_{t,\omega}(u)du \\ &= \left\langle x(u), g^*_{t,\omega}(u)\right\rangle\end{aligned} \tag{10.95}$$

where $*$ means taking conjugation. If the window function $g(u)$ is equal to the Gaussian function

$$g(u) = e^{-\pi u^2} \tag{10.96}$$

Then Eq. (10.94) becomes

$$\hat{t}_x\hat{\omega}_x = 1/2 \tag{10.97}$$

That is, the lower bound of Heisenberg's inequality can be reached at this point, where the product of time width and bandwidth is minimum. Gabor transform in time–frequency analysis method uses Gaussian function as window function, so Gabor transform is used for time–frequency analysis to improve time–frequency resolution. After completing the time–frequency analysis, the single source points are identified and all the single source points are clustered. In summary, the steps of the underdetermined mixing matrix estimation method based on the improved identification time–frequency SSP are as follows:

Step 1 Perform Gabor transformation on the observation signal to obtain time–frequency observation signal $\boldsymbol{X}(t,\omega)$, where (t,ω) is time–frequency point.

Step 2 Delete the low energy points in the time–frequency observation signal according to the following formula

$$\left\|\boldsymbol{X}\left(t',\omega'\right)\right\| \geq \lambda \max\|\boldsymbol{X}(t,\omega)\| \tag{10.98}$$

where, $0 < \lambda \leq 0.3$, (t,ω) are the time–frequency points of time–frequency analysis, and $\left(t',\omega'\right)$ is the time–frequency points after eliminating low-energy points.

Step 3 Identify the time–frequency single source points according to Eq. (10.99)

$$\left|\frac{\left(\mathrm{Re}\left\{\boldsymbol{X}\left(t',w'\right)\right\}^{\mathrm{T}}\mathrm{Im}\left\{\boldsymbol{X}\left(t',w'\right)\right\}\right)}{\|\mathrm{Re}\{\boldsymbol{X}(t',w)\}\|\;\|\mathrm{Im}\{\boldsymbol{X}(t',w')\}\|}\right| \geq \rho \tag{10.99}$$

where, $0.9 \leq \rho < 1$. The signal vector satisfying the formula (10.99) is a single source signal vector.

Step 4 Take out the real parts of all the single source points and normalize them to obtain the normalized single source points.

Step 5 Cluster the normalized single source points to get the mixing matrix estimation.

10.5.4 Computer Simulation and Analysis

In order to analyze the performance of the underdetermined mixing matrix estimation algorithm for insufficiently sparse source signals based on single source point detection, linear frequency modulation signal, nonlinear frequency modulation signal, frequency modulation signal are used as the source signals, and Gaussian random matrix is used as the mixing matrix. The estimation error of mixing matrix and the signal-to-interference ratio of mixing matrix estimation are used as the estimation performance indexes.

1. **Single Source Point Detection Experiment**

There are three observing channels and four source signals, as shown in Fig. 10.7. The first source signal is a linear frequency modulation pulse signal, the frequency changes from 0 Hz to 400 kHz; the second source signal is a nonlinear frequency modulation pulse signal, the frequency changes from 500 to 100 kHz; the third source signal is a linear frequency modulation pulse signal, which frequency changes from 0 to 500 kHz. The fourth source signal is a 2FSK communication signal, including two carrier frequencies of 50 and 100 kHz. The sampling frequency is 1 MHz and the sampling time is 0.003 s.

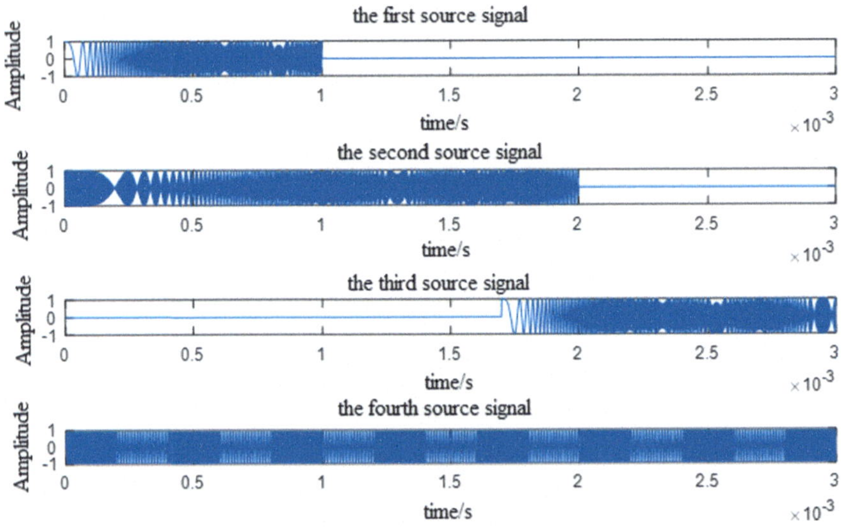

Fig. 10.7 Waveforms of source signals

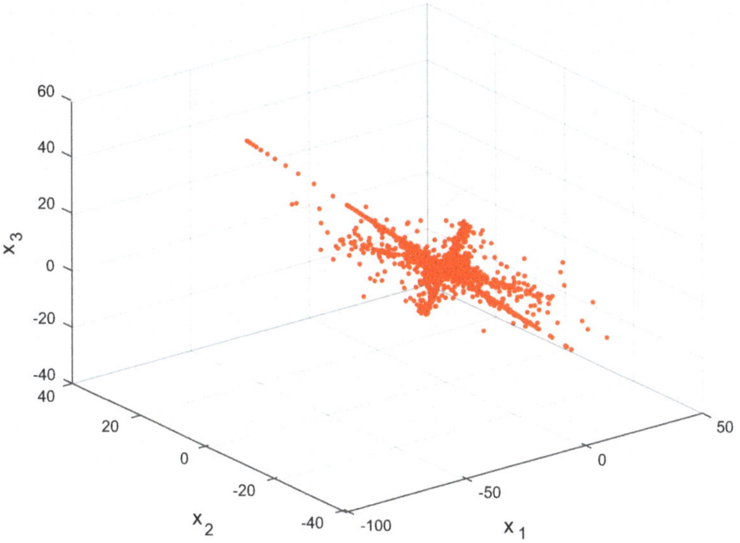

Fig. 10.8 Scatter plot of time–frequency observed signals after short-time Fourier transform

The scatter plot (real part) of the time–frequency observed signal after the short-time Fourier transform is shown in Fig. 10.8, which intuitively shows some linear clustering characteristics. But there are many interference points in some directions, which may produce large clustering bias during the clustering, thus reducing the estimation performance of the mixing matrix. The scatter plot after single source point detection is given below.

As shown in Fig. 10.9, after the single source point identification method introduced in Sect. 10.5.2 is used to find single source points, its scatter plot presents linear clustering characteristics, and the directions of four lines correspond to the four column vectors of the mixing matrix. After using the improved time–frequency SSP method, as shown in Fig. 10.10, the lines presented in the scatter plot are clearer and more conducive to clustering.

2. **Experiment of Mixing Matrix Estimation in Different SNR Environments**

In order to compare the anti-noise performance of mixing matrix estimation algorithm, the recognition time–frequency SSP method, TIFROM method and the improved recognition time–frequency method are simulated under different SNR environments, and the results are shown in Fig. 10.11. Because the time–frequency scatter plot of the observed signal does not satisfy the characteristics of the continuous distribution of the same single source point, the estimation error of the mixed matrix estimation using the mixed time–frequency ratio method is large. However, the method based on recognition time–frequency SSP only requires a certain number of single source points in the time–frequency domain, and does not require that their distribution be continuous. Therefore, the estimation error is significantly reduced

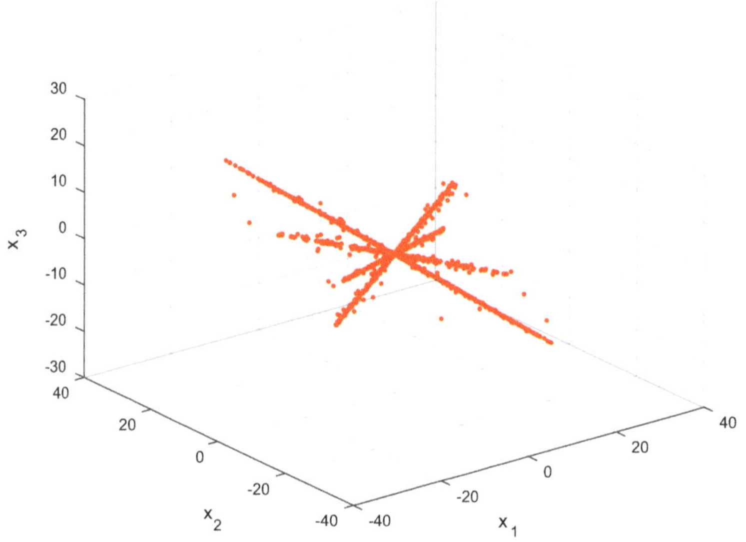

Fig. 10.9 Scatter plot of time–frequency observed signal after recognition SSP

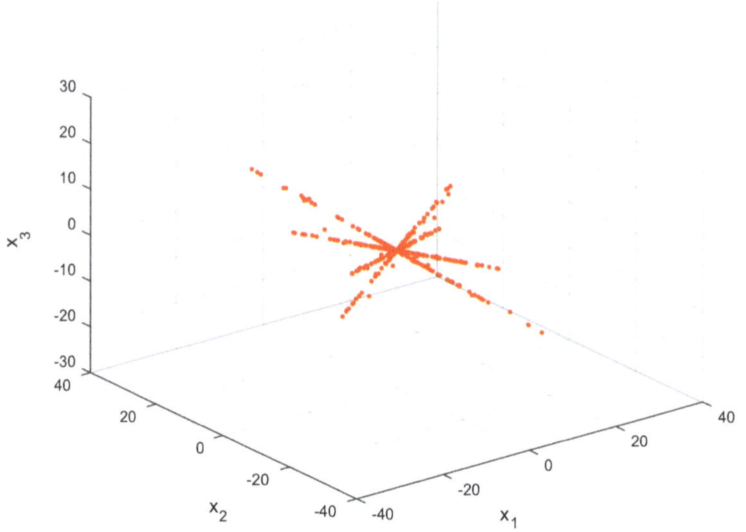

Fig. 10.10 Scatter plot of time–frequency observed signal after improved recognition SSP

when the method of recognition time–frequency SSP is used for detection. In the case of high signal-to-noise ratio (SNR), such as SNR = 45 dB, the estimation

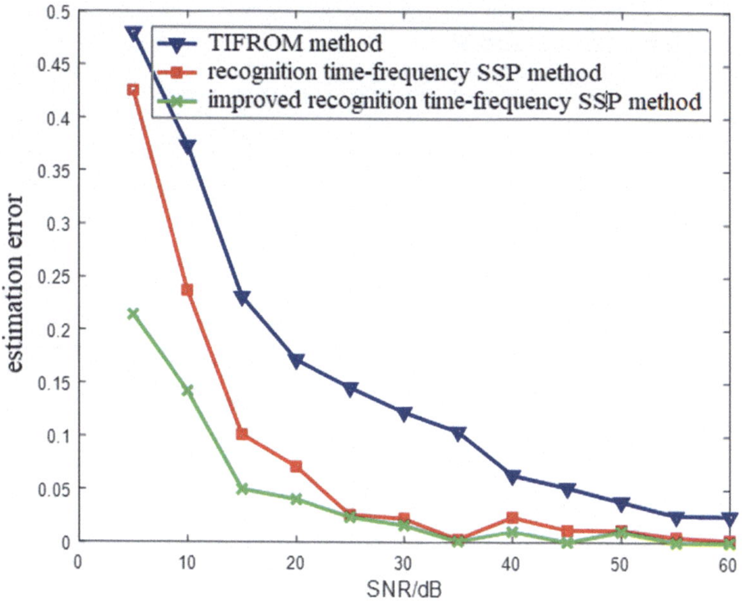

Fig. 10.11 Comparison of mixing matrix estimation performance in different SNR environments

error corresponding to TIFROM method is 0.0518, while the estimation error corresponding to the recognition time–frequency SSP method and the improved recognition time–frequency SSP method is less than 0.01. In a low SNR environment, it can be seen from Fig. 10.11 that the estimation errors of the two methods based on single source point recognition are significantly lower than that of TIFROM method. For example, when SNR = 15 dB, the estimation error of the improved recognition time–frequency SSP method is 50.65% lower than that of the recognition time–frequency SSP method. The recognition time–frequency SSP method is 56.16% lower than TIFROM method.

In the case of high SNR, because the estimation errors of the recognition time–frequency SSP method and the improved recognition time–frequency SSP method are close to 0 (in fact, the estimation errors of the two methods have difference on orders of magnitude), their performance curves almost coincide, and it is difficult to compare them intuitively. Therefore, the signal-to-interference ratio (SIR) is used as a performance indicator for comparison. As shown in Fig. 10.12, the SIR of the improved recognition time–frequency SSP is significantly higher than that of recognition time–frequency SSP and TIFROM method. For example, when SNR = 40 dB, the SIR of the improved recognition time–frequency SSP method is 61.69 dB, the recognition time–frequency SSP method and the mixed time–frequency ratio method are 51.4 dB and 23.89 dB, respectively. When SNR = 10 dB, the SNR of the improved recognition time–frequency SSP method is 26.19 dB, while the SIR of the other two methods are lower than 20 dB. It can be seen that the mixing matrix

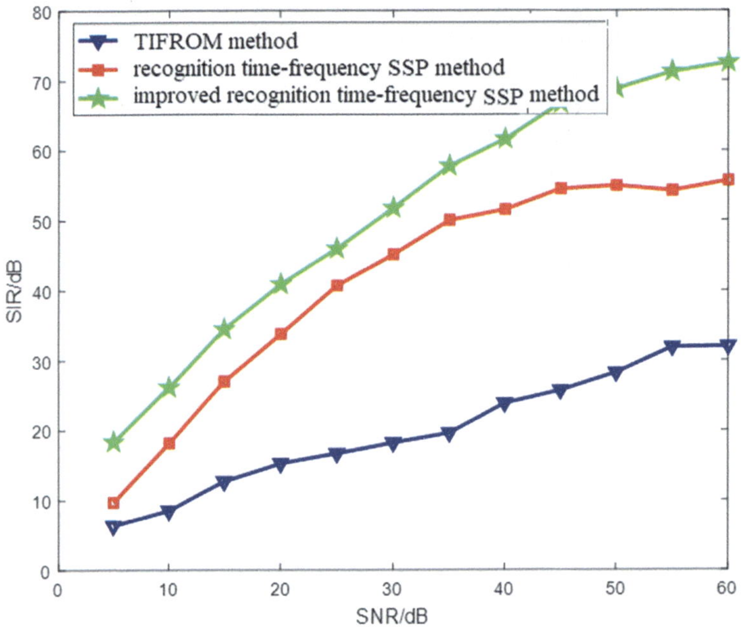

Fig. 10.12 The variation of SIR with SNR of different algorithms

estimation method based on improved recognition time–frequency SSP improves the estimation accuracy to some extent.

Summary

In this chapter, underdetermined mixing matrix estimation algorithms are studied when the source signal is not sufficiently sparse. Firstly, several mixing matrix estimation algorithms based on subspace estimation are studied. Because when the source signal is not sufficiently sparse, the scatterplot of the observed signal shows the characteristics of planar clustering, and the column vectors of the mixing matrix are at the intersection of these clustering planes. Therefore, the column vectors of mixing matrix can be obtained by estimating these clustering planes or subspaces. This chapter introduces the mixing matrix estimation algorithms based on k-dimension subspace, parameter estimation, plane clustering, and homogeneous polynomial representation, and introduces the principles and implementation steps of these algorithms in detail. The performance of these algorithms is simulated and analyzed.

In addition, several underdetermined mixing matrix estimation algorithms based on single source detection are introduced, including TIFROM method, recognition time–frequency SSP method and improved recognition time–frequency SSP method. The principles and implementation steps of the algorithms are given, and the performance of the algorithms in different environments is simulated and analyzed.

Chapter 11
Source Signal Recovery Algorithm Based on Compression Reconstruction

The second step to solve the problem of underdetermined blind source separation (UBSS) is to use the estimated mixing matrix to recover the source signal. Since the source signals in UBSS are generally sparse signals, we find that the mathematical model of source signal recovery in UBSS is consistent with the mathematical model of compressed signal reconstruction. Therefore, the idea of compression reconstruction can be used to recover the source signal in the problem of UBSS.

This chapter mainly introduces the source signal recovery algorithm based on compression reconstruction. Firstly, the mathematical model of compressed sensing is analyzed, and then the relationship between UBSS and compressed sensing is analyzed. Then, UBSS signal recovery algorithm based on greedy thought, the UBSS signal recovery algorithm based on L1 norm and the source signal recovery algorithm based on smooth L0 norm (SL0) are introduced respectively, and the principle and steps of the algorithm are introduced respectively. The performance of the algorithm is simulated and analysed [52].

11.1 Mathematical Model of Compressed Sensing

Candes and Donoho proposed the concept of compressed sensing (CS) in 2006. The theory of CS has brought a revolutionary breakthrough for signal sampling technology. The theory shows that when the signal is sparse, we can sample the signal at a frequency much lower than the Nyquist sampling rate, and accurately recover the source signal through the reconstruction algorithm.

Let $\boldsymbol{x}$ be the N-dimensional digital signal obtained by traditional sampling. If it is a K-sparse signal (only K non-zero values or can be represented by K non-zero values after some reversible transformation), the M-dimensional compressed signal $\boldsymbol{y}$ ($M < N$) can be directly obtained through the compressed sensing process, the relationship between $\boldsymbol{y}$ and $\boldsymbol{x}$ is as follows:

W. Fu, *Principle and Application of Blind Source Separation Technology*,
https://doi.org/10.1007/978-981-96-0022-9_11

$$\boldsymbol{y} = \boldsymbol{\Phi x} \tag{11.1}$$

When the signal $\boldsymbol{x}$ is a sparse signal, the signal $\boldsymbol{x}$ can be reconstructed with high precision according to the observed signal $\boldsymbol{y}$ and the measurement matrix $\boldsymbol{\Phi}$ when $\boldsymbol{\Phi}$ satisfies the restricted isometry property (RIP). However, many common signals are not sparse signals in the time domain, but they are sparse in the transform domain, which requires a sparse representation of the source signal, $\boldsymbol{x} = \boldsymbol{\Psi s}$, then the formula (11.1) can be rewritten into the following form:

$$\boldsymbol{y} = \boldsymbol{\Phi x} = \boldsymbol{\Phi \Psi s} = \boldsymbol{\Theta} s \tag{11.2}$$

where matrix $\boldsymbol{\Theta}$ is the product of measurement matrix $\boldsymbol{\Phi}$ and sparse transformation basis matrix $\boldsymbol{\Psi}$. Compressed sensing can be divided into two steps, namely the design of the measurement matrix and the reconstruction of the source signal. In order for the receiver to accurately recover the source signal, the designed measurement matrix $\boldsymbol{\Phi}$ needs to make matrix $\boldsymbol{\Theta}$ satisfy the Null Space Property (NSP) or Restricted Isometry Property (RIP) criterion. The NSP criterion is applicable to the measurement matrix design in the case of no noise, and the RIP is applicable to the case of noise, which is a broader criterion. The NSP criterion is stated as follows:

For any observed signal $\boldsymbol{y} \in R^M$, if and only if $spark(\boldsymbol{\Theta}) > 2K$, there is at most one signal $\boldsymbol{s} \in \sum_K$ such that $\boldsymbol{y} = \boldsymbol{\Theta} s$.

In the NSP criterion, $\sum_K$ represents the set of all K-sparse signals, and $spark(\boldsymbol{\Theta})$ represents the minimum number of linearly related columns in the matrix $\boldsymbol{\Theta}$. It can be seen from the NSP criterion that in order to reconstruct sparse signals as accurately as possible, we should make $spark(\boldsymbol{\Theta})$ as large as possible. Because of $spark(\boldsymbol{\Theta}) \in [2, M+1]$, the matrix we design should be able to satisfy $spark(\boldsymbol{\Theta}) = M+1$, and Gaussian matrices usually meet this requirement. NSP criterion only applies when there is no noise. When there is noise, we use RIP criterion. RIP criterion is expressed as follows:

If there is a constant $\delta \in (0, 1)$ such that for any $\boldsymbol{s} \in \sum_K$, it satisfies:

$$(1 - \delta_K)\|\boldsymbol{s}\|_2^2 \le \|\Theta s\|_2^2 \le (1 + \delta_K)\|\boldsymbol{s}\|_2^2 \tag{11.3}$$

We say that the matrix $\boldsymbol{\Theta}$ satisfies the K-order RIP criterion. If matrix $\boldsymbol{\Theta}$ meets the K-order RIP criterion, we can recover the K-sparse source signal with great probability.

11.2 Relationship Between Underdetermined Blind Source Separation and Compressed Sensing

By comparing Eqs. (2.1) and (11.2), we find that if the matrix Θ in the compressed sensing model is equal to $\boldsymbol{A}$ and the observed vector $\boldsymbol{y}$ is equal to $\boldsymbol{x}$, then the UBSS model and the compressed sensing model have the same mathematical expression. In addition, the purpose of compressed sensing reconstruction and UBSS source signal recovery is to find sparse signal $\boldsymbol{s}$ when the sensor matrix (or mixing matrix) is known, so the algorithm of signal reconstruction in compressed sensing can be applied to the problem of UBSS source signal recovery.

Pando Georgiev proposed a sparse component analysis (SCA) theory for the problem of UBSS, that is, when the following three conditions are met, the received signal can be used to accurately recover the source signal.

(1) Mixing matrix $\boldsymbol{A} \in R^{M \times N}$ is row full rank matrix, that is, any $M \times N$-dimensional submatrix in $\boldsymbol{A}$ is non-singular;
(2) Each column of the source signal matrix $\boldsymbol{s}$ has at most $M - 1$ non-zero elements;
(3) For any index set of $N - M + 1$ elements $I = \{i_1, i_2, \ldots, i_{N-M+1}\} \subset \{1, 2, \ldots, N\}$ there exist at least M column vectors of source signal matrix $\boldsymbol{s}$ such that each of them has zero elements in places with indexes I and each $m - 1$ of them are linearly independent.

In the above three conditions, condition (1) is the requirement for the matrix, condition (2) is the condition that each signal column vector should meet, and condition (3) requires that the received signal during the entire sampling period should carry all the information of the mixing matrix $\boldsymbol{A}$, so that the mixing matrix can be recovered. This theory proposed by Pando Georgiev has played a great role in promoting the development of UBSS.

The above SCA theory tells us that if conditions (1) and (3) are satisfied, the source signal can be completely recovered as long as the number of non-zero values in the source signal does not exceed $M - 1$. But the NSP criterion in compressed sensing tells us that in the case of $spark(\boldsymbol{A}) = M + 1$, we can accurately reconstruct the sparse signal with $\lfloor M/2 \rfloor$ non-zero values at most. At this point, the two conclusions are inconsistent. For example, for a mixing matrix with 3 rows and 4 columns, according to the theory proposed by Pando Georgiev, we can completely reconstruct the source signal containing two non-zero values, but according to the NSP criterion, we can only completely reconstruct the signal containing one non-zero value, and the two conclusions get different results. The NSP criterion has been proved mathematically, while the three conditions of the SCA theory proposed by Pando Georgiev have not been proved mathematically. In fact, in some special cases, when the mixing matrix and source signal in UBSS meet the three conditions of SCA theory, we may not be able to completely recover the source signal.

Here is a special case: Suppose the source signal satisfies conditions (2) and (3), and the mixing matrix is

$$A = \begin{bmatrix} 1\,0\,0\,1 \\ 0\,1\,0\,1 \\ 0\,0\,1\,1 \end{bmatrix} \tag{11.4}$$

A satisfies condition (1) in SCA. For source signals $s_1 = \begin{bmatrix} -1\,0\,0\,1 \end{bmatrix}^{\mathrm{T}}$, $s_2 = \begin{bmatrix} 0\,1\,0\,0 \end{bmatrix}^{\mathrm{T}}$, it can be obtained that $As_1 = As_2$, different source signals have the same received signal. At this time, the received signal vector falls on the intersection line of the subspace composed of matrix column vectors, and we cannot reconstruct the source signal without other information.

Through the above analysis, we can get the following conclusion: For UBSS source signal recovery problem that meets the three conditions of SCA, we can accurately recover the source signal vector whose number of non-zero values is less than or equal to $\lfloor M/2 \rfloor$, and recover the source signal vector whose number of non-zero values is in the interval $[\lfloor M/2 \rfloor + 1, M - 1]$ with a high probability.

11.3 Underdetermined Blind Separation Source Signal Recovery Based on Greedy Thoughts

Greedy algorithm is the most widely studied algorithm in sparse signal reconstruction (SCA). When applied to the problem of UBSS signal recovery, greedy algorithm shows better performance in the case of sparse source signals. Therefore, greedy algorithm is suitable for the case of sufficiently sparse source signals. Greedy algorithms can be subdivided into three categories: matching pursuit (MP) algorithms, complementary matching pursuit (CMP) algorithms, and gradient pursuit (GP) algorithms. The CMP algorithm has high precision but also high time complexity, while the MP algorithm and the GP algorithm have low time complexity but also low accuracy. To solve this problem, we can introduce the idea of subspace pursuit (SP) algorithm into the CMP algorithm to reduce the time complexity of the algorithm, and then use the smooth L0 norm to replace the L2 norm in the CMP algorithm to measure the sparsity of the algorithm and improve the accuracy of the algorithm.

11.3.1 Matching Pursuit Series Algorithms

1. **Matching Pursuit (MP)**

Matching Pursuit (MP) is the first algorithm proposed in the matching pursuit series of algorithms. The basic idea of MP algorithm is to first build a basic atom library $\Lambda = \{a_i\}$ (for UBSS, the atom library is a set of some column vectors of a mixing matrix), and all vectors in the library have unit norm. An atom a_i that best matches the signal x to be analyzed is selected from the library (the so-called best match means that the absolute value of the inner product of $\boldsymbol{x}$ and $\boldsymbol{a}_i$ is the largest). $\boldsymbol{x}$ is decomposed in the $\boldsymbol{a}_i$ direction, that is:

$$\boldsymbol{x} = \langle \boldsymbol{a}_i, \boldsymbol{x} \rangle \cdot \boldsymbol{a}_i + \boldsymbol{R}_1 \tag{11.5}$$

where, $\boldsymbol{R}_1$ is the residual after matching, and the component of the corresponding position of the recovered signal is $\langle \boldsymbol{a}_i, \boldsymbol{x} \rangle$. The atom that best matches $\boldsymbol{R}_1$ is then selected from the remaining atom library, the decomposition is performed again, and so on until the 2 norm of the remaining residual is less than a threshold. When this algorithm is applied to the UBSS signal recovery problem, it only needs to make the observed signal $\boldsymbol{x}$ as the signal to be analyzed, and the column vectors of the mixing matrix $\boldsymbol{A}$ are unitized as an atomic library. After matching, the decomposed value is added to the corresponding position of the recovered signal. When the residual is less than the given threshold, the recovered signal can be obtained.

The advantages of MP algorithm are simple algorithm and low time complexity, but the disadvantages are low recovery accuracy, especially when dealing with the problem of UBSS when the source signal is not sufficiently sparse, the recovery effect will be poor. In order to improve the reconstruction accuracy of the algorithm, there have been many improved algorithms based on the MP algorithm, including the famous Orthogonal Matching Pursuit (OMP), Subspace Pursuit (SP) and Sparsity Adaptive Matching Pursuit (SAMP).

2. **Orthogonal Matching Pursuit (OMP) Algorithm**

Because the atoms selected by MP algorithm are not optimal when the mixing matrix is not orthogonal, in order to overcome the shortcomings of MP algorithm, some scholars proposed orthogonal matching pursuit (OMP) algorithm on the basis of MP algorithm. The OMP algorithm still uses the matching degree between the observed signal and the atom to select the atom. The difference is that the OMP algorithm adds the step of orthogonalization of the selected atom, which improves the accuracy of the algorithm to a certain extent.

The basic idea of the OMP algorithm is to treat the mixing matrix as a redundant dictionary, each column vector of the matrix as an atom, and in each iteration process, select one atom that has the strongest correlation with the residual, so that the observed signal will have the smallest new residual after removing this correlation part. Iterate until the residual is less than an artificially set threshold. In this way, after many iterations, a matching set of atoms can be selected from the original redundant dictionary, so that the observed signal can accurately reconstruct the source signal using the atoms in this matching set.

In summary, the steps of the OMP algorithm can be summarized as follows:

Input: Mixing matrix $\boldsymbol{A}$, received signal $\boldsymbol{x}$;
Output: K-sparse approximation $\hat{\boldsymbol{s}}$ of the source signal.
Initialization: residual $\boldsymbol{r}_0 = \boldsymbol{x}$, index set $\boldsymbol{\Lambda}_0$ is an empty set, reconstruction matrix Φ_0 is an empty matrix, the number of iterations $k = 1$;

Step 1: Find the residual r_{k-1} and subscript λ corresponding to the column vector with the largest inner product in the mixing matrix $\boldsymbol{A}$, that is, $\lambda = \arg\max_{j=1,2,\ldots,N} |\langle \boldsymbol{r}_{k-1}, \boldsymbol{A}_j \rangle|$;

Step 2: Update index set $\boldsymbol{\Lambda}_k = \boldsymbol{\Lambda}_k \cup \{\lambda\}$, and update reconstruction matrix $\boldsymbol{\Phi}_k = [\boldsymbol{\Phi}_{k-1} \quad A_\lambda]$;

Step 3: Solve $\hat{s} = \arg\min_{\hat{s}} ||\boldsymbol{x} - \boldsymbol{\Phi}_k \hat{s}||_2$ by least square method to get $\hat{s} = (\boldsymbol{\Phi}_k^{\mathrm{T}} \boldsymbol{\Phi}_k)^{-1} \boldsymbol{\Phi}_k^{\mathrm{T}} \boldsymbol{x}$;

Step 4: Update the residual $\boldsymbol{r}_k = \boldsymbol{x} - \boldsymbol{\Phi}_k \hat{s}$ and increase the number of iterations $k = k + 1$;

Step 5: Determine whether the number of iterations k has reached the maximum number of iterations or the residual $\|r_{k-1}\|_2$ is small enough. If so, the iteration is stopped and $\hat{s}$ is output. If not, go to Step 1.

Because of its simplicity, low complexity and high recovery accuracy, OMP has become a classical algorithm in compressed sensing sparse signal reconstruction.

3. **Stagewise Orthogonal Matching Pursuit (StOMP)**

StOMP algorithm is an improved algorithm proposed by later scholars for OMP algorithm, which is different in the way of selecting atoms. Since OMP algorithm can only select one atom that best matches the current residual to add it to the support set each time, which will inevitably reduce the reconstruction speed. StOMP algorithm changes this atomic selection method, thus improving the reconstruction speed of the algorithm compared with OMP algorithm. StOMP algorithm first sets a threshold and then adds atoms that meet the threshold condition to the support set. If only one atom can be added to the support set during each iteration, the StOMP algorithm degenerates into OMP algorithm. OMP algorithm is a special case of the StOMP algorithm. We should avoid this situation, and try to choose a threshold that allows multiple atoms to enter the support set during each iteration. It can be seen that the selected threshold has an important impact on the reconstruction speed of StOMP algorithm.

The specific steps of StOMP algorithm are as follows:

Input: Mixing matrix $\boldsymbol{A} \in \boldsymbol{R}^{m \times n}$, observed signal $\boldsymbol{x} \in \boldsymbol{R}^m$, sparsity K, threshold α;

Output: Recovered signal $\hat{s}$;

Initialization: residual $\boldsymbol{r}_0 = \boldsymbol{x}$, index set $\boldsymbol{\Lambda}_0$ is an empty set, reconstruction matrix Φ_0 is an empty matrix, the number of iterations $k = 1$;

Step 1: Calculate the inner product of the current residual $\boldsymbol{r}_{k-1}$ and each column $\boldsymbol{A}_j$ in the mixing matrix $\boldsymbol{A}$, denoted by $\langle \boldsymbol{r}_{k-1}, \boldsymbol{A}_j \rangle$. Find the set $\boldsymbol{\lambda}_k$ of subscripts j for all columns whose $\langle \boldsymbol{r}_{k-1}, \boldsymbol{A}_j \rangle$ is greater than the threshold α, that is $\lambda_k = \{j | \langle \boldsymbol{r}_{k-1}, \boldsymbol{A}_j \rangle | > \alpha\}$.

Step 2: Update the index set $\boldsymbol{\Lambda}_k = \boldsymbol{\Lambda}_k \cup \{\lambda_k\}$, and the reconstruction matrix $\boldsymbol{\Phi}_k = [\boldsymbol{\Phi}_{k-1} \quad A_{\lambda_k}]$;

Step 3: Solve $\hat{s} = \arg\min_{\hat{s}} ||\boldsymbol{x} - \boldsymbol{\Phi}_k\hat{s}||_2$ by the least square method, get $\hat{s} = (\boldsymbol{\Phi}_k^{\mathrm{T}}\boldsymbol{\Phi}_k)^{-1}\boldsymbol{\Phi}_k^{\mathrm{T}}\boldsymbol{x}$;

Step 4: Update the residual $\boldsymbol{r}_k = \boldsymbol{x} - \boldsymbol{\Phi}_k\hat{s}$ and increase the number of iterations $k = k + 1$;

Step 5: Determine whether the number of iterations k has reached the maximum number of iterations K or the residual $|\boldsymbol{r}_k||_2$ is small enough. If so, the iteration is stopped and $\hat{s}$ is output. If not, go back to Step 1.

Whether it is OMP algorithm or StOMP algorithm, once an atom is selected to be added to the support set, no matter whether the atom is accurate in the subsequent iteration process, the atom must always exist in the support set. If the atom is not accurate, it can only remain in the support set and cannot be removed from the support set. This will certainly affect the recovery accuracy of the algorithm.

SP algorithm can overcome this problem effectively by combining the idea of backtracking. Assuming that the signal sparsity is k, the idea of SP algorithm is as follows: Instead of selecting one atom for each iteration, k atoms that match best are selected and added to the support set. Then, the measured signal is projected orthogonally into the multi-dimensional space formed by the atoms in the support set. The k components with the largest absolute value are identified from these results, which will be retained and the atoms corresponding to the remaining components are removed from the support set. Therefore, the k atoms that match best are always retained in the support set, and those atoms with low matching degree in the iterative process will be eliminated. By constantly updating the support set, the accuracy of the atoms in the support set is ensured, and the reconstruction accuracy of the algorithm can be improved to a certain extent. In addition, the initial step of SP algorithm is to select k atoms into the support set, instead of only selecting one at a time, which can also improve the reconstruction speed of the algorithm.

SAMP algorithm is a sparsity adaptive algorithm. Compared with OMP algorithm, StOMP algorithm and SP algorithm, SAMP algorithm does not need to know the sparsity of signal in advance in the process of reconstructing original signal, while the previously introduced algorithm must know the sparsity of signal in advance, especially SP algorithm. The accuracy of the reconstructed original signal will be much lower. In addition, the SAMP algorithm also selects multiple atoms to enter the support set in each iteration, and the number of atoms selected in each iteration is not fixed. When certain conditions are met, the number of atoms selected in the next iteration should be increased, and the increase rate is determined by the step size; otherwise, the number of atoms selected in the next iteration will remain unchanged. Obviously, the accuracy of SAMP algorithm is related to the step size. When the step size is large, the number of iterations is small, the algorithm complexity is low, but the accuracy is also low. When the step size is small, the recovery accuracy is higher, but the number of iterations increases and the algorithm complexity increases. Therefore, when the step size is appropriate, the ideal reconstruction result can be obtained, but when the step size is not appropriate, the reconstruction effect is poor.

11.3.2 Gradient Pursuit Series Algorithm

Gradient Pursuit algorithms are also a kind of greedy algorithms which are widely used. The main idea of the GP algorithm is to obtain an atom set by using the atom selection method in the MP algorithm in each iteration, and then get the estimated source signal by using the atom set and the idea of gradient descent. Compared with MP series algorithms, GP series algorithms generally have the advantage of low time complexity, but the disadvantage is that the recovery accuracy is not high enough. Well-known Gradient Pursuit series algorithms include Gradient Pursuit (GP), Conjugate Gradient Pursuit (CGP) and Newton Pursuit (NP).

In the OMP algorithm, when we select a set Γ_k from a set of atoms, we use the least square method to solve for the estimation of the source signal. In the MP series algorithm, the method of pseudo-inverse of the submatrix is used to solve the least square solution, which will have high time complexity in the case of large matrix. In order to reduce the time complexity, the GP series algorithm first choose a direction that causes the value of function

$$f(\boldsymbol{s}_k) = \|\boldsymbol{x} - \boldsymbol{A}_{\Gamma_k}\boldsymbol{s}_k\|_2^2 \tag{11.6}$$

to decrease, and then select the appropriate step size, and gradually approximate the optimal value of the function $f(\boldsymbol{s}_k)$ through iteration, the formula of iterative approximation to the optimal value is

$$\boldsymbol{s}_k = \boldsymbol{s}_{k-1} + \alpha_k \boldsymbol{d}_{\Gamma_k} \tag{11.7}$$

where, $\boldsymbol{A}_{\Gamma_k}$ is a matrix composed of column vectors in the mixing matrix $\boldsymbol{A}$ with elements in the set Γ_k as subscripts, $\boldsymbol{d}_{\Gamma_k}$ is the iterative direction chosen at the k-th iteration, and α_k is the iterative step size at the k-th iteration. The iterative direction chosen by GP algorithm in each iteration is the negative gradient direction of function $f(\boldsymbol{s}_k)$, that is:

$$\boldsymbol{d}_{\Gamma_k} = -\nabla f(s_k) = \boldsymbol{A}_{\Gamma_k}^{\mathrm{T}}(x - \boldsymbol{A}_{\Gamma_k} s_k) \tag{11.8}$$

Since the search direction chosen each time is the negative gradient direction of the problem $\min f(s_k)$ to be solved, the update of the search direction is equivalent to the update of the search direction in the fastest descent method. Next, the generation method of searching step size in GP algorithm is introduced.

The searching step α_k of the k-th iteration in GP algorithm is obtained by exact one-dimensional search, that is, the optimal step size is obtained by a minimization problem, which is as follows:

$$\min f(\boldsymbol{s}_k + \alpha_k \boldsymbol{d}_k) \tag{11.9}$$

Taking the derivative of equation $f(\boldsymbol{s}_k + \alpha_k \boldsymbol{d}_k)$ so that it is equal to 0, the best step size α_k can be solved as follows:

$$\alpha_k = \frac{\langle \boldsymbol{r}_{k-1}, \boldsymbol{A}_{\Gamma_k} \boldsymbol{d}_{\Gamma_k} \rangle}{\left\| \boldsymbol{A}_{\Gamma_k} \boldsymbol{d}_{\Gamma_k} \right\|_2^2} \tag{11.10}$$

where $\boldsymbol{r}_k$ is the residue of the k-th iteration, $\boldsymbol{r}_k = \boldsymbol{x} - \boldsymbol{A}_{\Gamma_k} \boldsymbol{s}_k^{k-1}$, $\langle \boldsymbol{r}_{k-1}, \boldsymbol{A}_{\Gamma_k} \boldsymbol{d}_{\Gamma_k} \rangle = (\boldsymbol{r}_{k-1})^{\mathrm{T}} \boldsymbol{A}_{\Gamma_k} \boldsymbol{d}_{\Gamma_k}$. Thus, the signal can be updated by using formula (11.7).

In summary, the steps of GP algorithm are as follows:

Input: Received signal $\boldsymbol{x} \in \boldsymbol{R}^m$, mixed matrix $\boldsymbol{A} \in \boldsymbol{R}^{m \times n}$, observed signal, termination threshold ε;

Output: The recovered signal $\boldsymbol{s}$;
Initialization: residual $\boldsymbol{r}_0 = \boldsymbol{x}$, support set $\Gamma_0 = \emptyset$, $\boldsymbol{s}_0 = \boldsymbol{0}$, current iteration number $k = 1$;

Step 1: Calculate the inner product of the current residual $\boldsymbol{r}_{k-1}$ and each column A_j in the matrix $\boldsymbol{A}$, and compare these inner products to find out the subscript $\boldsymbol{\lambda}_k$ of the column corresponding to the maximum value in the result, i.e.$\boldsymbol{\lambda}_k = \arg\max_{j=1\ldots n} \left| \langle \boldsymbol{r}_{k-1}, A_j \rangle \right|$, and take it as the selected atom within the current iteration;

Step 2: Update the support set by adding the selected atom within the current iteration, that is, $\Gamma_k = \Gamma_{k-1} \cup \{\lambda_k\}$;

Step 3: Solve the iteration direction within the current iteration $\boldsymbol{d}_{\Gamma_k} = -\nabla f(s_k) = \boldsymbol{A}_{\Gamma_k}^T (x - \boldsymbol{A}_{\Gamma_k} s_k)$;

Step 4: Calculate the iterative step size α_k of the current iteration by using formula (11.10);

Step 5: Update the signal, that is $\boldsymbol{s}_k = \boldsymbol{s}_{k-1} + \alpha_k \boldsymbol{d}_{\Gamma_k}$;

Step 6: Update the residual, that is, $\boldsymbol{r}_k = \boldsymbol{r}_{k-1} - \alpha_k \boldsymbol{A}_{\Gamma_k} \boldsymbol{d}_{\Gamma_k}$, and $k = k + 1$, if the condition $\|\boldsymbol{r}_k\|_2 \geq \varepsilon$ is met, then continue the iterative process and return to step 1; Otherwise, stop iteration.

Since the GP algorithm searches according to the negative gradient direction of the function during each iteration, and the iterative step size selected of each iteration is the optimal step size, the directions of two iterations are orthogonal. This phenomenon causes the initial point to follow a zigzag path to the optimal value point, which is equivalent to taking many detours. This phenomenon is mathematically known as "sawtooth". Due to the existence of "sawtooth", the convergence speed of GP algorithm is slow and the time complexity is high.

Lemma 11.1 *The direction of the previous search and that of the next search are always perpendicular and orthogonal to each other.*

Proof Let

$$\phi(\alpha_k) = f(\boldsymbol{x}_k + \alpha_k \boldsymbol{d}_k) \tag{11.11}$$

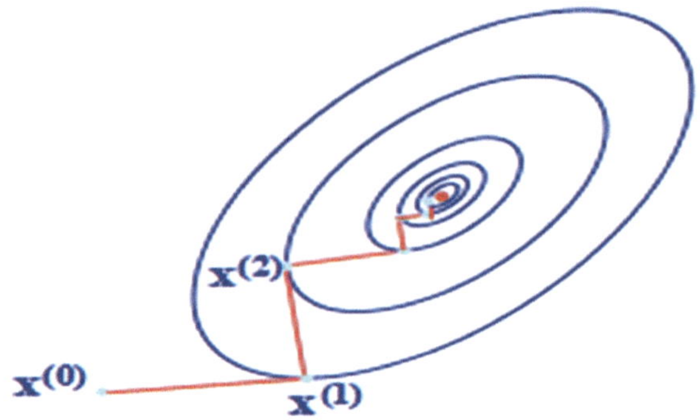

Fig. 11.1 Sawtooth phenomenon of the fastest descent method

Using exact one-dimensional search, take the derivative of $\phi(\lambda)$ and let it be 0, then:

$$\phi'(\alpha_k) = \nabla f(\boldsymbol{x}_k + \alpha_k \boldsymbol{d}_k)^{\mathrm{T}} \boldsymbol{d}_k = 0 \tag{11.12}$$

Equation (11.12) can be further reduced to Eq. (11.13):

$$0 = \nabla f(\boldsymbol{x}_k + \alpha_k \boldsymbol{d}_k)^{\mathrm{T}} \boldsymbol{d}_k = \nabla f(\boldsymbol{x}_{k+1})^{\mathrm{T}} \boldsymbol{d}_k = -\boldsymbol{d}_{k+1}^{\mathrm{T}} \boldsymbol{d}_k \tag{11.13}$$

where $\boldsymbol{d}_k$ is the search direction of the k-th iteration, α_k is the iterative step size during the k-th iteration, and $\boldsymbol{d}_{k+1}$ is the search direction of the $(k+1)$-th iteration. It can be seen that, in the fastest descent method, the search direction of the previous and the search direction of the last are perpendicular and orthogonal to each other, which leads to a problem that the process of using the fastest descent method to approach the optimal solution is tortuous. This problem of the fastest descent method is called sawtooth phenomenon, that is, the process of approaching the minimum point is "zigzag", as shown in Fig. 11.1.

Therefore, when using the fastest descent method to approach the minimum point, no matter from which initial point, it can be quickly approached to the minimum point. However, as the subsequent approximation step size becomes smaller and smaller, the moving speed becomes slower and slower. This will cause the speed of the fastest descent method to be slow when approaching the minimum point near the minimum point. Due to the defect of the fastest descent, the reconstruction speed of GP algorithm is slow when approaching the optimal solution near the optimal solution. Therefore, the overall reconstruction speed of GP algorithm is slow.

In order to overcome the shortcomings of GP algorithm and reduce the time complexity of the algorithm, conjugate gradient pursuit (CGP) algorithm uses the conjugate gradient direction instead of the negative gradient direction in GP algorithm to approximate the optimal value. Conjugation is defined as follows. If two vectors $\boldsymbol{d}_k$ and $\boldsymbol{d}_{k-1}$ satisfy $\boldsymbol{d}_k^T \boldsymbol{G} \boldsymbol{d}_{k-1} = 0$ with respect to an N-dimensional square matrix, the vectors $\boldsymbol{d}_k$ and $\boldsymbol{d}_{k-1}$ are said to be conjugate symmetric with respect to the matrix $\boldsymbol{G}$. The direction of the two iterations before and after are conjugate, so that the original point can approximate the optimal point along a relatively straight path, which effectively eliminates the problem of slow convergence caused by "sawtooth phenomenon" in GP algorithm, and significantly reduces the time complexity of the algorithm.

11.3.3 Complementary Matching Pursuit (CMP) Series Algorithm

In addition to MP series algorithms and GP series algorithms, another common greedy algorithm is the complementary matching pursuit (CMP) algorithm. The core idea of CMP algorithm is to find the optimal solution of source signal by using the complementary property of source signal in matrix space and orthogonal space of matrix. Here, we first introduce complementary characteristic of signals. In underdetermined blind source separation, since the M row vectors of the mixing matrix $\boldsymbol{A} \in \boldsymbol{R}^{M\times N}$ satisfy the independence of each other, then the columns of $\boldsymbol{A}^{\mathrm{T}}$ can span an M-dimensional subspace $\boldsymbol{S}_A$ in an N-dimensional space, which must have an $N-M$ dimensional orthogonal subspace $\boldsymbol{S}_A^{\perp}$. And $\tilde{\boldsymbol{s}} = \boldsymbol{A}^{\mathrm{T}}(\boldsymbol{A}\boldsymbol{A}^{\mathrm{T}})^{-1}\boldsymbol{x}$ is a solution of the equation $\boldsymbol{x} = \boldsymbol{A}\boldsymbol{s}$, and $\tilde{\boldsymbol{s}}$ must be in the space expressed by $\boldsymbol{S}_A$. The source signal is represented by $\boldsymbol{s}^0$, assuming that it can be decomposed into $\boldsymbol{s}^c$ and $\tilde{\boldsymbol{s}}$, i.e. $\boldsymbol{s}^0 = \tilde{\boldsymbol{s}} + \boldsymbol{s}^c$. Because of $\boldsymbol{A}\boldsymbol{s}^0 = \boldsymbol{A}\tilde{\boldsymbol{s}} = \boldsymbol{x}$, we have $\boldsymbol{A}\boldsymbol{s}^c = \boldsymbol{0}$, and the physical meaning of this equation is that $\boldsymbol{s}^c$ lies in the subspace $\boldsymbol{S}_A^{\perp}$. Now assume that only one element of the source signal vector $\boldsymbol{s}^0$ is nonzero, and the remaining $N-1$ elements are equal to zero. Because of $\boldsymbol{s}^0 = \tilde{\boldsymbol{s}} + \boldsymbol{s}^c$ and $\tilde{\boldsymbol{s}} = \boldsymbol{A}^{\mathrm{T}}(\boldsymbol{A}\boldsymbol{A}^{\mathrm{T}})^{-1}\boldsymbol{x}$, if we can solve the $\boldsymbol{s}^c$, problem will be solved. From the previous analysis, we already know that $\boldsymbol{s}^c$ lies in $\boldsymbol{S}_A^{\perp}$, and $\boldsymbol{G}$ is used to represent an $N\times(N-M)$-dimensional matrix, whose column vectors satisfy orthogonality and span subspace $\boldsymbol{S}_A^{\perp}$, and the matrix $\boldsymbol{G}$ can be obtained by QR decomposition or SVD decomposition of $\boldsymbol{A}^{\mathrm{T}}$. The result obtained by QR decomposition is as follows:

$$\boldsymbol{A}^{\mathrm{T}} = \boldsymbol{Q}\boldsymbol{R} = [\boldsymbol{Q}_1^{N\times M}\,\boldsymbol{Q}_2^{N\times(N-M)}]\begin{bmatrix}\boldsymbol{R}_1^{M\times M}\\ \boldsymbol{0}^{(N-M)\times M}\end{bmatrix} \tag{11.14}$$

where $\boldsymbol{Q}$ and $\boldsymbol{R}$ are $\boldsymbol{N}\times\boldsymbol{N}$ matrices, and $\boldsymbol{R}_1$ is an upper triangular matrix, then the subspaces corresponding to $\boldsymbol{Q}_1$ and $\boldsymbol{Q}_2$ are $\boldsymbol{S}_A$ and $\boldsymbol{S}_A^{\perp}$ respectively, and the columns of $\boldsymbol{Q}_1$ and $\boldsymbol{Q}_2$ are orthogonal to each other, then

$$\boldsymbol{G} = \boldsymbol{Q}_2 \tag{11.15}$$

Now suppose that the l-th component $\boldsymbol{s}^0(l)$ in $\boldsymbol{s}^0$ is an unknown non-zero component, and the remaining $N-1$ components of $\boldsymbol{s}^0$ are zero, which is denoted by $\bar{\boldsymbol{s}}^0(l)$, then $\bar{\boldsymbol{s}}^0(l) = \boldsymbol{0}$. $\boldsymbol{s}^c(l)$ represents the l-th component of $\boldsymbol{s}^c$, and the remaining $N-1$ elements are represented by $\bar{\boldsymbol{s}}^c(l)$. Because $\boldsymbol{s}^c$ is on the space $\boldsymbol{S}_A^{\perp}$, there must be a unique $N - M$-dimensional column vector z, such that:

$$Gz = s^c \tag{11.16}$$

Formula (11.16) can be rewritten as:

$$\begin{bmatrix} \overline{G_l} \\ g_l \end{bmatrix} z = \begin{bmatrix} \overline{s}^c(l) \\ s^c(l) \end{bmatrix} \tag{11.17}$$

where $\overline{G_l}$ is a matrix of row vectors in G except for row l, and g_l is the l-th row vector of G. So

$$\overline{G}_l z = \overline{s}^c(l) \tag{11.18}$$

$$g_l z = s^c(l) \tag{11.19}$$

The matrix $\overline{G}_l$ is column full rank and $\overline{s}^c(l)$ is also column full rank, so we can get $z = \left(\overline{G}_l^{\mathrm{T}}\overline{G}_l\right)^{-1}\overline{G}_l^{\mathrm{T}}\overline{s}^c(l)$. Furthermore, since $\overline{s}^0(l) = \mathbf{0}$, it follows that $\overline{s}^c(l) = -\overline{\tilde{s}}(l)$. Consequently, $s^c(l)$ and $s^0(l)$ can be calculated according to the following formula:

$$s^c(l) = g_l\left(\overline{G}_l^{\mathrm{T}}\overline{G}_l\right)^{-1}\overline{G}_l^{\mathrm{T}}\overline{s}^c(l) = -g_l\left(\overline{G}_l^{\mathrm{T}}\overline{G}_l\right)^{-1}\overline{G}_l^{\mathrm{T}}\overline{\tilde{s}}(l) \tag{11.20}$$

$$s^0(l) = \tilde{s}(l) + s^c(l) = \tilde{s}(l) - g_l\left(\overline{G}_l^{\mathrm{T}}\overline{G}_l\right)^{-1}\overline{G}_l^{\mathrm{T}}\overline{\tilde{s}}(l) \tag{11.21}$$

When the number of active source signals at a certain time is 1, the CMP algorithm takes the method of sequentially traversing each atom, while checking and judging whether the sum of the s^c and $\tilde{s}$ components corresponding to the remaining $N - 1$ atoms is 0. If the index of the active source signal is found, the formula $s^0(l) = \tilde{s}(l) - g_l\left(\overline{G}_l^{\mathrm{T}}\overline{G}_l\right)^{-1}\overline{G}_l^{\mathrm{T}}\overline{\tilde{s}}(l)$ can be used to calculate the active source signal at a certain time. In summary, the implementation steps of CMP algorithm are as follows:

Input: Mixed signal x, mixing matrix A;

Output: Estimation of the source signal s^0;

Step 1: Perform QR decomposition on the mixing matrix A, and obtain G according to Eqs. (11.14) and (11.15);

Step 2: Let $p_0 = \tilde{s} = A^{\mathrm{T}}(AA^{\mathrm{T}})^{-1}x$, set of subscripts $\mathrm{I} = \{1, 2, \ldots N\}$, and use $\overline{p}_j(i)$ to represent the vector consisting of the remaining elements after removing the i-th element from vector p_j. j represents the number of iterations, with the initial values $j = 1$, $i = 1$, and $s^0 = \mathbf{0}$;

Step 3: Calculate $e_j = p_{j-1} - G\left(\overline{G}_i^{\mathrm{T}}\overline{G}_i\right)^{-1}\overline{G}_i^{\mathrm{T}}\overline{p}_{j-1}(i)$, where $i \in \boldsymbol{I}$, and $\overline{G}_i$ represents a matrix of all rows in G except row i;

Step 4: Let $c_j = e_j(i)$, then $s^0(i) = s^0(i) + c(j)$;

Step 5: Find the new i: $i = \arg\min_{k \in I} \left\| \overline{e}_j(k) \right\|_2$, where $\overline{e}_j(k)$ is the vector obtained by removing the k-th element from $\boldsymbol{e}_j$;

Step 6: Update the complementary residual $\boldsymbol{p}_j = \boldsymbol{I}_{\bar{i}} \boldsymbol{e}_j$, where $\boldsymbol{I}_{\bar{i}}$ is a transformation matrix of an N-dimensional identity matrix, and the i-th diagonal element of the $\boldsymbol{I}_{\bar{i}}$ matrix is 0;

Step 7: Calculate the residual difference $\boldsymbol{R}_j = \boldsymbol{x} - \boldsymbol{A}\boldsymbol{s}^0$. When the 2-norm of $\boldsymbol{R}_j$ is smaller than the set threshold, the algorithm iteration is considered to be terminated; otherwise, let $j := j + 1$ and go back to Step 3 to continue the iteration.

The $\boldsymbol{s}^0$ obtained at the end of the algorithm is the source signal vector recovered by the CMP algorithm.

The OCMP algorithm follows the CMP algorithm which uses the complementary characteristics of signals to select atoms. Different from CMP algorithm, OCMP algorithm adopts orthogonal processing method in the process of approximating the source signal by using the selected atom set.

The steps of the OCMP algorithm are summarized as follows:

Input: Mixed signal $\boldsymbol{x}$, mixed matrix $\boldsymbol{A}$;

Output: Estimation of the source signal $\boldsymbol{s}^0$;

Step 1: Perform QR decomposition on the mixing matrix $\boldsymbol{A}$, and obtain $\boldsymbol{G}$ according to Eqs. (11.14) and (11.15);

Step 2: Initialize $\boldsymbol{p}_0 = \tilde{\boldsymbol{s}} = \boldsymbol{A}^{\mathrm{T}}(\boldsymbol{A}\boldsymbol{A}^{\mathrm{T}})^{-1}\boldsymbol{x}$, and set of subscripts $\mathrm{I} = \{1, 2, \ldots N\}$. The set of selected atoms is $\mathrm{J}_1 = \{1\}$, with $j = 1$, $i = 1$, and $\boldsymbol{s}^0 = \boldsymbol{0}$;

Step 3: Compute $\boldsymbol{e}_j = \boldsymbol{p}_j - \boldsymbol{G}(\overline{\boldsymbol{G}}_J^{\mathrm{T}} \overline{\boldsymbol{G}}_J)^{-1} \overline{\boldsymbol{G}}_J^{\mathrm{T}} \overline{\boldsymbol{p}}_j(\mathrm{J}_j)$, where $\overline{\boldsymbol{G}}_J$ represents a matrix consisting of row vectors in $\boldsymbol{G}$ other than the row vectors corresponding to the set J_j.

Step 4: Find out that the new i, $i = \arg\min_{i \in I} \left\| \overline{e}_j(i) \right\|_2$, where $\overline{e}_j(i)$ is the vector obtained by removing the i-th element from $\boldsymbol{e}_j$;

Step 5: Update the atom set: $\mathrm{J}_{j+1} = \mathrm{J}_j \cup i$;

Step 6: Update the values corresponding to the components of the selected atom: $s^0(\mathrm{J}_j) := s^0(\mathrm{J}_j) + \boldsymbol{e}_j(\mathrm{J}_j)$, where $s^0(\mathrm{J}_j)$ and $\boldsymbol{e}_j(\mathrm{J}_j)$ represent vectors consisting of all the $i(i \in \mathrm{J}_j)$-th elements in $\boldsymbol{s}^0$ and $\boldsymbol{e}_j$ respectively;

Step 7: Update the complementary residual $\boldsymbol{p}_j = \boldsymbol{I}_k \boldsymbol{e}_j$, where $\boldsymbol{I}_k$ represents a matrix that changes the diagonal component corresponding to set J_j in an N-dimensional identity matrix to 0;

Step 8: Calculate the residual $\boldsymbol{R}_j = \boldsymbol{x} - \boldsymbol{A}\boldsymbol{s}^0$. Stop the process when the norm of the residual is less than the set threshold, otherwise let $j = j + 1$, go back to step 3.

11.3.4 Subspace Complementary Matching Pursuit Algorithm

When there are multiple non-zero elements in the source signal, the method of CMP algorithm is to use the idea of cyclic iteration, using the minimum L2 norm to find a non-zero element in each iteration process, by increasing the number of iterations to find the index of the non-zero atom, and recover the source signal according to the selected index set. The result is that the algorithm operation speed is slow and the algorithm time complexity is high. If you can search multiple elements at a time, it will greatly improve the speed of the algorithm. Based on this idea, a Subspace Complementary Matching Pursuit (SCMP) algorithm is proposed. In the CMP algorithm, we need to find the coordinate z of the vector $\tilde{s}$ on the basis of the column vector of $\boldsymbol{Q}_2$. We randomly select $N-M$ row vectors from the matrix $\boldsymbol{Q}_2$ to form matrix $\boldsymbol{Q}_2[\Phi] \in R^{(N-M)\times(N-M)}$ (Φ represents a subset of any $N-M$ elements in the set $\{1, 2, \ldots, N\}$). According to $\boldsymbol{Q}_2[\Phi]$ and the corresponding $\tilde{s}[\Phi] \in R^{(N-M)}$, we get

$$z = (\boldsymbol{Q}_2[\Phi])^{-1}\tilde{s}[\Phi] = -(\boldsymbol{Q}_2[\Phi])^{-1}\boldsymbol{s}^c[\Phi] \tag{11.22}$$

Then

$$\tilde{s} = \boldsymbol{Q}_2 z = -\boldsymbol{Q}_2(\boldsymbol{Q}_2[\Phi])^{-1}\boldsymbol{s}^c[\Phi] \tag{11.23}$$

$$\boldsymbol{e} = \boldsymbol{s}^c[\Phi] - \boldsymbol{Q}_2(\boldsymbol{Q}_2[\Phi])^{-1}\boldsymbol{s}^c[\Phi] \tag{11.24}$$

In this way, there are C_N^{N-M} of these Φs, and the sparsity of $\boldsymbol{e}$ is measured by the norm, where the set $\tilde{\Phi}$ corresponding to $\boldsymbol{e}$ with the smallest norm is the index of $N-M$ zero elements in the source signal $\boldsymbol{s}$. The recovery signal can then be obtained as follows.

$$\boldsymbol{s} = \boldsymbol{s}^c + \tilde{s} = \boldsymbol{s}^c - \boldsymbol{Q}_2(\boldsymbol{Q}_2[\tilde{\Phi}])^{-1}\boldsymbol{s}^c[\tilde{\Phi}] \tag{11.25}$$

In CMP algorithm, L2 norm is used to measure the sparsity of $\boldsymbol{e}$. L2 norm is simple to calculate, but L2 norm cannot accurately reflect the sparsity of the signal. Although the L0 norm can accurately reflect the sparsity of the signal, it is very sensitive to noise and cannot be applied to the case with noise. Therefore, the signal sparsity can be measured using the approximate L0 norm. The approximate L0 norm is as follows.

$$f_\delta(s) = \frac{2}{\pi}\arctan\left(\frac{s^2}{2\delta^2}\right) \tag{11.26}$$

$$F_\delta(s) = \sum_{i=1}^{n} f_\delta(s_i) \tag{11.27}$$

Since $f_\delta(s)$ has the following properties:

$$\lim_{\delta \to 0} f_\delta(s) = \begin{cases} 1 & s \neq 0 \\ 0 & s = 0 \end{cases} \tag{11.28}$$

Then

$$\lim_{\delta \to 0} F_\delta(s) = ||\boldsymbol{s}||_0 \tag{11.29}$$

Therefore, the L0 norm can be approximated by the function $F_\delta(s)$. Where, the smaller the value of δ in Eq. (11.26), the more accurately it can reflect the sparsity of the signal, but at the same time, the more sensitive it is to noise. Therefore, δ should not be too small, and the value of δ can be max $\frac{|x(t)|}{5}$. According to the above analysis, the final SCMP algorithm steps are as follows:

Step 1: Perform QR decomposition on the transpose $\boldsymbol{A}^{\mathrm{T}}$ of the mixing matrix: $\boldsymbol{A}^{\mathrm{T}} = \boldsymbol{QR} = [\boldsymbol{Q}_1^{N\times K}\ \boldsymbol{Q}_2^{N\times(N-K)}]\begin{bmatrix}\boldsymbol{R}_1^{K\times K}\\ \boldsymbol{0}^{(N-K)\times K}\end{bmatrix}$, and get $\boldsymbol{s}_1 = \boldsymbol{A}^{\mathrm{T}}(\boldsymbol{A}\boldsymbol{A}^{\mathrm{T}})^{-1}\boldsymbol{x}(t)$;

Step 2: Find $f_i = F_\delta(\boldsymbol{s}_1 - \boldsymbol{Q}_2(\boldsymbol{Q}_2[\Phi_i])^{-1}\boldsymbol{s}_1[\Phi_i])$, $i = 1, 2, \ldots, C_N^{N-M}$, where Φ_i represents a subset of any $N - M$ elements in the set $\{i = 1, 2, \ldots N\}$, and $\boldsymbol{Q}_2[\Phi_i]$ represents a matrix of $N - M$ row vectors corresponding to the set Φ_i in $\boldsymbol{Q}_2$;

Step 3: Find the index set $j = \arg\min_i f_i$, $\tilde{\Phi} = \Phi_j$;

Step 4: The recovery signal $\hat{\boldsymbol{s}} = \boldsymbol{s}_1 - \boldsymbol{Q}_2(\boldsymbol{Q}_2[\tilde{\Phi}])^{-1}\boldsymbol{s}_1[\tilde{\Phi}]$ is obtained.

11.3.5 Algorithm Performance Simulation and Analysis

In this section, the proposed SCMP algorithm is simulated with OMP algorithm, CMP algorithm and OCMP algorithm, and the performances of the algorithms are compared from two aspects of algorithm accuracy and algorithm time complexity. In the experiment, we use Gaussian sparse signal as the source signal, and the mixing matrix is Gaussian random matrix. There are 5 source signals and 3 observed signals, and the number of sampling points is 1000, that is, $T = 1000$, $n = 5$, $m = 3$. In order to evaluate the accuracy of the algorithm, the correlation coefficient between the recovery signal and the source signal is used to measure the accuracy of the algorithm. The correlation coefficient is defined as follows:

$$e(\boldsymbol{s}, \hat{\boldsymbol{s}}) = \frac{|\langle \boldsymbol{s}, \hat{\boldsymbol{s}} \rangle|}{||\boldsymbol{s}||_2 ||\hat{\boldsymbol{s}}||_2} \tag{11.30}$$

where s and $\hat{s}$ represent the source signal vector and recovery signal vector respectively. If e is closer to 1, the correlation between s and $\hat{s}$ is stronger, indicating that the algorithm is more accurate, and the maximum value of correlation is 1. In the following simulation experiments, SCMP algorithm is compared with OMP, CMP and OCMP algorithm respectively in terms of source signal sparsity and SNR.

1. **Influence of Source Signal Sparsity on Algorithm Performance**

In this experiment, when the SNR is 10, 20 and 30 dB, the curve of the correlation coefficient between the recovered signal and the source signal changing with the sparsity of the source signal and the curve of the running time of each algorithm changing with the sparsity of the source signal are simulated. Figure 11.2 shows the curve of correlation coefficient changing with sparsity when the SNR is 10 dB. Figure 11.3 shows the curve of the change of running time with sparsity when the SNR is 10 dB.

The curves of correlation coefficient and running time changing with sparsity when the SNR is 20 dB are shown in Figs. 11.4 and 11.5 respectively.

Figure 11.6 shows the curve of correlation coefficient changing with sparsity when the SNR is 30 dB. Figure 11.7 shows the curve of the change of running time with sparsity when the SNR is 30 dB.

According to the simulation results in Figs. 11.2, 11.3, 11.4, 11.5, 11.6 and 11.7, when the SNR is not too high, such as SNR = 10 dB, the four algorithms of OMP, CMP, OCMP and SCMP have little difference in accuracy performance. When the

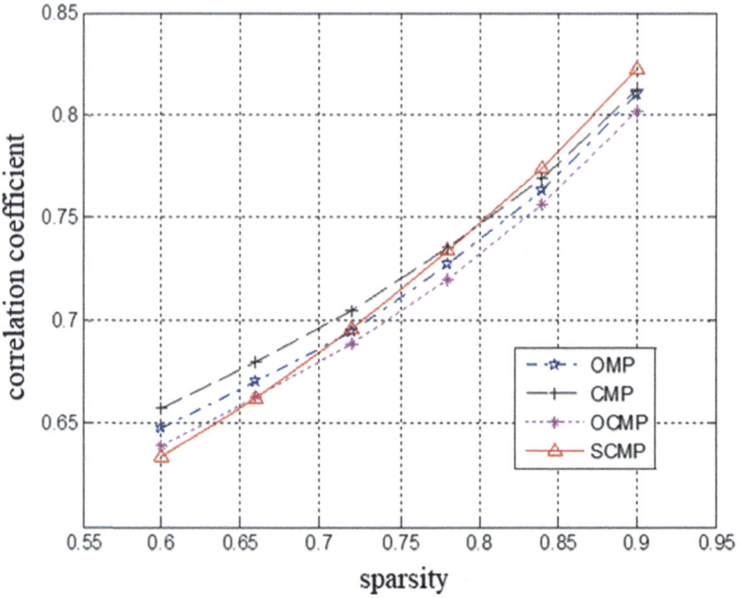

Fig. 11.2 Curves of correlation coefficient changing with sparsity when SNR is 10 dB

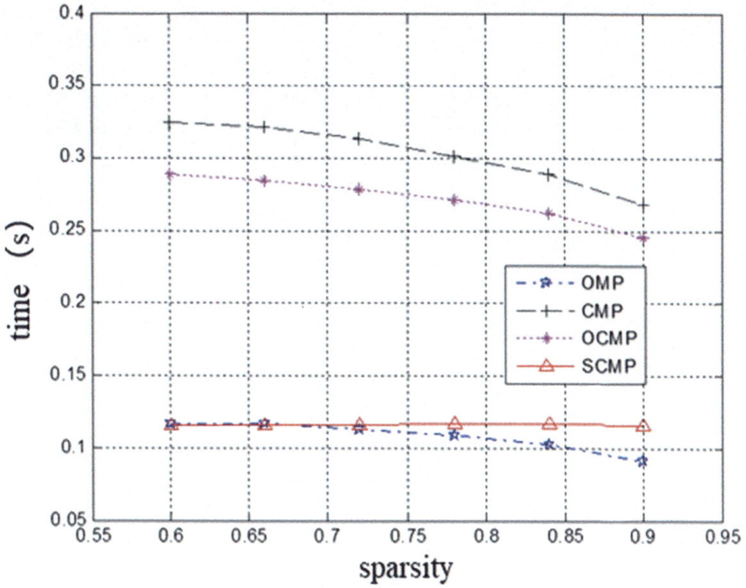

Fig. 11.3 Curves of the running time changing with sparsity when the SNR is 10 dB

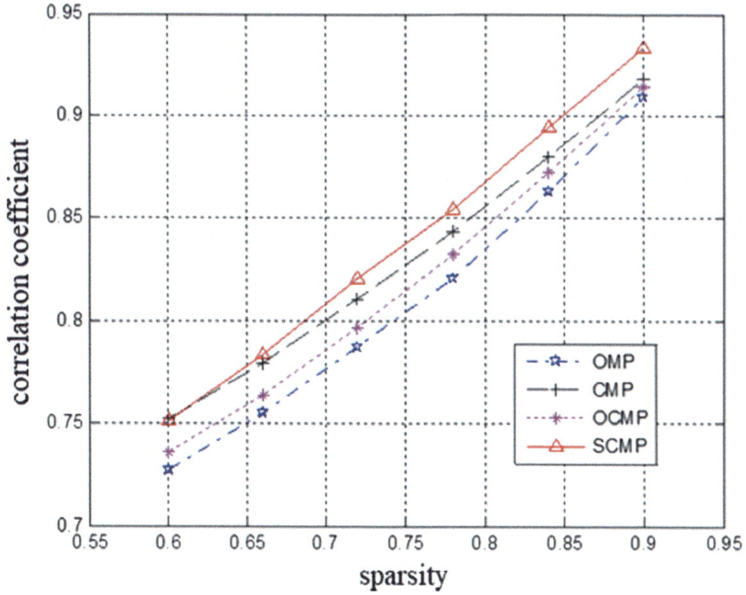

Fig. 11.4 Curves of correlation coefficient changing with sparsity when SNR is 20 dB

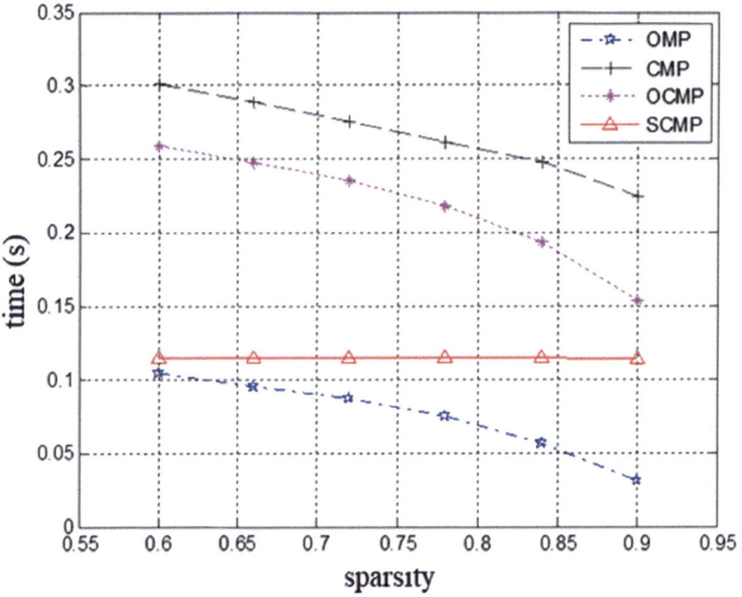

Fig. 11.5 Curves of the running time changing with sparsity when the SNR is 20 dB

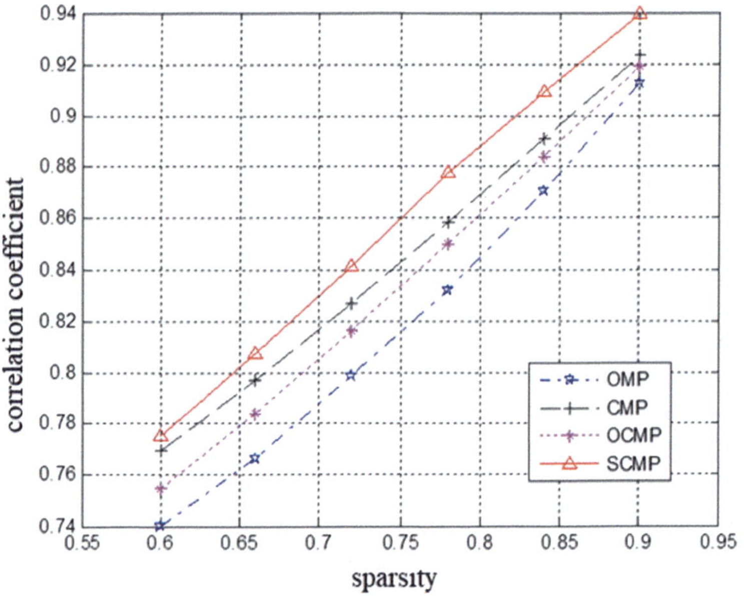

Fig. 11.6 Curves of correlation coefficient changing with sparsity when SNR is 30 dB

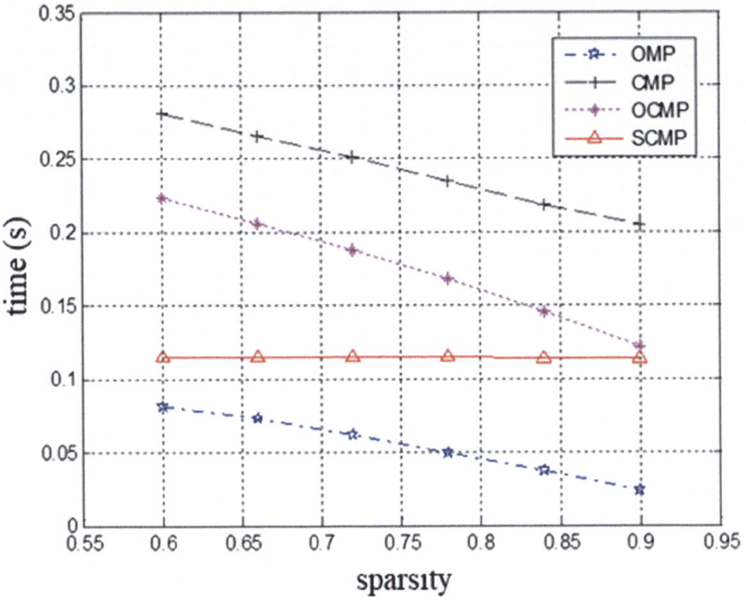

Fig. 11.7 Curves of the running time changing with sparsity when the SNR is 30 dB

signal sparsity is small, the correlation coefficient of SCMP algorithm is slightly lower than that of other three algorithms; when the signal sparsity is large, SCMP algorithm is slightly better than the other three algorithms in terms of accuracy. When the SNR is large, such as SNR = 30 dB, the SCMP algorithm is always superior to other algorithms in terms of recovery accuracy regardless of the source signal sparsity. In terms of time complexity, since the number of iterations in the SCMP algorithm has nothing to do with the source signal sparsity, the running time of the SCMP algorithm is a horizontal straight line when the source signal sparsity changes. The iteration times of OMP, CMP and OCMP are all related to the sparsity of the source signal. Therefore, the running time of these three algorithms shows a decreasing trend with the increase of sparsity. As can be seen from Figs. 11.3, 11.5 and 11.7, the running time complexity of SCMP algorithm is significantly smaller than that of CMP algorithm and OCMP algorithm. In the case of small source signal sparsity, the running time of SCMP algorithm is similar to that of OMP algorithm, and the running time of SCMP is slightly longer than that of OMP in the case of large source signal sparsity.

2. **Influence of SNR on Algorithm Performance**

In this experiment, when the source signal sparsity is 0.7, 0.8 and 0.9, the effects of SNR on the performance of the algorithm are simulated respectively. Figures 11.8 and 11.9 respectively show the curves of correlation coefficient and running time of each algorithm changing with SNR when source signal sparsity is 0.7.

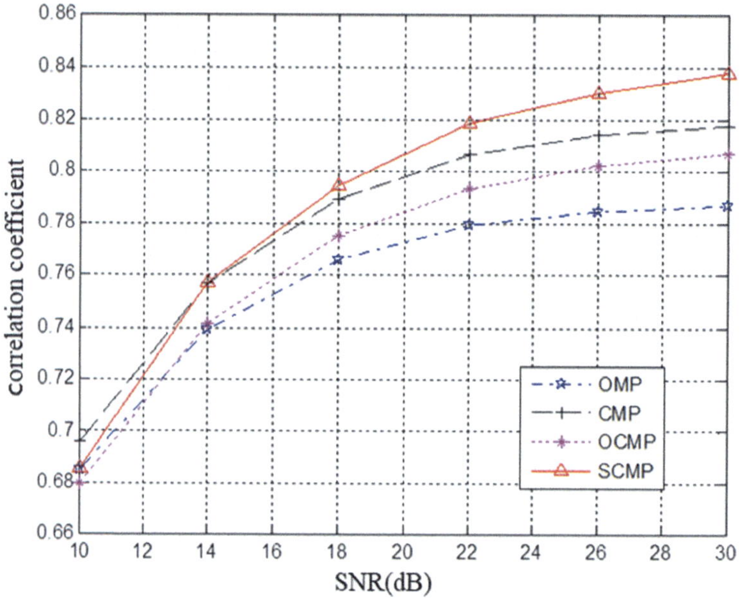

Fig. 11.8 Curves of correlation coefficient changing with SNR when source signal sparsity is 0.7

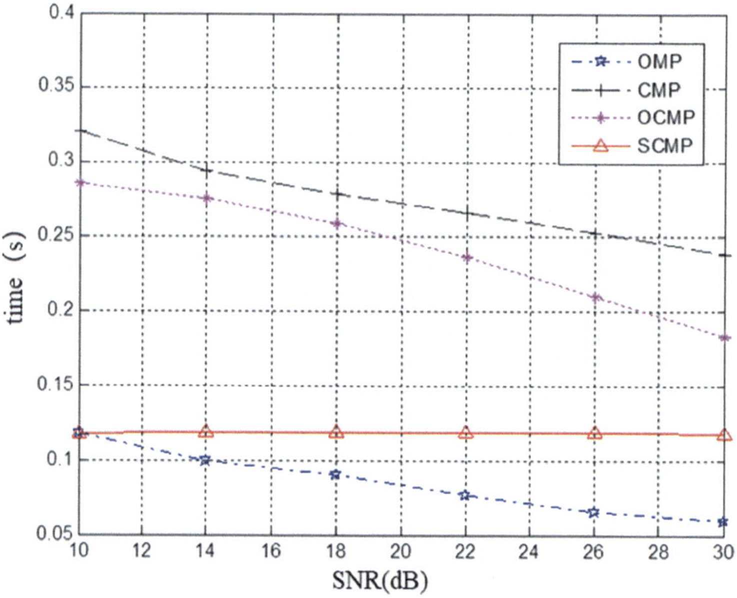

Fig. 11.9 Curves of running time changing with SNR when source signal sparsity is 0.7

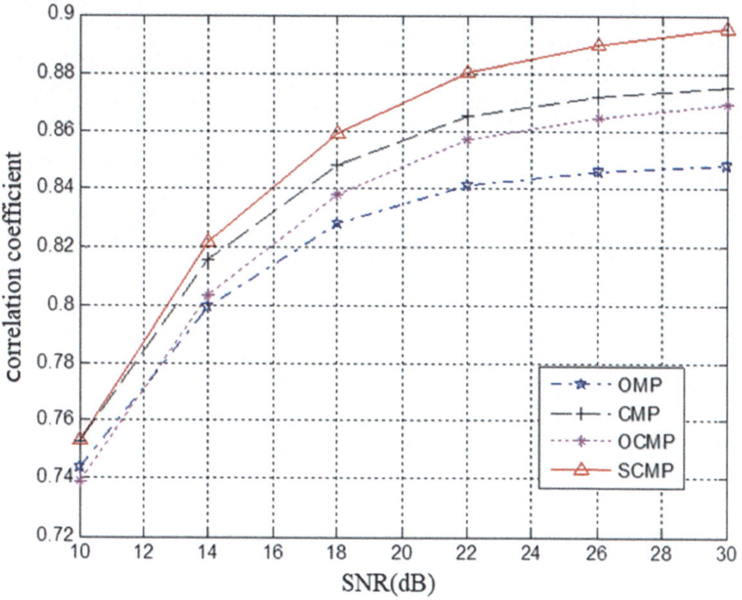

Fig. 11.10 Curves of correlation coefficient changing with SNR when source signal sparsity is 0.8

Figures 11.10 and 11.11 respectively show the curves of correlation coefficient and running time changing with SNR when source signal sparsity is 0.8.

Figures 11.12 and 11.13 respectively show the curves of correlation coefficient and running time with SNR when source signal sparsity is 0.9.

As can be seen from the above simulation results, when the SNR is small (such as SNR = 10 dB), the recovery accuracy of SCMP is similar to that of OMP, CMP and OCMP algorithms. When the SNR is large (such as SNR = 30 dB), the recovery accuracy of SCMP algorithm is higher than that of other three algorithms. In terms of running time, the running time of SCMP algorithm does not change with the change of SNR, while the other three algorithms decrease with the increase of SNR. When the SNR is small (for example, SNR = 10 dB), the running time of the SCMP algorithm is similar to that of the OMP algorithm. When the SNR is large (SNR = 30 dB), the running time of SCMP algorithm is higher than that of OMP algorithm, similar to OCMP algorithm, but lower than CMP algorithm.

Through the above two experiments, it can be seen that compared with OMP algorithm with low operational complexity, SCMP algorithm only increases the operation time a little, but significantly improves the accuracy of the algorithm. Compared with CMP and OCMP algorithms with relatively high algorithm time complexity, SCMP has improved both in the running time and algorithm accuracy.

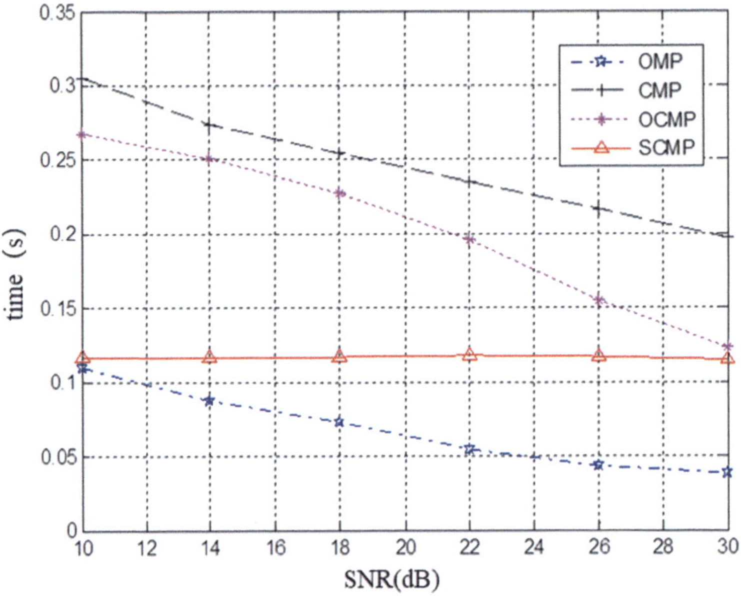

Fig. 11.11 Curves of running time changing with SNR when source signal sparsity is 0.8

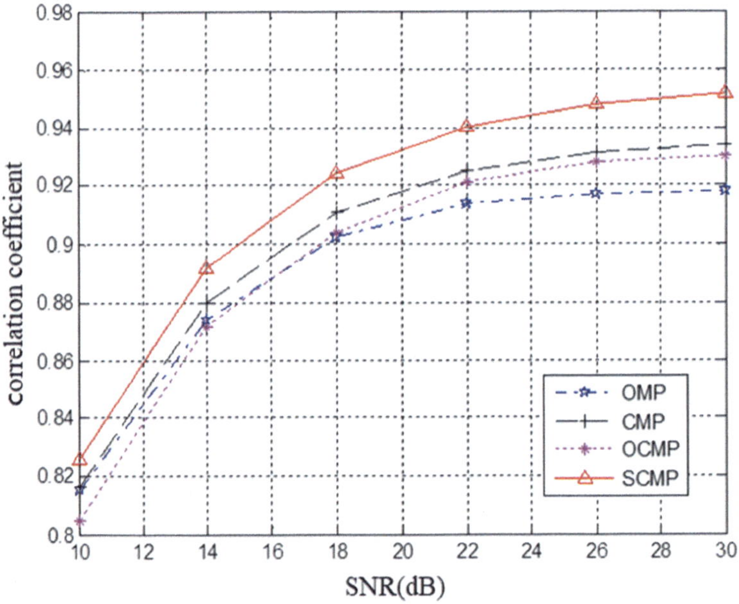

Fig. 11.12 Curves of correlation coefficient changing with SNR when source signal sparsity is 0.9

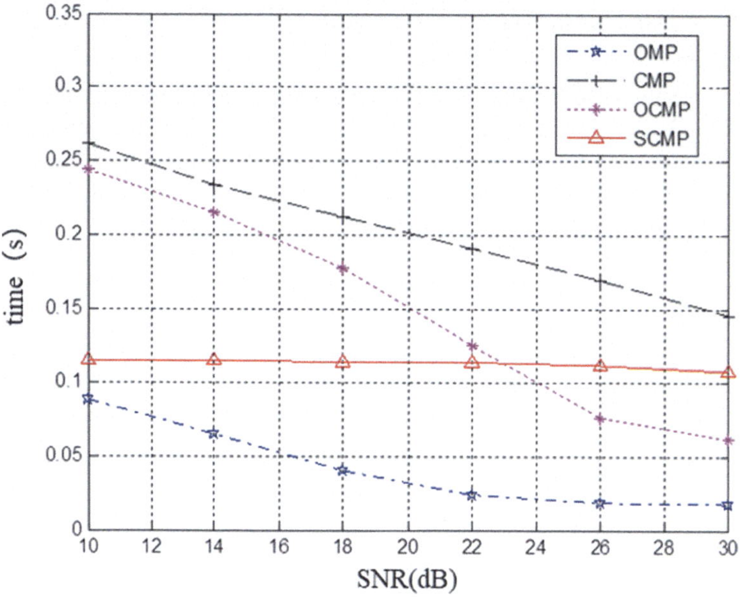

Fig. 11.13 Curves of running time changing with SNR when source signal sparsity is 0.9

11.4 Underdetermined Blind Separation Source Signal Recovery Based on ℓ_1 Norm

Because ℓ_1 norm is a good measure of signal sparsity, sparse signal reconstruction algorithms based on ℓ_1 norm have been widely paid attention. Compared with the greedy algorithm described in Sect. 11.3, the sparse signal reconstruction algorithm based on ℓ_1 norm shows better characteristics when the source signal is not sufficiently sparse, and is suitable for UBSS under the condition of the source signal is not sufficiently sparse. The existing sparse signal reconstruction algorithms based on ℓ_1 norm mainly include basis pursuit (BP) algorithm and gradient projection for sparce reconstruction (GPSR) algorithm. BP algorithm is usually solved by internal point method, which has high precision, but high time complexity, so it is difficult to be applied to practice. GPSR algorithm can obtain sparse estimation of source signal by solving optimization problems. Compared with BP algorithm, GPSR algorithm has a significant reduction in time complexity, but it is not satisfactory in recovery accuracy. To solve the problems in the above algorithms, the idea of complementary matching pursuit (CMP) can be integrated into the ℓ_1 norm algorithm and solved by iterative method. This method not only significantly reduces the time complexity of the algorithm, but also achieves a recovery accuracy similar to that of BP algorithm, showing a good recovery effect.

11.4.1 Basis Pursuit (BP) Algorithm

The basis pursuit (BP) algorithm is a very famous algorithm based on the minimum ℓ_1 norm. The advantages of BP algorithm are high precision and strong robustness, but the disadvantages are high complexity. The basic idea of BP algorithm is to transform the Eq. (11.31) into a linear programming problem through variable substitution, and find the optimal solution of the problem through linear programming.

$$\min ||\boldsymbol{s}||_1, \text{ s.t.} \boldsymbol{As} = \boldsymbol{x} \tag{11.31}$$

The following introduces the basic idea of BP algorithm. First, let $\boldsymbol{s} = \boldsymbol{u} - \boldsymbol{v}$, where $\boldsymbol{u}$, $\boldsymbol{v}$ are non-negative vectors, $\boldsymbol{u} = [u_i]_{N\times 1} \geq 0$, $\boldsymbol{v} = [v_i]_{N\times 1} \geq 0$, then Eq. (11.31) can be converted to the following linear optimization problem with non-negative constraints:

$$\begin{aligned} &\min \sum_{i=1}^{N} (u_i + v_i) \\ &\text{s.t.} \quad [\boldsymbol{A}, -\boldsymbol{A}]\begin{bmatrix} \boldsymbol{u} \\ \boldsymbol{v} \end{bmatrix} = \boldsymbol{x}, u_i \geq 0, v_i \geq 0, i = 1, 2, \ldots \end{aligned} \tag{11.32}$$

If $\boldsymbol{z} = \begin{bmatrix} \boldsymbol{u} \\ \boldsymbol{v} \end{bmatrix}$, $\boldsymbol{B} = [\boldsymbol{A}, -\boldsymbol{A}]$, $\boldsymbol{c} = [1, 1, \ldots, 1]^T \in R^{2N\times 1}$, then formula (11.32) can be written as:

$$\begin{aligned} &\min \ \boldsymbol{c}^{\mathrm{T}}\boldsymbol{z} \\ &\text{s.t.} \quad \boldsymbol{Bz} = \boldsymbol{x} \end{aligned} \tag{11.33}$$

For the solution of the Eq. (11.33), firstly, a solution $\boldsymbol{z}^{(0)}$ satisfying the equation is found according to the equation $\boldsymbol{Bz} = \boldsymbol{x}$, and then the optimal solution is gradually approached by searching along a direction that reduces the value of the function $f = \boldsymbol{c}^{\mathrm{T}}\boldsymbol{z}$. The negative gradient direction of the function $f = \boldsymbol{c}^{\mathrm{T}}\boldsymbol{z}$ is the direction that decreases the objective function, but if we directly use the negative gradient direction $-\boldsymbol{c}$ as the search direction, the new iteration point may not satisfy $\boldsymbol{Bz} = \boldsymbol{x}$. In order to make the newly searched iteration point $\boldsymbol{z}^{(k)}$ still satisfy the equation $\boldsymbol{Bz}^{(k)} = \boldsymbol{x}$, the projection $\boldsymbol{d}^{(k)}$ of $-\boldsymbol{c}$ on the orthogonal space of matrix $\boldsymbol{B}$ should be used as the search direction. The null space of matrix $\boldsymbol{B}$ is defined as:

$$\mathbb{N} = \{\boldsymbol{x} | \boldsymbol{Bx} = 0\} \tag{11.34}$$

Since the projection $\boldsymbol{d}^{(k)} \in \mathbb{N}$ is satisfied:

$$\boldsymbol{Bd}^{(k)} = \boldsymbol{0} \tag{11.35}$$

then

$$Bz^{(k+1)} = B(z^{(k)} + \alpha_k d^{(k)}) = Bz^{(k)} = x \tag{11.36}$$

The condition $z^{(k+1)} > 0$ can be guaranteed by controlling the search step size.

BP algorithm can recover the original sparse signal through the global optimal solution. Compared with greedy algorithm, BP algorithm has the advantages of higher precision and stronger robustness. However, BP algorithm involves multiple matrix multiplication and matrix inversion in the operation process, so the time complexity of the algorithm is high, and it is difficult to be applied to practice.

11.4.2 Gradient Projection for Sparse Recovery (GPSR) Algorithm

GPSR algorithm transforms the optimization problem of Eq. (11.31) into an unconstrained convex optimization problem. Compared with BP algorithm, the time complexity of GPSR algorithm is significantly reduced, and the accuracy of GPSR algorithm is also improved when the source signal is not sufficiently sparse. However, the disadvantage is that the accuracy of the algorithm is significantly lower than that of the previous greedy algorithm and BP algorithm when the source signal is very sparse.

GPSR algorithm is divided into GPSRBasic algorithm and GPSRBB algorithm. The main difference between the two algorithms is that the two algorithms choose different ways to reduce the value of the objective function. GPSRBasic chooses the direction of negative gradient as the direction in which the value of the objective function decreases. GPSRBB calculates the direction in which the value of the objective function decreases according to a modified Newton method.

GPSRBasic reformulates the constrained optimization problem (11.31) into the unconstrained convex optimization problem shown in Eq. (11.37) to solve sparse signals.

$$\min \frac{1}{2}||x - As|| + \tau||s||_1 \tag{11.37}$$

Equation (11.37) still cannot be solved directly. In order to solve it more easily, Eq. (11.37) should be further transformed into a quadratic programming problem. Here we need to express s as the subtraction of two non-negative vectors, as shown in Eq. (11.38):

$$s = u - v, \quad u \geq 0, v \geq 0 \tag{11.38}$$

where

$$u_i = (s_i)_+, \quad v_i = (-s_i)_+, \quad i = 1, 2, \ldots, N \tag{11.39}$$

$(x)_+$ means taking the larger value between x and 0, that is, $(x)_+ = \max(0, x)$. Therefore, we have $||\boldsymbol{s}||_1 = \boldsymbol{I}_N^{\mathrm{T}}\boldsymbol{u} + \boldsymbol{I}_N^{\mathrm{T}}\boldsymbol{v}$, where $\boldsymbol{I}_N = [1, 1, \ldots, 1]^{\mathrm{T}}$ is a vector containing N ones. Therefore, Eq. (11.37) can be rewritten as the following bounded quadratic programming problem:

$$\min \frac{1}{2}||\boldsymbol{x} - \boldsymbol{A}(\boldsymbol{u} - \boldsymbol{v})||_2^2 + \tau\boldsymbol{I}_N^{\mathrm{T}}\boldsymbol{u} + \tau\boldsymbol{I}_N^{\mathrm{T}}\boldsymbol{v}, \quad \text{s.t.}\, \boldsymbol{u} \geq 0, \boldsymbol{v} \geq 0 \tag{11.40}$$

Equation (11.40) can be rewritten as a more standard bounded quadratic programming problem:

$$\min_{z} F(z) = \min_{z} \boldsymbol{c}^{\mathrm{T}} z + \frac{1}{2} z^{\mathrm{T}} Bz, \quad \text{s.t.} \quad z \geq 0 \tag{11.41}$$

where:

$$z = \begin{bmatrix} \boldsymbol{u} \\ \boldsymbol{v} \end{bmatrix} \tag{11.42}$$

$$\boldsymbol{b} = \boldsymbol{A}^{\mathrm{T}}\boldsymbol{x} \tag{11.43}$$

$$\boldsymbol{c} = \tau\boldsymbol{I}_{2N} + \begin{bmatrix} -\boldsymbol{b} \\ \boldsymbol{b} \end{bmatrix} \tag{11.44}$$

$$\boldsymbol{B} = \begin{bmatrix} \boldsymbol{A}^{\mathrm{T}}\boldsymbol{A} & -\boldsymbol{A}^{\mathrm{T}}\boldsymbol{A} \\ -\boldsymbol{A}^{\mathrm{T}}\boldsymbol{A} & \boldsymbol{A}^{\mathrm{T}}\boldsymbol{A} \end{bmatrix} \tag{11.45}$$

GPSRBasic uses a gradual approximation method to find the optimal value of the formula (11.41). First, select a scalar parameter $\alpha_k > 0$, let

$$\boldsymbol{w}^{(k)} = (z^{(k)} - \alpha_k \nabla F(z^{(k)}))_+ \tag{11.46}$$

where

$$\nabla F(z^{(k)}) = \boldsymbol{c} + \boldsymbol{B}z^{(k)} \tag{11.47}$$

Then, select the second parameter $\lambda_k \in (0, 1]$ and let:

$$z^{(k+1)} = z^{(k)} + \lambda_k(\boldsymbol{w}^{(k)} - z^{(k)}) \tag{11.48}$$

During the approximation, GPSRBasic searches in the direction of negative gradient $-\nabla F(z^{(k)})$ during each iteration. In order to ensure that the feasible solution is always in the non-negative quadrant, it is necessary to project $-\nabla F(z^{(k)})$ in the non-negative quadrant. Let the projection vector be $\boldsymbol{g}^k$ and the i-th component g_i^k of $\boldsymbol{g}^k$ be

$$g_i^k = \begin{cases} (\nabla F(z^k))_i & \text{if} z_i^k > 0 \text{ or } (\nabla F(z^k))_i < 0 \\ 0 & \text{otherwise} \end{cases} \tag{11.49}$$

GPSRBasic selects the initial α_0 by solving the optimal step size:

$$\alpha_0 = \arg\min_{\alpha} F(z^k - \alpha \boldsymbol{g}^k) \tag{11.50}$$

Through the above formula, it can be accurately calculated as follows.

$$\alpha_0 = \frac{(\boldsymbol{g}^k)^{\mathrm{T}}\boldsymbol{g}^k}{(\boldsymbol{g}^k)^{\mathrm{T}}\boldsymbol{B}\boldsymbol{g}^k} \tag{11.51}$$

In order to avoid the value of α_0 being too large or too small, the GPSRBasic algorithm sets the value of α_0 within a range of $[\alpha_{\min}, \alpha_{\max}]$, here $0 < \alpha_{\min} < \alpha_{\max}$.

The steps of GPSRBasic algorithm are as follows:

Input: Observed signal vector $\boldsymbol{x}$, mixing matrix $\boldsymbol{A}$;
Output: Estimated source signal $\boldsymbol{s}$;
Step 1: Initialize the estimation of the source signal $\boldsymbol{s}_0 = \boldsymbol{A}^{\dagger}\boldsymbol{x}$; Select parameters $\beta \in (0, 1)$, $\mu \in (0, 1/2)$ and the number of iterations $k = 1$;
Step 2: Get z according to Eqs. (11.39) and (11.42);
Step 3: According to Eqs. (11.43), (11.44) and (11.45), get $\boldsymbol{B}$ and $\boldsymbol{c}$;
Step 4: Calculate $\nabla F(z^{(k)}) = \boldsymbol{c} + \boldsymbol{B}z^{(k)}$ and find $\boldsymbol{g}^k$ according to Eq. (11.49);
Step 5: Calculate $\alpha^* = \frac{(\boldsymbol{g}^k)^{\mathrm{T}}\boldsymbol{g}^k}{(\boldsymbol{g}^k)^{\mathrm{T}}\boldsymbol{B}\boldsymbol{g}^k}$, and find $\alpha_0 = \mathrm{mid}(\alpha_{\min}, \alpha^*, \alpha_{\max})$, where mid() means to find the middle value;
Step 6: Select the appropriate α_k so that α_k is the first number that satisfy $F((z^{(k)} - \alpha_k \nabla F(z^{(k)}))_+) \le F(z^{(k)}) - \mu \nabla F(z^{(k)})^{\mathrm{T}}[z^{(k)} - (z^{(k)} - \alpha_k \nabla F(z^{(k)}))_+]$ in sequence $\alpha_0, \beta\alpha_0, \beta^2\alpha_0, \ldots$, then update $z^{(k+1)} = (z^{(k)} - \alpha_k \nabla F(z^{(k)}))_+$;
Step 7: If $\|z^{(k+1)} - z^{(k)}\|_2^2 < \varepsilon$, then the iteration is stopped, and $z^{(k+1)}$ is the estimate of z, go to Step 8; Otherwise $k \leftarrow k + 1$, return to Step 3.
Step 8: Calculate the final estimation of the source signal $\boldsymbol{s}$ according to Eqs. (11.38) and (11.42).

GPSRBasic algorithm has the advantages of strong robustness, higher accuracy than BP algorithm in the case of large noise, i.e. low SNR, and is suitable for sparse signal reconstruction under the condition that the source signal is not sufficiently sparse. However, the disadvantages are also obvious, that is, when the noise intensity is small, that is, the SNR is large, the recovery accuracy is significantly lower than that of BP algorithm. In this regard, GPSRBB algorithm combines the idea of BB approximation algorithm, reducing the complexity of the algorithm to a certain extent and improving the recovery accuracy of the algorithm.

The steps of GPSRBB algorithm are as follows:

Input: Observed signal vector $\boldsymbol{x}$, mixing matrix $\boldsymbol{A}$;
Output: Estimated source signal $\boldsymbol{s}$;
Step 1: Initialize the estimate of the source signal $s_0 = \boldsymbol{A}^{\dagger}\boldsymbol{x}$, select parameter $\alpha_{\min}, \alpha_{\max}, \varepsilon$, and the number of iterations $k = 1$;
Step 2: Get z according to Eqs. (11.39) and (11.42);
Step 3: According to Eqs. (11.43), (11.44) and (11.45), get $\boldsymbol{B}$ and $\boldsymbol{c}$;
Step 4: Calculate $\nabla F(z^{(k)}) = \boldsymbol{c} + \boldsymbol{B}z^{(k)}$ and get $\boldsymbol{g}^k$ according to Eq. (11.49);
Step 5: Calculate $\alpha^* = \frac{(\boldsymbol{g}^k)^{\mathrm{T}}\boldsymbol{g}^k}{(\boldsymbol{g}^k)^{\mathrm{T}}\boldsymbol{B}\boldsymbol{g}^k}$, and find $\alpha_1 = \text{mid}(\alpha_{\min}, \alpha^*, \alpha_{\max})$, where mid$(\cdot)$ means to find the middle value;
Step 6: Calculate $\delta_k = (z^{(k)} - \alpha_k \nabla F(z^{(k)}))_+ - z^{(k)}$;
Step 7: Find the λ_k that minimizes $F(z^{(k)} + \lambda_k \boldsymbol{\delta}_k)$ over $\lambda_k \in [0, 1]$, that is, $\lambda_k = -\frac{\boldsymbol{\delta}_k^{\mathrm{T}}\boldsymbol{B}z^{(k)} + \boldsymbol{c}^{\mathrm{T}}\boldsymbol{\delta}_k}{\boldsymbol{\delta}_k^{\mathrm{T}}\boldsymbol{B}\boldsymbol{\delta}_k}$;
Step 8: Update:$z^{(k+1)} = z^{(k)} + \lambda_k \boldsymbol{\delta}_k$;
Step 9: Update α: Calculate $\gamma_k = (\boldsymbol{\delta}_k)^{\mathrm{T}}\boldsymbol{B}\boldsymbol{\delta}_k$. If $\gamma_k = 0$, then make $\alpha_{k+1} = \alpha_{\max}$; otherwise, set $\alpha_{k+1} = \text{mid}(\alpha_{\min}, \frac{\|\boldsymbol{\delta}_k\|_2^2}{\gamma_k}, \alpha_{\max})$;
Step 10: If $\|z^{(k+1)} - z^{(k)}\|_2^2 < \varepsilon$, stop iteration, and $z^{(k+1)}$ is the estimation of z, go to Step 11; otherwise $k \leftarrow k + 1$, return to Step 3.
Step 11: Calculate the final estimation of the source signal $\boldsymbol{s}$ according to Eqs. (11.38) and (11.42).

Compared with GPSRBasic algorithm, GPSRBB improves the performance of the algorithm to a certain extent, but compared with BP algorithm, GPSRBB algorithm still has unsatisfactory recovery accuracy in the case of high SNR. Therefore, we propose a complementary matching pursuit algorithm based on L1 norm.

11.4.3 Complementary Matching Pursuit Based on L1 Norm (L1CMP) Algorithm

1. **Algorithm Principle**

In underdetermined blind source separation, since the mixing matrix $\boldsymbol{A} \in \boldsymbol{R}^{M \times N}$ is a full row rank matrix, the matrix $\boldsymbol{A}^{\mathrm{T}}$ spans an M-dimensional subspace in N-dimensional space, which we call S_A. Corresponding to S_A there is an orthogonal subspace of $N - M$ dimension, denoted here as $S_A^{\perp}$. The matrix corresponding to $S_A^{\perp}$ can be obtained by QR decomposition of $\boldsymbol{A}^{\mathrm{T}}$:

$$\boldsymbol{A}^{\mathrm{T}} = \boldsymbol{Q}\boldsymbol{R} = [\boldsymbol{Q}_1^{N \times M} \quad \boldsymbol{Q}_2^{N \times (N-M)}] \begin{bmatrix} \boldsymbol{R}_1^{N \times M} \\ 0^{(N-M) \times N} \end{bmatrix} \tag{11.52}$$

where Q_1 spans the space S_A, Q_2 spans the space $S_A^{\perp}$, and the column vectors of Q_1 are orthogonal to the column vectors of Q_2. Thus, the source signal s can be expressed in the form of a two-part sum: $s = s_1 + s_2$, where s_1 falls on space S_A and s_2 falls on space $S_A^{\perp}$, so there are $As_2 = 0$, $As = As_1 = x$. s_1 can be get directly with the minimized L_2 norm:

$$s_1 = A^{\mathrm{T}}(A^{\mathrm{T}}A)^{-1}x \tag{11.53}$$

Although s_2 cannot be found directly, since s_2 falls on the plane $S_A^{\perp}$, s_2 can be expressed as:

$$s_2 = -Q_2 z \tag{11.54}$$

where z is the coefficient vector. Therefore, the source signal s can be expressed as:

$$s = s_1 - Q_2 z \tag{11.55}$$

If the L_1 norm is used to measure the sparsity of the signal, to find the sparsest solution of the source signal s is equivalent to solving the following unconstrained optimization problem:

$$\min_z = |s_1 - Q_2 z|_1 \tag{11.56}$$

Let's set

$$F(z) = ||s_1 - Q_2 z||_1 \tag{11.57}$$

The derivative of $F(z)$ with respect to z is

$$\frac{\partial F(z)}{\partial z} = -Q_2^{\mathrm{T}} sign(s_1 - Q_2 z) \tag{11.58}$$

Letting the derivative equal to 0, we have:

$$Q_2^{\mathrm{T}} sign(s_1 - Q_2 z) = 0 \tag{11.59}$$

This is a nonlinear equation set that cannot be solved directly. Here, we rewrite the formula (11.59) by an iterative approach as:

$$\frac{\partial F(z)}{\partial z} = -Q_2^{\mathrm{T}} sign(s_1 - Q_2 z) = -Q_2^{\mathrm{T}} U(z)(s_1 - Q_2 z) \tag{11.60}$$

where

$$\boldsymbol{U}(z) = \begin{bmatrix} \frac{1}{s_1[1]-\boldsymbol{Q}_2[1]z} & 0 & \cdots & 0 \\ 0 & \frac{1}{s_1[2]-\boldsymbol{Q}_2[2]z} & \cdots & 0 \\ \vdots & \vdots & \ddots & \vdots \\ 0 & 0 & \cdots & \frac{1}{s_1[2]-\boldsymbol{Q}_2[2]z} \end{bmatrix} \tag{11.61}$$

$\boldsymbol{Q}_2[i]$ represents the i-th row of the matrix $\boldsymbol{Q}_2$, and $s_1[i]$ represents the i-th element of the signal $\boldsymbol{s}_1$. By setting $\frac{\partial F(z)}{\partial z} = 0$, we get:

$$z = (\boldsymbol{Q}_2^{\mathrm{T}}\boldsymbol{U}(z)\boldsymbol{Q}_2)^{-1}\boldsymbol{Q}_2^{\mathrm{T}}\boldsymbol{U}(z)\boldsymbol{s}_1 \tag{11.62}$$

By replacing z to the right of the equals sign with $z^{(k)}$, and z to the left of the equals sign with $z^{(k+1)}$, we can get the recursion formula from $z^{(k)}$ to $z^{(k+1)}$:

$$z^{(k+1)} = (\boldsymbol{Q}_2^{\mathrm{T}}\boldsymbol{U}(z)\boldsymbol{Q}_2)^{-1}\boldsymbol{Q}_2^{\mathrm{T}}\boldsymbol{U}(z^{(k)})\boldsymbol{s}_1 \tag{11.63}$$

When the square $||z^{(k+1)} - z^{(k)}||_2^2$ of the 2-norm of the difference of the coefficient vectors obtained by two adjacent iterations is less than the threshold ε, we consider that the iterative process has reached the optimal value, then stop the iteration, and obtain the estimated value of the source signal as follows.

$$\hat{\boldsymbol{s}} = \boldsymbol{s}_1 - \boldsymbol{Q}_2 z \tag{11.64}$$

In summary, the algorithm steps of CMP algorithm based on L_1 norm (L_1CMP) are summarized as follows.

Step 1: Initialization: iteration number $k = 0$, $z^{(k)} = 0^{(N-M)\times 1}$, $\boldsymbol{s}_1 = \boldsymbol{A}^{\mathrm{T}}(\boldsymbol{A}^{\mathrm{T}}\boldsymbol{A})^{-1}\boldsymbol{x}$, QR decomposition of $\boldsymbol{A}^T$:

$$\boldsymbol{A}^{\mathrm{T}} = \boldsymbol{QR} = [\boldsymbol{Q}_1^{N\times M} \quad \boldsymbol{Q}_2^{N\times(N-M)}]\begin{bmatrix} \boldsymbol{R}_1^{N\times M} \\ \boldsymbol{0}^{N\times(N-M} \end{bmatrix}$$

Step 2: Update: $z^{(k)} = (\boldsymbol{Q}_2^{\mathrm{T}}\boldsymbol{U}(z^{k-1})\boldsymbol{Q}_2)^{-1}\boldsymbol{Q}_2^{\mathrm{T}}\boldsymbol{U}(z^{(k-1)})\boldsymbol{s}_1$;
Step 3: If $||z^{(k+1)} - z^{(k)}||_2^2 < \varepsilon$, go to Step 4, otherwise, go back to step 2;
Step 4: Get the estimated signal $\hat{\boldsymbol{s}} = \boldsymbol{s}_1 - \boldsymbol{Q}_2 z^{(k)}$.

Compared with traditional algorithms based on minimal L_1 norm (such as BP algorithm and GPSR algorithm), the L_1CMP algorithm proposed in this section searches in $N - M$-dimensional space, while the traditional reconstruction algorithm based on minimal L_1 norm searches in $2N$-dimensional space. Therefore, the algorithm proposed in this section significantly reduces the time and space complexity of the algorithm. On the other hand, compared with greedy algorithms, L_1CMP algorithm obtains the sparse solution by solving the global optimal, and its objective function is the L_1 norm of the signal, that is, the sparsity of the signal. When the source signal is sparse, its optimal solution can approach the source signal more accurately.

Moreover, because greedy algorithms approach global optimal solutions through local optimal solutions, and need to find matching atoms in the iterative process, it is found in experiments that when the source signal is not sufficiently sparse, greedy algorithms often cannot accurately select matching atoms. Therefore, the accuracy of the L_1CMP algorithm is higher than that of the greedy algorithm, especially when the source signal is not sufficiently sparse.

2. **Algorithm Convergence Proof**

It is proved that the L_1CMP algorithm converges when the initial value $z^{(0)} = \mathbf{0}$, that is, when the initial iteration point is selected as the minimum 2-norm point. Since a monotonic sequence with both upper and lower bounds must converge, we prove the convergence of the sequence from two aspects of monotony and boundedness. The proof is given here only for $\boldsymbol{Q}_2 \in R^N$.

Because $\boldsymbol{Q}_2^T \boldsymbol{s}_1 = 0$, that is:

$$\sum_{i=1}^{N} s_1[i]\boldsymbol{Q}_2[i] = 0 \tag{11.65}$$

So

$$\frac{s_1[1]}{\boldsymbol{Q}_2[1]}, \frac{s_1[2]}{\boldsymbol{Q}_2[2]}, \ldots, \frac{s_1[N]}{\boldsymbol{Q}_2[N]} (\boldsymbol{Q}_2[i] \neq 0, i = 1, 2, \ldots, N) \tag{11.66}$$

Can't be positive or negative simultaneous. let

$$\Gamma_+ = \left\{ \frac{s_1[i]}{\boldsymbol{Q}_2[i]} > 0, i = 1, 2, \ldots, N \right\} \tag{11.67}$$

$$\Gamma_- = \left\{ \frac{s_1[i]}{\boldsymbol{Q}_2[i]} < 0, i = 1, 2, \ldots, N \right\} \tag{11.68}$$

The minimum value in Γ_+ is $\frac{s_1[p]}{Q_2[p]}$, and the maximum value in Γ_- is $\frac{s_1[n]}{Q_2[n]}$. When $z_k \in \left(\frac{s_1[n]}{Q_2[n]}, \frac{s_1[p]}{Q_2[p]}\right)$, there are:

$$s_1[i] - \boldsymbol{Q}_2[i]z_k \neq 0, i = 1, 2, 3 \tag{11.69}$$

That is, when $\frac{s_1[n]}{Q_2[n]} < z_k < \frac{s_1[p]}{Q_2[p]}$, $s_1[i] - \boldsymbol{Q}_2[i]z_k$ is always positive or always negative. So, when $z_k \in \left(\frac{s_1[n]}{Q_2[n]}, \frac{s_1[p]}{Q_2[p]}\right)$, $\boldsymbol{Q}_2^{\mathrm{T}} sign(s_1 - \boldsymbol{Q}_2 z_k)$ is a constant value because of

$$z_{k+1} = (\boldsymbol{Q}_2^{\mathrm{T}}\boldsymbol{U}(z_k)\boldsymbol{Q}_2)^{-1}\boldsymbol{Q}_2^{\mathrm{T}}\boldsymbol{U}(z_k)\boldsymbol{s}_1$$
$$= \frac{\boldsymbol{Q}_2^{\mathrm{T}} sign(s_1 - \boldsymbol{Q}_2 z_k)}{\frac{(\boldsymbol{Q}_2[1])^2}{|s_1[1]-\boldsymbol{Q}_2[1]z_k|} + \frac{(\boldsymbol{Q}_2[2])^2}{|s_1[2]-\boldsymbol{Q}_2[1]z_k|} + \frac{(\boldsymbol{Q}_2[3])^2}{|s_1[3]-\boldsymbol{Q}_2[1]z_k|}} + z_k \tag{11.70}$$

Therefore, the sequence z_{k+1} increases or decreases monotonically in the interval $\left(\frac{s_1[n]}{Q_2[n]}, \frac{s_1[p]}{Q_2[p]}\right)$.

And because:

$$\lim_{z_k \to \left(\frac{s_1[p]}{Q_2[p]}\right)^-} z_{k+1} = \lim_{z_k \to \left(\frac{s_1[p]}{Q_2[p]}\right)^-} (\boldsymbol{Q}_2^{\mathrm{T}}\boldsymbol{U}(z_k)\boldsymbol{Q}_2)^{-1}\boldsymbol{Q}_2^{\mathrm{T}}\boldsymbol{U}(z_k)\boldsymbol{s}_1$$
$$= \lim_{z_k \to \left(\frac{s_1[p]}{Q_2[p]}\right)^-} \frac{\frac{Q_2[1]s_1[1]}{s_1[1]-Q_2[1]z_k} + \frac{Q_2[2]s_1[2]}{s_1[2]-Q_2[2]z_k} + \cdots + \frac{Q_2[N]s_1[N]}{s_1[N]-Q_2[N]z_k}}{\frac{Q_2^2[1]}{s_1[1]-Q_2[1]z_k} + \frac{Q_2^2[2]}{s_1[2]-Q_2[2]z_k} + \cdots + \frac{Q_2^2[N]}{s_1[N]-Q_2[N]z_k}}$$
$$= \lim_{z_k \to \left(\frac{s_1[p]}{Q_2[p]}\right)^-} \frac{\frac{Q_2[p]s_1[p]}{s_1[p]-Q_2[p]z_k}}{\frac{Q_2^2[p]}{s_1[p]-Q_2[p]z_k}} = \frac{s_1[p]}{Q_2[p]} \tag{11.71}$$

In a similar way,

$$\lim_{z_k \to \left(\frac{s_1[n]}{Q_2[n]}\right)^+} z_{k+1} = \lim_{z_k \to \left(\frac{s_1[n]}{Q_2[n]}\right)^+} (\boldsymbol{Q}_2^{\mathrm{T}}\boldsymbol{U}(z_k)\boldsymbol{Q}_2)^{-1}\boldsymbol{Q}_2^{\mathrm{T}}\boldsymbol{U}(z_k)\boldsymbol{s}_1 = \frac{s_1[n]}{\boldsymbol{Q}_2[n]} \tag{11.72}$$

From the above analysis, it can be seen that the sequence z_{k+1} is monotonic in the interval $\left(\frac{s_1[n]}{Q_2[n]}, \frac{s_1[p]}{Q_2[p]}\right)$ and has both upper and lower bounds. Considering z_{k+1} as a function of z_k, we can see from the above proof that for $z_k \in \left(\frac{s_1[n]}{Q_2[n]}, \frac{s_1[p]}{Q_2[p]}\right)$, there is $z_{k+1} \in \left(\frac{s_1[n]}{Q_2[n]}, \frac{s_1[p]}{Q_2[p]}\right)$, and z_k is a monotone sequence. Therefore, when $z_0 = 0$, the sequence $z_{k+1} = (\boldsymbol{Q}_2^T\boldsymbol{U}(z_k)\boldsymbol{Q}_2)^{-1}\boldsymbol{Q}_2^T\boldsymbol{U}(z_k)\boldsymbol{s}_1$ must be a convergent sequence, and the convergence of the algorithm is proved.

11.4.4 Algorithm Simulation and Analysis

In the experiment, we used 5 Gaussian sparse signals as source signals and Gaussian random matrix as the mixing matrix to obtain 3 observed signals. The number of sampling points is 1000, that is, $T = 1000$, $n = 5$, $m = 3$. L_1CMP algorithm, BP algorithm based on L_1 norm, GPSRBasic algorithm and GPSRBB algorithm are used to recover the source signal. In order to evaluate the accuracy of the algorithm, we use the correlation coefficient between the recovery signal and the source signal.

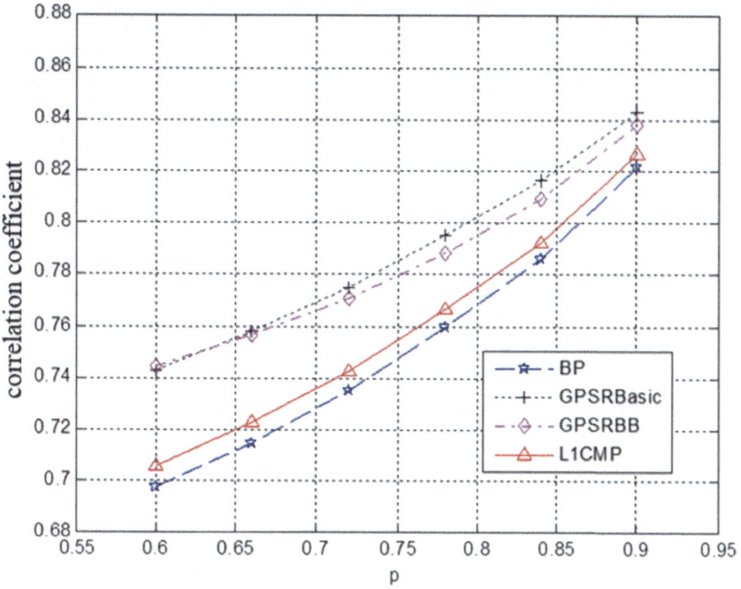

Fig. 11.14 Curves of correlation coefficients of different algorithms changing with sparsity when the SNR is 10 dB

1. **Influence of Sparsity on Algorithm Performance**

In this experiment, the curves of the correlation coefficient between the recovery signal and the source signal changing with the sparsity of each algorithm and the curves of the running time of each algorithm changing with the sparsity of the source signal are simulated respectively when the SNR is 10 dB, 20 dB and 30 dB.

Figures 11.14 and 11.15 respectively show the curves of correlation coefficient and running time of different algorithms changing with sparsity when the SNR is 10 dB.

Figures 11.16 and 11.17 respectively show the curves of correlation coefficient and running time of different algorithms changing with sparsity when the SNR is 20 dB.

Figures 11.18 and 11.19 respectively show the curves of correlation coefficient and running time of different algorithms changing with sparsity when the SNR is 30 dB.

As can be seen from the simulation results in Figs. 11.14, 11.15, 11.16, 11.17, 11.18 and 11.19, when the SNR is not very high (such as SNR = 10 dB), the accuracy of BP and L_1CMP algorithms is lower than that of GPSRBasic and GPSRBB algorithms. When the SNR is high (such as SNR = 20 dB or 30 dB), L_1CMP algorithm is significantly better than GPSRBasic and GPSRBB algorithm and slightly better than BP algorithm in terms of accuracy. In terms of time complexity, L_1CMP algorithm is lower than GPSRBasic algorithm. Compared with GPSRBB algorithm, L_1CMP

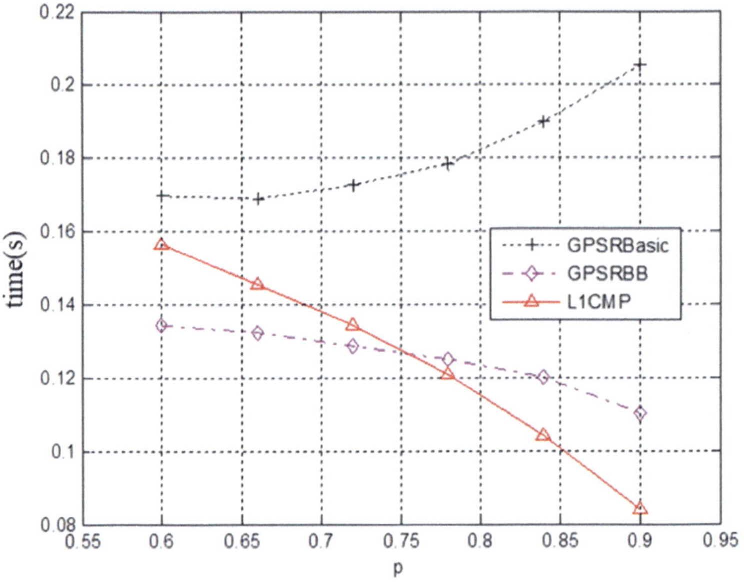

Fig. 11.15 Curves of the running time of different algorithms changing with the sparsity when the SNR is 10 dB

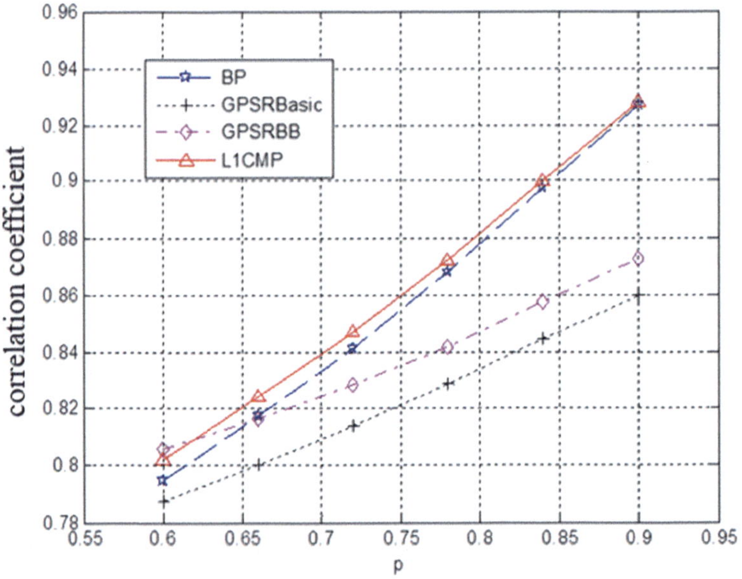

Fig. 11.16 Curves of correlation coefficients of different algorithms changing with sparsity when the SNR is 20 dB

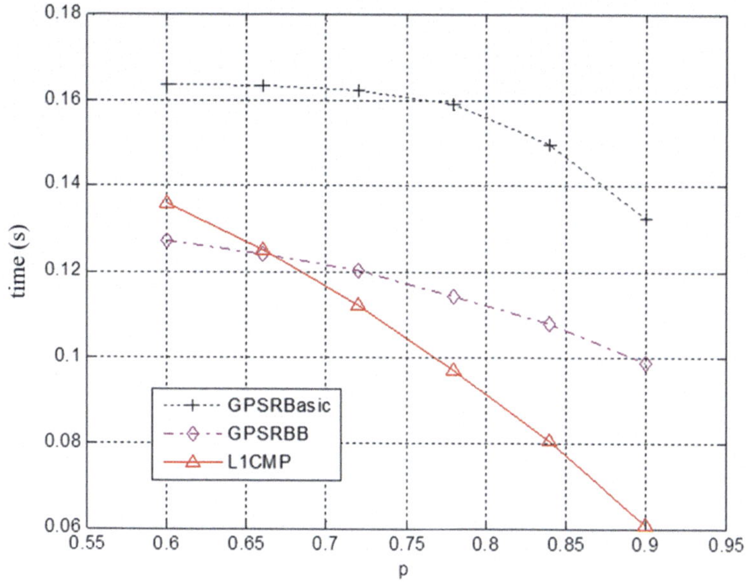

Fig. 11.17 Curves of the running time of different algorithms changing with the sparsity when the SNR is 20 dB

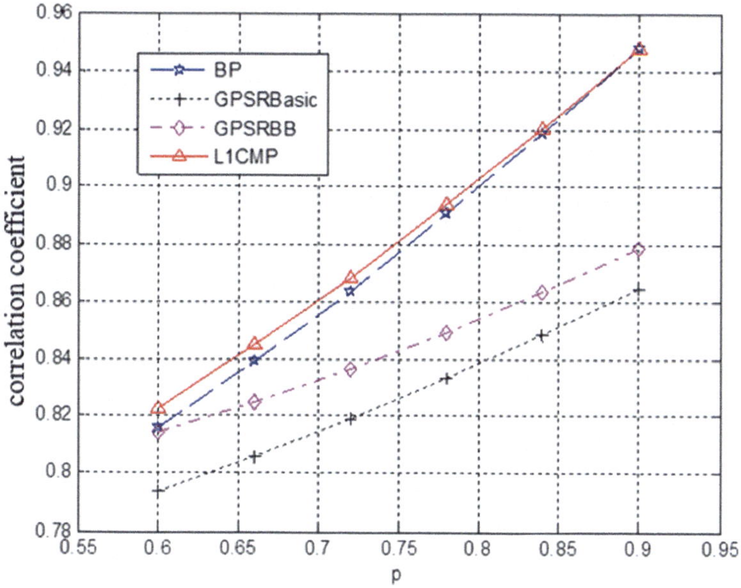

Fig. 11.18 Curves of correlation coefficients of different algorithms changing with sparsity when the SNR is 30 dB

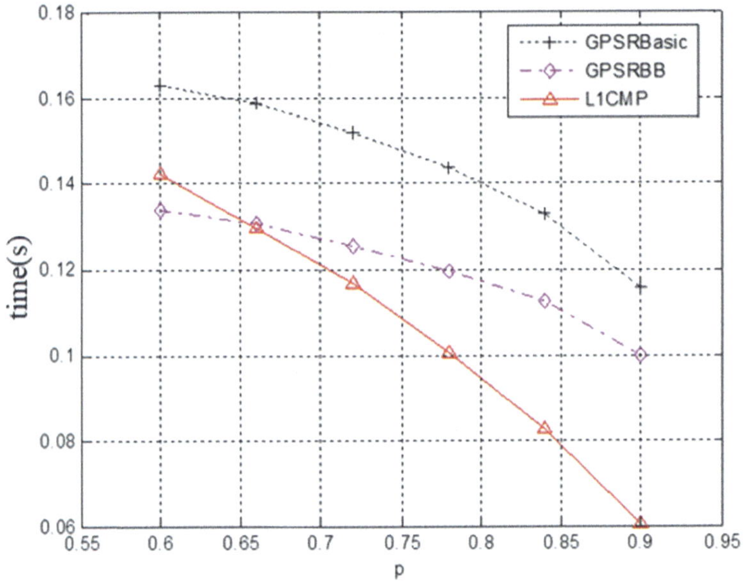

Fig. 11.19 Curves of the running time of different algorithms changing with the sparsity when the SNR is 30 dB

algorithm is slightly higher than GPSRBB algorithm in the case of small sparsity, and lower than GPSRBB algorithm in the case of large sparsity. In general, L_1CMP algorithm has better performance than other algorithms in recovery accuracy and complexity.

2. **Impact of Noise on Algorithm Performance**

In this experiment, when the source signal sparsity is 0.7, 0.8 and 0.9, the curves of the correlation coefficient between the recovery signal and the source signal changing with the SNR and the curves of the running time of different algorithms changing with the SNR are simulated respectively.

Figures 11.20 and 11.21 show the curves of correlation coefficient and running time of different algorithms changing with SNR when source signal sparsity is 0.7.

Figures 11.22 and 11.23 show the curves of correlation coefficient and running time of different algorithms changing with SNR when source signal sparsity is 0.8.

Figures 11.24 and 11.25 show the curves of correlation coefficient and running time of different algorithms changing with SNR when source signal sparsity is 0.9.

As can be seen from the above simulation results, in the case of a given source signal sparsity, when the SNR is small, for example, when the SNR is equal to 10 dB, the accuracy of the L_1CMP algorithm is lower than that of GPSRBasic and GPSRBB algorithms. When the SNR is high, the accuracy of L_1CMP algorithm is higher than that of GPSRBasic and GPSRBB algorithm. Compared with BP algorithm, the accuracy of L_1CMP algorithm is slightly higher than BP algorithm regardless of

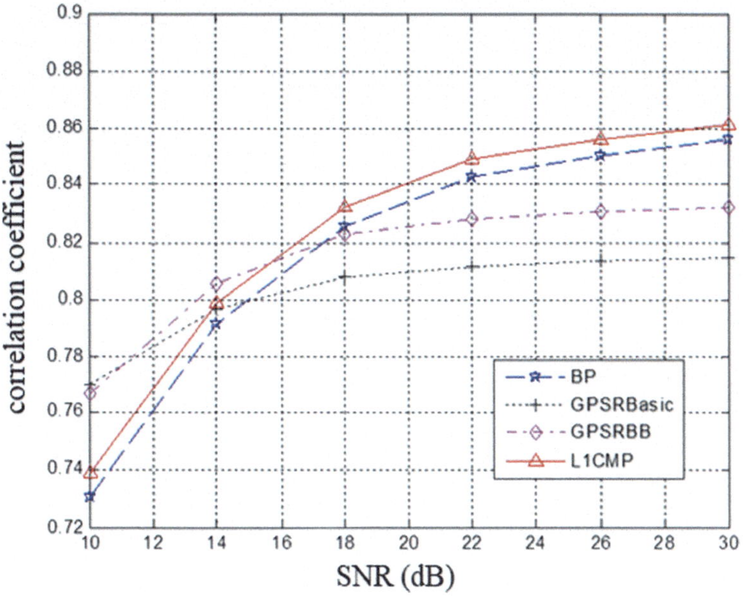

Fig. 11.20 Curves of the correlation coefficient of different algorithms changing with the SNR when the source signal sparsity is 0.7

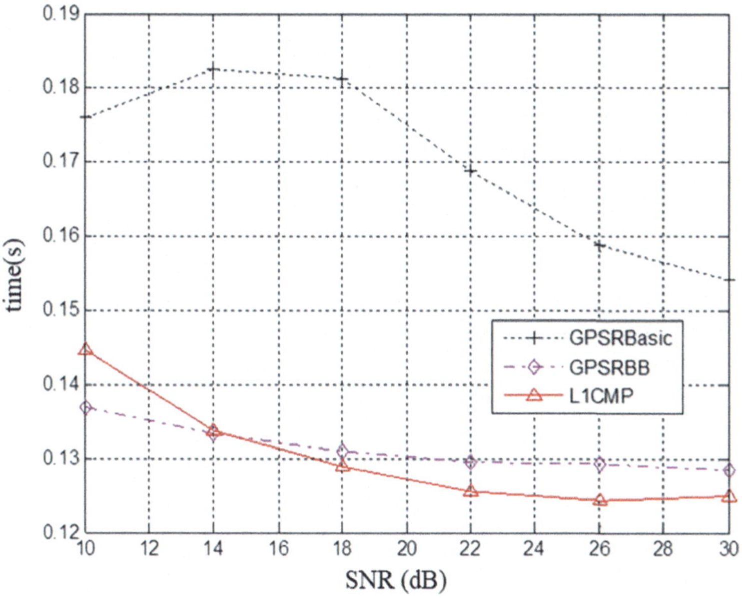

Fig. 11.21 Curves of the running time of different algorithms changing with the SNR when the source signal sparsity is 0.7

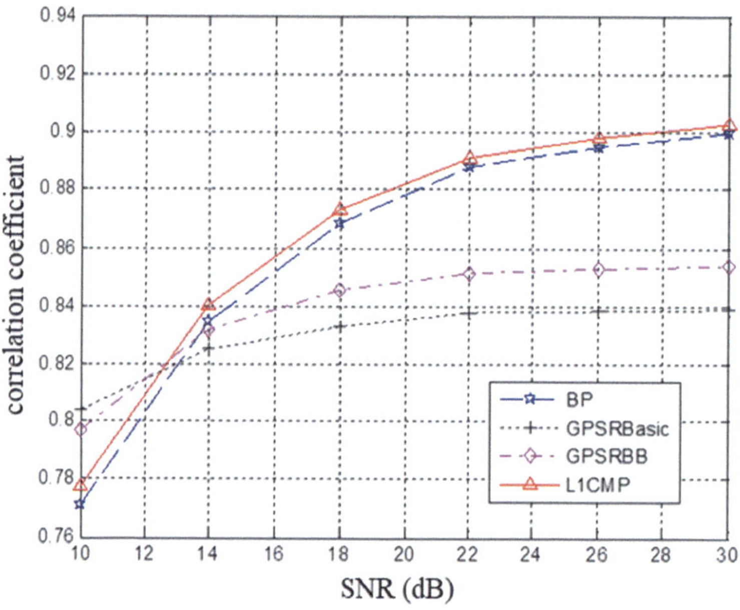

Fig. 11.22 Curves of the correlation coefficient of different algorithms changing with the SNR when the source signal sparsity is 0.8

SNR. Compared with GPSRBasic algorithm, time complexity of L_1CMP algorithm is significantly less than that of GPSRBasic algorithm. Compared with GPSRBB algorithm, when the SNR is small, the time complexity of L_1CMP algorithm is close to that of GPSRBB; when the SNR is large, the time complexity of L_1CMP algorithm is significantly smaller than that of GPSRBB algorithm.

11.5 Source Signal Recovery Algorithm Based on Smooth L0 (SL0) Norm and Its Improvement

The sparse signal reconstruction algorithm based on smoothing L_0 norm (SL_0) approximates the L_0 norm by using a smoothing function, and then solves the minimum L_0 norm solution by using the method dealing with convex optimization problems. Compared with greedy algorithm, the SL_0 algorithm has lower requirement for source signal sparsity.

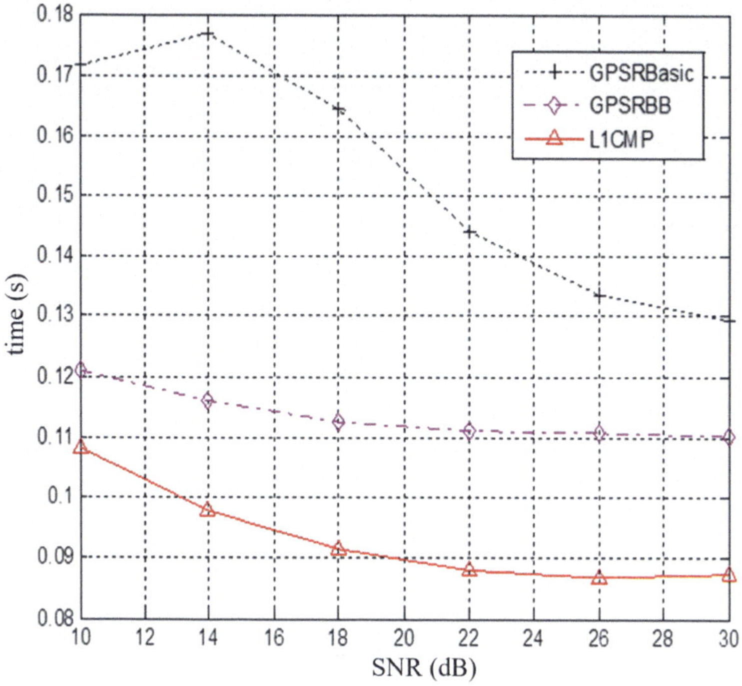

Fig. 11.23 Curves of the running time of different algorithms changing with the SNR when the source signal sparsity is 0.8

11.5.1 Smoothing L0 (SL0) Norm Algorithm

The smooth L_0 norm algorithm uses a smooth function $F(s) = N - \sum_{i=1}^{N} f(s_i)$ to approximate the L_0 norm, and then searches the optimal solution of the function along the direction of the negative gradient to obtain the sparse solution of the source signal. In SL0 algorithm, Gaussian function is used to approximate L_0 norm. That is

$$f_\sigma(s_i) = \exp\left(-\frac{s_i^2}{2\sigma^2}\right) \tag{11.73}$$

where σ is a relatively small parameter. Note that

$$\lim_{\sigma \to 0} f_\sigma(s_i) = \begin{cases} 1 & \text{if} \quad s_i = 0 \\ 0 & \text{if} \quad s_i \neq 0 \end{cases} \tag{11.74}$$

So

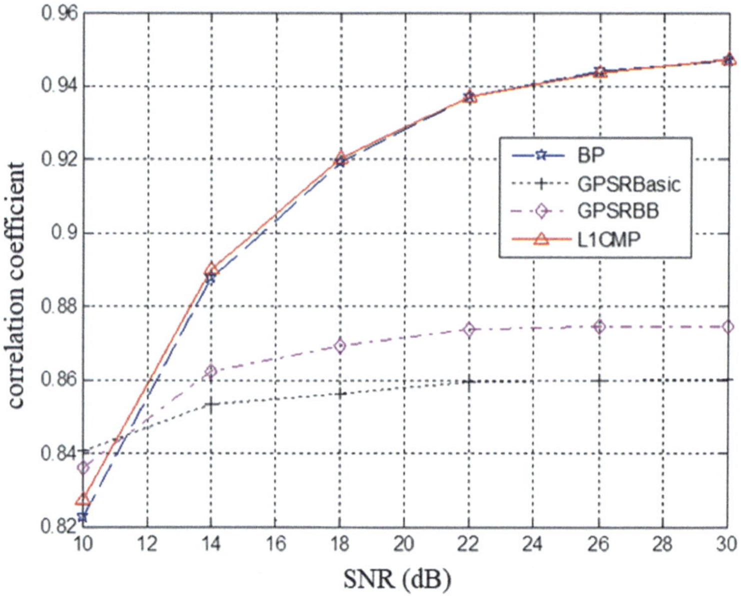

Fig. 11.24 Curves of the correlation coefficient of different algorithms changing with the SNR when the source signal sparsity is 0.9

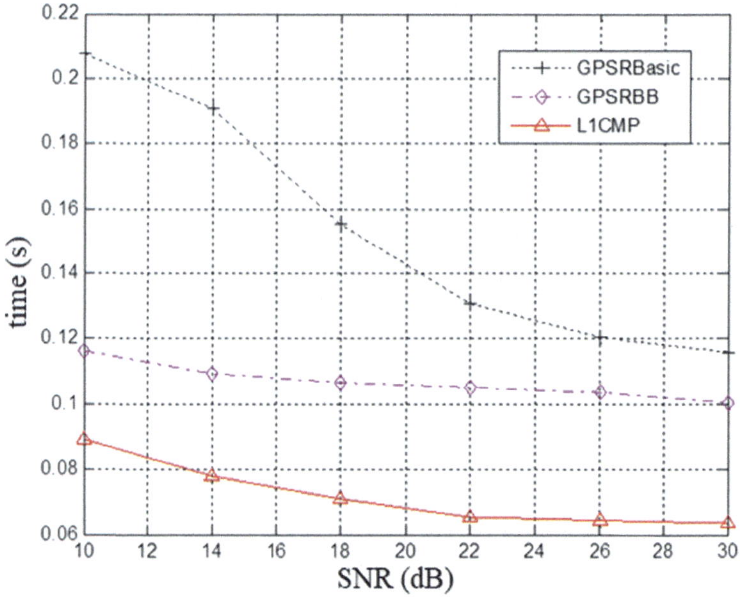

Fig. 11.25 Curves of the running time of different algorithms changing with the SNR when the source signal sparsity is 0.9

$$||\boldsymbol{s}||_0 \approx F_\sigma(\boldsymbol{s}) = N - \sum_{i=1}^{N} f_\sigma(s_i) \tag{11.75}$$

That is, the L_0 norm can be approximated by $F_\sigma(\boldsymbol{s})$. For parameter σ, if the value of σ is small, $F_\sigma(\boldsymbol{s})$ is a non-smooth function with more local minima, so it is difficult to solve the global minimum of $F_\sigma(\boldsymbol{s})$, and the optimal solution obtained by gradient descent is probably not the global optimal solution but the local optimal solution. When σ is large, $F_\sigma(\boldsymbol{s})$ is a relatively smooth function with few local minima, and it is relatively easy to solve the global minima of $F_\sigma(\boldsymbol{s})$, but the convergence speed will be slow when using gradient descent to solve the optimal solution of the function. The SL_0 algorithm first initializes a large σ, and then gradually decreases the value of σ during each iteration. The advantages of this approach are: on the one hand, in each iteration, the initialization value used by the algorithm is a solution that is relatively close to the optimal value, so it is not easy for the algorithm to converge to a local optimal value; On the other hand, with the decrease of σ value, the convergence speed of the function becomes faster and faster, which reduces the complexity of the algorithm to a certain extent.

The steps of the SL_0 algorithm are as follows.

Step 1: Let $\boldsymbol{s}_0$ be the least squares solution of $\boldsymbol{x} = \boldsymbol{A}\boldsymbol{s}$, that is, $\boldsymbol{s}_0 = \boldsymbol{A}^{\dagger}\boldsymbol{x}$;
Step 2: Select a descending sequence for parameter σ: $[\sigma_1, \sigma_2, \ldots, \sigma_J]$;
Step 3: For $j = 1, 2, \ldots, J$, repeat Steps 4–6.
Step 4: Let $\sigma = \sigma_j$;
Step 5: Calculate the maximum value of the function F_σ through L iterations using the fastest descent method:

(5.1) Initialize $\boldsymbol{s} = \boldsymbol{s}_{j-1}$;
(5.2) For $l = 1, 2, \ldots, L$, repeat (a)–(c);

(a) Let $\Delta \boldsymbol{s} = \left[s_1 e^{-\frac{s_1^2}{2\sigma^2}} \quad s_2 e^{-\frac{s_2^2}{2\sigma^2}} \quad \cdots \quad s_N e^{-\frac{s_N^2}{2\sigma^2}} \right]^T$;
(b) Update: $\boldsymbol{s} \leftarrow \boldsymbol{s} - \mu \Delta \boldsymbol{s}$ (where μ is a smaller constant number);
(c) Project $\boldsymbol{s}$ onto space $R(\boldsymbol{A}^{\mathrm{T}})$: $\boldsymbol{s} \leftarrow \boldsymbol{s} - \boldsymbol{A}^{\mathrm{T}}(\boldsymbol{A}\boldsymbol{A}^{\mathrm{T}})^{-1}(\boldsymbol{A}\boldsymbol{s} - \boldsymbol{x})$;

Step 6: Let $\boldsymbol{s}_j = \boldsymbol{s}$;
Step 7: Get the final recovery signal $\hat{\boldsymbol{s}} = \boldsymbol{s}$.

SL0 algorithm uses gradient descent method to solve the optimal value, which is easy to produce "sawtooth" phenomenon and convergence speed is slow. At the same time, the algorithm needs to set a step size μ in the iterative process. When the step size is not accurate, the recovery accuracy will be poor.

11.5.2 Robust SL0 (RSL0) Algorithm

In the case of noise, the relationship between the source signal and the observed signal is as follows.

$$\boldsymbol{x} = \boldsymbol{A}\boldsymbol{s} + \boldsymbol{v} \tag{11.76}$$

If the range of noise energy is known, that is, $||\boldsymbol{v}||_2 < \xi$, then $||\boldsymbol{x} - \boldsymbol{A}\boldsymbol{s}||_2 < \xi$ can be obtained. The RSL0 algorithm is an algorithm that uses the optimization method to solve the sparse solution of the signal under the assumption that the range of noise energy can be determined. RSL0 follows the fastest descent algorithm in SL0, and uses $||\boldsymbol{x} - \boldsymbol{A}\boldsymbol{s}||_2 < \xi$ instead of $\boldsymbol{x} = \boldsymbol{A}\boldsymbol{s}$ in SL0 to constrain the optimal solution. When the noise energy is known, RSL0 can improve the accuracy of the algorithm to some extent. The steps of RSL0 algorithm are as follows.

Step 1: Initialize $\boldsymbol{s}_0 = \boldsymbol{A}^{\dagger}\boldsymbol{x}$, and step size μ_0;
Step 2: Select a set of descending sequences for parameter σ: $[\sigma_1, \sigma_2, \ldots, \sigma_K]$, let $k = 1$;
Step 3: Let $\sigma = \sigma_k, \boldsymbol{s} = \boldsymbol{s}_k$;
Step 4: For $l = 1, 2, \ldots, L$, loop (4.1)–(4.3);

(4.1) Let $\Delta \boldsymbol{s} = \begin{bmatrix} s_1 e^{-\frac{s_1^2}{2\sigma^2}} & s_2 e^{-\frac{s_2^2}{2\sigma^2}} & \cdots & s_N e^{-\frac{s_N^2}{2\sigma^2}} \end{bmatrix}^{\mathrm{T}}$;
(4.2) $\boldsymbol{s} \leftarrow \boldsymbol{s} - \mu_0 \Delta \boldsymbol{s}$;
(4.3) If $||\boldsymbol{A}\boldsymbol{s} - \boldsymbol{x}||_2 > \xi$, project $\boldsymbol{s}$:

$$\boldsymbol{s} \leftarrow \boldsymbol{s} - \boldsymbol{A}^{\mathrm{T}}(\boldsymbol{A}\boldsymbol{A}^{\mathrm{T}})^{-1}(\boldsymbol{A}\boldsymbol{s} - \boldsymbol{x})$$

Step 5: If $k < K$, $k = k + 1$, return to step 3; else $\boldsymbol{s}_K = \boldsymbol{s}$, go to step 6.
Step 6: Get the recovery signal $\hat{\boldsymbol{s}} = \boldsymbol{s}_K$.

11.5.3 Hybrid Optimization SL0 (HOSL0) Algorithm

SL0 uses the fastest descent method to find the sparsest solution of the optimization problem. Although the gradient descent method is simple, it needs to choose a suitable iteration step size. When the step size is small, the obtained solution is close to the optimal solution, and the algorithm precision is high, but the disadvantage is that the algorithm convergence speed is slow. When the step size is large, the convergence speed of the algorithm will be faster, but the accuracy of the algorithm will be low.

Compared with the fastest descent method, Newton's method has the advantage of quadratic convergence. When the initial point selected in the iterative process is close to the optimal solution, Newton's method can approach the optimal solution at a faster speed. The difference between modified Newton method (NSL0) and SL0 algorithm is that NSL0 no longer uses the fastest descent method to approximate the optimal solution, but adopts a modified Newton method to approximate the optimal solution step by step. They are only different in the search direction, and the search direction of Newton's method is as follows:

$$\boldsymbol{d} = -\boldsymbol{G}^{-1}\boldsymbol{g} \tag{11.77}$$

where $\boldsymbol{g}$ is the gradient of the objective function $F_\sigma(\boldsymbol{s})$, that is $\boldsymbol{g} = \nabla F_\sigma(\boldsymbol{s})$, and $\boldsymbol{G}$ is the Hessen matrix of the objective function $F_\sigma(x)$, that is

$$\boldsymbol{G} = \nabla^2 F_\sigma(\boldsymbol{s}) \tag{11.78}$$

where

$$\begin{aligned}\boldsymbol{g} = \nabla F_\sigma(\boldsymbol{s}) &= \left[\frac{\partial f_\sigma(s_1)}{\partial s_1}, \frac{\partial f_\sigma(s_2)}{\partial s_2}, \ldots, \frac{\partial f_\sigma(s_N)}{\partial s_N}\right]^{\mathrm{T}} \\ &= \left[\frac{s_1}{\sigma^2} e^{-\frac{s_1^2}{2\sigma^2}}, \frac{s_2}{\sigma^2} e^{-\frac{s_2^2}{2\sigma^2}}, \ldots, \frac{s_N}{\sigma^2} e^{-\frac{s_N^2}{2\sigma^2}}\right]^{\mathrm{T}}\end{aligned} \tag{11.79}$$

$$\begin{aligned}\boldsymbol{G} = \nabla^2 F_\sigma(\boldsymbol{s}) &= \begin{bmatrix} \frac{\partial^2 f_\sigma(s_1)}{\partial s_1^2} & 0 & 0 \cdots 0 \\ 0 & \frac{\partial^2 f_\sigma(s_2)}{\partial s_2^2} & 0 \cdots 0 \\ \vdots & \vdots & \vdots \cdots 0 \\ 0 & 0 & 0 \cdots \frac{\partial^2 f_\sigma(s_N)}{\partial s_N^2} \end{bmatrix} \\ &= \begin{bmatrix} (\frac{\sigma^2 - s_1^2}{\sigma^4}) e^{-\frac{s_1^2}{2\sigma^2}} & 0 & 0 \cdots 0 \\ 0 & (\frac{\sigma^2 - s_2^2}{\sigma^4}) e^{-\frac{s_2^2}{2\sigma^2}} & 0 \cdots 0 \\ \vdots & \vdots & \vdots \ddots \vdots \\ 0 & 0 & 0 \cdots (\frac{\sigma^2 - s_N^2}{\sigma^4}) e^{-\frac{s_N^2}{2\sigma^2}} \end{bmatrix}\end{aligned} \tag{11.80}$$

To ensure that the Newtonian direction $\boldsymbol{d}$ is the descending direction of $F_\sigma(\boldsymbol{s})$, $\nabla^2 F_\sigma(\boldsymbol{s})$ is required to be a positive definite matrix, it can be simply proved as follows.

Let $\boldsymbol{D} = (\nabla F_\sigma(\boldsymbol{s}))^{\mathrm{T}}\boldsymbol{d} = -(\nabla F_\sigma(\boldsymbol{s}))^{\mathrm{T}}[\nabla^2 F_\sigma(\boldsymbol{s})]^{-1}\nabla F_\sigma(\boldsymbol{s})$. For the Newtonian direction $\boldsymbol{d}$ to be the descending direction of $F_\sigma(\boldsymbol{s})$, $\boldsymbol{D}$ must be less than 0, and thus $(\nabla F_\sigma(\boldsymbol{s}))^{\mathrm{T}}[\nabla^2 F_\sigma(\boldsymbol{s})]^{-1}\nabla F_\sigma(\boldsymbol{s}) > 0$, which means that $\nabla^2 F_\sigma(\boldsymbol{s})$ is a positive definite matrix.

However, the search direction $\boldsymbol{d}$ obtained by Eq. (11.77) may not be the direction of descending. Only when Hessen matrix is positive definite, $\boldsymbol{d}$ can be guaranteed to be the direction of descending. In order to solve this problem, the Hessen matrix of the objective function in Newton's method can be modified to ensure that it is positive definite, and then to ensure that the search direction $\boldsymbol{d}$ is the direction of descending, which is usually modified as follows.

$$G = \nabla^2 F_\sigma(s) + v \tag{11.81}$$

where:

$$v = \begin{bmatrix} \frac{2s_1^2}{\sigma^4} e^{-\frac{s_1^2}{2\sigma^2}} & 0 & \cdots & 0 \\ 0 & \frac{2s_2^2}{\sigma^4} e^{-\frac{s_2^2}{2\sigma^2}} & \cdots & 0 \\ \vdots & \vdots & \ddots & 0 \\ 0 & 0 & \cdots & \frac{2s_N^2}{\sigma^4} e^{-\frac{s_N^2}{2\sigma^2}} \end{bmatrix} \tag{11.82}$$

Then the diagonal element of the modified Hessen matrix is $(\frac{\sigma^2+s_n^2}{\sigma^4})e^{-\frac{s_n^2}{2\sigma^2}}$, which obviously meets the conditions of positive definite matrix. Put the modified Hessen matrix into the Newton direction formula, and the modified Newton direction can be obtained after simplification as follows.

$$d = -G^{-1}g = \left[\frac{-\sigma^2 s_1}{\sigma^2 + s_1^2}, \frac{-\sigma^2 s_2}{\sigma^2 + s_2^2}, \ldots, \frac{-\sigma^2 s_N}{\sigma^2 + s_N^2}\right]^{\mathrm{T}} \tag{11.83}$$

Therefore, the modified Newton method iteration formula is as follows:

$$s \leftarrow s + d \tag{11.84}$$

Newton method has a second-order convergence speed, but the choice of the initial point is very important. If the initial point is close to the minimum point, it can quickly converge to the optimal point; If the initial point is far from the minimum point, Newton method may not converge, or converge to a wrong optimal solution. In view of the shortcomings, it can be considered to find a good initial point for Newton method, and this initial point should be near the global optimal point, which can ensure that Newton method converges to the optimal point with high precision and fast speed. Consider combining these two algorithms to enhance strengths and avoid weaknesses, the proposed algorithm not only converges to the optimal solution, but also has a fast convergence speed, which is hybrid optimization SL0 algorithm (HOSL0). The basic idea is summarized as follows:

(1) First, the fastest descent method is used to find a good initial point in a large range for modified Newton algorithm, which makes use of the advantages of the fastest descent method, such as low requirements for the initial point, large initial step size and fast convergence.
(2) Take the point found in the previous step as the initial point, and use the modified Newton algorithm to improve the approximation speed and accuracy.

The steps of the HOSL0 algorithm are as follows.

Step 1: Set the initial value $s = A^{\mathrm{T}}(AA^{\mathrm{T}})^{-1}x$ and the maximum number of iterations K;

Step 2: For $l = 1{:}\,L$ (L is the number of iterations of the fastest descent method), execute (2.1)–(2.4);

(2.1) Select the decreasing sequence $[\sigma_1, \sigma_2, \ldots, \sigma_K]$ of parameter σ, and let $k = 1$;
(2.2) Select $\sigma = \sigma_k$ and calculate the negative gradient direction:

$$\boldsymbol{d}_1 = -\left[\frac{s_1}{\sigma^2} e^{-\frac{s_1^2}{2\sigma^2}}, \frac{s_2}{\sigma^2} e^{-\frac{s_2^2}{2\sigma^2}}, \ldots, \frac{s_N}{\sigma^2} e^{-\frac{s_N^2}{2\sigma^2}}\right]^{\mathrm{T}};$$

(2.3) Get $\boldsymbol{s} = \boldsymbol{s} + \mu\sigma^2 \boldsymbol{d}_1$ by using the fastest descent method;
(2.4) Get $\boldsymbol{s} = \boldsymbol{s} - \boldsymbol{A}^{\mathrm{T}}(\boldsymbol{A}\boldsymbol{A}^{\mathrm{T}})^{-1}(\boldsymbol{A}\boldsymbol{s} - \boldsymbol{x})$ by gradient projection;

Step 3: For $p = 1{:}\,P$ (P is the number of Newton method iterations), execute (3.1)–(3.3);

(3.1) Newton direction $\boldsymbol{d}$ is calculated by using Eq. (11.80):
(3.2) Get $\boldsymbol{s} := \boldsymbol{s} + \boldsymbol{d}$ by Newton method;
(3.3) Get $\boldsymbol{s} := \boldsymbol{s} - \boldsymbol{A}^{\mathrm{T}}(\boldsymbol{A}\boldsymbol{A}^{\mathrm{T}})^{-1}(\boldsymbol{A}\boldsymbol{s} - \boldsymbol{x})$ by gradient projection;

Step 4: If $k < K$, let $k = k + 1$, and go to step 2, otherwise stop iteration, and get the optimal solution.

11.5.4 Radial Basis Function Cascade Network for Sparse Signal Recovery (RASR) Algorithm

RASR algorithm uses a combination of minimum L0 norm and minimum L2 norm. That is, RASR algorithm obtains the optimal solution by jointly solving the following two optimization problems [53]:

$$\min_s E = \min_s \frac{1}{2} ||\boldsymbol{x}_j - \boldsymbol{A}_j \boldsymbol{s}||_2^2, j = 1, 2, \ldots, M \tag{11.85}$$

$$\min_s F_\sigma(\boldsymbol{s}) = \min_s \sigma^2 \left(N - \sum_{i=1}^{N} \exp\left(-\frac{s_i^2}{2\sigma^2}\right)\right) \tag{11.86}$$

For the optimization problem (11.85), the least square method of gradient descent can be used. However, the error function defined in the Eq. (11.85) is a scalar value, and the gradient of the error E spans the n-dimensional subspace $\frac{\partial E}{\partial s(k)} \in R^N$. And there are M gradient equations because each row of the mixing matrix $\boldsymbol{A}$ is considered separately.

$$\left[\frac{\partial E}{\partial s(k)}\right]_j = -\boldsymbol{A}_j^{\mathrm{T}}(\boldsymbol{x}_j - \boldsymbol{A}_j \boldsymbol{s}(k)), j = 1, 2, \ldots, M \tag{11.87}$$

$$\left[\frac{\partial E}{\partial s_i(k)}\right]_j = -A_{ji}(x_j - A_j s(k)), i = 1, 2, \ldots, N, j = 1, 2, \ldots, M \tag{11.88}$$

Therefore, the minimum recovery error criterion is used to optimize $s(k)$ to obtain the following M equations.

$$s(k+1) = s(k) - \mu \frac{\partial E}{\partial s(k)} = s(k) + \mu A_j^{\mathrm{T}}(x_j - A_j s(k)), j = 1, 2, \ldots, M \tag{11.89}$$

where μ is a constant. Its optimal value is determined by computing the new gradient corresponding to $s(k+1)$ and equating the new gradient to the Eq. (11.89). Define $s(k+1)$ in Eq. (11.89) as $\hat{s}(k)$ and $\hat{E} = \frac{1}{2}||x_j - A_j\hat{s}||_2^2$ compute new update equation.

$$\begin{aligned} s(k+1) &= \hat{s}(k) - \mu \frac{\partial E}{\partial \hat{s}(k)} \\ &= \hat{s}(k) + \mu A_j^{\mathrm{T}}(x_j - A_j \hat{s}(k)) \\ &= \hat{s}(k) - \mu A_j^{\mathrm{T}} A_j \hat{s}(k) + \mu A_j^{\mathrm{T}}(x_j) \\ &= (I - \mu A_j^{\mathrm{T}} A_j)\hat{s}(k) + \mu A_j^{\mathrm{T}}(x_j) \\ &= (I - \mu A_j^{\mathrm{T}} A_j)[s(k) + \mu A_j^{\mathrm{T}}(x_j - A_j s(k))] + \mu A_j^{\mathrm{T}}(x_j) \\ &= s(k) + \mu A_j^{\mathrm{T}}(x_j - A_j s(k))(2I - \mu A_j^{\mathrm{T}} A_j) \end{aligned} \tag{11.90}$$

Here μ is kept constant throughout the iterations and comparing the two gradient Eqs. (11.89) and (11.90), it is evident that

$$(2I - \mu A_j^{\mathrm{T}} A_j) := I \tag{11.91}$$

Then

$$\mu = \frac{1}{||A_j||^2} \tag{11.92}$$

So the iterative formula from $s(k)$ to $s(k+1)$ is:

$$s(k+1) = s(k) + \frac{A_j^{\mathrm{T}}}{||A_j||^2}(x_j - A_j s(k)), j = 1, 2, \ldots, M \tag{11.93}$$

For the formula (11.86), the gradient descent method is directly used to solve it, and the iterative formula is as follows:

$$s_i(k+1) = s_i(k) - \eta s_i(k) \exp\left(\frac{-s_i^2(k)}{2\sigma^2}\right), \quad i = 1, \ldots, N \tag{11.94}$$

where η is the step size and is determined by the initial value of $\boldsymbol{s}_0$. Similar to the decreasing parameter sequence in the SL0 algorithm, the parameter σ gradually decreases during iteration.

$$\sigma_{k+1} = \sigma_k \delta \tag{11.95}$$

The parameters can be initialized as follows. $\boldsymbol{s}_k$ has an initial value of $\boldsymbol{s}(0) = \boldsymbol{A}^{\dagger}\boldsymbol{x} = \boldsymbol{A}^{\mathrm{T}}(\boldsymbol{A}\boldsymbol{A}^{\mathrm{T}})^{-1}\boldsymbol{x}$, and δ is a scale parameter less than 1, which can take a value of 0.6. The smaller the value of σ, the more $F_\sigma(\boldsymbol{s})$ can approximate L0 norm of vector $\boldsymbol{s}$. However, as the value of σ decreases, the smoothness of $F_\sigma(\boldsymbol{s})$ becomes worse and worse, and the number of local extremum points increases. Therefore, in the process of alternating optimization, the value of σ should be gradually reduced, σ is initialized to $\sigma_0 = 2\max\{\boldsymbol{s}_0\}$, and the minimum value is $\sigma_{\min}$.

In summary, the steps of the RASR algorithm are summarized as follows:

Step 1: Initialize the source signal $\boldsymbol{s}(0) = \boldsymbol{A}^{\dagger}\boldsymbol{x}$, and the number of iterations $k = 1$;
Step 2: Set parameters $\sigma_0 = 2\max\{\boldsymbol{s}(0)\}$, $\eta = 1.2\max\{\boldsymbol{s}(0)\}$, $0.6 \le \delta < 1$;
Step 3: while $\sigma_k \ge \sigma_{\min}$ do

(1) minimize $F_\sigma(s)$:
for $i = 1, 2, \ldots, N$

$$\Delta_i = s_i(k) - \eta s_i(k)\exp\left(\frac{-s_i^2(k)}{2\sigma^2}\right)$$

end
update the source signal to be recovered: $\boldsymbol{s}'(k) = [\Delta_1, \Delta_2, \ldots, \Delta_N]^{\mathrm{T}}$;

(2) minimize E
for $j = 1, 2, \ldots, M$

$$\boldsymbol{s}''(k) = \boldsymbol{s}'(k) + \frac{\boldsymbol{A}_j^{\mathrm{T}}}{\|\boldsymbol{A}_j\|_2^2}\left(x_j - \boldsymbol{A}_j\boldsymbol{s}(k)\right);$$

$$\boldsymbol{s}'(k) \leftarrow \boldsymbol{s}''(k);$$

end
update the source signal to be recovered: $\boldsymbol{s}(k+1) \leftarrow \boldsymbol{s}'(k)$;

(3) update parameter: $\sigma_{k+1} = \delta\sigma_k$;

(4) update iterations: $k \leftarrow k + 1$;

end do

Step 4: Output $\boldsymbol{s}(k)$.

11.5.5 Modified Newton RASR Algorithm

The RASR algorithm realizes the source signal recovery by solving the two optimization problems of formulas (11.85) and (11.86). In the process of solving the formula (11.86), the RASR algorithm uses the fastest descent method. The disadvantage of this method is that the accuracy of the algorithm is related to the iteration step size. Larger step size can improve the computation speed and reduce the computation complexity of the algorithm, but it will reduce the precision of the recovery. Smaller step size can improve the recovery accuracy to a certain extent, but it will increase the number of algorithm iterations and increase the computation complexity of the algorithm. In order to enhance the robustness of the algorithm, the accuracy and complexity of the algorithm are not affected by the step size, the modified Newton optimization method is introduced into the RASR algorithm.

It can be seen from Sect. 11.5.3 that when modified Newton method is used to solve the optimization problem (11.86), the recursion formula for the kth to $(k+1)$th iterations is as follows:

$$\boldsymbol{s}(k+1)=\boldsymbol{s}(k)+\boldsymbol{d}_k \tag{11.96}$$

where:

$$\boldsymbol{d}_k=\left[-\frac{\sigma_k^2 s_1}{\sigma_k^2+s_1^2}\ -\frac{\sigma_k^2 s_2}{\sigma_k^2+s_2^2}\ \cdots\ -\frac{\sigma_k^2 s_N}{\sigma_k^2+s_N^2}\right]^{\mathrm{T}} \tag{11.97}$$

In order to make $F_\sigma(\boldsymbol{s})$ more and more close to the L0 norm of the source signal vector $\boldsymbol{s}$, a decreasing σ is selected at each iteration, the minimum of σ is $\sigma_{\min}$, that is, $\sigma_{\min}<\sigma_{k+1}<\sigma_k$.

The solution method for optimization problem (11.85) is the same as in Sect. 11.5.4, so the iterative formula is the same as Eq. (11.93).

In summary, the steps of source signal recovery algorithm based on modified Newton RASR algorithm (NRASR) is as follows.

Step 1: Initialize the source signal $\boldsymbol{s}(0)=\boldsymbol{A}^{\mathrm{T}}(\boldsymbol{A}\boldsymbol{A}^{\mathrm{T}})^{-1}\boldsymbol{x}$, scale parameters $\delta=0.6$, $\sigma_0=2\max\{\boldsymbol{s}_0\}$, $\sigma_{\min}=10^{-5}$, and the number of iterations $k=0$;

Step 2: If $\sigma_k>\sigma_{\min}$, perform Step 3 to Step 7; otherwise, go to Step 8.

Step 3: for $j=1,2,\ldots M$ (M is the number of observed signals)

(1) $\nabla_k=\boldsymbol{A}_j^{\mathrm{T}}(x_j-\boldsymbol{A}_j\boldsymbol{s}(k))/\|\boldsymbol{A}_j\|_2$, where $\boldsymbol{A}_j$ is the j-row of $\boldsymbol{A}$;
(2) update $\boldsymbol{s}_k$: $\boldsymbol{s}(k)\leftarrow\boldsymbol{s}(k)+\nabla_k$;

End;

Step 4: Calculate the corrected Newtonian direction $\boldsymbol{d}_k$:

$$\boldsymbol{d}_k=\left[-\frac{\sigma_k^2 s_1(k)}{\sigma_k^2+s_1^2(k)}\ -\frac{\sigma_k^2 s_2(k)}{\sigma_k^2+s_2^2(k)}\ \cdots\ -\frac{\sigma_k^2 s_N(k)}{\sigma_k^2+s_N^2(k)}\right]^{\mathrm{T}};$$

Step 5: Update: $\boldsymbol{s}(k)\leftarrow\boldsymbol{s}(k)+\boldsymbol{d}_k$;

Step 6: Update parameters: $\sigma_{k+1}=\delta\sigma_k$;

Step 7: Update the source signal to be recovered: $\boldsymbol{s}(k+1)=\boldsymbol{s}(k)$; Update iteration times: $k\leftarrow k+1$;

Step 8: Output $\boldsymbol{s}(k)$.

11.5.6 Algorithm Performance Simulation and Analysis

In order to compare the performance of different source signal recovery algorithms, SL0, RSL0, RASR, HOSL0 and NRASR algorithms are simulated in this section [50]. The mixing matrix is a Gaussian random matrix with 3 rows and 5 columns, that is, there are 3 received signals and 5 source signals. The source signals are random sparse signals, and the number of sampling points is 1000. Figures 11.26 and 11.27 show the curves of correlation coefficients and running time of the five algorithms changing with the SNR when the source signal sparsity is 0.7.

Figures 11.28 and 11.29 show the curves of correlation coefficient and running time of the five algorithms changing with the SNR when the source signal sparsity is 0.8.

Figures 11.30 and 11.31 show the curves of correlation coefficients and running time of the five algorithms changing with the SNR when the source signal sparsity is 0.9.

Figures 11.32 and 11.33 show the curves of correlation coefficients and running time of the five algorithms changing with the sparsity when the SNR is 10 dB.

Figures 11.34 and 11.35 show the curves of correlation coefficients and running time of the five algorithms changing with the sparsity when the SNR is 20 dB.

Figures 11.36 and 11.37 show the curves of correlation coefficients and running time of the five algorithms changing with the sparsity when the SNR is 30 dB.

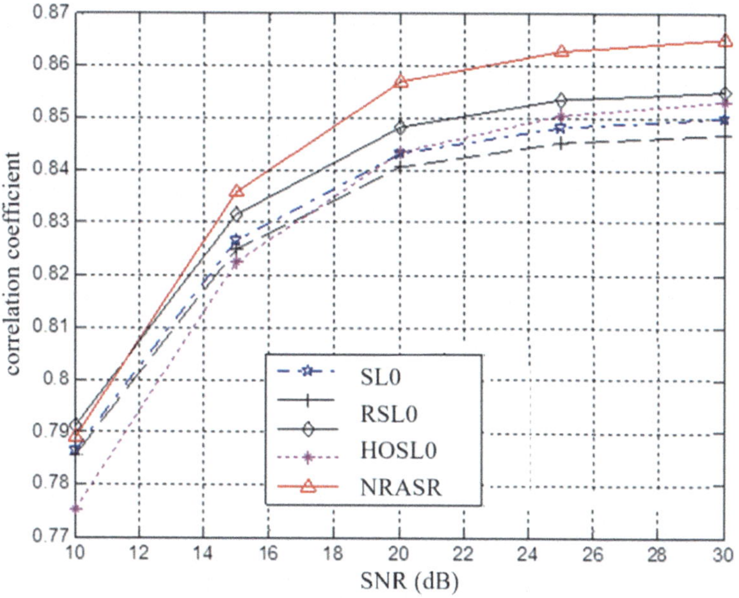

Fig. 11.26 Curves of the correlation coefficients of five algorithms changing with the SNR when source signal sparsity is 0.7

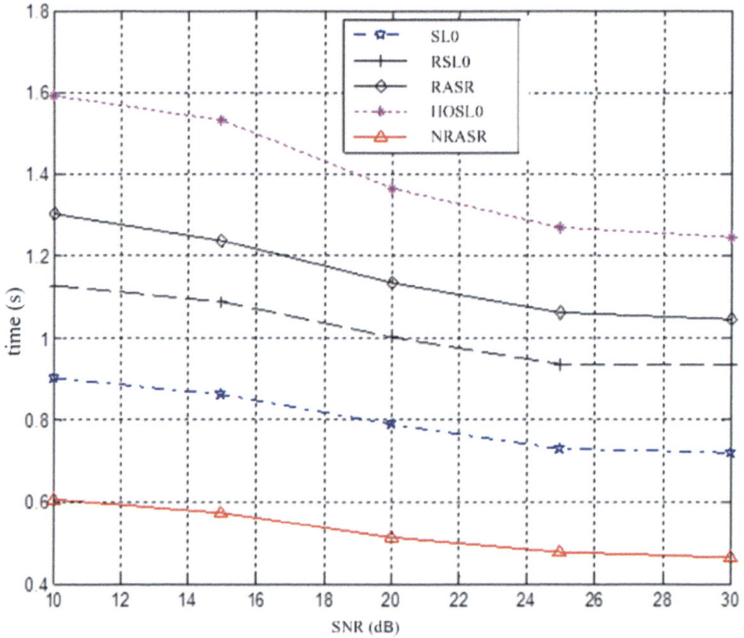

Fig. 11.27 Curves of the running time of five algorithms changing with the SNR when the source signal sparsity is 0.7

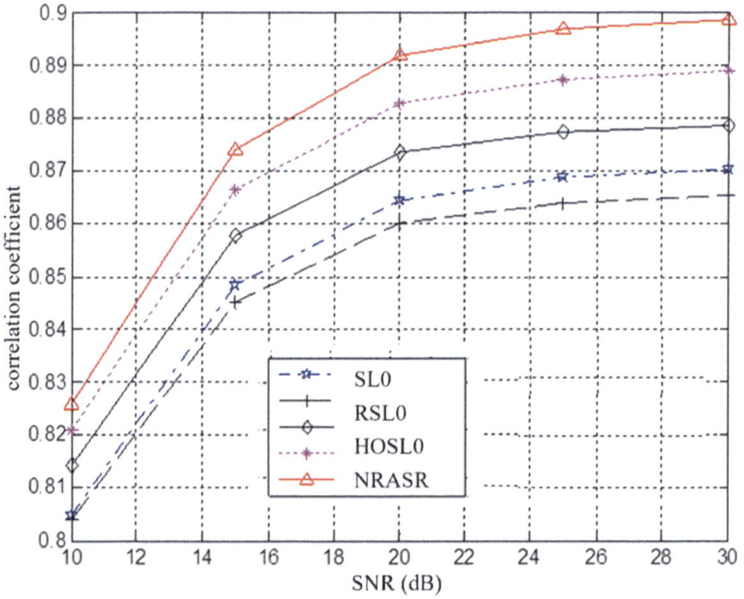

Fig. 11.28 Curves of the correlation coefficient of five algorithms changing with the SNR when source signal sparsity is 0.8

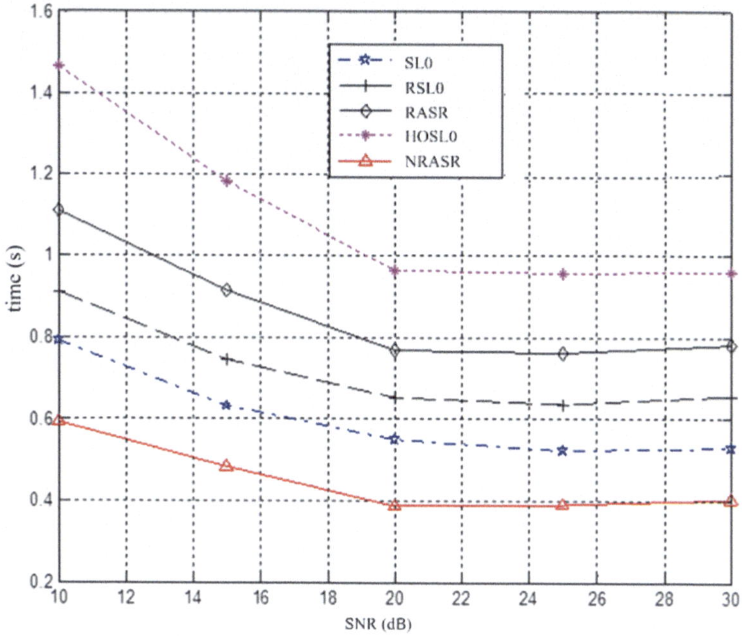

Fig. 11.29 Curves of the running time of five algorithms changing with the SNR when the source signal sparsity is 0.8

As can be seen from Figs. 11.26, 11.28 and 11.30, in the case of small SNR, such as when the SNR is less than 10 dB, correlation coefficient of NRASR algorithm is close to the other four algorithms. In the case of large SNR, NRASR algorithm has higher correlation coefficient than other four algorithms. This means that, compared with the other four algorithms, when SNR is small, NRASR algorithm is similar to the other four algorithms in terms of signal recovery accuracy. When the SNR is large, the proposed NRASR algorithm is better than the other four algorithms in terms of signal recovery accuracy.

As can be seen from Figs. 11.32, 11.34 and 11.36, in the case of a fixed SNR, when the sparsity is relatively small (such as less than 0.7), that is, when the source signal is not sufficiently sparse, NRASR algorithm has similar recovery accuracy to SL0, RSL0 and RASR algorithms, and is superior to HOSL0 algorithm. When the source signal sparsity is large (such as greater than 0.8), the NRASR algorithm has better performance in terms of recovery accuracy than the other four algorithms. This shows that compared with HOSL0 algorithm, the recovery accuracy of NRASR algorithm is always higher than HOSL0 algorithm regardless of the sparsity. Compared with SL0, RSL0 and RASR algorithms, NRASR algorithm has obvious advantages in the case of large sparsity.

From the remaining 6 figures, it can be seen that NRASR algorithm is significantly lower than the other four algorithms in terms of running time. In general,

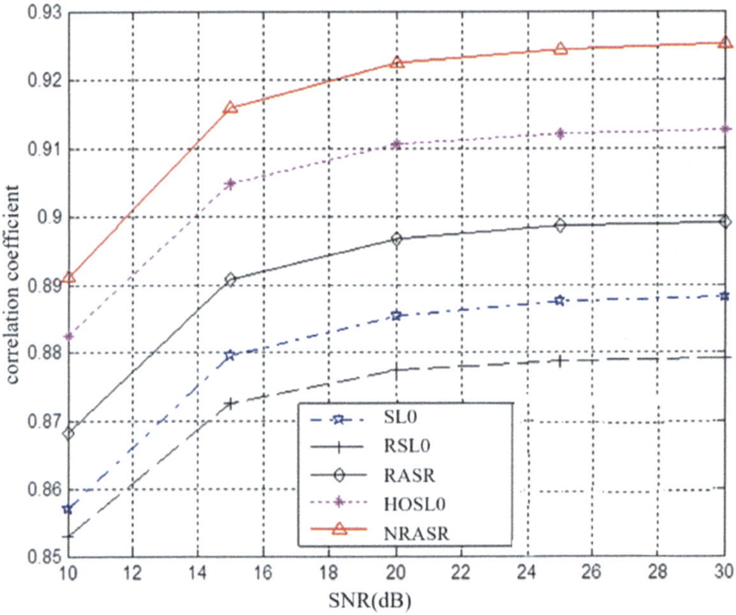

Fig. 11.30 Curves of the correlation coefficient of five algorithms changing with the SNR when source signal sparsity is 0.9

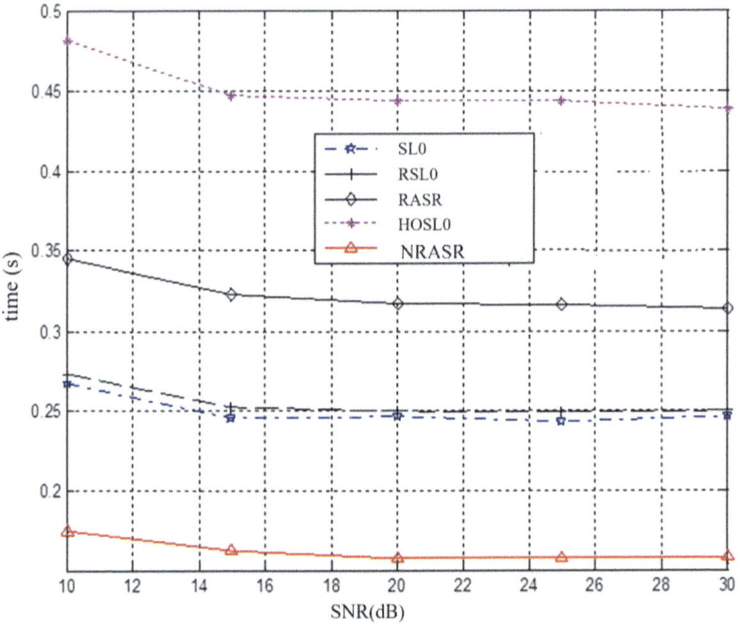

Fig. 11.31 Curves of the running time of five algorithms changing with the SNR when the source signal sparsity is 0.9

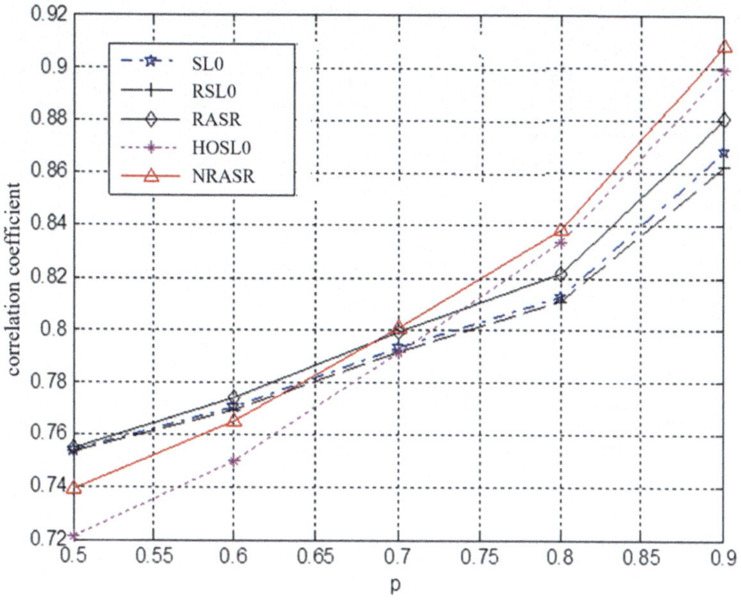

Fig. 11.32 Curves of the correlation coefficient of five algorithms changing with sparsity when the SNR is 10 dB

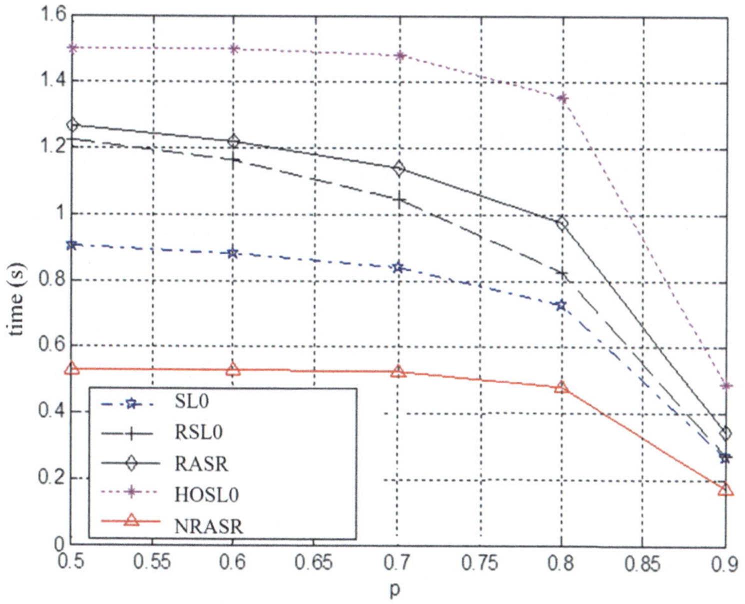

Fig. 11.33 Curves of the running time of five algorithms changing with sparsity when the SNR is 10 dB

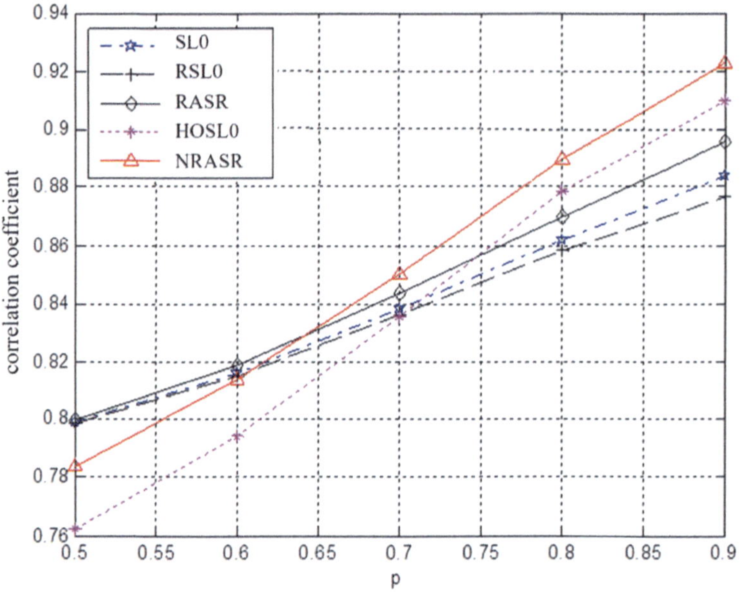

Fig. 11.34 Curves of the correlation coefficient of five algorithms changing with sparsity when the SNR is 20 dB

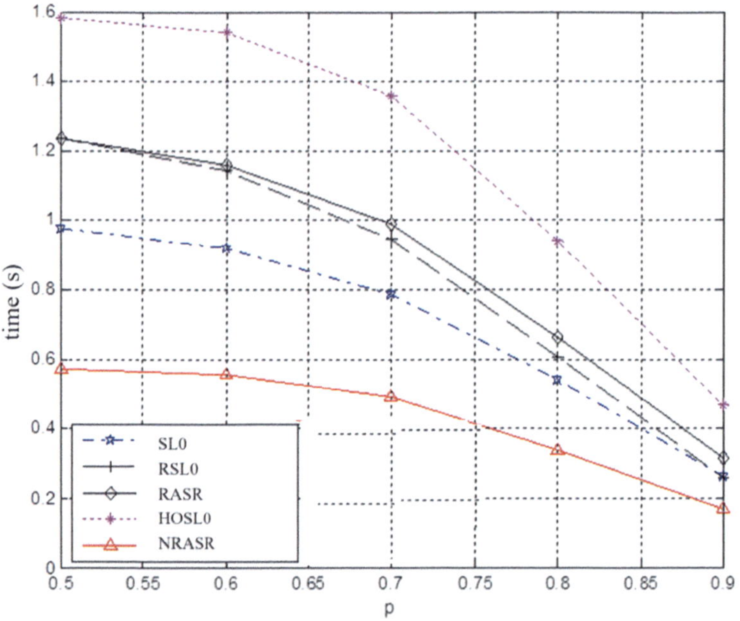

Fig. 11.35 Curves of the running time of five algorithms changing with sparsity when the SNR is 20 dB

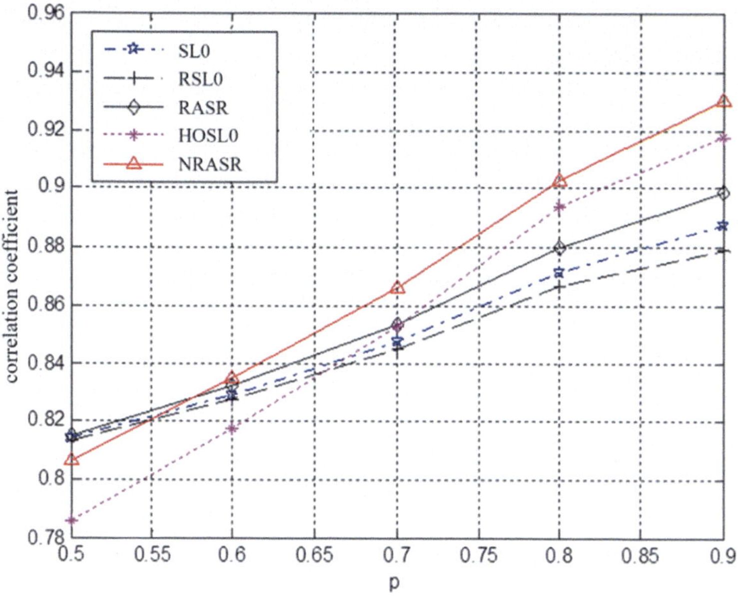

Fig. 11.36 Curves of the correlation coefficient of five algorithms changing with sparsity when the SNR is 30 dB

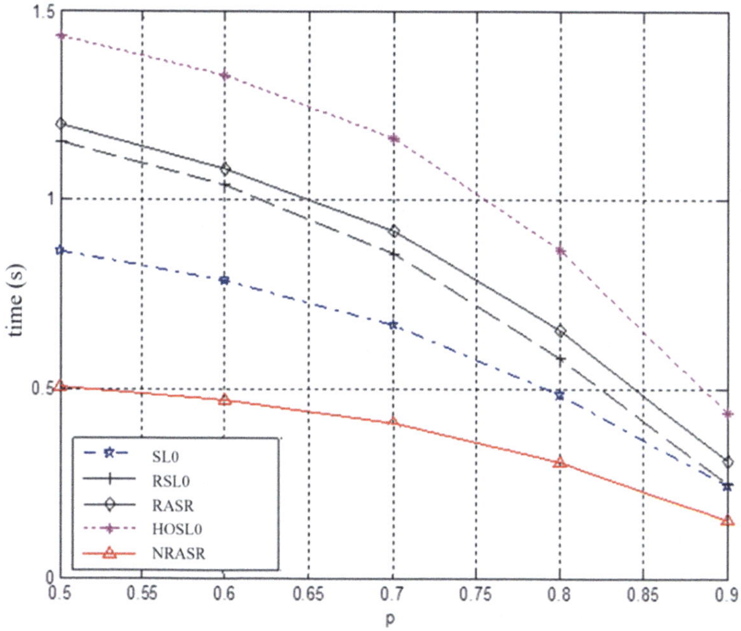

Fig. 11.37 Curves of the running time of five algorithms changing with sparsity when the SNR is 30 dB

NRASR algorithm performs better than other algorithms in recovery accuracy and computation complexity.

Summary

This chapter introduces a variety of underdetermined source signal recovery algorithms based on compression reconstruction, including greedy algorithm, algorithms based on L1 norm, algorithms based on L0 norm, algorithms based on radial basis function and so on. Based on the analysis of algorithm principle, the algorithm implementation steps are given, and the algorithm performance is simulated and analyzed.

Chapter 12
Other Algorithms for Source Signal Recovery

In this chapter, other algorithms for source signal recovery are studied. Firstly, a source signal recovery algorithm based on the shortest path is introduced, and then the problems existing in the source signal recovery algorithm based on statistical sparse decomposition are analyzed, and an improved source signal recovery algorithm based on statistical sparse decomposition is presented [54]. Finally, the performance of several algorithms is analyzed and simulated.

12.1 Source Signal Recovery Algorithm Based on Shortest Path

The shortest path method is the simplest source signal recovery algorithm in the underdetermined blind source separation, which is suitable for the source signal recovery when the number of observed signals is 2 and the source signals are sufficiently sparse. According to the sparse component analysis theory, the problem of recovering source signals is generally converted to solving the following optimization problems.

$$\begin{cases} \min\limits_{s(t)} \sum\limits_{i=1}^{N} |s_i(t)| \\ \boldsymbol{x}(t) = \boldsymbol{A}\boldsymbol{s}(t) = \sum\limits_{i=1}^{N} a_i s_i(t) \end{cases} \tag{12.1}$$

where N is the number of source signals, $\boldsymbol{s}(t) = [s_1(t), s_2(t), \ldots, s_N(t)]^T$, $s_i(t)$ is the i-th ($i = 1, 2, \ldots, N$) sample value of source signal at time t, $\boldsymbol{x}(t) = [x_1(t), x_2(t), \ldots, x_M(t)]^{\mathrm{T}}$ is the observed signal vector, and M is the number of observed signals. For the case of $M = 2$, the solution idea of the above problem

W. Fu, *Principle and Application of Blind Source Separation Technology*,
https://doi.org/10.1007/978-981-96-0022-9_12

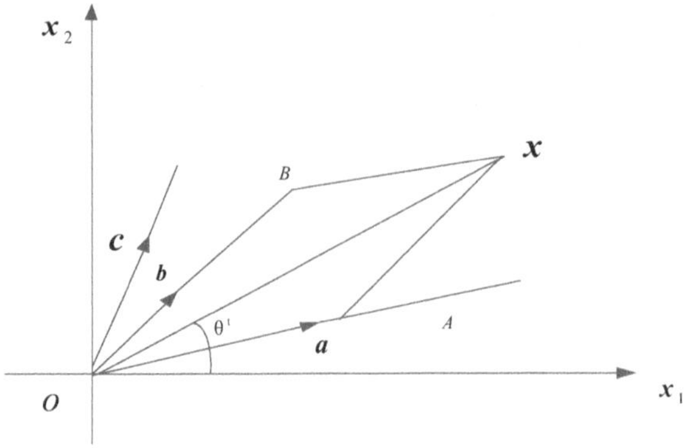

Fig. 12.1 Schematic diagram of the shortest path method when the observed signal is 2

is shown in Fig. 12.1. In this case, to minimize $\sum_{i=1}^{N} |s_i(t)|$ is to do a linear decomposition of the observed signal along the direction of two columns of the mixing matrix to find the shortest path from the origin to the observed signal. As can be seen from Fig. 12.1, the shortest path from the origin to the observed signal is the two vectors $\boldsymbol{a}$ and $\boldsymbol{b}$ that are closest to the Angle of $\boldsymbol{x}$.

When the number of observed signals is greater than 2, the original shortest path method can be improved, and the improved algorithm is suitable for source signal recovery when the number of observed signals is greater than 2. The improvement idea is to select the first observed signal, and then iterate to select the remaining $M - 1$ observed signals, that is, the observed signal to be processed each time is the combination of the first observed signal and any other observed signals, and there is a total of $M - 1$ combinations. Then we take two signals as observed signals, and use the shortest path method to recover the source signal. The recovered source signals have $M - 1$ groups and the $M - 1$ groups of source signals are averaged, and the source signals to be recovered are obtained.

The steps of shortest path method are as follows.

Step 1: Select the first observed signal, and then the remaining $M - 1$ signals are selected in turn. We can get $M - 1$ signal combinations $\tilde{X}_m(t) = [x_1(t), x_m(t)]^{\mathrm{T}}$, $(m = 2, 3, \ldots, M)$ with only 2 observed signals.

Step 2: Each group of observed signals in step 1 is preprocessed by removing the column vector with all observed signals being zero and then normalizing it.

Step 3: Calculate the angles of each column vector of the mixing matrix: $\theta_{m,j}^{A} = \tan^{-1}(a_{m,j}/a_{1,j})$, $(m = 2, 3, \ldots, M)$, where $\theta_{m,j}^{A}$ represents the angle corresponding to the m-th row and j-th column in the mixing matrix.

Step 4: for $t = 1$: T do

Step 5: for $m = 2$: M do

(5.1) Calculate the angle of the m-th observed signal relative to the first observed signal at time t

$$\theta_m^t = \tan^{-1}(\boldsymbol{x}_m(t)/\boldsymbol{x}_1(t)) \tag{12.2}$$

(5.2) Find the angle of the two mixing matrix column vectors closest to the observed signal vector angle and record the corresponding column serial number j_1, j_2;

(5.3) Let $\tilde{\boldsymbol{a}}_{j_1} = \left[a_{1,j_1}, a_{m,j_1}\right]^{\mathrm{T}}$, $\tilde{\boldsymbol{a}}_{j_2} = \left[a_{1,j_2}, a_{m,j_2}\right]^{\mathrm{T}}$, where $\tilde{\boldsymbol{a}}_{j_1}$, $\tilde{\boldsymbol{a}}_{j_2}$ are the two column vectors closest to the direction of the observation signal combination $\tilde{X}_m$;

(5.4) Let $\tilde{\boldsymbol{A}}_m = \left[\tilde{\boldsymbol{a}}_{j_1}, \tilde{\boldsymbol{a}}_{j_2}\right] = \begin{bmatrix} a_{1,j_1}, a_{1,j_2} \\ a_{m,j_1}, a_{m,j_2} \end{bmatrix}$, and $\boldsymbol{W}_m = \tilde{\boldsymbol{A}}_m^{-1}$;

(5.5) Calculate the source signals recovered from the observed signals at time t

$$\begin{cases} \left[\hat{\boldsymbol{s}}_{j1}^m(t), \hat{\boldsymbol{s}}_{j2}^m(t)\right]^{\mathrm{T}} = \boldsymbol{W}_m \tilde{X}_m(t) \\ \hat{\boldsymbol{s}}_j^m(t) = 0 \ \text{ for } \ j \neq j_1, j_1, \end{cases} \tag{12.3}$$

End step 5.

Step 6: Calculate the average of source signals recovered at time t

$$\hat{\boldsymbol{s}}(t) = \frac{1}{M-1} \sum_{m=2}^{M} \hat{\boldsymbol{s}}^m(t) \tag{12.4}$$

where $\hat{\boldsymbol{s}}^m(t) = \left[\hat{\boldsymbol{s}}_1^m(t), \hat{\boldsymbol{s}}_2^m(t), \ldots, \hat{\boldsymbol{s}}_N^m(t)\right]$.

End step 4.

12.2 Source Signal Recovery Algorithm Based on Statistical Sparse Decomposition

Statistical Sparse Decomposition (SSDP) algorithm firstly clustered the observed signals to obtain the mixing matrix, and then selected the columns of the mixing matrix corresponding to the leading source signals according to the criterion of the minimum mutual correlation of the source signals in the same time period. On this basis, the source signals were recovered in the same time period under the noise environment.

$$\boldsymbol{C}_x = \boldsymbol{E}((\boldsymbol{x}(t) - \boldsymbol{E}_x)(\boldsymbol{x}(t) - \boldsymbol{E}_x)^{\mathrm{T}}) \tag{12.5}$$

where $\boldsymbol{x}(t)$ denotes the mean value of the mixed signal at time t, and correspondingly, the covariance matrix of the source signal within is defined as

$$\boldsymbol{C}_s = \boldsymbol{E}((\boldsymbol{s}(t) - \boldsymbol{E}_s)(\boldsymbol{s}(t) - \boldsymbol{E}_s)^{\mathrm{T}}) \tag{12.6}$$

where $s(t)$ represents the source signal at time t, and E_s is the mean of $s(t)$.

In a given time interval, the covariance matrix is estimated as follows:

$$C_x \approx \frac{1}{\Delta t} \sum_{t=t_o}^{t_1} \left[x(t) - \frac{1}{\Delta t} \sum_{t=t_o}^{t_1} x(t) \right] \left[x(t) - \frac{1}{\Delta t} \sum_{t=t_o}^{t_1} x(t) \right]^{\mathrm{T}} \tag{12.7}$$

where $\Delta t = t_1 - t_0 + 1$. Due to the independence of source signals, when the number of observed signals is M and the number of source signals is N, within Δt time interval, the source signal covariance matrix can be expressed as

$$C_s \approx \begin{bmatrix} C_s^{sub} & 0_{M \times (N-1)} \\ 0_{(N-M) \times M} & 0_{(N-M) \times (N-M)} \end{bmatrix} \tag{12.8}$$

where $C_s^{sub} \in R^{M \times M}$ is a diagonal matrix, and C_s^{sub} can be estimated according to the following formula.

$$C_s^{sub} = A_{j_1, j_2, \ldots, j_M}^{-1} C_x \left(A_{j_1, j_2, \ldots, j_M}^{-1} \right)^{\mathrm{T}} \tag{12.9}$$

Therefore, the problem of recovering the source signal is converted to finding suitable $j_1, j_2, \ldots j_M$ so that C_s^{sub} is a diagonal matrix, which can be calculated by the following formula:

$$[j_1, j_2, \ldots, j_M] = \underset{j_1, \ldots j_M = 1, \ldots N}{\operatorname{argmin}} \frac{\sum_{i=1}^{M} \sum_{j>i} \left| C_s^{sub}(i, j) \right|}{\sqrt{C_s^{sub}(1,1), \ldots C_s^{sub}(M, M)}} \tag{12.10}$$

Once the appropriate $j_1, j_2, \ldots j_M$ is found, the expression used to reconstruct the signal is described as follows:

$$\begin{cases} \left[s_{j1}(t), s_{j2}(t) \ldots, s_{jM}(t) \right]^{\mathrm{T}} = A_{j_1 j_2, \ldots j_M}^{-1} [x_1(t), x_2(t), \ldots x_M(t)]^{\mathrm{T}} \\ s_j(t) = 0, j \neq j_1, j_2 \ldots, j_M \end{cases} \tag{12.11}$$

Research and simulation show that SSDP algorithm requires that the number of active source signals M' must be equal to the number of observed signals M in each processing time window. But in practice, within any Δt time, the number of active source signals is not always equal to M, in this case, we need to make some improvements to the algorithm.

12.3 Improved SSDP Source Signal Recovery Algorithm

The research and simulation show that the performance of SSDP algorithm is poor, and the algorithm requires that the number of active source signals M' must be equal to the number of observed signals M in each processing time window. When $M' < M$ for example, in period of time Δt, $s_1, s_2, \ldots, s_{M'}$ are not zero, and the other $N - M'$ source signals are zero, that is, the formula (12.8) is no longer a diagonal matrix, but has the following form.

$$\boldsymbol{C}_s^{sub} = \begin{bmatrix} \tilde{\boldsymbol{C}}_s^{sub} & 0_{M' \times (N-M')} \\ 0_{(N-M') \times M'} & 0_{(N-M') \times (N-M')} \end{bmatrix} \tag{12.12}$$

There are only M' diagonal elements in $\boldsymbol{C}_s^{sub}$ that are non-zero, and the other $N - M'$ diagonal elements should be 0 in theory. In particular, when $M' = 1$, ${C_s^{sub}}_{M \times M}$ theoretically has only one non-zero diagonal element, and all the rest are zero. It can be seen that it is obviously wrong to solve the optimal combination of column vectors according to Eq. (12.10).

According to the above analysis, it can be seen that the current SSDP algorithm assumes that the number of active source signals M' within each processing interval Δt is equal to the number of observed signals M. When this condition is not satisfied, the SSDP algorithm cannot correctly select the appropriate mixing matrix column vectors. At this time, M optimal column vectors are found through the following formula.

$$\left[j_1, j_2, \ldots, j_M\right] = \underset{j_1,\ldots j_M = 1,\ldots N}{\operatorname{argmin}} \frac{\prod\limits_{i=1}^{M} \prod\limits_{j>i} \boldsymbol{C}_s^{sub}(i,j)}{\sum\limits_{l=1}^{M'} \lambda_l} \tag{12.13}$$

$$\boldsymbol{C}_s^{sub} = \boldsymbol{A}_{j_1, j_2, \ldots, j_M}^{-1} \boldsymbol{C}_x \left(\boldsymbol{A}_{j_1, j_2, \ldots, j_M}^{-1}\right)^{\mathrm{T}} \tag{12.14}$$

where λ_l is the l-th element obtained after sorting the diagonal elements of $\boldsymbol{C}_s^{sub}$ from largest to smallest. This method is called the improved sparse statistical decomposition method.

The following describes the estimation method of the number of active source signals at a certain time.

Assuming that $M'(M' \leq M)$ source signals are active in time interval Δt, the observed signal can be expressed as $\boldsymbol{x}(t) = \left[\boldsymbol{a}_{j_1}, \boldsymbol{a}_{j_2}, \ldots, \boldsymbol{a}_{j_{M'}}\right] \times \left[s_{j_1}(t)\ s_{j_2}(t)\ s_{j_3}(t)\ \ldots s_{j_{M'}}(t)\right]^{\mathrm{T}} + \boldsymbol{v}(t)$. Finding the correlation matrix of the observed signal, we get $\boldsymbol{C}_x = \boldsymbol{A}^{M'} \boldsymbol{C}_s^{M'} \left(\boldsymbol{A}^{M'}\right)^{\mathrm{T}} + \boldsymbol{C}_v$, where $\boldsymbol{x}(t) \in R^{M \times 1}$ is the observed signal vector, $\boldsymbol{a}_{j_{M'}} \left(M' = 1, 2, \ldots M\right)$ represents column $j_{M'}$ in matrix $\boldsymbol{A}$, and $s_{j_{M'}}(t)$ is the $j_{M'}$-th source. $\boldsymbol{v}(t)$ represents a vector composed of Gaussian white noise received by each receiving channel, and $A^{M'} = \left[a_{j_1}, a_{j_2}, \ldots, a_{j_{M'}}\right] \in R^{M \times M'}$ represents a submatrix composed of M' column vectors in A. C_x is the observed signal covariance matrix in time interval Δt, and C_v is the noise covariance matrix. After the eigenvalue decomposition of C_x, M eigenvalues and corresponding eigenvectors are obtained. Among these M eigenvalues, there should be M' large eigenvalues $\lambda_1, \lambda_2, \ldots, \lambda_{M'}$, and their corresponding eigenvectors form the signal subspace; in addition, there are $M - M'$ small eigenvalues $\lambda_{m'+1}, \lambda_2, \ldots, \lambda_M$, and their corresponding eigenvectors form the noise subspace. The eigenvalues of C_x satisfies $\lambda_1 > \lambda_2 > \lambda_{M'} \gg \lambda_{M'+1} > \cdots > \lambda_M$, you can see, when $M' < M$, $\lambda_{M'}/\lambda_{M'+1} \gg 1$. According to this feature, the number of active source signals M' can be estimated, that is, the ratio of adjacent eigenvalues is calculated, that is, let $b_{M'} = \lambda_{M'}/\lambda_{M'+1}\left(M' = 1, 2, \ldots, M - 1\right)$, and the corresponding minimum M' is the number of active source signals when $b_{M'}$ is greater than a certain threshold value (such as 20). When this condition is not satisfied, the number of active source signals is considered to be M.

In the above processing method, although M' may be less than M, M column vectors are still found. As long as the M column vectors contain M' real column vectors, the source signal can be well recovered. Therefore, as long as the number of source signals M is not incorrectly estimated as a smaller number than M, the estimation error of the number of source signals has no impact on the processing result of the algorithm. In the estimation of the number of source signals, the probability of the above errors is relatively small, so this processing method is relatively good.

Then recover the source signals according to the following formula.

$$\begin{cases} \left[s_{j1}(t), s_{j2}(t) \ldots, s_{jM}(t)\right]^{\mathrm{T}} = A_{j_1, j_2, \ldots, j_M}^{-1} \left[x_1(t), x_2(t), \ldots x_M(t)\right]^{\mathrm{T}} \\ s_j(t) = 0, j \neq j_1, j_2 \ldots, j_M \end{cases} \tag{12.15}$$

When the number of active source signals is less than or equal to the number of observed signals at the same time, improved statistical sparse decomposition (ISSDP) can recover the source signals better.

The specific implementation steps of ISSDP algorithm are as follows.

Step 1: The observed signal with sampling number T is divided into L segments according to the sampling time. The length of each segment is ΔT, and the segment sequence number $i (i \leq L)$ of the observed signal to be processed is initialized to 1, where $\Delta T = T/L$.

Step 2: The covariance matrix of the observed signal in the i-th segment in the period of $((i - 1) \cdot \Delta T + 1) : (i \cdot \Delta T)$ is calculated as follows.

$$C_{x^i} = E\left(\left(x^i - E_{x^i}\right)\left(x^i - E_{x^i}\right)^{\mathrm{T}}\right)$$

Step 3: Estimate the number of active source signals in time period of $((i - 1) \cdot \Delta T + 1) : (i \cdot \Delta T)$, and the specific method is as follows. Firstly, the eigenvalue decomposition of the covariance matrix of the observed signal in i-th segment is carried out to generate M eigenvalues, which are arranged in descending order $\lambda_1 > \lambda_2 > \lambda_{M'} \gg \lambda_{M'+1} > \cdots > \lambda_M$. Then the ratio of the sorted adjacent eigenvalues $b = \lambda_{M'}/\lambda_{M'+1}$ $\left(M' = 1, 2, \ldots M - 1\right)$ is calculated. Lastly, determine the minimum M' when the ratio b is greater than 20, and the minimum M' is the estimated number of active source signals in the time period of $((i-1)\cdot \Delta T + 1) : (i \cdot \Delta T)$.

Step 4: According to Eq. (12.13), the columns corresponding to the sparse decomposition matrix are calculated.

Step 5: According to Eq. (12.15), the source signal recovered in the time period of $((i-1)\cdot \Delta T + 1) : (i \cdot \Delta T)$ is calculated and the segmented serial number i of the observed signal is updated $i := i + 1$.

Step 6: Determine whether $i \le L$ is true. If yes, go to Step 2. If no, output the final recovered source signal matrix as

$$s = \left[\left[s_{j_1}^1, s_{j_2}^1 \ldots s_{j_M}^1\right]^{\mathrm{T}}, \left[s_{j_1}^2, s_{j_2}^2 \ldots s_{j_M}^2\right]^{\mathrm{T}}, \ldots, \left[s_{j_1}^i, s_{j_2}^i \ldots s_{j_M}^i\right]^{\mathrm{T}}, \ldots \left[s_{j_1}^L, s_{j_2}^L \ldots s_{j_M}^L\right]^{\mathrm{T}}\right]$$

where $s_{j_m}^i$ represents the source signal vector recovered according to the j_m-th $(m = 1, 2, \ldots, M)$ column of the mixing matrix during the i-th $(i = 1, 2, \ldots, L)$ time period ΔT. $s_{j_m}^i = \left[s_{j_m}((i-1)\cdot \Delta T + 1), s_{j_m}((i-1)\cdot \Delta T + 2), \ldots, s_{j_m}(i \cdot \Delta T)\right]$.

12.4 Performance Simulation and Analysis of Algorithm

Simulation 1

In order to reflect the advantages of SSDP algorithm and ISSDP algorithm, the performance of the two algorithms is simulated and compared under different circumstances. The mixing matrix used in the simulation is a randomly generated 5×6 dimensional Gaussian mixing matrix, and the source signal used is 6 radar signals. Figures 12.2, 12.3, 12.4 and 12.5 show the time-domain waveforms of the source signal when at most 1, 2, 4 and 5 source signals are active at a given time respectively [54].

The types of source signals shown in Figs. 12.2, 12.3, 12.4 and 12.5 are the same, but the signal distribution is different. The source signals are 6 radar signals, the sampling frequency of the signals is 30×10^6 Hz, and the sampling number is 3000. The first signal and the second signal are conventional radar signals with carrier frequencies of 3 MHz and 5 MHz respectively. The third signal is a linear FM radar signal with a carrier frequency of 2 MHz, and the fourth signal is a nonlinear FM radar signal with a carrier frequency of 3 MHz. The fifth and sixth signals are sinusoidal phase-modulated radar signals with carrier frequencies of 5 MHz and 2 MHz respectively.

Figures 12.6 and 12.7 show the performance curves of SSDP algorithm when the number of active source signals is different.

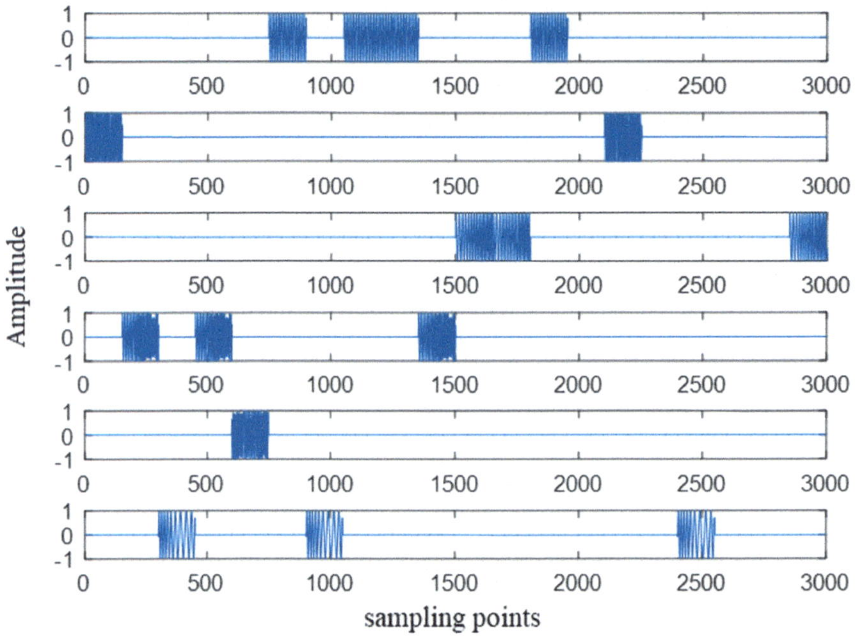

Fig. 12.2 Time domain waveforms of the source signals when only one source signal is active at the same time

As can be seen from Figs. 12.6 and 12.7, SSDP algorithm performs well only when the number of active source signals is equal to the number of observed signals. When there are five source signals in action, SIR and correlation coefficient of the SSDP algorithm gradually increase with the increase of the input SNR. When the input SNR is 10 dB, the SIR of the SSDP algorithm reaches 10 dB and the correlation coefficient reaches 0.95.

Figures 12.8 and 12.9 show the performance curves of ISSDP algorithm when the number of active source signals is different.

According to Figs. 12.8 and 12.9, it can be seen that the performance of ISSDP algorithm is better regardless of whether the number of active source signals is equal to the number of observed signals at the same time, and the sparser the source signals, the better its recovery accuracy. When the input SNR is 10 dB and five source signals are active, the SIR of separated signal is 9.5 dB and the correlation coefficient is 0.95. When 4 source signals are active, the SIR of separated signal is 12.7 dB and the correlation coefficient reaches 0.97. When two source signals are active, the SIR of separated signal 14 dB and the correlation coefficient reaches 0.98; when one source signal is active, the SIR of separated signal 15.6 dB and the correlation coefficient reaches 0.985.

As shown in Figs. 12.6, 12.7, 12.8 and 12.9, when the number of active source signals is equal to the number of observed signals, the recovery accuracy of the SSDP algorithm is slightly higher than that of the ISSDP algorithm. When the input SNR

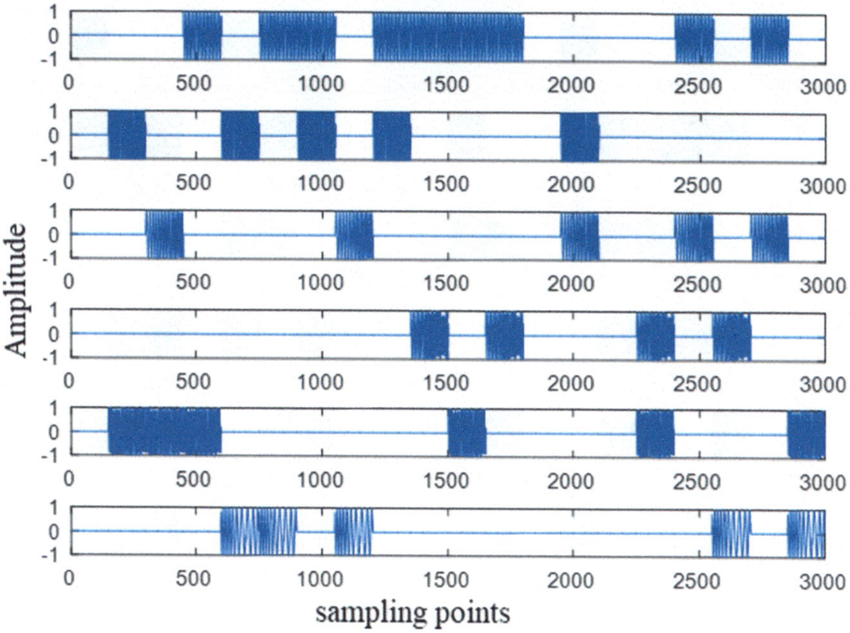

Fig. 12.3 Time domain waveforms of the source signals when two source signals are active at the same time

is 10 dB, the SIR of separated signals of the SSDP algorithm is 10 dB, and that of the ISSDP algorithm is 9.5 dB, and the correlation coefficient of the two algorithms is 0.95. However, when the number of active source signals is less than the number of observed signals, the performance of ISSDP algorithm is much better than that of SSDP algorithm.

Simulation 2

The source signal in simulation 2 is of the same type as the source signal in simulation 1, which are 6 radar signals. However, the number of active source signals in simulation 2 is not exactly the same at different times, and there may be 1, 2, 3, 4 or 5 source signals at some sampling points. The mixing matrix is a randomly generated dimensional Gaussian mixing matrix. Figure 12.10 is the time-domain waveforms of the source signal, Fig. 12.11 is the SIR curves of SSDP algorithm and ISSDP algorithm, and Fig. 12.12 is the correlation coefficient change curves of SSDP algorithm and ISSDP algorithm.

From Figs. 12.10, 12.11 and 12.12, it can be seen that during the sampling period, when the number of active source signals is not always equal to the number of observed signals, the SSDP algorithm can hardly recover the source signals, while the ISSDP algorithm can still achieve a relatively ideal recovery effect. When the input SNR is 10 dB, the SIR of the SSDP algorithm is only 3 dB and the correlation

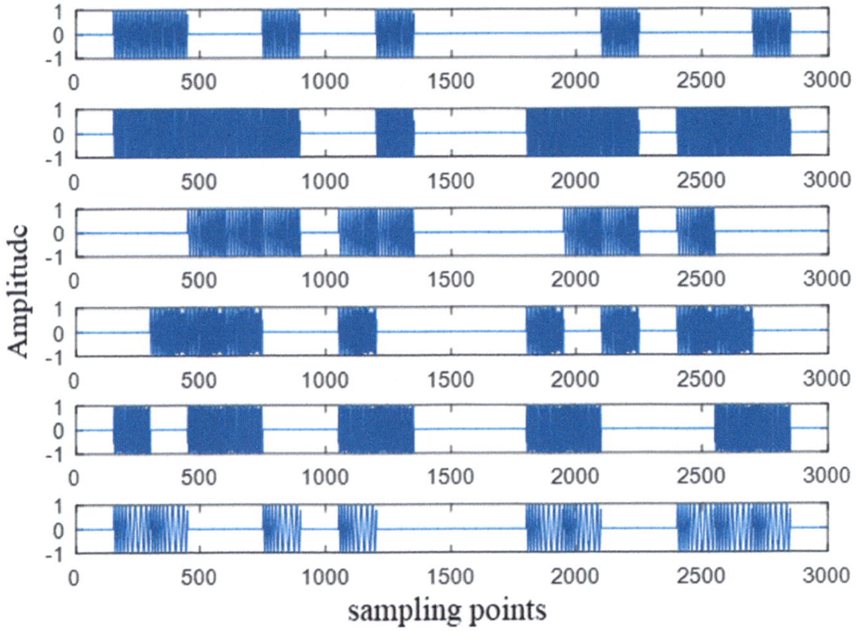

Fig. 12.4 Time domain waveforms of the source signals when four source signals are active at the same time

coefficient is 0.78, but the SIR of ISSDP algorithm reaches 13.6 dB and the correlation coefficient reaches 0.98.

According to simulation 1 and simulation 2, SSDP algorithm can realize source signal recovery only when the number of active source signals is equal to the observed signals, while ISSDP algorithm can realize source signal recovery when the number of active source signals is less than or equal to the number of observed signals.

Summary

This chapter introduces other three algorithms of underdetermined blind separation source signal recovery, including source signal recovery algorithm based on shortest path, source signal recovery algorithm based on statistical Sparse decomposition (SSDP) and improved Statistical sparse decomposition (ISSDP) source signal recovery algorithm. The performance of SSDP and ISSDP algorithm is simulated and compared. The results show that ISSDP algorithm has better performance.

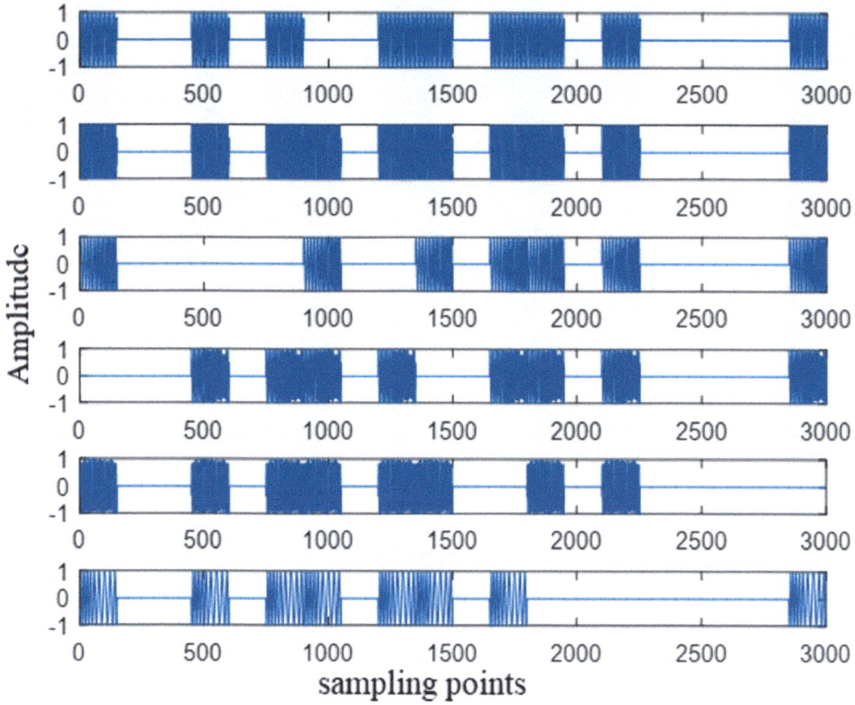

Fig. 12.5 Time domain waveforms of the source signals when five source signals are active at the same time

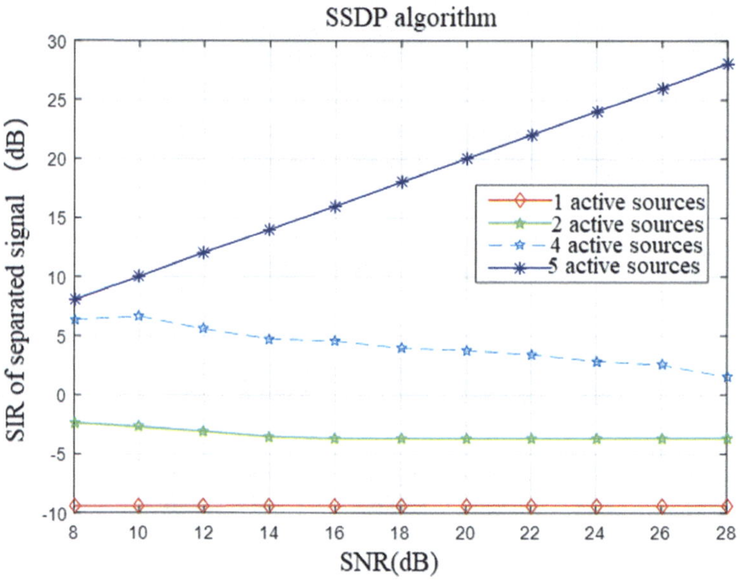

Fig. 12.6 SIR curves of separated signals of SSDP algorithm

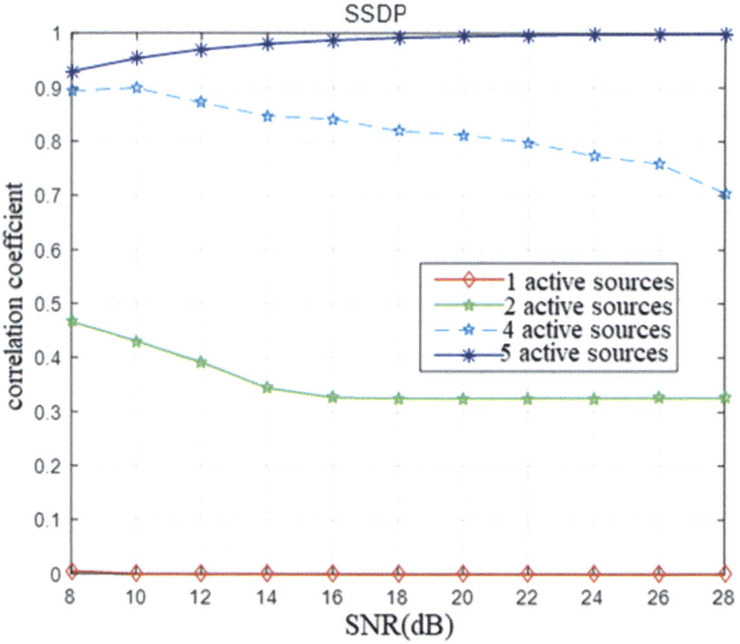

Fig. 12.7 Correlation coefficient curves of SSDP algorithm

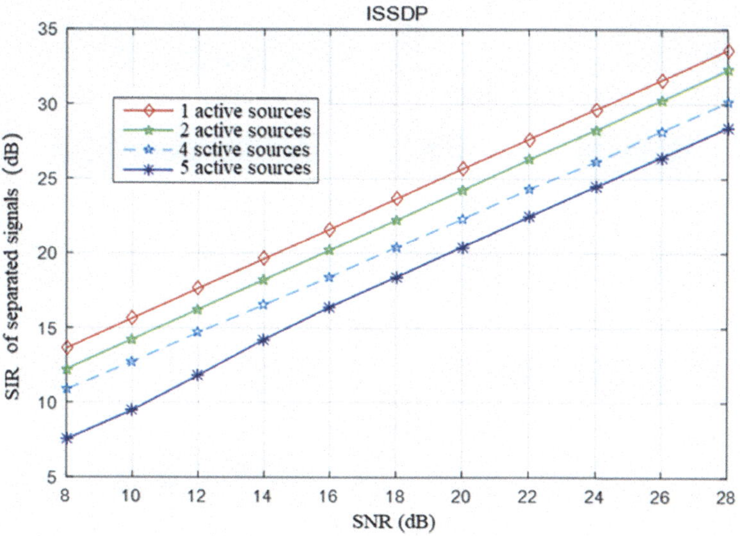

Fig. 12.8 SIR curves of separated signals of ISSDP algorithm

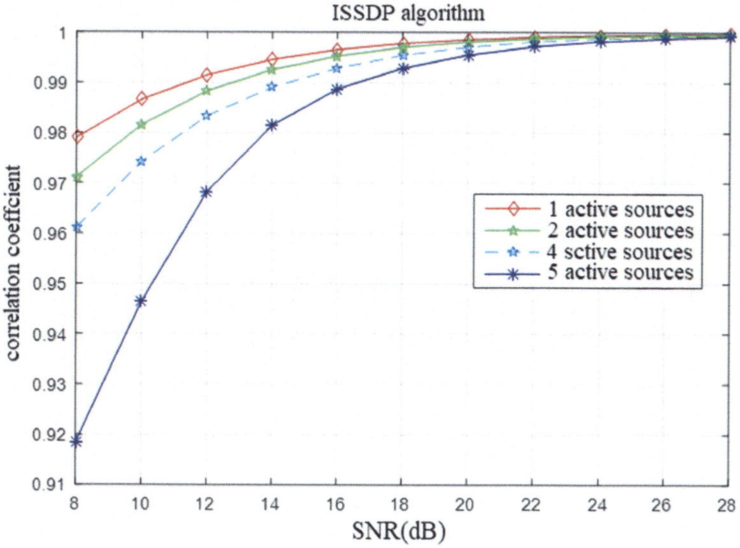

Fig. 12.9 Correlation coefficient curves of ISSDP algorithm

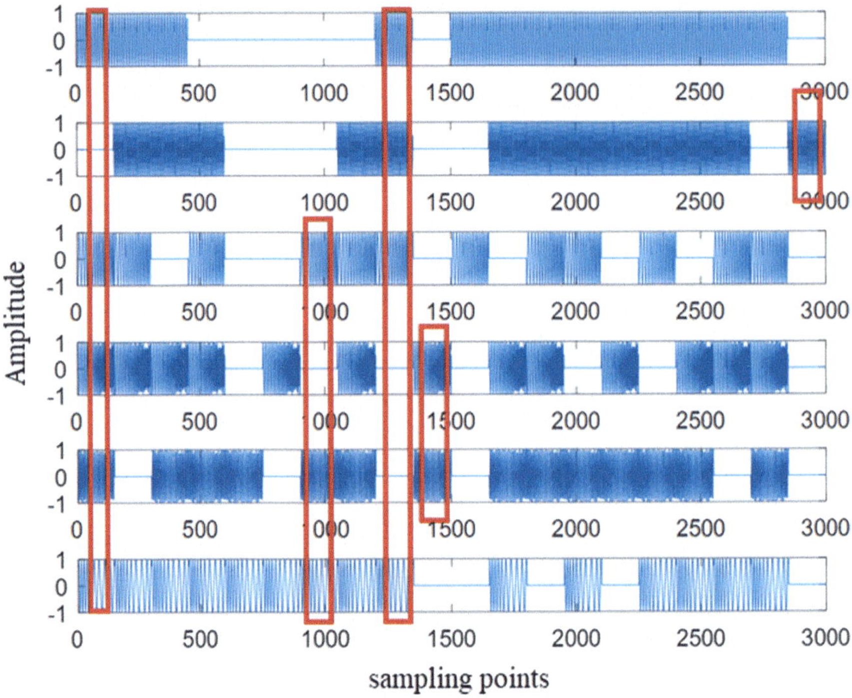

Fig. 12.10 Time domain waveforms of source signals

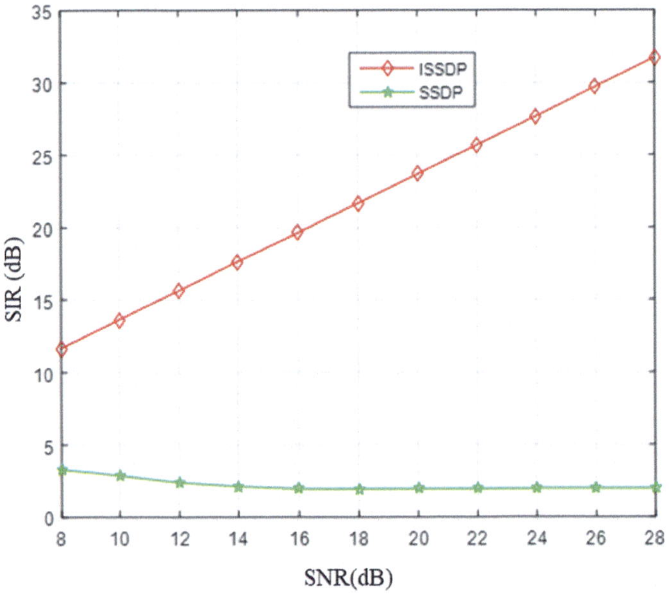

Fig. 12.11 The SIR curves of SSDP algorithm and ISSDP algorithm

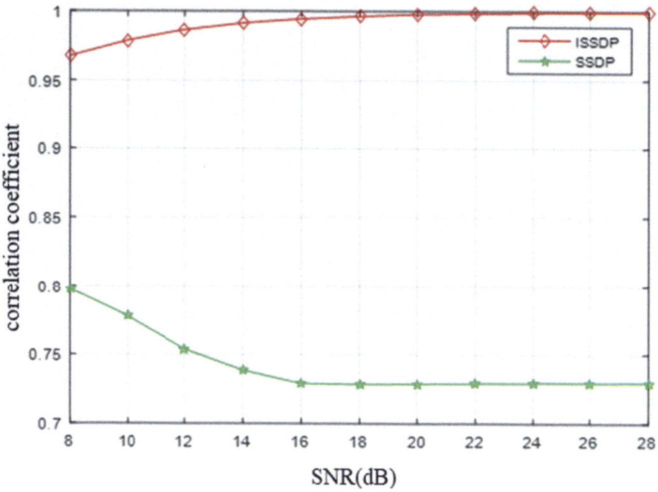

Fig. 12.12 The correlation coefficient change curves of SSDP algorithm and ISSDP algorithm

Chapter 13
Application of Blind Source Separation in Military Communication

Blind source separation technology has a wide application prospect in military communication. This chapter introduces the application examples of blind source separation technology in military communication signal processing. Firstly, a DS-CDMA pseudo-code estimation and multi-user separation technique based on BSS is introduced, the principle of the algorithm is given, and the performance of the algorithm is verified by simulation. Secondly, we study the frequency hopping signal sorting and splicing technology based on BSS, and verify its effectiveness. Thirdly, the blind source separation of the spectrum aliasing communication signals is simulated, and the feasibility of the blind source separation technology to extract useful signals from aliasing signals in the communication countermeasure is verified. Finally, the communication anti-jamming technology based on BSS is studied, and the simulation proves that the BSS technology can resist various types of interference [25].

13.1 DS-CDMA Signal Pseudo-Code Estimation and Multi-user Separation Based on BSS

Some of the existing multi-user detection technologies of CDMA signal need to know the spread spectrum code information of all users, and some even need strict timing and channel information, which will be greatly limited in practical applications. The analysis of DS-CDMA system model shows that the DS-CDMA system model is consistent with the linear instantaneous mixing model in blind source separation, so BSS technology can be used for the multi-user detection of DS-CDMA system. This method does not need to know all the user's spread spectrum code information, and does not need to perform channel estimation. Only according to the received signal, the original bit information sent by the user can be estimated. Because BSS makes full use of the independences of different user information, its performance is better

W. Fu, *Principle and Application of Blind Source Separation Technology*,
https://doi.org/10.1007/978-981-96-0022-9_13

than the traditional multi-user detection algorithm. In addition, the BSS technique is used to estimate the spread spectrum code information of each user, which has broad application prospects in communication reconnaissance. In communication reconnaissance, we do not know any user's spread spectrum code information about the intercepted enemy signal, and we can estimate the pseudo-code of CDMA signal by using the blind estimation algorithm introduced in this chapter, and then demodulate the information code element [55].

13.1.1 DS-CDMA Signal Model

1. **DS-CDMA Signal Model of Synchronous Single Path Channel**

In the downlink of DS-CDMA system, when the multipath environment is not considered, for synchronous DS-CDMA system, the received data is

$$r(t) = \sum_{m=0}^{N-1} \sum_{k=1}^{K} b_{k,m} a_k s_k(t - mT) + n(t) \tag{13.1}$$

where a_k is the fading factor for the k-th user. In general, for all users in the downlink, we have $\{\{a_1 = a_2 = \cdots a_K = a\}\}$. $b_{k,m}$ is the m-th symbol transmitted by the k-th user, T is the symbol period, and $s_k(\cdot)$ is the user's chip sequence. When $t \in [0, T)$, then $s_k(t) = \{-1, +1\}$. And $s_k(t)$ is 0 in other conditions. $n(t)$ is the noise, N is the number of symbols transmitted by the user during the observing time, and K is the number of users.

Let C be the length of the spread spectrum code sequence and the chip rate be C/T. To sample the received data at the chip rate, we have

$$\begin{aligned} r(p \cdot \Delta t) = r\left(p \cdot \frac{T}{C}\right) &= \sum_{m=0}^{N-1} \sum_{k=1}^{K} b_{k,m} \cdot a \cdot s_k\left(p \cdot \frac{T}{C} - mT\right) + n\left(p \cdot \frac{T}{C}\right) \\ &= \sum_{m=0}^{N-1} \sum_{k=1}^{K} b_{k,m} \cdot a \cdot s_k\left[(p - mC)\frac{T}{C}\right] + n\left(p \cdot \frac{T}{C}\right) \end{aligned} \tag{13.2}$$

Let $\tilde{r}(p) \triangleq r(p \cdot \Delta t), \tilde{s}_k(p) \triangleq s_k(p \cdot \Delta t), \tilde{n}(p) \triangleq n(p \cdot \Delta t)$ represent discrete sample of $r(t), s_k(t), n(t)$ respectively. We know from the definition of $s_k(t)$ that

$$\tilde{s}_k(p) = \begin{cases} -1 \text{ or}, +1, & p \in [0, C) \\ 0, & \text{others} \end{cases} \tag{13.3}$$

Then Eq. (13.2) can be written as:

$$\tilde{r}(p) = \sum_{\substack{m=0 \\ p-mC\in[0,C)}}^{N-1} \sum_{k=1}^{K} b_{k,m} \cdot a \cdot \tilde{s}_k\big[(p - mC)\big] + \tilde{n}(p) \tag{13.4}$$

Let

$$\tilde{\boldsymbol{r}}_m = [\tilde{r}[mC], \tilde{r}[mC+1]], \ldots, \tilde{r}[(m+1)C-1]^{\mathrm{T}} \tag{13.5}$$

$$\tilde{n}_m = [n(mC), n(mC+1), n[(m+1)C-1]]^{\mathrm{T}} \tag{13.6}$$

Then

$$\tilde{\boldsymbol{r}}_m = \sum_{k=1}^{K} b_{k,m} \cdot a \cdot \boldsymbol{g}_k + \tilde{\boldsymbol{n}}_m \tag{13.7}$$

where

$$\boldsymbol{g}_k = [as_k[1], as_k[2], \ldots, as_k[C]]^{\mathrm{T}} \tag{13.8}$$

Let $\boldsymbol{G} = \left[\boldsymbol{g}_1\ \boldsymbol{g}_2\ \ldots\ \boldsymbol{g}_K\right]$, $\boldsymbol{b}_m = \left[b_{1,m}\ b_{2,m}\ \ldots\ b_{K,m}\right]^{\mathrm{T}}$, and Eq. (13.7) can be written as

$$\boldsymbol{r}_m = \boldsymbol{G} \times \boldsymbol{b}_m + \boldsymbol{n}_m \tag{13.9}$$

Equation (13.9) is the signal model of synchronous DS-CDMA system without considering multipath environment.

2. **DS-CDMA Signal Model of Multipath Channel**

When considering the multipath environment, for a synchronous DS-CDMA system, the received data of DS-CDMA system downlink is

$$r(t) = \sum_{m=0}^{N-1}\sum_{k=1}^{K}\sum_{l=1}^{L} b_{k,m} \cdot a_{kl} \cdot s_k(t - mT - \tau_{kl}) + n(t) \tag{13.10}$$

where a_{kl} is the fading factor of the k-th user and l-th path, it is generally assumed that $a_{1l} = a_{2l} = \cdots a_{Kl} = a_l$. $b_{k,m}$ is the m-th symbol transmitted by the k-th user, T is the symbol period, and $s_k(\cdot)$ is the user's chip sequence, when $t \in [0, T)$, $s_k(t) = \{-1, +1\}$, and in other cases $s_k(t)$ is 0. τ_{kl} is the delay of k-th user and l-th path. For all users, there is $\tau_{1l} = \tau_{2l} = \cdots \tau_{Kl} = \tau_l$, $0 < \tau_l < T/2$. $n(t)$ is the noise, N is the number of symbols transmitted by the user during the observing time, and K is the number of users. Then

$$r(t) = \sum_{m=0}^{N-1} \sum_{k=1}^{K} \sum_{l=1}^{L} b_{k,m} \cdot a_l \cdot s_k(t - mT - \tau_l) + n(t) \tag{13.11}$$

Similarly, C is the length of the spread spectrum code sequence, and the chip rate is C/T. Then the received data $r(t)$ is sampled with the chip rate, then we get

$$r\left(p \cdot \frac{T}{C}\right) = \sum_{m=0}^{N-1} \sum_{k=1}^{K} \sum_{l=1}^{L} b_{k,m} \cdot a_l \cdot s_k\left(p \cdot \frac{T}{C} - mT - \tau_l\right) + n\left(p \cdot \frac{T}{C}\right) \tag{13.12}$$

Let $\Delta t = T/C$, $d_l = \lceil \tau_l \cdot C/T \rceil$, here $[\cdot]$ represents taking the integer closest to the variable, d_l is the discrete delay of the l-th path, $d_l \in \{0, 1, \ldots, (C-1)/2\}$, then the above formula can be written as

$$r(p \cdot \Delta t) = \sum_{m=0}^{N-1} \sum_{k=1}^{K} \sum_{l=1}^{L} b_{k,m} \cdot a_l \cdot s_k\big[(p - mC - d_l)\Delta t\big] + n(p \cdot \Delta t) \tag{13.13}$$

Take its discrete value, then:

$$\tilde{r}(p) = \sum_{m=0}^{N-1} \sum_{k=1}^{K} \sum_{l=1}^{L} b_{k,m} a_l \tilde{s}_k\big[p - mC - d_l\big] + \tilde{n}(p) \tag{13.14}$$

Let

$$\tilde{\boldsymbol{r}}_m = [\tilde{r}[mC]\tilde{r}[mC+1]\cdots\tilde{r}[(m+1)C-1]\tilde{r}[(m+1)C]]^{\mathrm{T}} \tag{13.15}$$

$$\tilde{\boldsymbol{n}}_m = [\tilde{n}[mC]\tilde{n}[mC+1]\cdots\tilde{n}[(m+1)C-1]\tilde{n}[(m+1)C]]^{\mathrm{T}} \tag{13.16}$$

Then

$$\tilde{\boldsymbol{r}}_m = \sum_{k=1}^{K} \left[b_{k,m-1} \sum_{l=1}^{L} a_l \boldsymbol{g}_{k,l} + b_{k,m} \sum_{l-1}^{L} a_l \underline{\boldsymbol{g}}_{k,l} \right] + \tilde{\boldsymbol{n}}_m \tag{13.17}$$

where $\tilde{\boldsymbol{n}}_m$ represents the noise vector, then the "front" and "back" halves of the spread spectrum code vector are respectively

$$\boldsymbol{g}_{kl} = [s_k(C - d_l + 1) \ldots s_k(C)0 \ldots 0]^{\mathrm{T}} \tag{13.18}$$

$$\underline{\boldsymbol{g}}_{kl} = [0 \ldots 0 s_k(1) \ldots s_k(C - d_l)]^{\mathrm{T}} \tag{13.19}$$

Let $\boldsymbol{b}_m = [b_{1m}, b_{2m}, \ldots, b_{Km}]^T$, $\boldsymbol{G}_0 = \left[\sum_{l=1}^{L} a_l g_{1l}, \sum_{l=1}^{L} a_l g_{2l}, \ldots, \sum_{l=1}^{L} a_l g_{Kl}\right]$, $\boldsymbol{G}_1 = \left[\sum_{l=1}^{L} a_l \underline{g}_{1l}, \sum_{l=1}^{L} a_l \underline{g}_{2l}, \ldots, \sum_{l=1}^{L} a_l \underline{g}_{Kl}\right]$, and Eq. (13.17) can be expressed by the matrix as

$$\tilde{\boldsymbol{r}}_m = [\boldsymbol{G}_0\ \boldsymbol{G}_1]\begin{bmatrix}\boldsymbol{b}_m \\ \boldsymbol{b}_{m-1}\end{bmatrix} + \tilde{\boldsymbol{n}}_m = \boldsymbol{G}\hat{\boldsymbol{b}}_m + \tilde{\boldsymbol{n}}_m \tag{13.20}$$

where $\boldsymbol{G}_0, \boldsymbol{G}_1$ are $C \times K$ mixing matrices corresponding to source signal $\boldsymbol{b}_m$ and delay signal $\boldsymbol{b}_{m-1}$ respectively. The column vectors of $\boldsymbol{G}_0, \boldsymbol{G}_1$ are given by the "before" and "after" parts of the coding vector respectively. Formula (13.20) is the DS-CDMA signal model under multipath channel conditions.

13.1.2 Pseudo-Code Estimation of DS-CDMA Signal Based on BSS

It can be seen that the signal model of DS-CDMA system described in Eq. (13.9) is consistent with the linear mixing model in blind source separation. If the number of observed symbols is N, that is, in Eq. (13.9) $m = 1, 2, \ldots, N$, let $\boldsymbol{X} = [\boldsymbol{r}_1, \boldsymbol{r}_2, \ldots \boldsymbol{r}_N]$, $\boldsymbol{B} = [\boldsymbol{b}_1, \boldsymbol{b}_2, \ldots, \boldsymbol{b}_N]$, without considering the influence of noise, then

$$\boldsymbol{X} = \boldsymbol{G}\boldsymbol{B} \tag{13.21}$$

The correlation matrix for $\boldsymbol{X}$ is $\boldsymbol{R}_{XX} = E\{\boldsymbol{X}\boldsymbol{X}^{\mathrm{T}}\}$, and the eigenvalue decomposition of $\boldsymbol{R}_{XX}$ is

$$\boldsymbol{R}_{XX} = \boldsymbol{U}\boldsymbol{D}\boldsymbol{U}^{\mathrm{T}} \tag{13.22}$$

Suppose $\boldsymbol{U}_s \in R^{L\times K}$ is a matrix composed of K principal eigenvectors of $\boldsymbol{R}_{XX}$, then the space spanned by the column vectors of $\boldsymbol{U}_s^{\mathrm{T}} \in R^{K\times L}$ should belong to the same subspace as the subspace spanned by the column vectors of $\boldsymbol{G}^{\mathrm{T}}$, that is, there is a linear transformation relationship between them. Suppose the linear transformation is $\boldsymbol{A}$, then

$$\boldsymbol{U}_s^{\mathrm{T}} = \boldsymbol{A}\boldsymbol{G}^{\mathrm{T}} \tag{13.23}$$

Equation (13.23) is equivalent to a linear mixing model of BSS, where $\boldsymbol{U}_s^{\mathrm{T}}$ is the mixed (observed) signal matrix, $\boldsymbol{A}$ is the unknown mixing matrix, and $\boldsymbol{G}^{\mathrm{T}}$ is the source signal. Therefore, according to $\boldsymbol{U}_s^{\mathrm{T}}$, $\boldsymbol{G}^{\mathrm{T}}$ can be estimated by using the BSS algorithm. Assuming that the signal modulation mode is BPSK modulation, the separated signal after blind source separation is $\boldsymbol{Y}$, then the user's spread spectrum code can be estimated as follows.

$$\hat{G}^{\mathrm{T}} = sign(\boldsymbol{Y}) \tag{13.24}$$

In multipath environment, when the delay of main path is 0 and the gain of main path is much larger than other paths, the above method can also be used to estimate the user's spread spectrum code information.

13.1.3 Multi-user Separation and Demodulation of DS-CDMA Signal Based on BSS

As can be seen from Eq. (13.20), the multipath CDMA system model is consistent with the instantaneous linear mixing model in BSS, so we naturally think that BSS technology can be used to estimate user's information bits. For civil communication, generally in the downlink, users only know their own spread spectrum code and cannot know the spread spectrum code of other users, and the traditional multi-user detection requires each user to know the spread spectrum codes of all users. The BSS technology has no limitation on this, and it can estimate the information bits transmitted by each user without any spread spectrum code information. By using BSS technology, the channel estimation can also be omitted, and the user transmitted information can be estimated using only the mixed signals received from multiple users.

In addition, in civil communication, even if the downlink users know their own spread spectrum code information, they can use the matching filter method to detect the information of interest, but the simulation results show that this matching filter method is only the best in the synchronous single path CDMA system, and in the multipath environment, the bit error rate (BER) performance of signal detected by this method will deteriorate. Because the BSS technology makes use of the independence of different user information bits, the BER performance is greatly improved.

Because of the inherent ambiguity of BSS technology, if the initial conditions of BSS algorithm are not restricted, the corresponding relationship between the information detected by this method and the user cannot be determined. In order to solve this problem, we can use the method of training sequence to limit the initial vector of the algorithm, so as to ensure that the recovered information bit is the information bit of the user that we are interested in. The method is as follows.

Assuming that you are interested in the information of the first user, the initial value of the separating vector is

$$\boldsymbol{w}(0) = \boldsymbol{Y}_p \left[b_{11}\ b_{12}\ \ldots\ b_{1p}\right]^{\mathrm{T}} \tag{13.25}$$

where $b_{11}\ b_{12}\ \ldots\ b_{1p}$ is the first p bits of information (training sequence) transmitted by the interested user, and Y_p is the observed signal corresponding to the training sequence. After the observed signal matrix $\boldsymbol{R} = [\boldsymbol{r}_1\ \boldsymbol{r}_2\ \ldots\ \boldsymbol{r}_M]$ is whitened, the initial value $\boldsymbol{w}(0)$ of the separating vector represented by Eq. (13.25) can be used to separate the signal vector of interest, denoted by $\boldsymbol{y}_1$, by using the overdetermined BSS algorithm introduced in the previous chapter (such as EASI, FastICA, etc.). Assuming that the information modulation mode is BPSK modulation, then taking the sign operation on $\boldsymbol{y}_1$ can estimate the bit information transmitted by the interested user.

In communication reconnaissance, we cannot know any spread spectrum code information about the enemy's CDMA signal, so it is obviously impossible to estimate the original information by using traditional matching filter or decorrelation algorithm. Therefore, in communication reconnaissance, BSS technology is used to detect the multi-user CDMA signals, so as to realize non-cooperative demodulation of enemy CDMA signal user's information, and its application prospect will be broader.

13.1.4 Computer Simulation

1. **Simulation Results of Multi-user Spread Spectrum Code Estimation**

The number of users is 4, the spread spectrum code is a 64-bit Walsh sequence, the modulation mode of CDMA signal is BPSK modulation, and the SNR is -10 dB. The algorithm proposed in Sect. 13.1.3 is used to estimate the spread spectrum code. Figure 13.1 shows the simulation results in a single path channel environment. Figure 13.2 shows the results of spread spectrum code estimation in a multipath environment with three paths. The delay and gain of paths respectively are delay = [0 10 20], gain = [1 0.1 0.01].

As can be seen from Figs. 13.1 and 13.2, the spread spectrum codes of the four users can be correctly estimated [56].

In order to study the performance of using blind source separation algorithm to estimate spread spectrum code under different user numbers, Figs. 13.3 and 13.4 show the number of spread spectrum codes of all users can be completely estimated correctly after 50 simulations under single path and multipath environment respectively, where N is the number of users.

From the above simulation results, we can see that in single path or multipath environment, when the length of spread spectrum code is 64, if the number of users is less than 6, as long as the SNR is above -6 dB, the method given in this paper can estimate the spread spectrum code of all users correctly. When the number of users is smaller than 6, the spread spectrum code of all users can be correctly estimated when the SNR is only -8 dB. If the length of the spread spectrum code is fixed, when the number of users increases to a certain number, it will be difficult to estimate the user spread spectrum code completely and correctly in all experiments. Therefore, the length of the spread spectrum code is fixed, and the number of users supported by the algorithm has an upper limit.

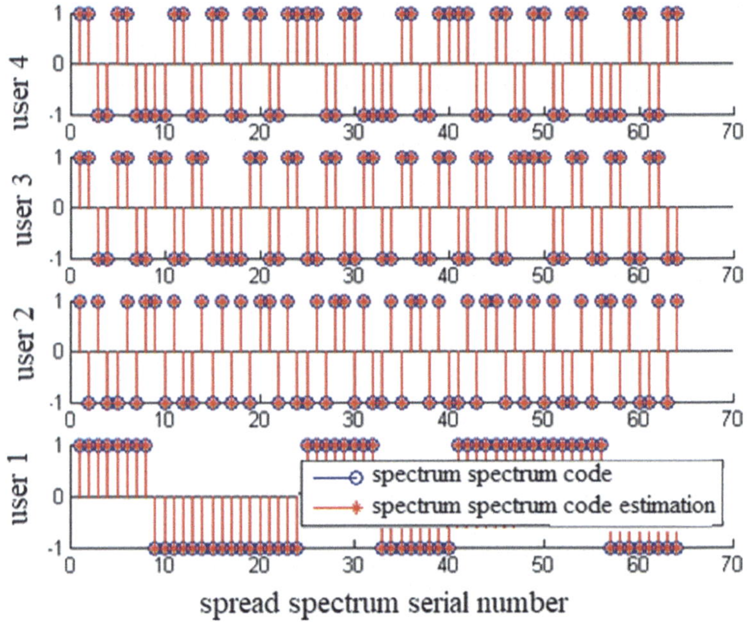

Fig. 13.1 Spread spectrum sequence estimation in single path channel environment

In addition, the simulation shows that if the length of spread spectrum code increases, the number of users supported by the algorithm will also increase. In other words, under the condition of the same number of users and the same SNR, the longer the length of spread spectrum code, the greater the proportion that can be correctly estimated. For example, when the length of the spread spectrum code is 128, the number of users is 7, and the SNR is −6 dB, the spread spectrum code of all users can be correctly estimated in 50 experiments, and when the length is 64, only some simulation experiments can correctly estimate the spread spectrum code of all users.

2. **Simulation Results of Multi-user Separation**

Assume that the number of users is 4, the spread spectrum code is a golden sequence of length 31, and the channel environment is a multipath channel with 4 multipaths, delay and gain of paths are: delay = [0 3 5 7], gain = [1 0.5 0.3 0.1]. The received signals are detected by blind source separation algorithm, matching filter and spatial decorrelation algorithm. The number of information bits transmitted by each user in each simulation is 10,000, and 100 Monte Carlo simulations are carried out. Figure 13.5 shows the relationship between the average bit error rate (BER) and SNR of the first user.

As can be seen from the figure, when BSS technology is used for multi-user detection, its BER performance is greatly improved compared with the matching filter and multi-user detection algorithm when the SNR is high. When the SNR is

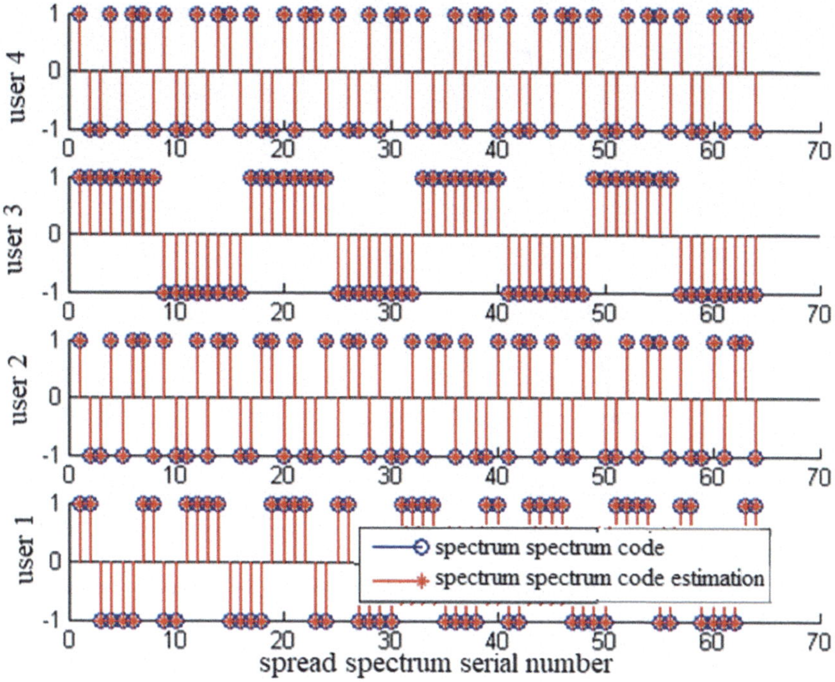

Fig. 13.2 Spread spectrum sequence estimation in multipath channel environment

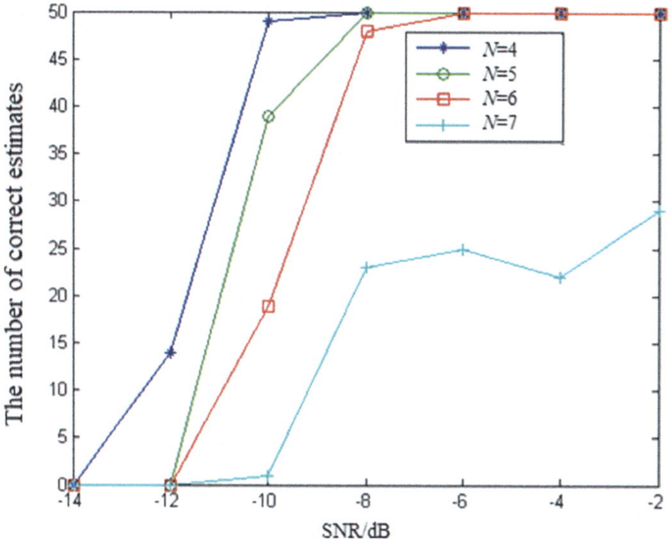

Fig. 13.3 The number of spread spectrum codes of all users can be correctly estimated in 50 experiments (single path)

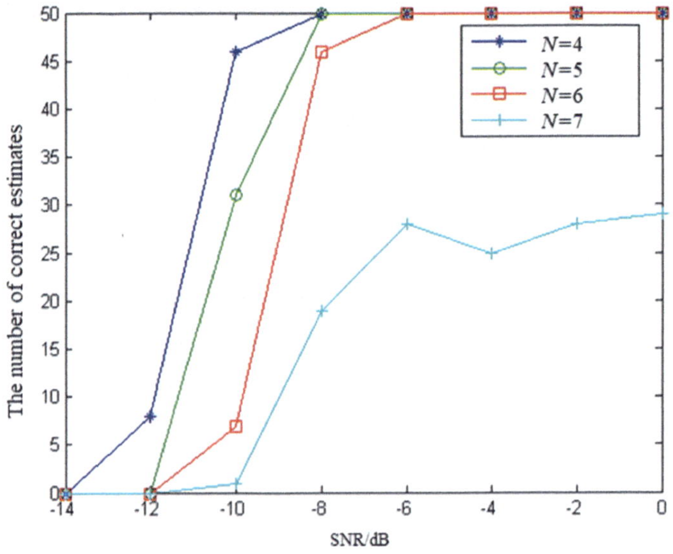

Fig. 13.4 The number of spread spectrum codes of all users can be correctly estimated in 50 experiments (multi-path)

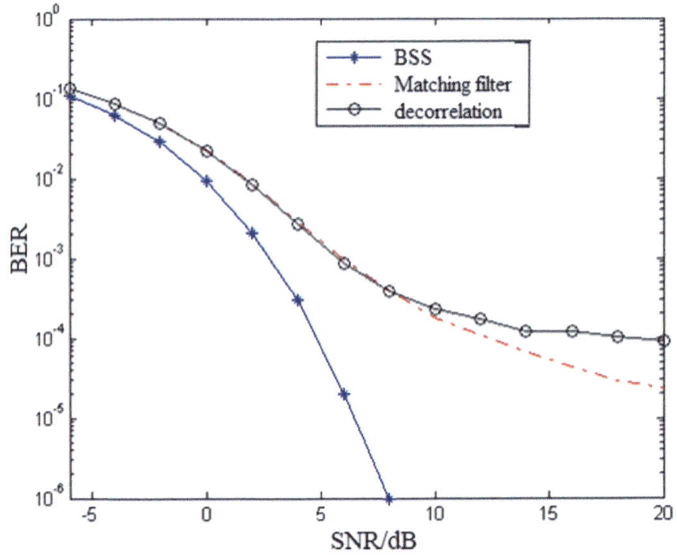

Fig. 13.5 BER performance comparison for the first user

6 dB, the BER is reduced by two orders of magnitude compared with the matching filter or the decorrelation algorithm. In addition, compared with the decorrelation technique, the BSS method has the advantage that it does not need to know the spread spectrum code information of all users, which is very important for communication reconnaissance.

By analyzing the DS-CDMA signal model, it is found that the DS-CDMA system model is consistent with the linear instantaneous mixing BSS model, and the BSS technology is proposed for the blind estimation of DS-CDMA signal spread spectrum code. In addition, the proposed multi-user detection algorithm based on BSS has better BER performance than the traditional matching filtering method and decorrelation algorithm. Moreover, the traditional decorrelation algorithm needs to know the spread spectrum code information of all users, while the BSS algorithm does not need to know this information, and can estimate the bit information transmitted by each user only based on the observed signal. Therefore, the multi-user detection algorithm based on BSS proposed by us not only has great advantages in civilian communication, but also has broad application prospects in communication reconnaissance.

13.2 Frequency Hopping Signal Sorting and Splicing Technology Based on BSS

The reconnaissance of frequency hopping (FH) signals (sorting, splicing, de-hopping and demodulation of FH signal) has always been a technical problem faced by communication countermeasures. The performance of traditional FH network station sorting and splicing methods based on arrival time deteriorates sharply under complex electromagnetic environment. Since the FH signals of different network stations have statistical independence, and there is correlation between different hopping points of the same FH signal, it is natural to think whether BSS technology can be used to realize the sorting and splicing of FH signal. Next, we demonstrate the feasibility of sorting and splicing of FH signal based on BSS technology through simulation.

Suppose that the frequency sets of the three FH signals are respectively:

$$f_{11} = 63\,\text{MHz}\; f_{12} = 60\,\text{MHz}\; f_{13} = 69\,\text{MHz}\; f_{14} = 66\,\text{MHz}$$

$$f_{21} = 67\,\text{MHz}\; f_{22} = 64\,\text{MHz}\; f_{23} = 67\,\text{MHz}\; f_{24} = 70\,\text{MHz}$$

$$f_{31} = 68\,\text{MHz}\; f_{32} = 65\,\text{MHz}\; f_{33} = 71\,\text{MHz}\; f_{34} = 62\,\text{MHz}$$

The data sampling frequency is 56 MHz, the rate of frequency hopping is 1000 hop/s, the data rate is 16 Kb/s, the duty ratio of the FH signal is 1/10, and the SNR is 20 dB. The signal incident angles of three FH signals are 0°, 20°, 40° respectively, and a uniform circular array of 6 elements is adopted for receiving. The radius of the circular array is 3/4 times the wavelength of the center frequency band. After the BSS algorithm is used to separate the observed signals, the ISR of the three separated signals within 4 frequency hopping time cycles is as follows:

$\text{ISR}_1 = [-37.1352\ -49.2193\ -49.0134]$; $\text{ISR}_2 = [-34.3432\ -36.4996\ -42.4762]$;

$\text{ISR}_3 = [-41.0309\ -31.8062\ -35.0421]$; $\text{ISR}_4 = [-32.4836\ -41.2423\ -36.3608]$;

Where ISR_i represents the ISR of the three separated signals in the i-th FH period. After convergence, the global matrices in the 4 FH period respectively are

$$|\boldsymbol{G}_1| = \begin{bmatrix} 0.0017 & 1.0515 & 0.0145 \\ 0.0008 & 0.0038 & 1.1200 \\ 1.0478 & 0.0037 & 0.0004 \end{bmatrix}; \quad |\boldsymbol{G}_2| = \begin{bmatrix} 0.0018 & 1.0417 & 0.0199 \\ 0.0061 & 0.0156 & 1.1167 \\ 1.0490 & 0.0076 & 0.0022 \end{bmatrix}$$

$$|\boldsymbol{G}_3| = \begin{bmatrix} 0.0014 & 1.0319 & 0.0090 \\ 0.0104 & 0.0269 & 1.1228 \\ 1.0439 & 0.0184 & 0.0014 \end{bmatrix}; \quad |\boldsymbol{G}_4| = \begin{bmatrix} 0.0016 & 1.0611 & 0.0252 \\ 0.0047 & 0.0084 & 1.1130 \\ 1.0461 & 0.0154 & 0.0041 \end{bmatrix}$$

It can be seen that the global matrix after the algorithm convergence is approximately a quasi-identity matrix with only one non-zero element in each row and column, so the blind separation algorithm in this paper can separate FH signals well. In order to see the waveform of the separated signal more directly, the time domain waveform of the three FH source signals and separated signals are shown in Figs. 13.6 and 13.8 respectively [25]. Figure 13.7 shows the time domain waveform diagrams of 6 observed (mixed) signals. As can be seen from Fig. 13.8, after separating the FH mixed signals, the separated signal is a good estimate of each source signal.

In order to more intuitively see the frequency correspondence between the separation signal and the corresponding source signal, their amplitude-frequency characteristics are given in Figs. 13.9, 13.10 and 13.11.

13.3 Blind Separation of Spectrum Aliasing Signal

In the communication countermeasure, sometimes the communication signal is often interfered by the enemy interference signal, and its spectrum is also covered with the interference signal, so how to extract useful communication signal from the interference signal is a problem that must be solved in the communication countermeasure. Since the interference signal and the communication signal come from different transmitting end, and they are statistically independent, so we can use BSS technology to accomplish the above tasks.

Assume that the three source signals: s1 is a complex signal (i.e. interference signal) with a frequency range of 64–76 MHz; s2 is a BPSK signal with carrier frequency of 71 MHz and a baud rate of 20 kB/s; s3 is a BPSK signal with carrier

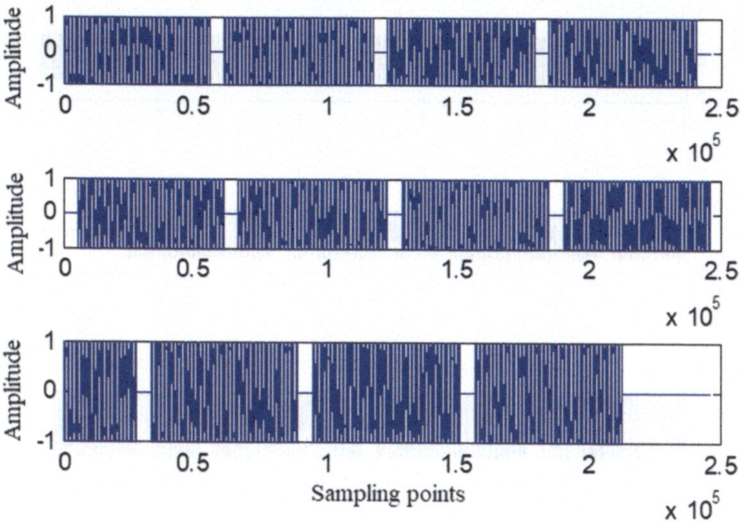

Fig. 13.6 Time domain waveforms of source signals

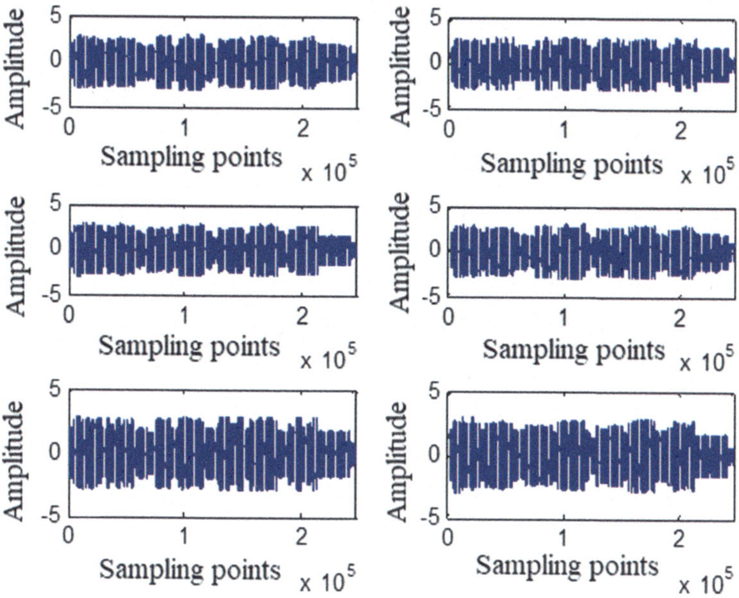

Fig. 13.7 Time domain waveforms of observed signals

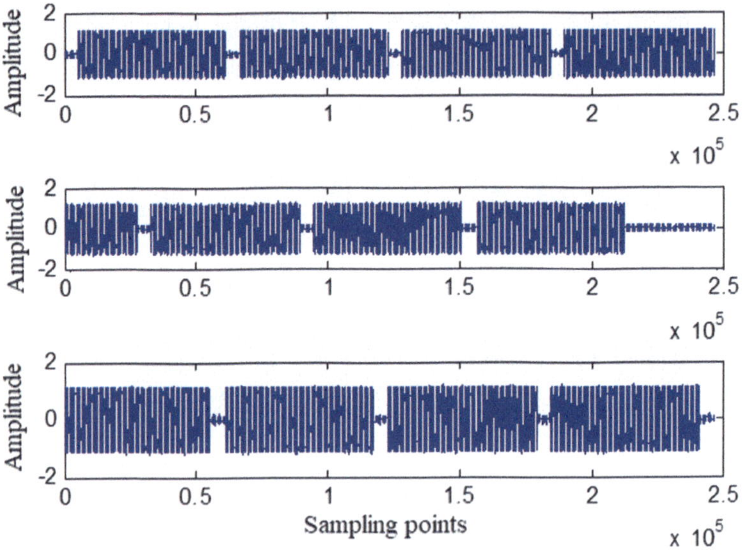

Fig. 13.8 Time domain waveforms of separated signals

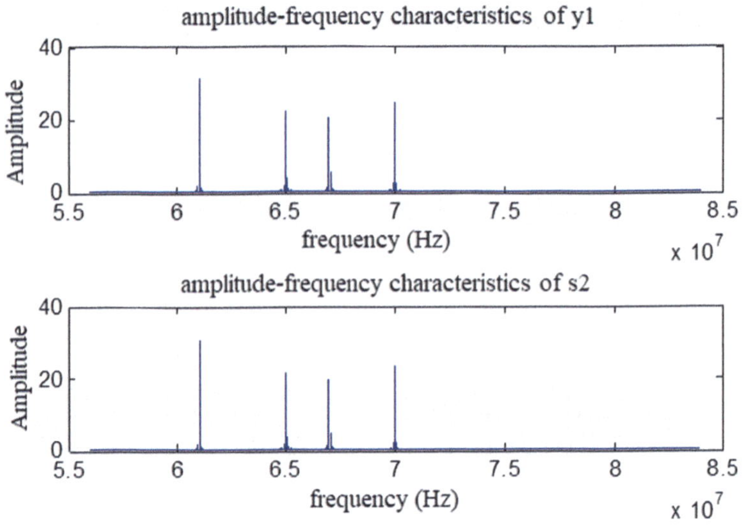

Fig. 13.9 Amplitude-frequency characteristics of separated signal ***y*1** and corresponding source signal s2

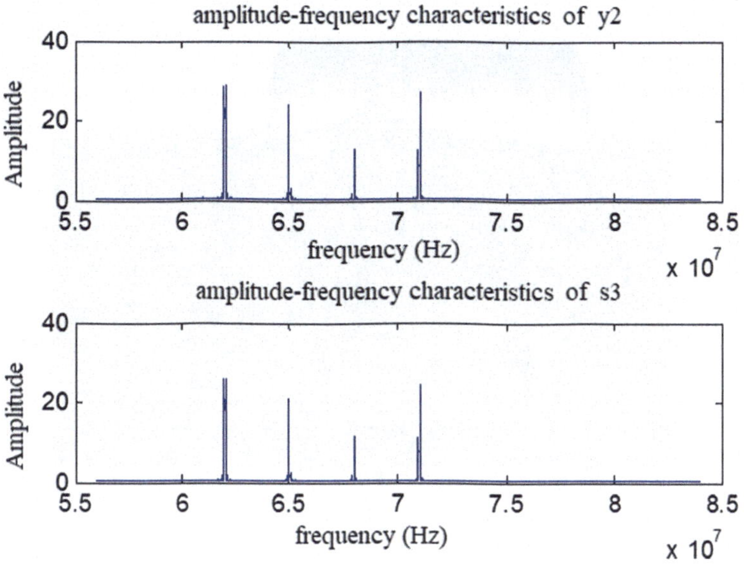

Fig. 13.10 Amplitude-frequency characteristics of separated signal ***y***2 and corresponding source signal s3

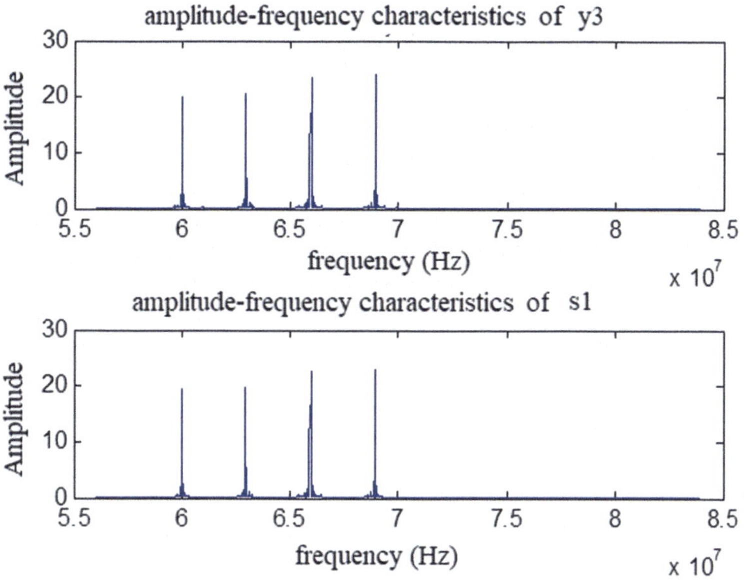

Fig. 13.11 Amplitude-frequency characteristics of separated signal ***y***3 and corresponding source signal s1

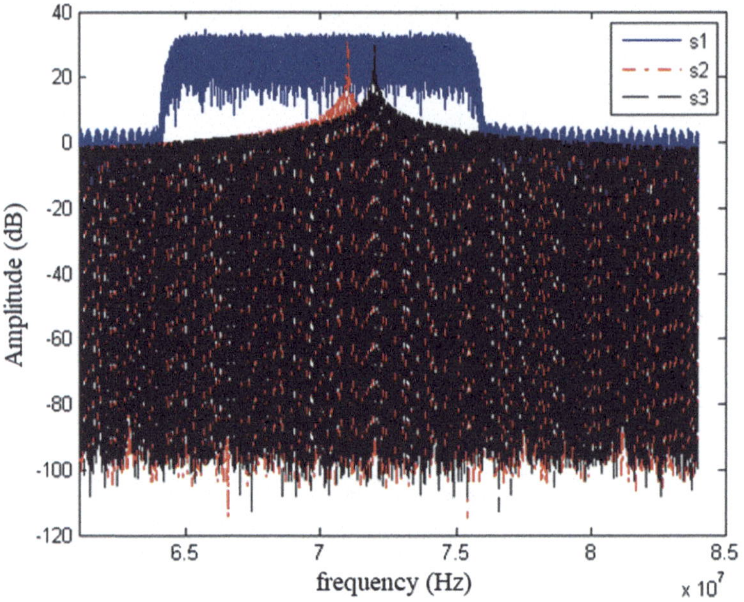

Fig. 13.12 Spectrum diagrams of source signals

frequency of 72 MHz and baud rate of 20 kB/s. Moreover, the spectrum of s2 and s3 is hidden in the spectrum of s1, and the spectrum diagrams of the three source signals are shown in Fig. 13.12 [25].

A uniform circular array of 4 elements is adopted to receive signals, and the mixing matrix is

$$\boldsymbol{H} = \begin{bmatrix} 0.5000 - 0.8660i & 0.8623 - 0.5065i & 0.8584 + 0.5129i \\ 0.7381 - 0.6746i & 0.5183 - 0.8552i & 0.9614 - 0.2752i \\ 1.0000 - 0.0000i & 0.6059 - 0.7955i & 0.5956 - 0.8033i \\ 0.7381 + 0.6746i & 0.9625 - 0.2715i & 0.5059 - 0.8626i \end{bmatrix}$$

The spectrum diagrams of the observed signals are shown in Fig. 13.13.

After the four observed signals are separated by EASI blind separation algorithm, the global matrix is

$$\boldsymbol{G} = \begin{bmatrix} -0.5715 + 0.9870i & -0.0975 - 0.0325i & 0.0684 + 0.0961i \\ 0.0030 + 0.0029i & -0.0001 + 0.0001i & -23.5606 - 23.5799i \\ 0.0017 + 0.0001i & 33.3226 + 0.8475i & -0.0002 + 0.0000i \end{bmatrix}$$

The modulus of G is

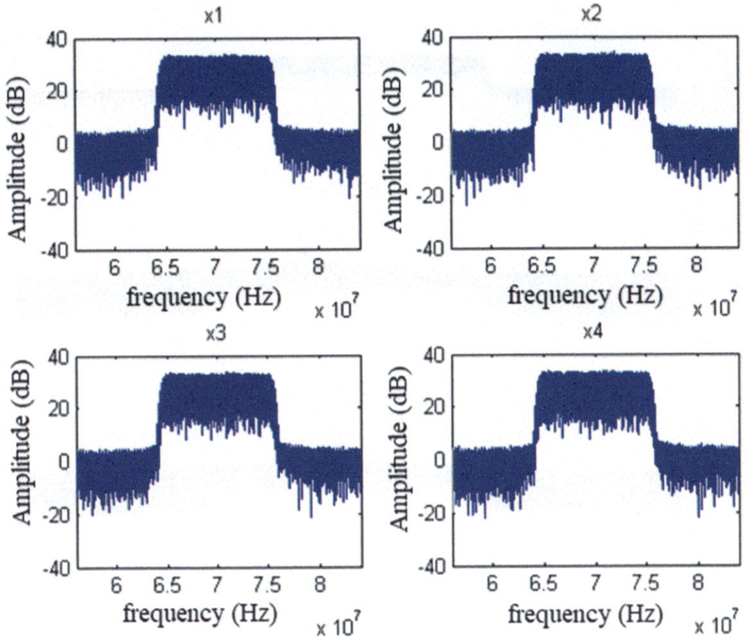

Fig. 13.13 Spectrum diagrams of observed signals

$$|G| = \begin{bmatrix} 1.1405 & 0.1028 & 0.1179 \\ 0.0042 & 0.0002 & 33.3334 \\ 0.0017 & 33.3334 & 0.0002 \end{bmatrix}$$

It can be seen from the results that the BSS technology can separate the source signals of spectrum aliasing, that is, it can extract useful communication signals from high-power interference signals.

In order to visually see the separation effect, the spectrum diagrams and time domain waveforms of the separated signals are given respectively in Figs. 13.14 and 13.15.

Figures 13.16 and 13.17 show the time domain waveforms of the observed signals and the source signal respectively.

When there is Gaussian white noise with a SNR of 20 dB, the global matrix of EASI algorithm is as follows.

$$\boldsymbol{G} = \begin{bmatrix} -0.5705 + 0.9865i & 0.4480 + 0.0141i & 0.1532 - 0.1538i \\ -0.0013 + 0.0013i & -0.4486 - 0.2050i & -35.1293 - 35.1314i \\ 0.0039 - 0.0037i & 49.1149 + 1.2509i & 0.4441 + 0.1619i \end{bmatrix}$$

The modulus of G is

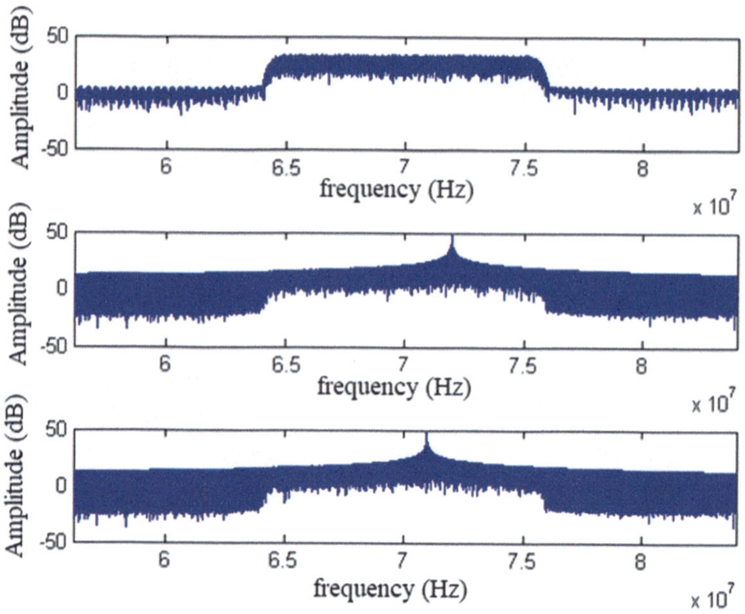

Fig. 13.14 Spectrum diagrams of separated signals by EASI algorithm

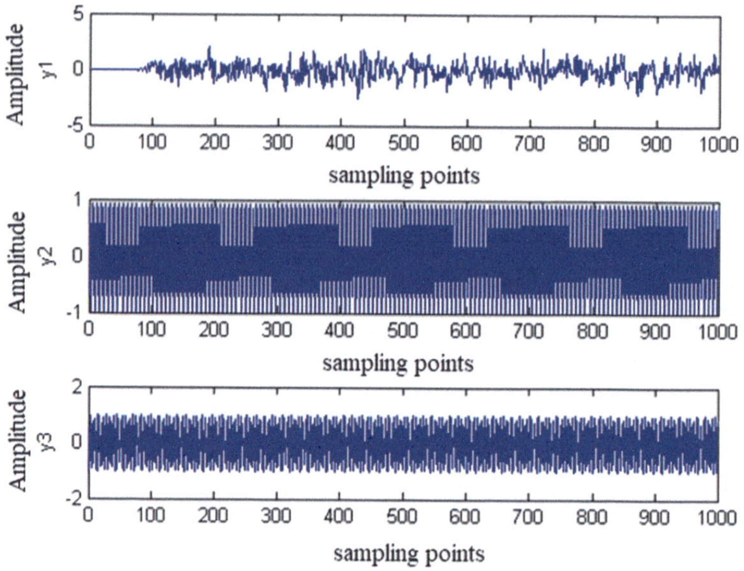

Fig. 13.15 Time domain waveforms of separated signals

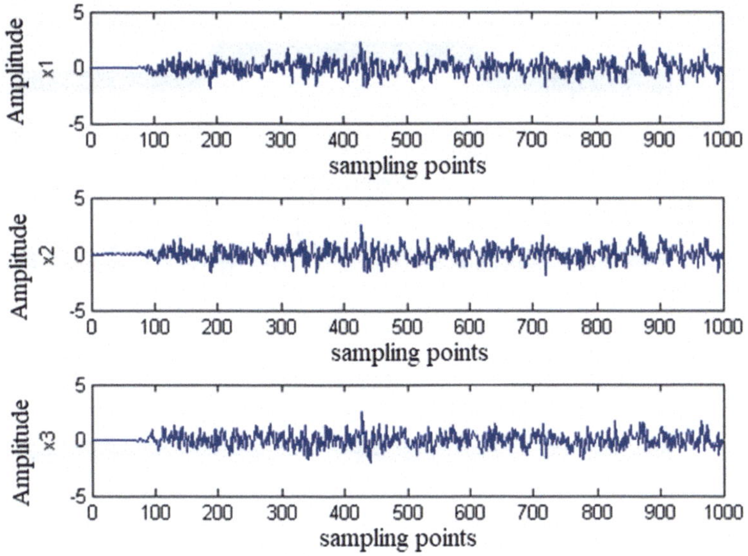

Fig. 13.16 Time domain waveform diagrams of the first 3 mixed signals

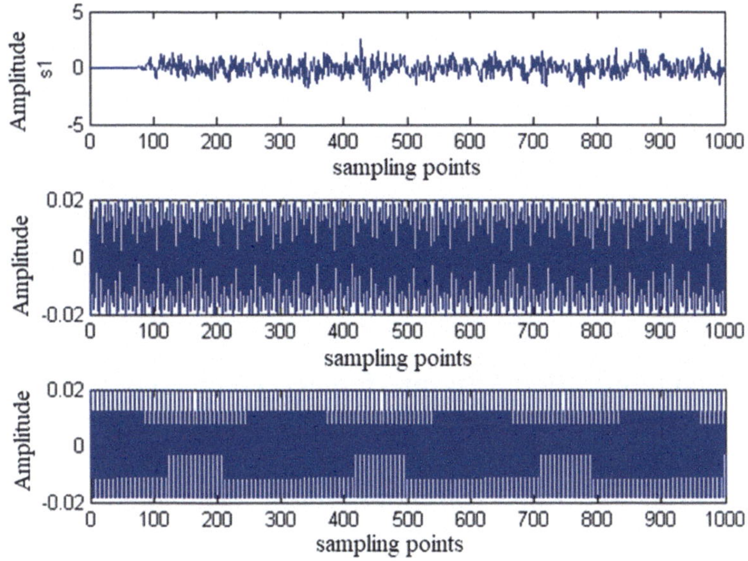

Fig. 13.17 Time domain waveform diagrams of source signal

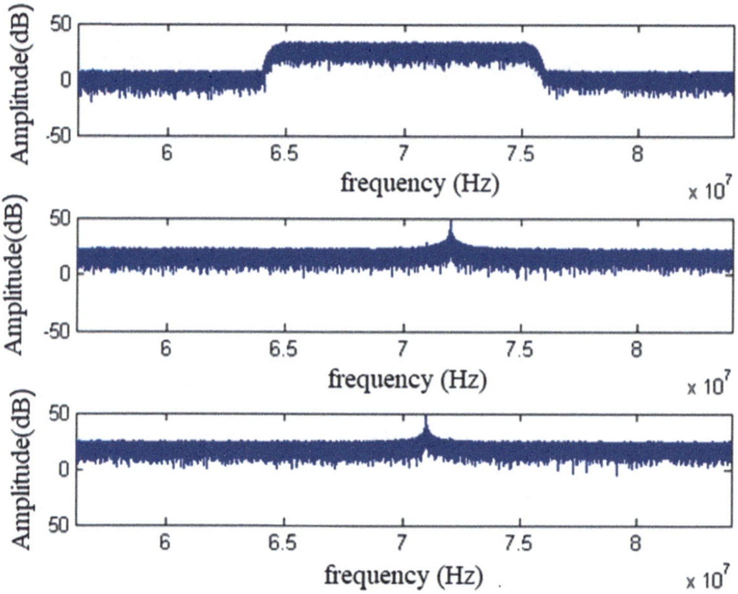

Fig. 13.18 Amplitude-frequency characteristics of separated signals when SNR is 20 dB

$$|\boldsymbol{G}| = \begin{bmatrix} 1.1395 & 0.4483 & 0.2171 \\ 0.0019 & 0.4933 & 49.6818 \\ 0.0054 & 49.1308 & 0.4727 \end{bmatrix}$$

Figures 13.18 and 13.19 give the spectrum diagrams and time domain waveforms of the separated signal respectively. It can be seen from the result that the source signals s2 and s3 can still be separated, but the separation effect has worsened.

13.4 Communication Anti-interference Technology Based on BSS

13.4.1 Principle of BSS Anti-interference Technology

In the modern high-tech war, the electronic battlefield is very serious, and various automatic, intelligent and multifunctional communication countermeasure systems with rapid response ability have brought increasingly serious threats to military communication. In order to ensure the safety and reliability of communication links, the communication system and equipment must have anti-interference, anti-detection, anti-direction finding and other anti-countermeasures capabilities. When developing communication equipment, all countries in the world pay great attention

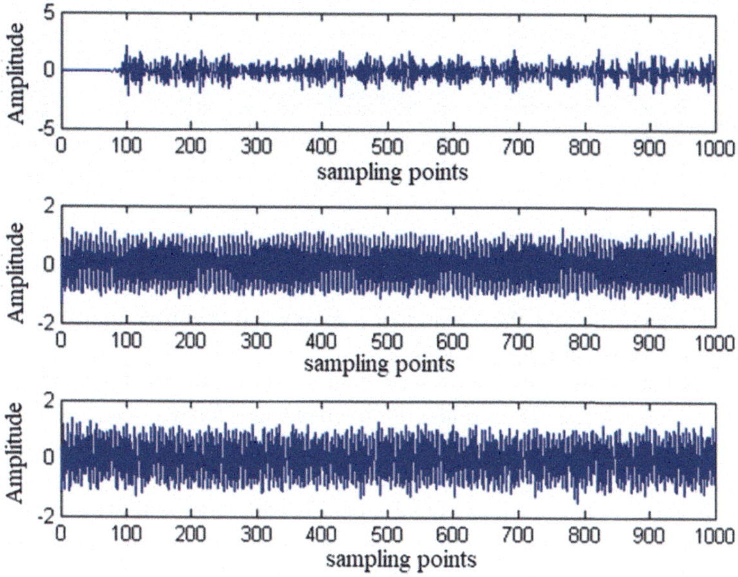

Fig. 13.19 Time-domain waveforms of separated signals when the SNR is 20 dB

to the development of communication anti-countermeasure technology to improve the timeliness, reliability and confidentiality of military communication.

The system, method and measure of communication anti-interference technology can be divided into four categories: (1) frequency domain anti-interference technology based on spread spectrum technology; (2) Time domain anti-jamming technology based on adaptive time varying and processing technology; (3) Spatial anti-jamming technology based on adaptive zeroing antenna; (4) Error correcting coding technology. In recent years, in order to win the victory of electronic warfare in the future, all countries in the world are stepping up the research and development of various anti-jamming technologies, and vigorously develop the integrated anti-jamming technology that combines multiple systems.

The development of electronic countermeasures technology makes the technical level of jamming and anti-jamming higher and higher. In order to ensure the smooth communication link, the new generation of communication equipment generally adopts a comprehensive anti-jamming technology. The jamming capability of modern military communication equipment and system has reached a very high level, but the development of anti-jamming technology is far from the peak. With the emergence and evolution of new, advanced, multifunctional, super-power, ultra-wideband interference systems, the electronic warfare battle will be more intense, and military communication stations and communication networks will face the challenge of four dimensional integration of the full frequency band and automated blocking jamming. In order to ensure the safe and reliable transmission of necessary information in the

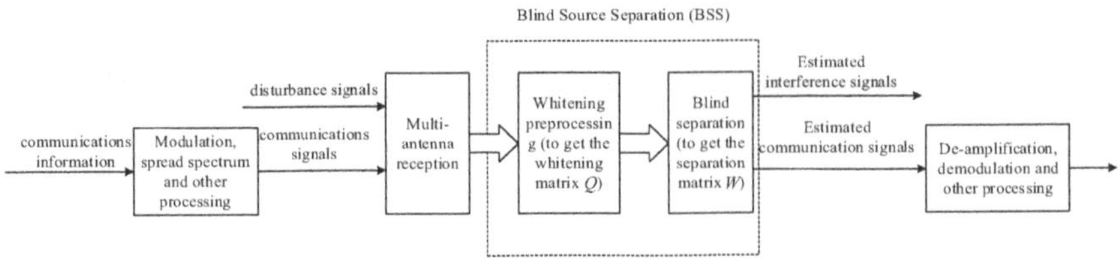

Fig. 13.20 Functional block diagram of anti-jamming system based on blind source separation

future high-tech battlefield, the research of communication anti-jamming technology will continue in a wider range of fields.

We know that no matter what form of jamming the enemy uses (same frequency interference, broadband interference, narrowband interference, etc.), the interference signal and communication signal should be statistically independent, so the received signal is often a mixture of independent source signals if multiple antennas are used to receive, the received signal model is consistent with the signal model of BSS. Therefore, we can use blind source separation technology to separate the interference signal and the communication signal from the mixed signal, which can greatly suppress the enemy interference signal, and achieve the purpose of anti-jamming. This is the anti-jamming technology based on blind source separation. This technology can be combined with the existing anti-jamming techniques such as spread spectrum, frequency hopping and time hopping to form a comprehensive anti-jamming system, so as to further improve the anti-jamming performance of the system. The functional block diagram of the anti-jamming system is shown in Fig. 13.20 [25].

13.4.2 Computer Simulation

1. Simulation of Anti-jamming Principle of Blind Source Separation

The communication signal is a 4QAM signal with a carrier frequency of 63 MHz, the symbol transmission rate is 56 K baud/s, the interference signal is a narrowband signal with a bandwidth of 10 kHz and the same frequency, the sampling frequency is 56 MHz, the power ratio of the interference signal and the communication signal is 20 dB, the signal incident angle is 40°, 80°, and the noise is Gaussian white noise with SNR = 10 dB. A uniform linear array of 4 elements is adopted for receiving. The EASI blind source separation algorithm is used to blind separate the received signals, and the ISR is −33.8 dB after separation. Figure 13.21 shows the constellation diagram of the separated communication signal.

It can be seen from the figure that the constellation diagram of the separated communication signal is basically consistent with that of the source signal except for the angular rotation. Therefore, the BSS technology can indeed recover the source communication signal from the communication signal interfered by the same

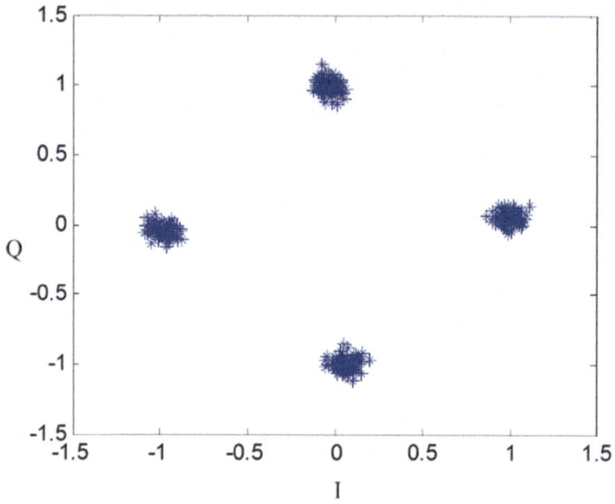

Fig. 13.21 Constellation diagram of communication signal after separation

frequency narrowband signal, which verifies the feasibility of anti-jamming by blind source separation technology.

2. **Simulation of Anti-jamming Performance of BPSK Signal**

(1) Same Frequency Narrowband Interference

The source communication signal is a BPSK signal with a carrier frequency of 630 kHz, symbol rate is 10 K baud/s, sampling frequency is 560 kHz, and the interference signal is a narrowband signal with a bandwidth of 2 kHz. The communication signal and interference signal incident angles are 40° and 80° respectively. The array radius is half the wavelength of the signal. Figure 13.22 shows the change curves of the ISR before separation (input ISR) and the ISR after separation (output ISR) under different SNR conditions.

When the input ISR is constant, the larger the SNR, the smaller the output ISR, that is, the better the anti-jamming performance. When the SNR increases to a certain extent, the output ISR no longer decreases, and the anti-jamming performance of the system reaches the limit. When the SNR is constant, the smaller the input ISR, the larger the output ISR and the worse the anti-jamming performance. Because the SNR defined here is the power ratio of communication signal and noise, the input ISR is small, means that the interference power is small, so the interference to noise power ratio (INR) is reduced. And in blind source separation, we treat the interference signal and communication signal as the source signals, so this is equivalent to the reduction of the SNR in BSS, the separation effect becomes worse, that is, the anti-jamming effect becomes worse. When the input ISR reaches a certain value, the output ISR no longer decreases, that is, the anti-jamming performance of the system also reaches the limit.

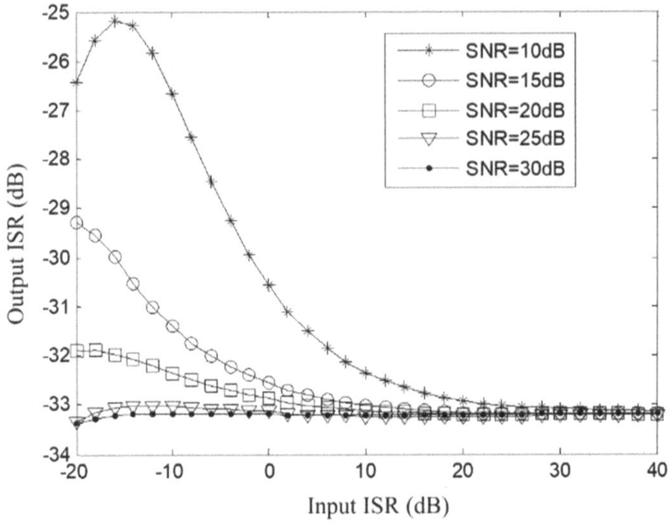

Fig. 13.22 Simulation results of anti same frequency narrowband interference performance of BPSK signal

It is noted that when the SNR is constant, such as SNR = 10 dB, and the input ISR is less than −16 dB, the output ISR will not increase but decrease with the decrease of the input ISR. This is because the input ISR is very small, and the degree of deterioration of BSS performance is smaller than the degree of decline of the input ISR, so the total output ISR decreases.

(2) Same frequency broadband interference

The source communication signal is a BPSK signal with a carrier frequency of 630 kHz, a symbol rate of 10 K baud/s, a sampling frequency of 560 kHz, and the interference signal is a broadband signal with a bandwidth of 30 kHz. The communication signal and interference signal incident angles are 40° and 80° respectively. The array radius is half the wavelength of the signal. Figure 13.23 shows the anti-interference performance curve. The same conclusion can be obtained from Fig. 13.23 as Fig. 13.22. In addition, the performance of the system against the same frequency broadband interference is better than that of the narrowband interference.

(3) Correlation Interference

The source communication signal is a BPSK signal with a carrier frequency of 630 kHz, and the symbol rate is 10 K baud/s. The carrier frequency of the interference signal differs by 1 kHz from that of the communication signal, and other parameters are the same. The incident angles of the two signals are 40° and 80° respectively, and a uniform linear array of 4 elements is adopted for receiving. Figure 13.24 shows the anti-correlation interference performance curves, from which the same conclusion can be drawn as Fig. 13.22.

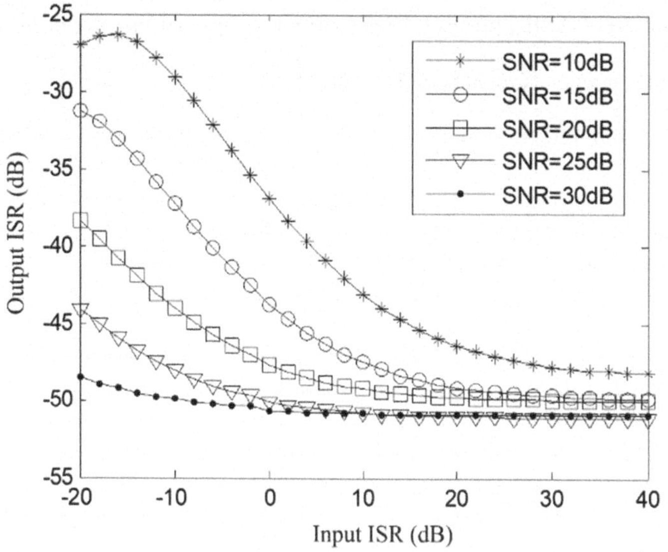

Fig. 13.23 Simulation results of anti same frequency broadband interference performance of BPSK signal

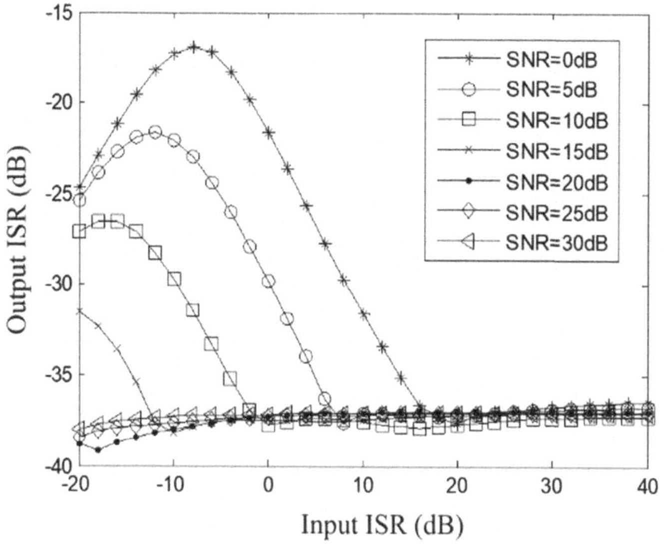

Fig. 13.24 Simulation results of anti-correlation interference performance of BPSK signal

3. **Simulation of Anti-jamming Performance of Spread Spectrum Signal**

(1) Same Frequency Narrowband Interference

The source communication signal is a DSSS signal, the spread spectrum code length is 64 bit Walsh code, adopts BPSK modulation, symbol rate is 1.25 K baud/s, signal carrier frequency is 630 kHz, sampling frequency is 560 kHz, and interference signal is a narrowband signal with a bandwidth of 2 kHz and the same frequency. The communication signal and interference signal incident angles are 40° and 80° respectively. Figure 13.25 shows the relationship between the output ISR and the input ISR.

As can be seen from Fig. 13.25, when the input ISR is constant, the larger the SNR, the better the anti-jamming performance. When the SNR is constant, the larger the input ISR, the smaller the output ISR, and the better the anti-jamming performance. But when the input ISR is increased to a certain extent, the anti-jamming performance of the system reaches the limit, and the output ISR no longer decreases. Comparing Figs. 13.22 and 13.25, we can see that the anti same frequency interference performance of the spread spectrum performance is stronger than the anti-interference performance of the BPSK signal.

(2) Same Frequency Broadband Interference

The source communication signal is a DSSS signal, the spread spectrum code length is 64 bit Walsh code, the BPSK modulation is adopted, the symbol rate is 1.25 K

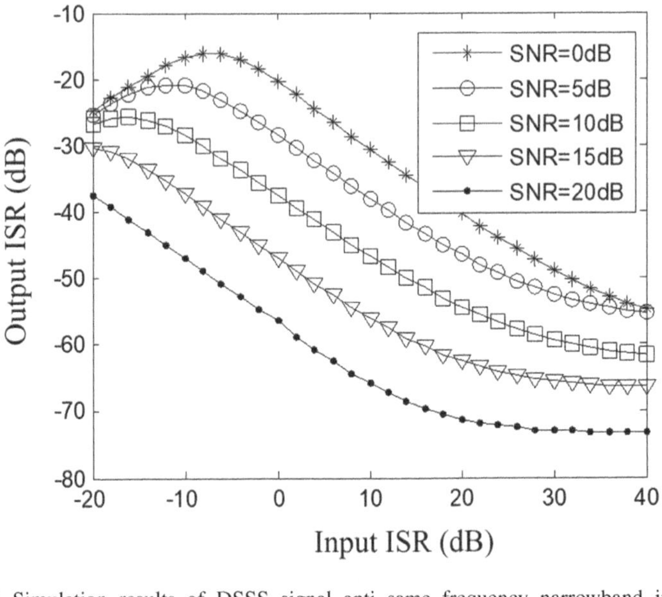

Fig. 13.25 Simulation results of DSSS signal anti same frequency narrowband interference performance

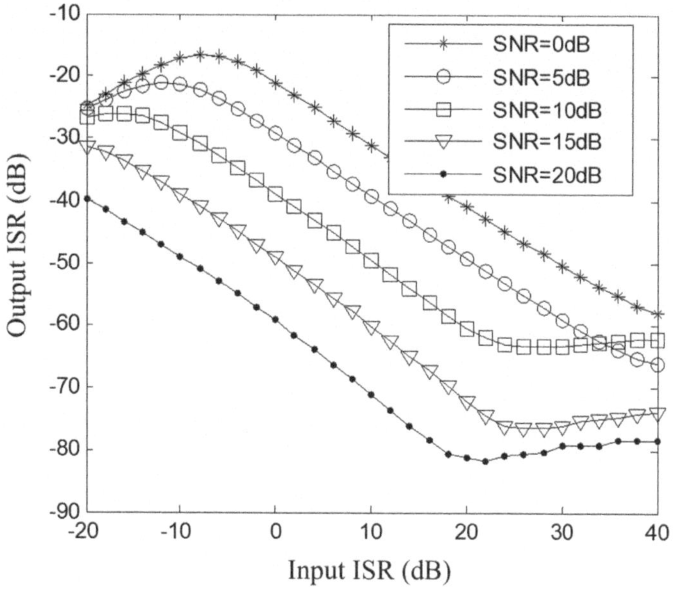

Fig. 13.26 Simulation results of DSSS signal against same frequency broadband interference performance

baud/s, the signal carrier frequency is 630 kHz, the sampling frequency is 560 kHz, and the interference signal is a broadband signal with a bandwidth of 160 kHz. The communication signal and interference signal incident angles are 40° and 80° respectively. Figure 13.26 shows the performance curve of DSSS signal against same frequency broadband interference. Comparing Figs. 13.23 and 13.26, we can see that the anti-jamming performance of spread-spectrum signal is stronger than that of BPSK signal.

(3) Correlation Interference

Both the source communication signal and the interference signal are DSSS signals and adopt BPSK modulation. The spread spectrum code length of the communication signal is 64 bit Walsh code, the symbol rate is 1.25 k baud/s, the signal carrier frequency is 630 kHz, the sampling frequency is 560 kHz, the interference signal carrier frequency is 1 kHz off from the communication signal, and other parameters are the same. The communication signal and interference signal incident angles are 40° and 80° respectively. Figure 13.27 shows the performance curves of the DSSS signal anti-correlation interference performance. The same conclusion as above diagrams can also be seen from this graph.

According to the condition that communication signal and enemy interference signal are independent each other, an anti-jamming communication system based on blind source separation is introduced. The system can effectively suppress narrowband interference and broadband interference in the same frequency. The computer

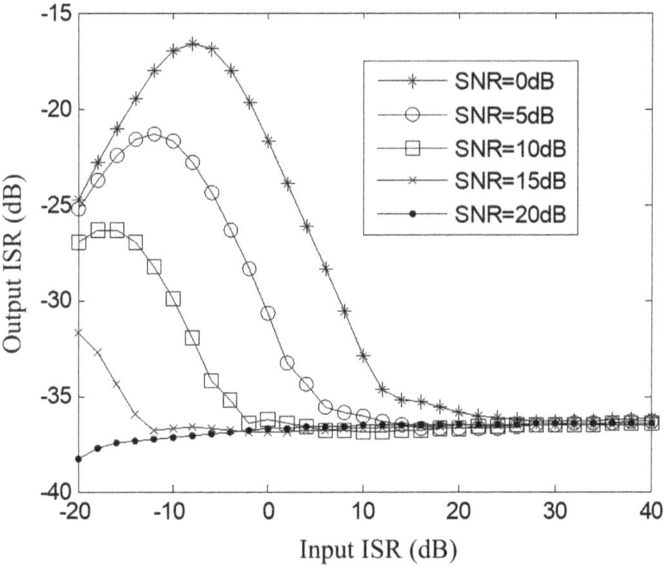

Fig. 13.27 Simulation results of anti-correlation interference performance of DSSS signal

simulation proves the effectiveness of the system. If this technology is combined with other anti-jamming technologies, the anti-jamming effect of the system will be further improved.

Summary

This chapter mainly introduces the application of BSS technology in communication signal processing, including CDMA signal multi-user detection technology based on BSS, frequency hopping signal sorting and splicing technology based on BSS, and communication anti-jamming technology based on BSS. At the same time, this chapter also simulates the blind separation of spectrum aliasing signals. The effectiveness of BSS technique for spectrum aliasing communication signal separation is verified.

Chapter 14
Experimental Experiment of BSS of Communication Signal

In order to verify the effectiveness of the BSS algorithm in practical applications, we set up an experimental system to collect some actual data, and then use BSS technology to separate the aliasing (or partial aliasing) signals in the time frequency domain, and carry out subsequent demodulation processing of the separated signals. The experimental results in this chapter prove the feasibility of the application of BSS technology in communication reconnaissance [45].

14.1 Composition of the Experimental System

The composition of the experimental system is shown in Fig. 14.1.

The receiving antenna array is a 5-element uniform circular array with a radius of 60 cm, but only four of the signals are used for processing, so the actual antenna array model is equivalent to a non-uniform circular array of 4 elements. The frequency range of the 8-channels broadband receiver is 20–1350 MHz. The receiver first performs down-conversion processing on the received signal and converts the signal to the intermediate frequency of 70 MHz with a bandwidth of 20 MHz and a sampling frequency of 56 MHz. After mixing, the multi-channel 70 MHz intermediate frequency (IF) signal is sent to the BSS processing module, which outputs the separated IF signal, and finally the multi-channel post-processing unit demodulates the separated signal.

W. Fu, *Principle and Application of Blind Source Separation Technology*,
https://doi.org/10.1007/978-981-96-0022-9_14

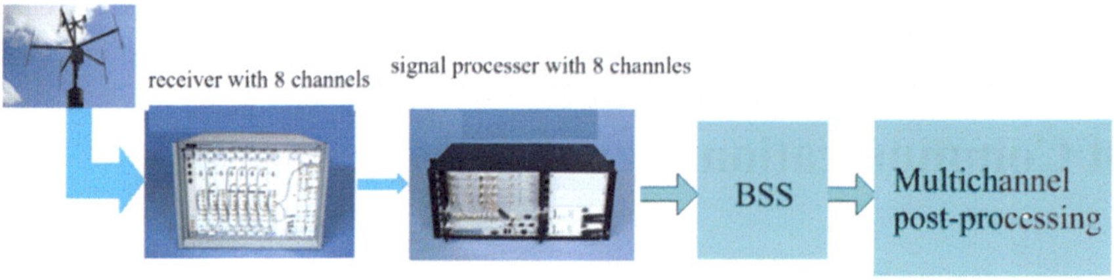

Fig. 14.1 Composition diagram of the experimental system

14.2 Principle Experiments of Blind Source Separation

14.2.1 BSS of Digital Modulated Signals

Experiment 1 four source signals, four receiving antennas—(different frequencies, different modulation)

The four source signals are: a 2FSK signal with a frequency of 1302 MHz, a code rate of 20 Kbps, and a power of −5 dBm; 16QAM signal with a frequency of 1304 MHz, a bit rate of 50 Kbps and a power of −20 dBm; A BPSK signal with a frequency of 1307 MHz, a bit rate of 30 Kbps and a power of −5 dBm; and a QPSK signal with a frequency of 1309 MHz, a bit rate of 100 Kbps and a power of −25 dBm.

Figures 14.2 and 14.3 show the amplitude frequency characteristics and time domain waveforms of the four observed (mixed) signals. EASI algorithm was used for BSS of the observed signal after whitening. Figures 14.4 and 14.5 show the amplitude frequency characteristics and time domain waveforms of the separated signals. It can be seen from the results that the BSS algorithm can separate the four source signals very well.

Experiment 2, four source signals, four antennas receive—(different frequencies, same modulation)

The four source signals are: BPSK signal with a frequency of 1302 MHz and a bit rate of 20 Kbps; BPSK signal with a frequency of 1304 MHz and a bit rate of 50 Kbps; A QPSK signal at 1307 MHz with a bit rate of 30 Kbps, and a QPSK signal at 1309 MHz with a bit rate of 100 Kbps.

Figures 14.6 and 14.7 show the amplitude frequency characteristics and time domain waveforms of the four observed (mixed) signals. EASI algorithm was used for blind separation of the observed signals after whitening. Figures 14.8 and 14.9 show the amplitude frequency characteristics and time domain waveforms of the separated signals. It can be seen from the separation results that the BSS algorithm can also separate the four source signals very well.

The above two experimental results show that the BSS algorithm can not only separate the source signals of different modulation modes, but also separate the source signals of the same modulation mode.

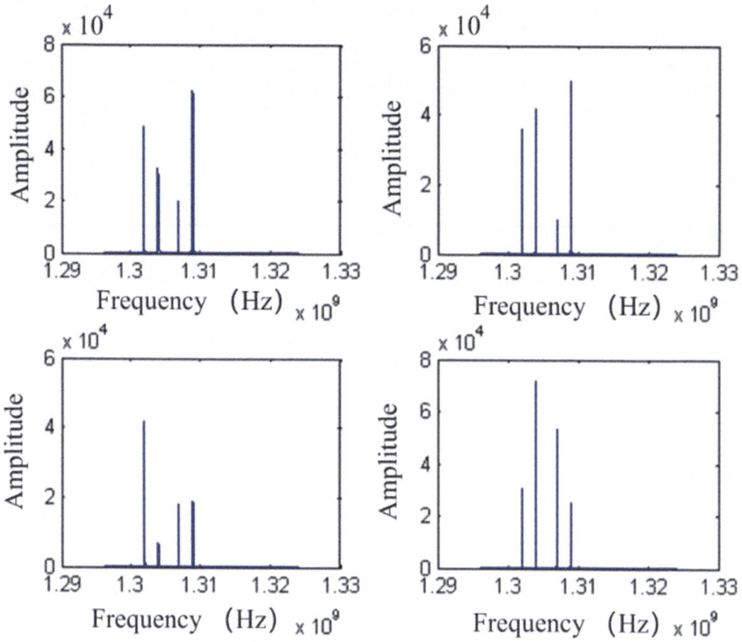

Fig. 14.2 Amplitude frequency characteristics of mixed signals

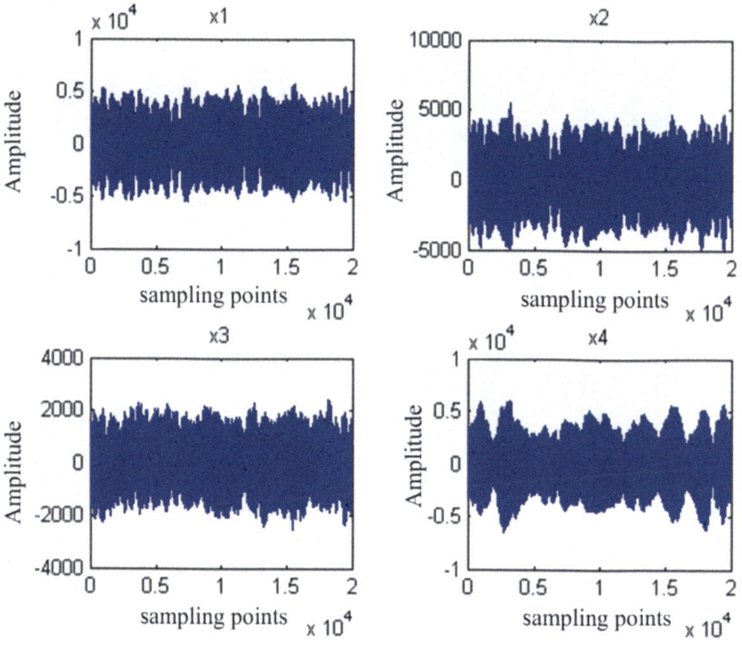

Fig. 14.3 Time domain waveforms of mixed signal

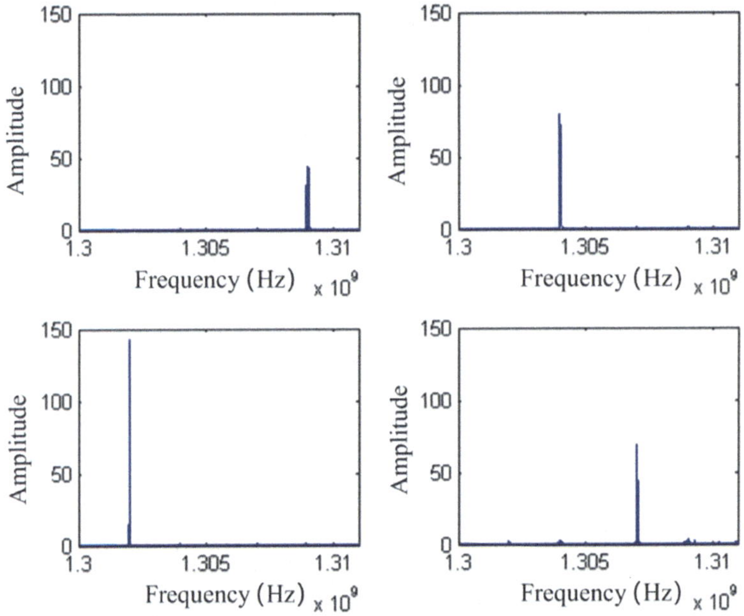

Fig. 14.4 Amplitude-frequency characteristics of separated signals by EASI algorithm

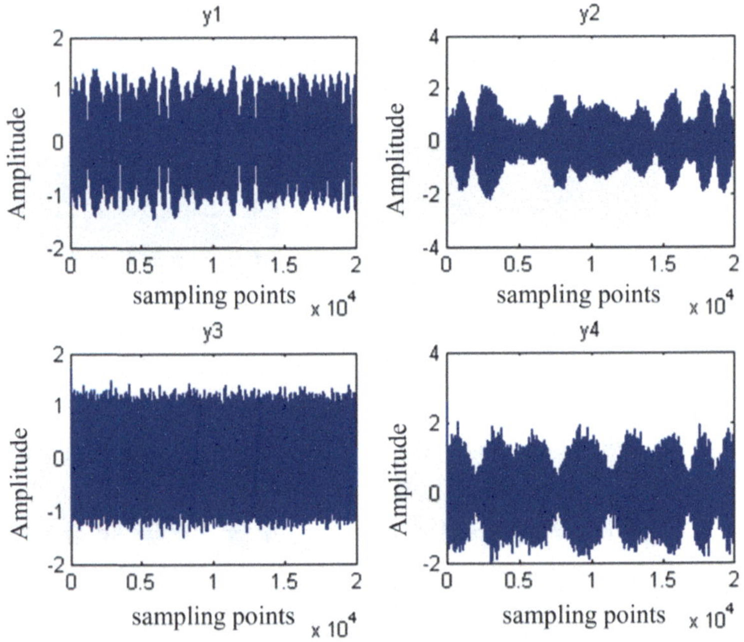

Fig. 14.5 Time domain waveforms of signal separated by EASI algorithm

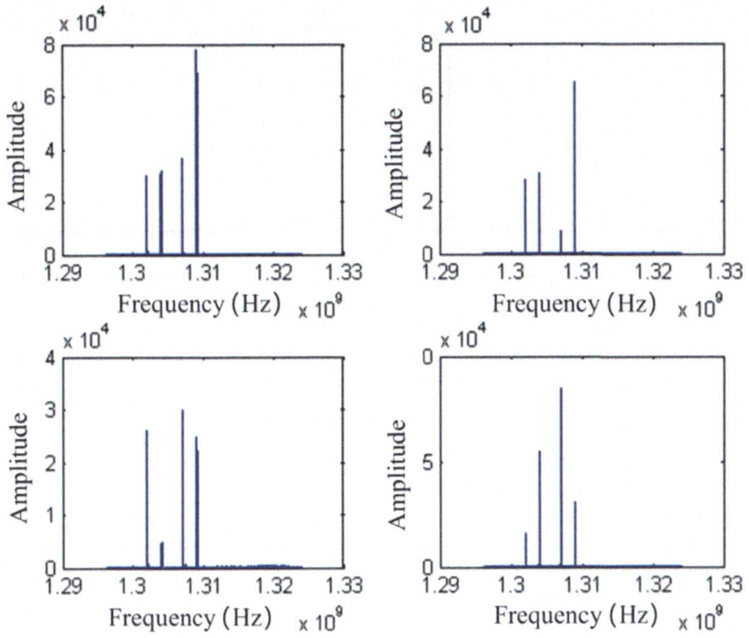

Fig. 14.6 Amplitude frequency characteristics of the observed signals

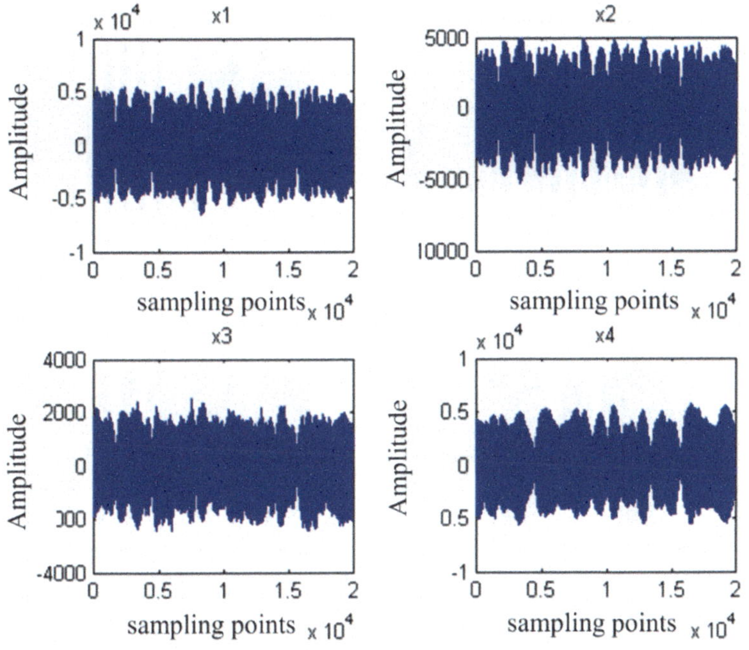

Fig. 14.7 Time domain waveforms of observed signals

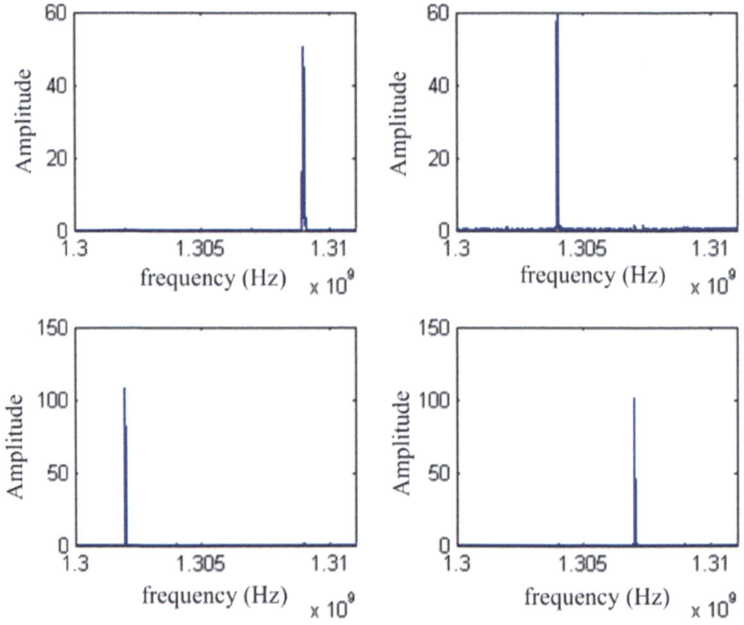

Fig. 14.8 Amplitude frequency characteristics of separated signals by EASI algorithm

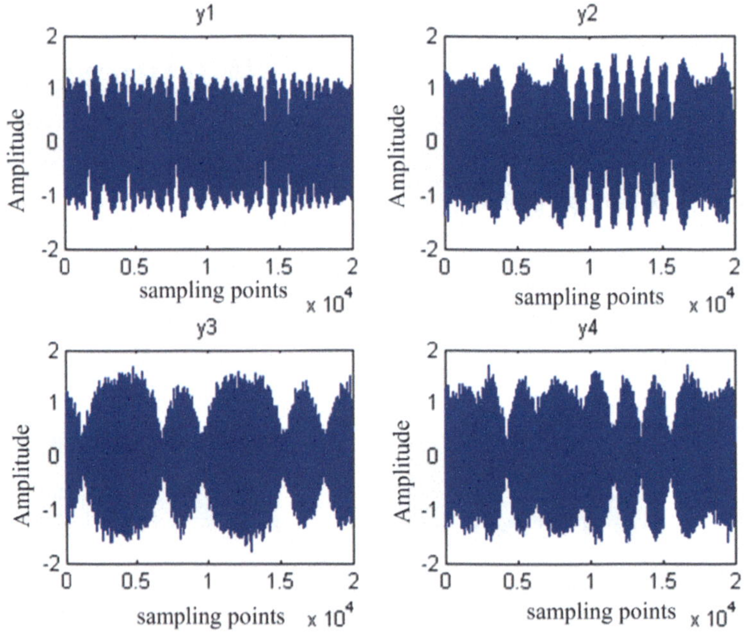

Fig. 14.9 Time domain waveforms of separated signals by EASI algorithm

14.2.2 Blind Separation of FM Broadcast Signals

One source signal is an FM broadcast signal with a carrier frequency of 88.2 MHz, and the other source signal (that is, interference signal) is a single tone FM signal with the same frequency, that is, the carrier frequency is 88.2 MHz, the modulated signal is a sine wave of 1 kHz, the frequency offset is 40 kHz, and the source signal power is 20 dBm. At the same time, in order to make the power of the interference signal cover the power of the broadcast station, the interference signal is amplified by power amplifier.

The receiving antenna is non-uniform circular array with 4 elements. The receiver bandwidth is 20 MHz, the center frequency is 70 MHz, A/D conversion is performed on the received four signals, the sampling frequency is 56 MHz, and then down conversion it to zero IF.

After the received four signals are de-averaged and whitened, the EASI algorithm is used for separation, and the separated signals are demodulated. The results are shown in Fig. 14.10. In the figure, the first line represents the waveform after direct demodulation of the received signals, the second and third lines represent the waveforms after demodulation of the two separated signals respectively. The y-axis in Fig. 14.10 is the value after A/D transformation, which only shows the change trend in signal amplitude, not the specific value of the signal. The following figures are the same as above.

It can be seen from the figure that after blind source separation, the FM broadcast signal can be recovered well. And through the sound equipment, before separation we

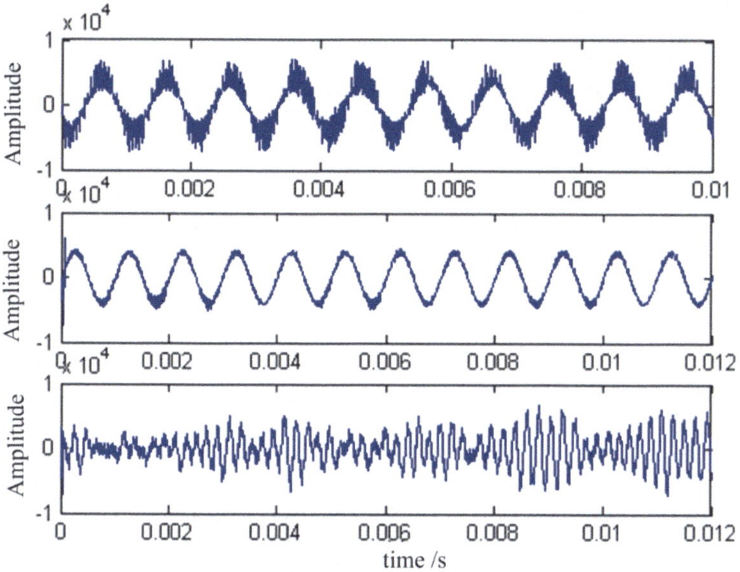

Fig. 14.10 Time domain waveform diagrams after signal demodulation before and after separation

can hear a noisy sound mixed with signals and speech. When the power of interference signal is relatively large, we can only hear the interference signal of single tone FM, and after BSS, the voice signal received is free from this interference, and we can hear the sound of the radio station very clearly.

14.3 Effect of Carrier Frequency Interval of Source Signal on Separation Performance

Experiment 1. Source signal is single tone frequency modulation signal and sine wave frequency modulation signal

The two source signals are:

Source signal 1 is a single tone frequency modulation signal, whose frequency offset is 25 kHz, and carrier frequency is 45 MHz. Modulated signal is sinusoidal signal with a frequency of 1 kHz, signal power is 20 dBm;

Source signal 2 is a square wave frequency modulation signal, whose frequency offset is 25 kHz, modulated signal is a square wave with a frequency of 500 Hz, signal power is 20 dBm, carrier frequency of source signal 2 is 45, 45.002, 45.01 MHz.

Figures 14.11, 14.12 and 14.13 show the separation results at different carrier frequencies respectively. The top of each figure represents the waveform after demodulating the signal before BSS, and the middle and the bottom of each figure represent the waveforms of demodulated signal after BSS.

We can see that, even if the carrier frequency of two signals is exactly the same, the blind source separation algorithm can separate them well, and the change of carrier frequency difference has little impact on the separation effect.

Experiment 2. Both source signals are square wave frequency modulation signals

The two source signals are:

Source signal 1 is a square wave frequency modulation signal, whose frequency offset is 25 kHz. Modulated signal is a square wave with a frequency of 1 kHz (period is 1 ms). The power of source signal 1 is 20 dBm and its carrier frequency is 45 MHz;

Source signal 2 is a square wave frequency modulation signal, whose frequency offset is 25 kHz. Modulated signal is square wave with a frequency of 400 Hz (period is 2.5 ms). The power of source signal 2 is 20 dBm and its carrier frequencies are 45.005 MHz, 45.008 MHz, 45.02 MHz, 45.04 MHz respectively.

Figures 14.14, 14.15, 14.16 and 14.17 show the separation results of two source signals with different carrier frequency intervals respectively. The top of each figure represents the waveform of demodulated signal before BSS, the middle and the bottom represent the waveforms of two demodulated signals after BSS.

It can be seen from the above results that the larger the carrier frequency interval of two square wave FM signals, the better the separation effect.

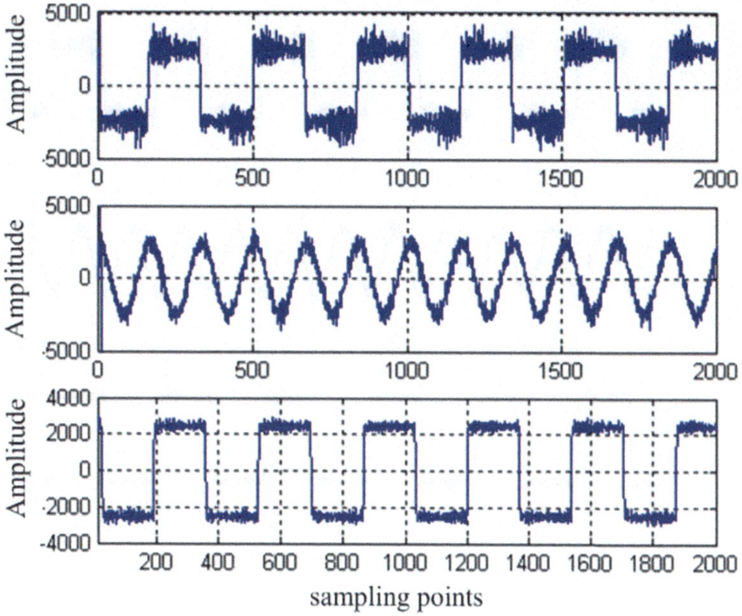

Fig. 14.11 Waveforms of demodulated signals before and after separation when the carrier frequencies are the same

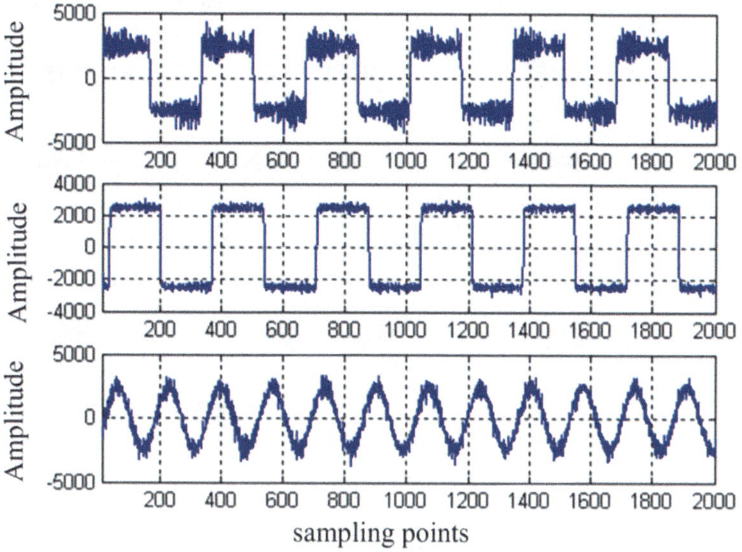

Fig. 14.12 Waveforms of demodulated signals before and after separation when carrier frequency difference is 2 kHz

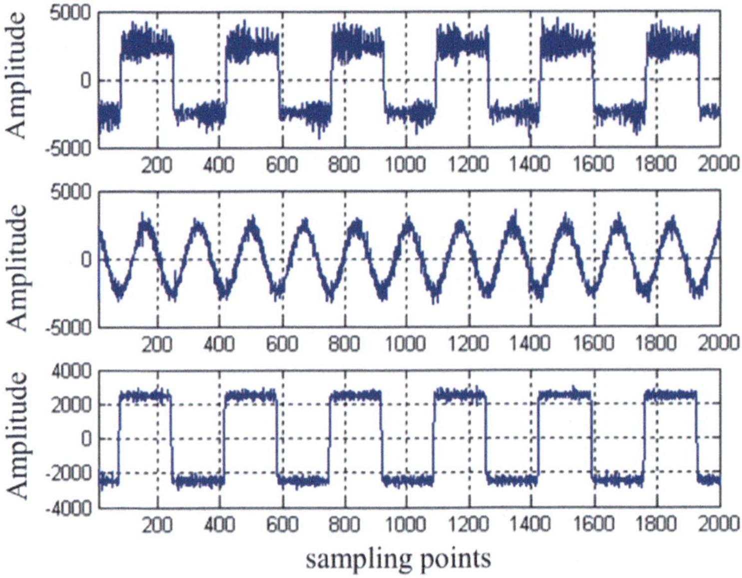

Fig. 14.13 Waveforms of demodulated signal before and after separation when carrier frequency difference is 10 kHz

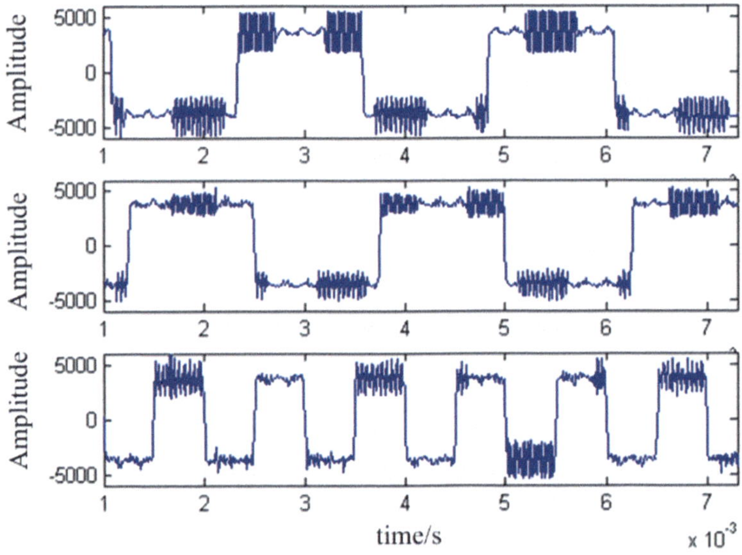

Fig. 14.14 Waveforms of demodulated signals before and after separation when carrier frequency difference is 5 kHz

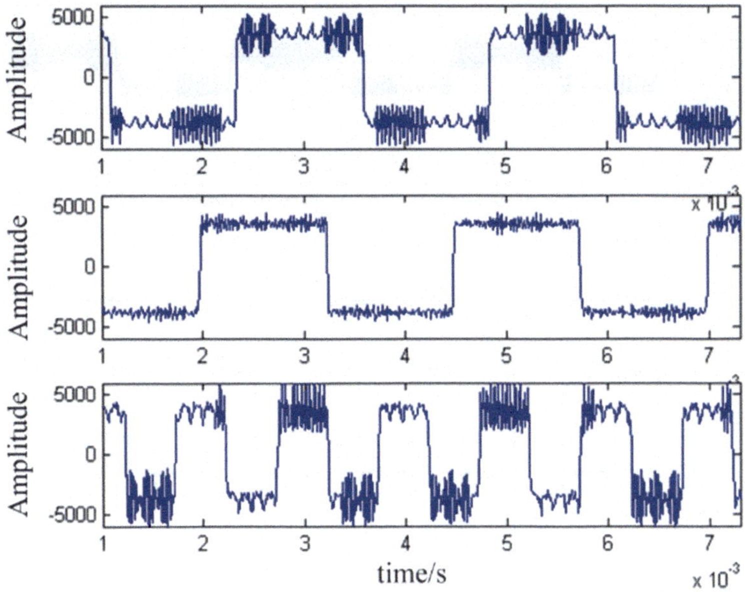

Fig. 14.15 Waveforms of demodulated signals before and after separation when carrier frequency difference is 8 kHz

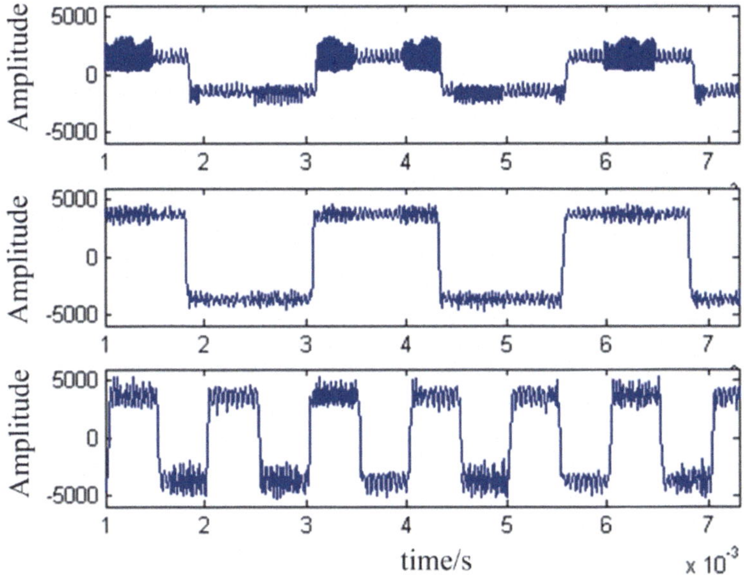

Fig. 14.16 Waveforms of demodulated signals before and after separation when carrier frequency difference is 20 kHz

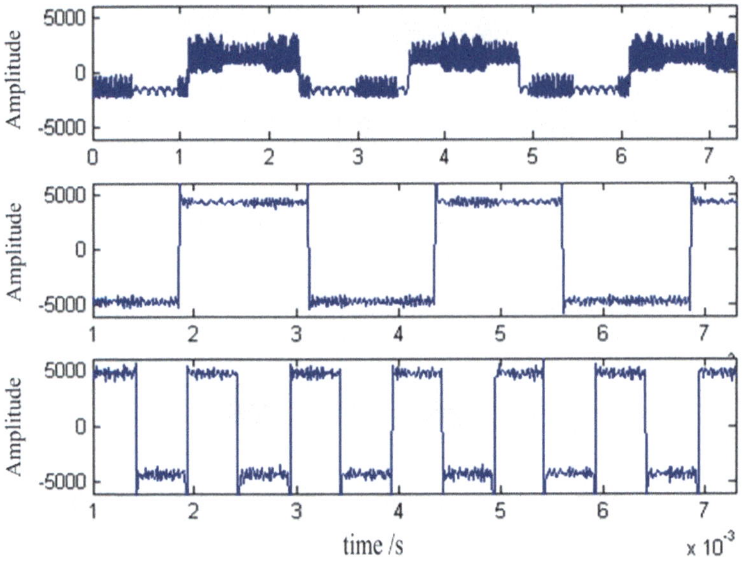

Fig. 14.17 Waveforms of demodulated signals before and after separation when carrier frequency difference is 40 kHz

Experiment 3. Two source signals are single tone FM signals

The two source signals are:

Source signal 1 is a single tone frequency modulation signal with a frequency offset of 25 kHz and carrier frequency 45 MHz. Modulated signal is a sinusoidal signal with a frequency of 10 kHz;

Source signal 2 is a single tone frequency modulation signal with a frequency offset of 25 kHz, and carrier frequency is 45.005 MHz, 45.02 MHz, 45.04 MHz, and 45.05 MHz respectively. Modulated signal is a sinusoidal signal with a frequency of 5 kHz.

Figures 14.18, 14.19 and 14.20 show the separation results at different carrier frequency intervals. Similarly, the top of each figure represents the waveform of demodulated signal before BSS, and the middle and the bottom represent the waveforms of two demodulated signals after BSS. As can be seen from the figures, when the carrier frequency interval is very small, the separation effect is relatively poor, and with the increase of the carrier frequency interval, the separation effect is getting better and better.

It can also be seen from the above results that under the condition of other parameters unchanged, the larger the carrier frequency interval of two single tone FM signals, the better the separation effect.

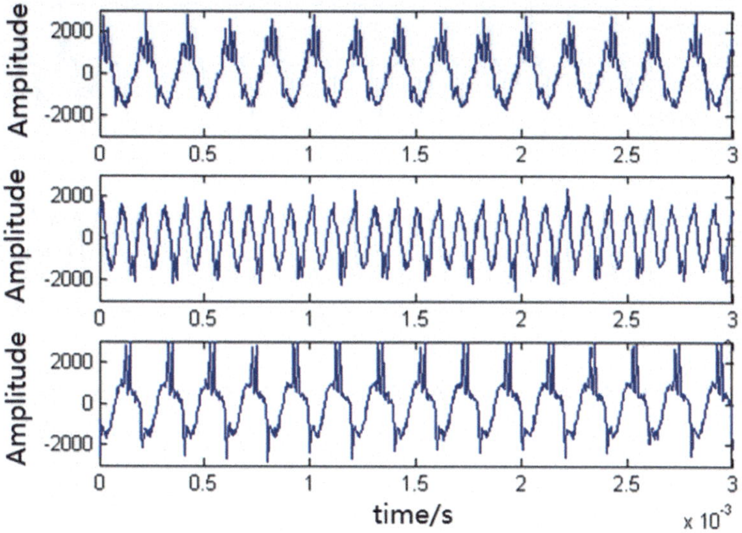

Fig. 14.18 Waveforms of demodulated signals before and after separation when the carrier frequency interval is 5 kHz

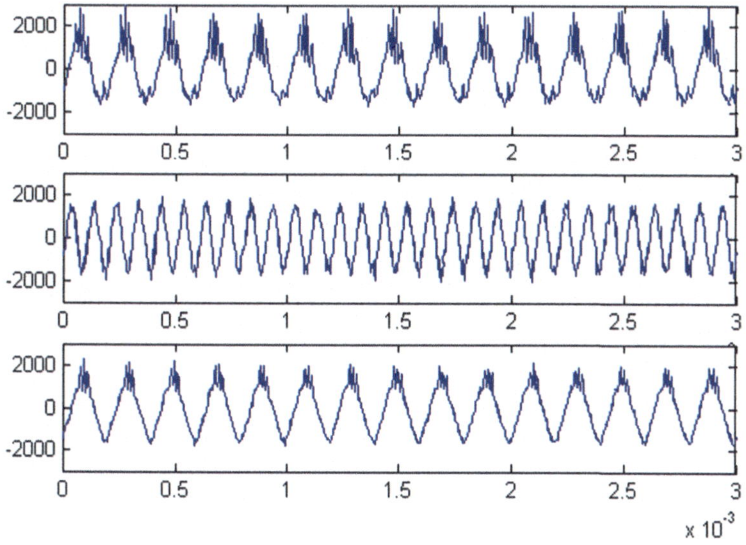

Fig. 14.19 Waveforms of demodulated signals before and after separation when the carrier frequency interval is 20 kHz

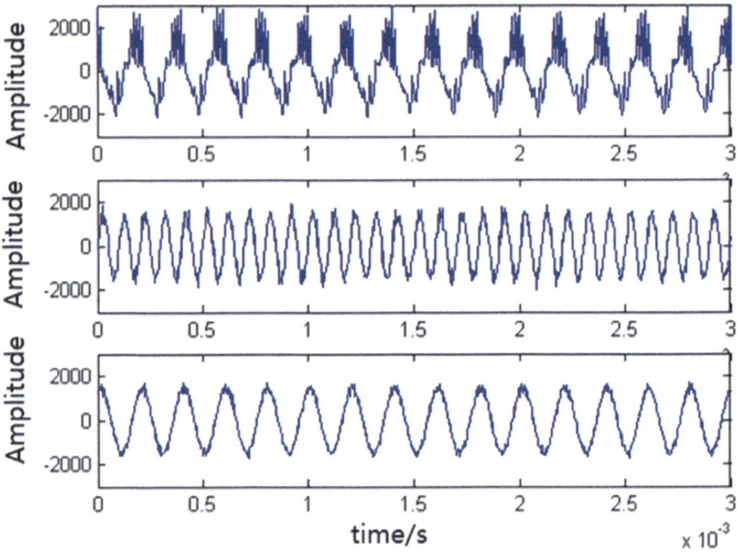

Fig. 14.20 Waveforms of demodulated signals before and after separation when the carrier frequency interval is 40 kHz

14.4 Effect of SNR on Separation Performance

Suppose two source signals are:

Source signal 1 is a single tone frequency modulation signal with a frequency offset of 25 kHz, and a carrier frequency of 45 MHz. Modulated signal is a sinusoidal wave with a frequency of 1 kHz.

Source signal 2 is a square wave frequency modulation signal with a frequency offset of 25 kHz, and a carrier frequency of 45 MHz. Modulated signal is a square wave with a frequency of 500 Hz (period is 2 ms).

The levels of source signals are 20 dBm, 18 dBm, 12 dBm, 5 dBm; Since the noise is unchanged, the reduction of the level of source signal is equivalent to the reduction of the SNR. Figures 14.21, 14.22, 14.23 and 14.24 show the results of blind separation under different signal levels. Similarly, the top of each figure represents the waveform of demodulated signal before separation, and the middle and the bottom of each figure represent the waveforms of two demodulated signal after separation.

As can be seen from the figures above, the smaller the signal level, that is, the lower the SNR, the worse the blind separation effect.

14.5 Spatial Resolution Experiment of BSS Algorithm

The two source signals are as follows.

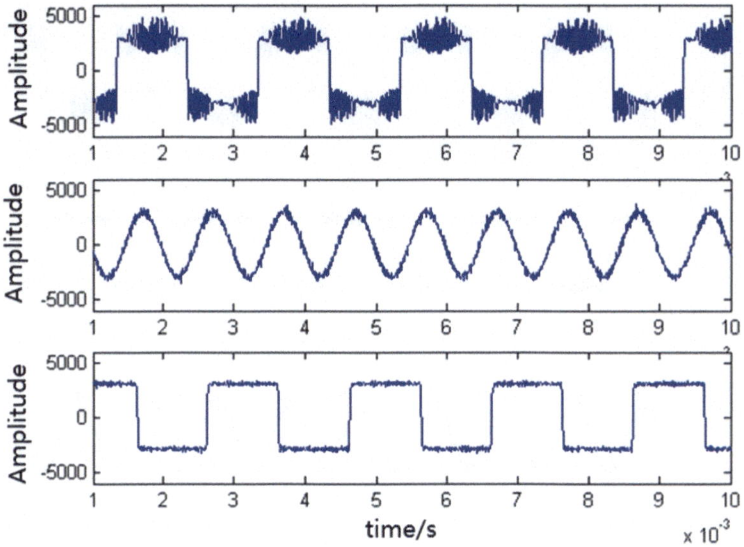

Fig. 14.21 Waveforms of demodulated signals before and after separation when source signal level is 20 dBm

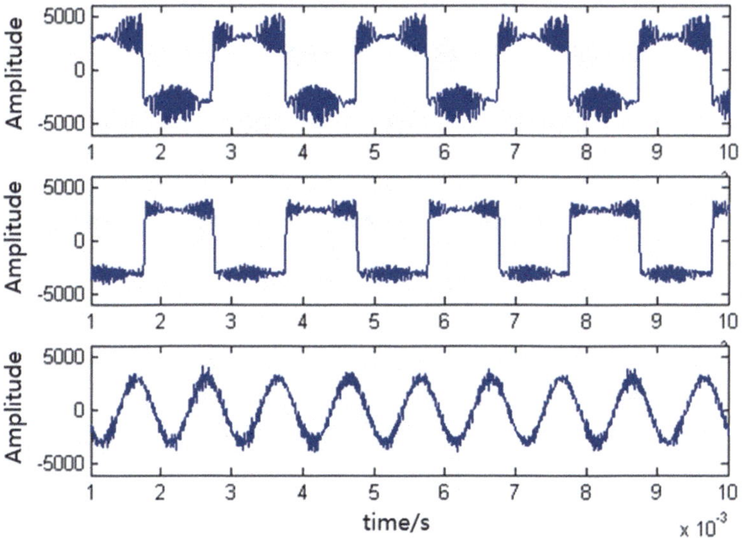

Fig. 14.22 Waveforms of demodulated signals before and after separation when source signal level is 18 dBm

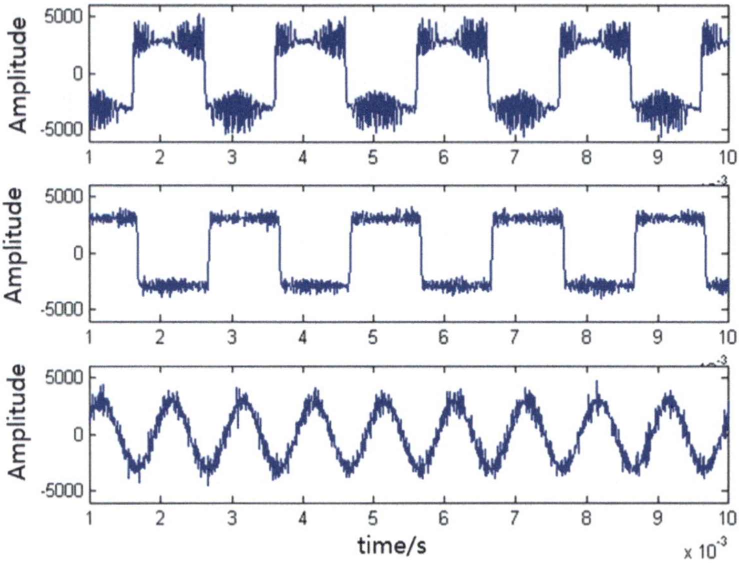

Fig. 14.23 Waveforms of demodulated signals before and after separation when source signal level is 12 dBm

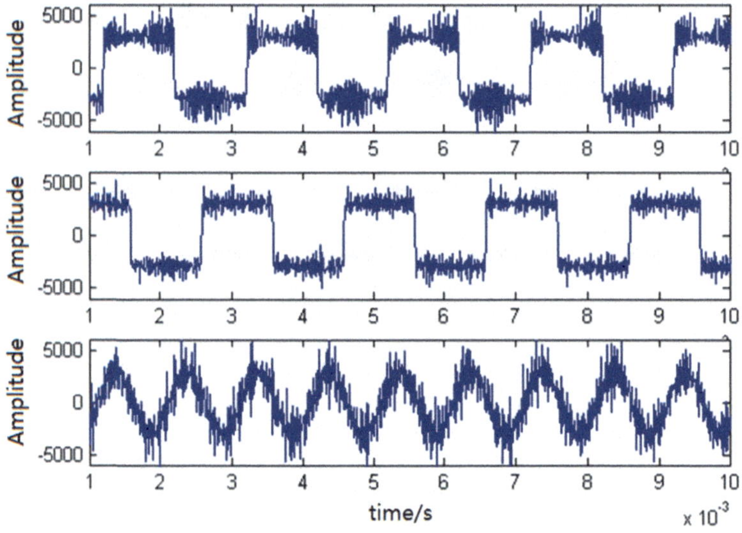

Fig. 14.24 Waveforms of demodulated signals before and after separation when source signal level is 5 dBm

Source signal 1 is a single tone frequency modulation signal with a carrier frequency of 400 MHz, frequency offset of 100 kHz. Modulated signal is a sinusoidal wave with a frequency of 5 kHz;

Source signal 2 is a square wave frequency modulation signal with a carrier frequency of 400 MHz, frequency offset of 100 kHz. Modulated signal is a square wave with a frequency of 2 kHz.

The difference of incident angle between two source signals is 0°12′, 0°37′, 2°30′, 4°47′, 8° respectively. And the SNR is close to 60 dB. The following Figures 14.25, 14.26, 14.27, 14.28 and 14.29 show the simulation results at different included angles. The first line in the figure represents the demodulated waveform of the signal without separation, and the second and third lines represent the demodulated waveforms of the two signals obtained after separation.

As can be seen from Fig. 14.25, when difference of incident angle is 0°12′, the waveform of demodulated signal obtained before and after separation is almost unchanged, and the separation algorithm cannot separate the source signal, because difference of incident angle is too small, and the mixing matrix is close to singular. The condition that the mixing matrix is non-singular is not satisfied, so the blind source separation algorithm fails. As can be seen from Fig. 14.26, when the difference of incident angle increases to 0°37′, the separation algorithm can basically separate the two source signals, but the separation effect is not particularly good. When the difference of incident angle continues to increase, the blind separation effect becomes better and better, as shown in Figs. 14.27, 14.28 and 14.29.

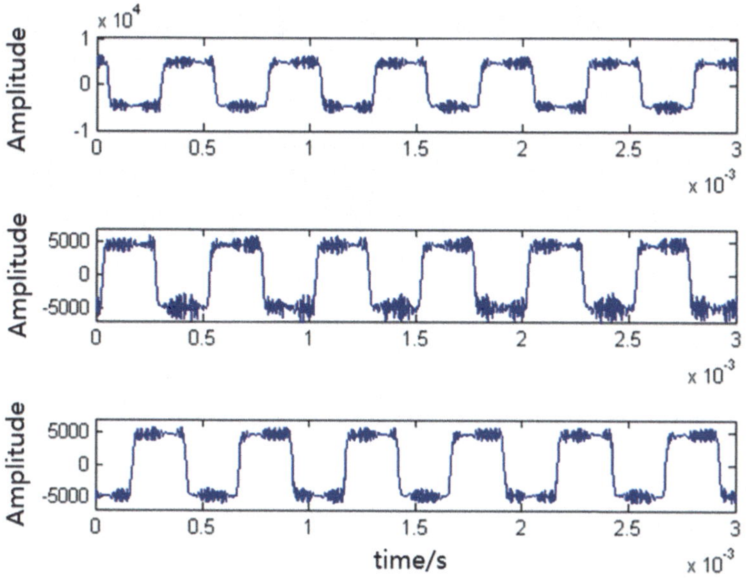

Fig. 14.25 Waveforms of demodulated signals before and after separation when difference of incident angle is 0°12′

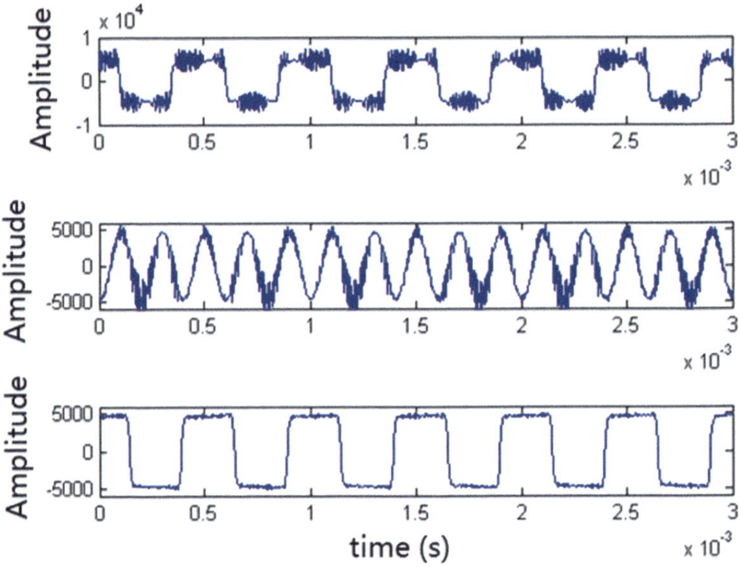

Fig. 14.26 Waveforms of demodulated signals before and after separation when difference of incident angle is 0°37′

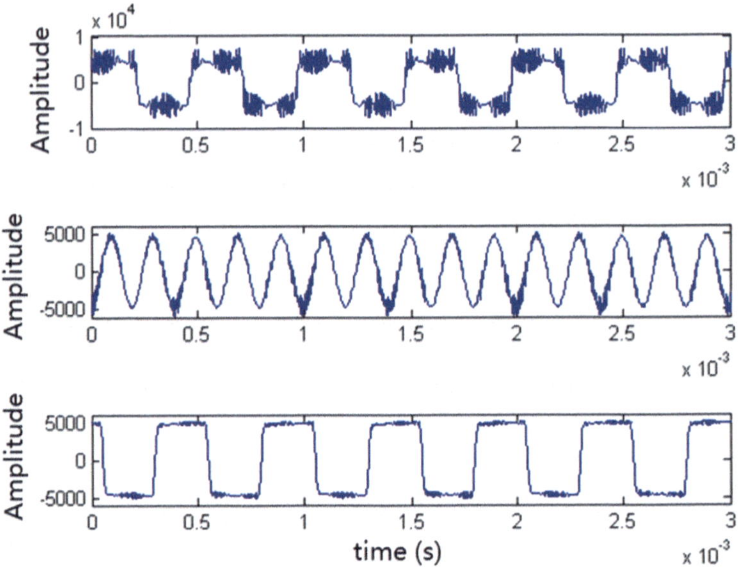

Fig. 14.27 Waveforms of demodulated signals before and after separation when difference of incident angle is 2°30′

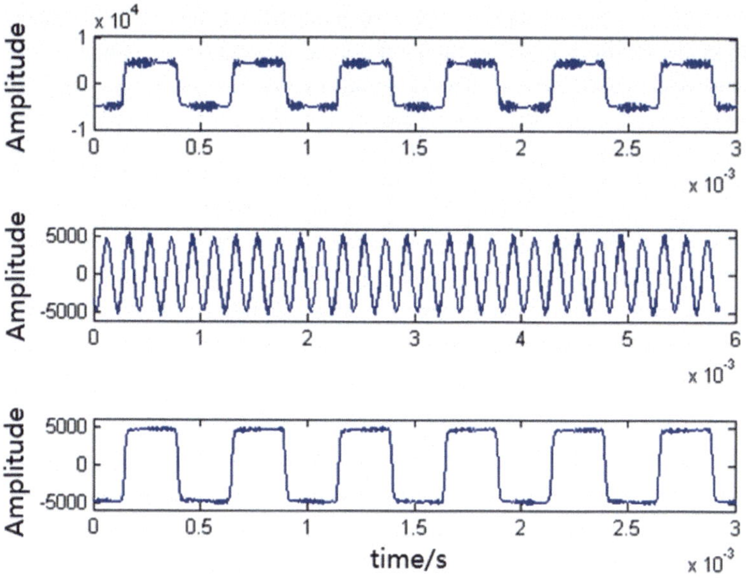

Fig. 14.28 Waveforms of demodulated signals before and after separation when difference of incident angle is 4°47′

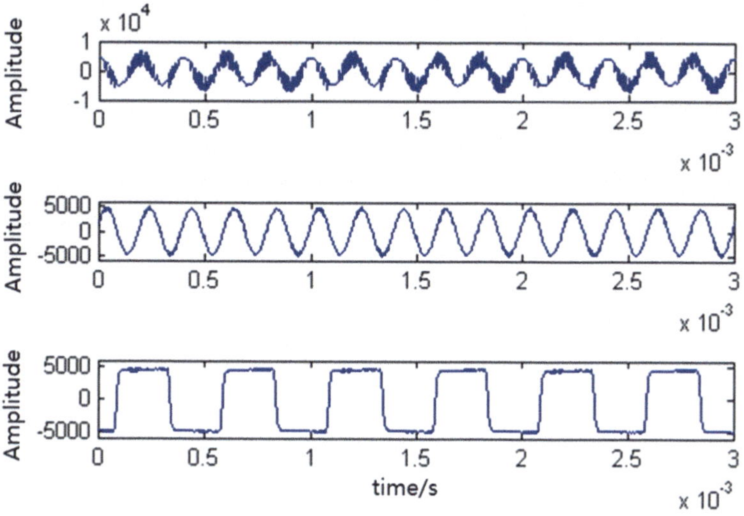

Fig. 14.29 Waveforms of demodulated signals before and after separation when difference of incident angle is 8°

From the above experiments, it can be seen that the spatial resolution of the blind separation algorithm is still very high. Of course, the main reason that such a good effect can be achieved here is that the signal's carrier frequency is large, then the wavelength is short, and the ratio of the array radius to the signal wavelength is large (it is 0.8 in this experiment). Therefore, even if the difference of incident angle is small, the condition number of the mixing matrix is still small. In other words, the non-singularity of the mixing matrix is good, so the blind separation algorithm can still separate the source signal well. If the ratio of the array radius to the signal wavelength decreases, the spatial resolution will also decrease. In addition, in this experiment, the number of source signals is only two. If the number of source signals is increased, the corresponding spatial resolution will also decrease.

Summary

In this chapter, the blind source separation technique is experimentally tested by using real transmitting and receiving equipment. In the case of changing different parameters, the BSS algorithm is used to separate the received mixed signals and demodulate the separated signals. At the same time, for the convenience of comparison, we compare the demodulated signal waveforms before and after separation. The experimental results show that when the two source signals are both single tone or square wave frequency modulation, when all other parameters being equal, the larger the carrier frequency interval, the better the separation effect; when other parameters remain unchanged, the larger the source signal power transmission, or the larger the SNR, the better the separation effect; when other parameters are unchanged, the greater the difference of incident angles between the two source signals, the better the separation effect.

References

1. Holland JG (1978) Behaviorism part of the problem or part of the solution, Journal of Applied Behavior Analysis 11(1):163–174.
2. Cardoso JF (1989) Blind identification of independent components with higher order statistics, Proc. Workshop on Higher-Order Spectral Analysis, pp 157–160.
3. Comon P (1989) Separation of stochastic processes, Proc. Workshop on Higher-Order Spectral Analysis, pp 174–179.
4. Tong L, Liu R, Soon VC, et al (1991) Indeterminacy and identifiability of blind identification. IEEE Transactions on Circuits and Systems 38(5):499–509.
5. Comon P (1994) Independent component analysis, a new concept.? Signal Processing 36(2):287–314.
6. Cardoso JF, and Laheld BH(1996) Equivariant adaptive source separation, IEEE Transactions of Signal Processing 44(12):3017–3030.
7. Amari S, Cichocki A (1998) Adaptive blind signal processing: neural network approaches. Proceeding of IEEE 86(10):2026–2047.
8. Amari S (1998) Natural gradient works efficiently in learning. Neural Computation 10(2):251–276.
9. Amari S, Cardosol F (1997) Blind source separation-semiparametric statistical approach. IEEE Transaction on Signal Processing 45(11):2692–2700.
10. Hyvarinen A, Oja E (1997) A fast fixed-point algorithm for independent component analysis. Neural Compotation 9(7):1483–1492.
11. Hyvarinen A (1997) One-unit contrast function for independent component analysis: a statistical analysis. In Proc. IEEE workshop on Neural Networks for Signal Processing, pp 388–397.
12. Abrard F, Deville Y (2005) A time–frequency blind signal separation method applicable to underdetermined mixtures of dependent sources. Signal Processing 85(7):1389–1403.
13. Georgiev P, Theis F, Cichocki A (2005) Sparse component analysis and blind source separation of underdetermined mixtures. IEEE Transactions on Neural Networks 16(4):992–996.
14. Donoho D, Elad M (2003) Optimally sparse representation in general (nonorthogonal) dictionaries via L1 minimization. Proceedings of the National Academy of Sciences 100(5):2197–2202. https://doi.org/10.1073/pnas.0437847100
15. Roweis S (1998) EM Algorithms for PCA and SPCA. Advances in Neural Information Processing Systems 10:626–632.
16. Xu H, Caramanis C, Sanghavi S (2012) Robust pca via outlier pursuit. IEEE Transactions on Information Theory 58(5):3047–3064.
17. Comon P (1994) Independent component analysis, a new concept? Signal Processing 36(2):287–314.
18. Hyvarinen A, Oja E (2000) Independent component analysis: algorithm and applications. Neural Networks 13:411–430.

W. Fu, *Principle and Application of Blind Source Separation Technology*,
https://doi.org/10.1007/978-981-96-0022-9

19. Hyvarinen A, Oja E (1997) A fast fixed-point algorithm for independent component analysis. Neural Compotation 9(7):1483–1492.
20. Hyvarinen A (1999) Fast and robust fixed-point algorithms for independent component analysis. IEEE Transactions on Neural Networks 10(3):626–634.
21. Nong B (2017) Research on mixing matrix estimation and source signal recovery for Undetermined Blind Source Separation, Dissertation, Xidian University (in Chinese).
22. Oja E (1997) The nonlinear PCA learning rule in independent component analysis. Neurocomputing 17(1):25–45.
23. Xu R, Wunsch DI (2005) Survey of clustering algorithms. IEEE Transactions on Neural Networks 16(3):645.
24. Fu WH, Ma LF, Li AL (2014) Blind estimation of underdetermined mixing matrix based on improved K-means clustering, Systems Engineering and Electrnics (11):2143–2148 (in Chinese).
25. Fu WH (2007) Research on blind source separation and its application in communication reconnaissance, Dissertation, Xidian University (in Chinese).
26. Xu XH, Gao J, Fan ZG (2007) Methods for estimation of the number of sources in blind source separation, Journal of Hefei University of the Technology 31(1):1–4 (in Chinese).
27. Akaike H (1974) A new look at the statistical model identification, IEEE Transactions on Automatic Control 19(6):716–723. https://doi.org/10.1109/TAC.1974.1100705
28. Rissanen J (1983) Information in prediction and estimation, The 22nd IEEE Conference on Decision and Control, San Antonio, TX, USA, pp 308–310.
29. Zhao WS (2023) Research on Single Channel Blind Source Separation Technology of Communication Signals, Dissertation, Xidian University (in Chinese).
30. Amari S (1998) Natural gradient works efficeiently in learning. Neural Computation 10:271–276.
31. Chambers JA, Jafari MG, McLaughlin S (2004) Variable step-size EASI algorithm for sequential blind source separation. Electronics Letters 40(6):393–394
32. Fu WH, Yang XN, Liu NA (2008) Novel algorithm for communication signal blind separation —EASIBSA Algorithm, Journal of Wuhan University of Technology 30(1):28–31
33. Fu WH, Yang XN, Liu NA (2008) Step-size Optimization Based EASI Algorithm for Blind Source Separation, Journal of Sichuan University (Engineering Science Edition) 40(1):910–913 (in Chinese).
34. Cichocki A, Amari S (2002) Adaptive blind signal and image processing: learning algorithms and applications. New York: John Wiley & Sons Ltd.
35. Hyvarinen A, Karhunen J, Oja E (2001) Independent Component analysis, A Wiley-Interscience Publication.
36. Jutten C, Herault J (1988) Independent component analysis versus PCA, Proceeding of European Signal Processing Conference. Grenoble: European Signal Processing Conference, pp 643–646.
37. Tichavsk P, Koldovsk Z, Oja E (2005) Asymptotic performance analysis of the fixed-point algorithm (Fast-ICA) for independent component analysis, IEEE/SP 13th Workshop on Statistical Signal Processing. https://doi.org/10.1109/SSP.2005.1628757
38. Fu WH, Yang XN, Liu NA (2008) Robust Algorithm for Communication Signal Blind Separation Fourth-Order-Cumulant-Based 30(8):1853–1856 (in Chinese).
39. Cardoso JF, Souloumiac A (1996) Jacobi angles for simultaneous diagonalization. SIAM Journal on Matrix Analysis and Application 17(1):161–164.
40. Zhang WT (2011) Diagonalization in blind source separation, Dissertation, Xidian University (in Chinese).
41. Zhang WT, Lou ST (2016) Multicriteria optimization for nonunitary joint block diagonalization, IEEE International Conference on Acoustics, Speech and Signal Processing. Shanghai: IEEE, pp 2509–2513.
42. Ghennioui H, Fadaili EM, Thirion-Moreau N, et al (2007) A Nonunitary Joint Block Diagonalization Algorithm for Blind Separation of Convolutive Mixtures of Sources. IEEE Signal Processing Letters 14(11):860–863.

43. Zhang C (2019) Research on Frequency Domain Algorithm for Convolutive Blind Source Separation, Dissertation, Xidian University (in Chinese).
44. Matsuoka K (2002) Minimal distortion principle for blind source separation, Proceedings of the 41st SICE Annual Conference, Osaka, Japan, 4:2138–2143. https://doi.org/10.1109/SICE.2002.1195729
45. Ikram MZ, Morgan DR (2002) A beamforming approach to permutation alignment for multichannel frequency-domain blind speech separation, IEEE International Conference on Acoustics, Speech, and Signal Processing, Orlando, FL, USA, pp I-881–I-884. https://doi.org/10.1109/ICASSP.2002.5743880
46. Fu WH, Zhang C (2018) Independent Vector Analysis Convolutive Blind Separation Algorithm Based on Step-size Adaptive, Journal of Electronic and Information Technology 40(9):2158–2164 (in Chinese).
47. Li AL (2014) Research on Mixing Matrix Estimation Algorithm for Underdetermined Blind Source Separation, Dissertation, Xidian University (in Chinese).
48. Zhou XB (2018) Research on mixing matrix estimation for underdetermined blind source separation, Dissertation, Xidian University (in Chinese).
49. Fu WH, Zhou XB, Nong B (2017) Estimation of Underdetermined Mixing Matrix Based on Similarity Measurement, Journal of Beijing University of Posts and Telecommunications 40(6):24–29 (in Chinese).
50. Nong B (2017) Research on mixing matrix estimation and source signal recovery for Undetermined Blind Source Separation, Dissertation, Xidian University (in Chinese).
51. Nong B, Fu WH (2018) Mixing matrix estimation in UBSS based on homogeneous polynomials, IET Signal Processing 12(9):1123–1130.
52. Chen JH (2015) Research on source signal recovery of Undetermined Blind Source Separation based on Compressed Sensing, Dissertation, Xidian University (in Chinese).
53. Fu WH, Nong B, Chen JH, et al (2017) Source recovery in underdetermined blind source separation based on RBF network, Journal of Beijing University of Posts and Telecommunications 40(1):94–98 (in Chinese).
54. Tian DY (2017) Research on Compressive Sensing Sparse Reconstruction Algorithm for Undetermined Blind Source Separation, Dissertation, Xidian University (in Chinese).
55. Fu WH, Yang XN, Li NA, et al (2009) Up-Link Multi-User Detection in CDMA System Based on Blind Source Separation, Journal of Beijing University of Posts and Telecommunications 32(2):111–114 (in Chinese).
56. Fu WH, Yang XN, Li NA (2008) The Multi-User Detection and Chip Sequence Estimation for CDMA System Based on the Blind Source Separation, Acta Electronica Sinica 36(7):1319–1323 (in Chinese).

The manufacturer's authorised representative in the EU is Springer Nature Customer Service Centre GmbH, Europaplatz 3, 69115 Heidelberg, Germany. If you have any concerns regarding our products, please contact ProductSafety@springernature.com

Printed and bound by CPI Group (UK) Ltd, Croydon, CR0 4YY
01/10/2026
02225930-0012